Progress in Power-to-Gas Energy Systems

Progress in Power-to-Gas Energy Systems

Editor

Johannes Schaffert

MDPI • Basel • Beijing • Wuhan • Barcelona • Belgrade • Manchester • Tokyo • Cluj • Tianjin

Editor
Johannes Schaffert
Fuel and Appliance Technology
Gas- und Wärme-Institut Essen e.V. (GWI)
Essen
Germany

Editorial Office
MDPI
St. Alban-Anlage 66
4052 Basel, Switzerland

This is a reprint of articles from the Special Issue published online in the open access journal *Energies* (ISSN 1996-1073) (available at: www.mdpi.com/journal/energies/special_issues/Progress_P2G).

For citation purposes, cite each article independently as indicated on the article page online and as indicated below:

LastName, A.A.; LastName, B.B.; LastName, C.C. Article Title. *Journal Name* **Year**, *Volume Number*, Page Range.

ISBN 978-3-0365-7557-5 (Hbk)
ISBN 978-3-0365-7556-8 (PDF)

Contents

About the Editor

Johannes Schaffert

After completing his doctorate in physics in 2013, Johannes Schaffert began researching a broad variety of topics related to the energy transition through the use of green gases or renewable electricity for the Gas- und Wärme-Institut Essen (GWI). His fields of expertise include hydrogen, power-to-gas, natural gas and hydrogen grids, bio-methane, appliance technology and energy system analysis. In addition, his work touches on combustion science, district heating and technical rules and regulations. Johannes' research focus is on interdisciplinary projects that prepare and accompany the energy transition towards a sustainable energy supply.

Preface to "Progress in Power-to-Gas Energy Systems"

Like many of you, dear readers, I plunged into the exciting field of hydrogen and power-to-gas, which promises us renewable energy storage and use in molecular forms—either gaseous or liquid—to decarbonise all energy sectors that currently depend on fossil fuels. Over the past decade in this topic, I have been privileged to witness a number of developments, trends and advancements that have enabled us to stand where we stand today. We find ourselves at a crossroads between theoretical groundwork and the actual, large-scale technical implementation of the energy transition.

This reprint brings together current research findings from highly diverse disciplines, all of which can make a valuable contribution to the success of the energy transition. I am convinced that this interdisciplinarity of the common global challenge and its cross-sectoral character are central challenges of the energy transition. There will not be one blueprint solution that can be applied across the globe; rather, the research results yield insights for possible partial solutions that can be implemented in national, regional or even local energy systems at different implementation levels and detail, depending on boundary conditions and specific demands.

The first contribution in this reprint is an editorial regarding the contents of the ten following research articles. It contains a table of the articles, including research fields, titles and methods. The reader can use it as a quick overview before turning to the detailed articles. I can therefore be brief and simply refer to the overarching themes covered here: You will find three contributions regarding combustion research, one contribution regarding electro catalysis; several contributions from the perspective of energy economy; articles on energy storage, the mobility sector, and several energy system analyses; and finally, a contribution on energy law and regulation.

As Guest Editor, I have been able to rely on competent co-authors, all of whom are recognised experts in their field and represent leading research institutions. The scope of this reprint can in no way do full justice to their wealth of knowledge and experience, but it offers exciting insights into their latest research findings. My sincere thanks go to them once again. Special thanks are also due to all colleagues not listed as co-authors in this reprint. Thank you for the lessons I was able to learn from working with you. It is my pleasure to share the results with interested readers worldwide.

Johannes Schaffert
Editor

energies

Editorial

Progress in Power-to-Gas Energy Systems

Johannes Schaffert

Gas- und Wärme-Institut Essen e.V. (GWI), Hafenstrasse 101, 45356 Essen, Germany; johannes.schaffert@gwi-essen.de

1. Introduction

Hydrogen is expected to become a key component in the decarbonized energy systems of the future. Its unique chemical characteristics make hydrogen a carbon-free fuel that is suitable to be used as broadly as fossil fuels are used today. Since hydrogen can be produced by splitting water molecules using electricity as the only energy input needed, hydrogen offers the opportunity to produce a fully renewable fuel if the electricity input also only stems from renewable sources. Once renewable electricity is converted into hydrogen, it can be stored over long periods of time and transported over long, even intercontinental, distances. Underground hydrogen storage, pipelines, compressors, liquefaction-units, and transportation ships are infrastructures and suitable technologies to establish a global hydrogen energy system. Several chemical synthesis routes exist to produce more complex products from green hydrogen to fulfil the demands of various end-users and industries. One exemplary power-to-gas product is methane, which can be used as a natural gas substitute. Furthermore, ammonia, alcohols, kerosene, and all other important products from hydrocarbon chemistry can be synthesized using green hydrogen.

In the light of the continuously exacerbating crisis of global warming, the urgency of deploying green technologies could not be more pressing. Researchers and industry innovators worldwide study and develop all aspects of power-to-gas technologies, ranging from hydrogen production, conversion, transport, handling, etc., to the broad variety of end-use options. In the meanwhile, it remains an open research question to what extent the portfolio of power-to-gas technologies and products will enter the markets. For the various end-use sectors, the future defossilized energy mix will develop into different optima depending on the availability, flexibility, and cost of energy, as well as technical implications for the end users and the specific legal and regulatory framework. As a result, we will see different local or regional energy mixes and fuel compositions around the globe.

A major unknown variable is the depths in which direct electrification of processes will be implemented. Direct electrification—in cases where it leads to similar product qualities in industry or comfort in the mobility or household sectors—promises higher efficiencies and makes the more complex synthesis routes of power-to-gas technologies obsolete. However, a fully electrified energy system will lack the large scale and long-lasting storage option, not cover the fuel demands by high-temperature industrial processes, and lack back-up power plants needed for weather-conditions with insufficient renewable electricity supply. Thus, the role of molecules as energy carriers is very crucial, but at the same time is challenged by the competing direct electrification in some market segments.

A recent Special Issue of the open access journal *Energies* published ten research articles from the field of hydrogen and power-to-gas energy. The Special Issue entitled 'Progress in Power-to-Gas Energy Systems' covers several aspects of hydrogen energy systems, including hydrogen production, transport/distribution, storage, end use, as well as legal and regulatory aspects. This editorial summarizes the key findings. Finally, a brief discussion will pick up the main ideas of the introduction and place the research results of the Special Issue in the wider context of the defossilization of energy sectors and the competition of electrons vs. molecules.

Citation: Schaffert, J. Progress in Power-to-Gas Energy Systems. *Energies* **2023**, *16*, 135. https://doi.org/10.3390/en16010135

Received: 13 October 2022
Accepted: 23 November 2022
Published: 23 December 2022

2. Special Issue Articles

The ten articles published in the Special Issue 'Progress in Power-to-Gas Energy Systems' are presented in Table 1, where an overview of the authors, research fields, titles and methodologies is given.

Table 1. Summary of the research fields, titles, and methodologies of the Special Issue articles.

Article	Authors	Research Field	Title	Methodology
[1]	Yan Zhao, Vince McDonell, Scott Samuelsen	Combustion Science	Residential Fuel Transition and Fuel Interchangeability in Current Self-Aspirating Combustion Applications: Historical Development and Future Expectations	Review Article
[2]	Jörg Leicher, Johannes Schaffert, Hristina Cigarida, Eren Tali, Frank Burmeister, Anne Giese, Rolf Albus, Klaus Görner, Stéphane Carpentier, Patrick Milin, Jean Schweitzer	Combustion Science	The Impact of Hydrogen Admixture into Natural Gas on Residential and Commercial Gas Appliances	Combustion Theory, Calculations, Experimental Investigations
[3]	Paul Glanville, Alex Fridlyand, Brian Sutherland, Miroslaw Liszka, Yan Zhao, Luke Bingham, Kris Jorgensen	Combustion Science	Impact of Hydrogen/Natural Gas Blends on Partially Premixed Combustion Equipment: NOx Emission and Operational Performance	Experimental Research
[4]	David Tetzlaff, Vasanth Alagarasan, Christopher Simon, Daniel Siegmund, Kai junge Puring, Roland Marschall, Ulf-Peter Apfel	Electro Catalysis	[NiFe]-(Oxy)Sulfides Derived from $NiFe_2O_4$ for the Alkaline Hydrogen Evolution Reaction	Experimental Research
[5]	Andreas Zauner, Karin Fazeni-Fraisl, Philipp Wolf-Zoellner, Argjenta Veseli, Marie-Theres Holzleitner, Markus Lehner, Stephan Bauer, Markus Pichler	Energy Economy, Energy Storage	Multidisciplinary Assessment of a Novel Carbon Capture and Utilization Concept including Underground Sun Conversion	Techno-economic assessment Life cycle assessment Legal assessment
[6]	Janos Lucian Breuer, Juri Scholten, Jan Christian Koj, Felix Schorn, Marc Fiebrandt, Remzi Can Samsun, Rolf Albus, Klaus Görner, Detlef Stolten, Ralf Peters	Mobility	An Overview of Promising Alternative Fuels for Road, Rail, Air, and Inland Waterway Transport in Germany	Mobility Sector Modeling
[7]	Hannu Karjunen, Eero Inkeri, Tero Tynjälä	Energy System Analysis	Mapping Bio-CO_2 and Wind Resources for Decarbonized Steel, E-Methanol and District Heat Production in the Bothnian Bay	Potential Analysis
[8]	Johannes Schaffert, Hans Christian Gils, Max Fette, Hedda Gardian, Christine Brandstätt, Thomas Pregger, Nils Brücken, Eren Tali, Marc Fiebrandt, Rolf Albus, Frank Burmeister	Energy System Analysis	Integrating System and Operator Perspectives for the Evaluation of Power-to-Gas Plants in the Future German Energy System	Energy System Modeling, Energy Economics, Assessment of Regulatory Framework
[9]	Yifei Lu, Thiemo Pesch, Andrea Benigni	Energy System Analysis	Simulation of Coupled Power and Gas Systems with Hydrogen-Enriched Natural Gas	Energy Grid Modeling
[10]	Lena Maria Ringsgwandl, Johannes Schaffert, Nils Brücken, Rolf Albus, Klaus Görner	Energy Law and Regulation	Current Legislative Framework for Green Hydrogen Production by Electrolysis Plants in Germany	Legal and Regulatory Assessment

The contributions bring together various research fields. The first item in Table 1 contains a comprehensive review paper on the historic development of residential combustion technologies and fuel transitions that have been realized in the past [1]. The following two articles [2,3] stem from the field of combustion science and analyze in detail the impact of hydrogen admixture in natural gas (or methane). Both pieces of research include theoretical and experimental approaches. In [3], the focus is on partially premixed combustion, which plays a dominating role in the US-American natural gas appliances. One contribution from the field of catalysis describes a production method for facilitating the Alkaline Hydrogen Evolution Reaction [4]. In [5], the role of underground gas storage as a potential large-scale methanation reactors is addressed. Hydrogen and hydrogen-derived fuels in various mobility sectors are studied in [6]. Three articles were published from the field of energy system analysis. In [7], a detailed case study of a region in Finland is modeled, while in [8], the German energy system was modeled and assessed from system and operator points of view. Gas and electricity energy networks were modeled in [9] with a focus on co-simulation and hydrogen/natural gas blends. Finally, an assessment of the current legal framework for green hydrogen production in Germany was published in [10].

In the following, brief summaries of the ten research papers are given.

The work by Zhao et al. [1] provides a comprehensive literature review of the historic development of residential fuel transition and fuel interchangeability in self-aspirating combustion applications. The researchers from University of California, Irvine, CA, USA, go back in time and reflect on the fuel transitions in the domestic heating sector in the past centuries in great detail. The paper includes the history of coal substituting fuelwood in England starting in the 1500s, and the introduction of manufactured gases (amongst others the so-called town gas) starting in the early 19th century, followed by the town-gas-to-natural-gas transition in the second half of the 20th century. In England, this fuel switch led to 35 million appliances with 200 million burners being converted. The authors also address safety issues related to carbon monoxide poisoning and the development of gas quality standards. The emergence of renewable gases since the 1980s is summarized for the case of biogas and renewable hydrogen, including an outlook on future developments. The competing options of electrifying homes versus adopting renewable gases is discussed as well. A separate chapter of this work discusses technical considerations of adopting renewable fuels in residential burners in detail, explaining relevant combustion characteristics and formulae. Subsequently, the burner performance indicators efficiency, emissions, flame characteristics, and ignition are discussed.

Leicher et al. [2] address the impact of hydrogen admixture into natural gas on residential and commercial gas appliances. Hydrogen admixture into existing natural gas grids is being discussed as a way for the gas industry to contribute to decarbonization efforts in the short-term. The work was carried out as part of the European Commission-funded research project THyGA, which currently investigates up to 100 appliances in laboratory tests with hydrogen blending levels up to 60% by volume [11–13].

The main findings of these theoretical considerations and the accompanying first measurements are that hydrogen admixture into natural gas can in many ways be treated as a conventional natural gas quality issue. Many effects of the changing fuel characteristics induced by hydrogen, e.g., in terms of laminar combustion velocities, combustion temperatures, or the formation of nitrogen oxides (NO_X), are largely compensated by shifts towards higher air excess ratios in uncontrolled residential appliances. In appliances with combustion control, the control systems were found to be unable to maintain a constant air excess ratio with increasing levels of H_2 admixture, at least at full load. This is, however, not a safety-relevant concern since air excess ratios increase with higher levels of hydrogen, making the formation of toxic carbon monoxide less likely. These findings are, to some extent, specific to certain technologies: fully premixed gas appliances, which are common in heating systems in the EU, are less sensitive to issues such as flame flash backs than partially premixed devices, which can be found in cooking applications and in American residential heating appliances.

The theoretical and experimental investigations in THyGA so far indicate that a hydrogen admixture of about 20 vol% into natural gas, as it is currently proposed, does not pose any safety-related challenges, e.g., in the form of flame flash backs, overheating, or increased pollutant emissions, to the appliance types found in the field today. There are additional aspects to consider, e.g., in the context of appliance adjustment in the field, which is a topic ongoing research within the THyGA project.

In Glanville et al. [3], six North American hydrogen/natural gas blending demonstrations were selected for a literature review that set the basis for the subsequent experimental approach. Through laboratory testing using purpose-built "simulators" and in situ tests and field sampling in a simulated operating environment, a series of short-term tests were performed on components and equipment. Performance, efficiency, emissions, and other factors were characterized as a function of hydrogen blending up to 30% by volume. In general, all appliances and their burners were able to tolerate this shift in fuel composition, without notable excursions in process temperatures or emissions, and anticipated trends were confirmed and further quantified for these appliances, ranging from the de-rating of heat input, to the increase in excess aeration, and to the NOx and CO emissions. For these partially premixed types of combustion appliances, the dominant impact of hydrogen blending without extra adjustments is the increase in excess air, often resulting in lower NOx emissions, surface temperatures, and other parameters. The most sensitive burners to hydrogen blending were of the "in-shot" variety, used by warm-air furnaces, tested in the laboratory. The flash back events observed were inconsistent and likely caused by either test procedures or sensitivities of the specific test stands used. Further investigation into these burners is recommended [3].

Tetzlaff et al. [4] focus on hydrogen production, more specifically, on precious-metal-free electrocatalysis, which is a key factor for industrial-scale hydrogen production. Using controlled (partial) sulfidation of Fe/Ni oxide nanoparticles, the authors describe a novel synthesis procedure for mixed Fe/Ni (oxy)sulfide materials. As a result, high overall activities of the synthesized electrocatalysts were reported for the electrochemical hydrogen evolution reaction and interpreted as a step forward towards designing transition metal chalcogenide catalyst materials for the hydrogen evolution reaction and efficient stoichiometric formulations of NiFe (oxy)sulfide-based catalysts. [4]

Zauner et al. [5] approach the Power-to-Gas topic and more specifically, the underground storage of renewable methane synthesized from hydrogen and carbon dioxide from economic, technical simulation, greenhouse gas emissions, and legal points of view. The so-called Underground Sun Conversion or geo-methanation process uses green hydrogen and carbon dioxide captured from industrial emitters. The gases are injected into depleted underground hydrocarbon reservoirs, where the methane synthesis is realized by biochemical processes. The resulting gas is cleaned to achieve natural gas grid compliant feed-in quality before it is used in industry, closing a carbon cycle. Results show, that the novel synthesis route for methane production may be at comparable or lower cost compared to conventional above-ground methane synthesis. However, it must be taken into account that in order to produce geomethane with an underground sun conversion plant, according to current knowledge, large quantities of carrier gas are required by the process and must be stored simultaneously with the hydrogen and carbon dioxide used for underground methanation. Therefore, the underground sun conversion technology is particularly suitable when large quantities of gas have to be stored already. This commonly occurs for reasons of system relevance or supply security. As an additional benefit of this storage process, geomethane can be produced from renewable hydrogen and carbon dioxide, thus contributing to the achievement of climate targets. However, a list of legal or regulatory barriers or gaps, along with technical uncertainties, preventing project realization on an industrial scale is reported [5].

Karjunen, Inkeri, and Tynjälä [7] performed a regional resource analysis for a potential hydrogen valley (i.e., a promising early adoption site for hydrogen). Open data sources were used to identify the existing local industrial facilities from the northern region of

Sweden and Finland. The steel industry alone could require 55 TWh of additional renewable electricity by 2045 for producing the required hydrogen for carbon-neutral steel production. The production of renewable methanol from electricity and CO_2 could also require up to 85 TWh of electricity annually if all of the available biogenic CO_2 resources in the area were to be utilized. To put these numbers in perspective, the current annual electricity consumption of Sweden and Finland is 127 and 81 TWh, respectively. The study also evaluated the role of wind power production as a source of additional renewable power generation in the region. Existing projects in the area already amounts to 16 TWh of annual electricity production, but even 100 TWh could be exceeded in the following decades. The study also performed an analysis of the utilization of residual electrolyzer heat in the district heat supply when primed with heat pumps. The required industrial volumes of hydrogen are so vast that eventually a significant oversupply of low-grade heat is to be expected, but first demonstrations still show great integration benefits. Future studies are recommended to assess the dynamic performance of the system, leading to a more concise implementation plan for the region.

Breuer et al. [6] review alternative fuels based on Power-to-Gas and Power-to-Liquid processes, as well as corresponding propulsions systems, to solve challenges of decarbonizing the road, rail, air, and inland waterway transport sectors. On the production side, the criteria of technical maturity, costs, as well as environmental impacts were evaluated. On the utilization side, possible blending with existing fossil fuels and the satisfaction of the required distances are composed. From today's perspective, the electrification of most long-distance, heavy-duty road, shipping, and aircraft transportation is highly unlikely. In conclusion, Methanol-to-Gasoline, Fischer–Tropsch diesel and kerosene, hydrogen, battery-electric propulsion, HVO, DME, and natural gas were identified as promising future fueling options. All the above-named alternative fuels can reach near-zero greenhouse gas emissions bounded to preconditions. The results of the cost value review highlight the insecurities around the regarded cost levels, production costs, cross-border prices, and end-user prices. The extracted interval sizes of cross-border prices for 2020 are 7 EURct/kWh_LHV for H_2, 10 EURct/kWh_LHV for SNG, and 8 EURct/kWh_LHV for unspecified PtL fuels. Cost insecurity increases for 2030 and 2050, as does the length of the value chains. At present, cost comparisons indicate that lower production costs of H_2 are almost compensated by higher transport costs in comparison to fuels that offer existing infrastructural compatibility.

In their energy system analysis, Schaffert et al. [8] study in what way, and in which sectors, renewable energy will be integrated in the German Energy System by 2030, 2040, and 2050 and what role hydrogen and methane from power-to-gas processes will play in this integration.

To address their research questions, techno-economic energy system modeling was performed. Evaluation of the resulting operation of energy technologies was carried out from system and business points of view. Special consideration of gas technologies, such as hydrogen production, transport, and storage, was taken as a large-scale and long-term energy storage option and a key enabler for the decarbonization of the non-electric sectors. The broad set of results gives insight into the entangled interactions of the future energy technology portfolio and its operation within a coupled energy system. Amongst other energy demands, CO_2 emissions, hydrogen production, and future power plant capacities are presented. One main conclusion is that integrating the first elements of a large-scale hydrogen infrastructure into the German energy system by 2030 is necessary for ensuring the supply of upscaling demands across all sectors. Within the regulatory regime of 2020, authors suggest that investment decisions for large scale hydrogen infrastructures may come too late and might jeopardize the chances of achieving transition targets within the 2050 horizon [8].

In a second contribution from the field of energy system analysis, Yifei et al. [9] simulate coupled power and gas energy systems, including the option of hydrogen-enriched natural gas. Their methodological approach can handle different gas compositions and is thus able to accurately analyze the impact of hydrogen injections into natural gas pipelines. An exemplary co-simulation of coupled power and gas networks proves the capabilities of the new model. The authors implemented a detailed description of the physical properties of the gas mixtures, which allows tackling co-simulation research questions in the future. The importance of detailed technical simulation is highlighted for a number of examples including the necessity of considering different and—importantly—varying gas compositions in simulations. [9]

The work by Ringsgwandl et al. [10] contributed to this Special Issue's topic from a complementary point of view by addressing the current legislative framework for implementing green hydrogen production plants. For the case of Germany, the authors analyze laws and ordinances to identify potential obstacles to the rollout of green hydrogen. Especially the implications on potential hydrogen production plant operators are focused on. Due to unbundling-related constraints, potential operators from the group of electricity transport system and distribution system operators lacking permission to operate hydrogen production plants. Moreover, ownership remains forbidden for them. The same applies to natural gas transport system operators. The case is less clear for natural gas distribution system operators, where explicit regulation is missing. It is finally analyzed if the production of green hydrogen production, in its competition with fossil hydrogen production, is currently supported, not only by the legal framework but also by the National Hydrogen Strategy and the Amendment of the Renewable Energies Act. It can be concluded that in recent amendments of German energy legislation, regulatory support for green hydrogen in Germany was found. The latest legislation has clarified crucial points concerning the ownership and operation of electrolyzers and the treatment of green hydrogen as a renewable energy carrier. This can be seen as a step forward towards a green hydrogen rollout; nevertheless, a number of clarifications are still needed to allow a swift, large-scale implementation of green hydrogen in a deeply decarbonized energy system. As an outlook, the already proposed amendment for the Renewable Energy Directive by the European Commission is expected to bring quotas for renewable hydrogen used in industry. This might become a game-changer and lead to a drastic acceleration of hydrogen deployment in European member states.

3. The Current Status of Research in the Field of Power-to-Gas Energy Systems

Power-to-Gas energy systems is a research field within Energy Science, that requires interdisciplinary approaches. This is especially due to the coupling functions that Power-to-Gas technologies perform, e.g., between the electricity and gas energy sectors. Technical development alone does not suffice to enable a mass roll-out of Power-to-Gas plants and Power-to-Gas products. Rather, complementary research fields need to work coherently to assess the strengths, weaknesses, chances, and risks of implementing Power-to-Gas technologies in today's energy systems.

The broad portfolio of research fields in energy science that intensely study the future prospects of hydrogen and related energy carriers is also reflected in the recent Special Issue of *Energies* [14], which is summarized in this editorial. We received contributions from combustion science [1–3], which is a field currently studying the hydrogen tolerability of today's gas applications. We also received a manuscript from the field of catalysis [4], a research field that focuses on the development of energy- and cost-effective materials for hydrogen production. Furthermore, one contribution deals with large-scale energy storage and hydrogen-to-methane conversion [5]. In addition, one paper that assesses various renewable fuel options for the mobility sector [6] has been included. Energy System Analysis is a broad topic itself, with three papers being published in this Special Issue. One of them modeled a region in Finland [7] as a case study for large-scale industrial energy applications. A second paper modeled the future German Energy System and assessed the results from techno-economic as well as operator perspective [8]. The third contribution focused on the technical modeling of power and gas networks including the option of hydrogen/natural gas blends [9]. Finally, one article is dedicated to the legal framework for implementing green hydrogen for the case of Germany [10].

Only by reflecting the progress of Power-to-Gas-related research from all relevant disciplines can the current status and remaining challenges of this transition be understood. The Special Issue discussed here can only yield punctual insights into current research. Many more crucial research fields remain untouched. As an example, the important topic of social acceptance of technologies and policies shall be mentioned here.

The technologies needed to supply economies with clean energy, to convert it into the desired form, to store, to distribute and to utilize it, are available and are constantly being optimized. Research in this field must bridge the gap between academic understanding of clean energy technologies on the one hand side and a very constructive applied science approach on the other hand side. By this re-focussing on the implementation of the various energy applications, the remaining knowledge gaps must be identified and filled to facilitate the mass roll-out of clean energy technologies.

Let us have a more detailed look at the research topics mentioned above. The three publications from the field of **combustion science** describe in high technical detail the chances of blending hydrogen into existing natural gas distribution networks and using it with todays' end-use appliances. Highly depending on the exact design of appliances, i.e., geometrical, burners, fuel–air mixing, combustion controls, etc., some hydrogen admixture may be tolerable without changing the appliances in the field. Operation safety is the dominating concern when debating the use of existing appliances with new gas properties, while in general, the distribution of (defossilized) gaseous fuels distributed through existing grid infrastructure appears advantageous for quick decarbonization effects in the heating sector. For high hydrogen contents in distributed gas mixtures, e.g., above 20 vol-% or 30 vol-%, and especially for the case of a pure hydrogen distribution on the household level, new appliance types designed for this purpose will need to be installed. Fuel cells would be a high-efficiency option here. Also, a hydrogen combustion-based heat supply in households can be realised. The electric-driven heat pump is in the lead so far. For well-insulated buildings, it can be a very cost-effective option. In industry, however, especially in high-temperature processes, fuels will be needed, e.g., for energy-intensive melting, blast furnaces or direct reduction of iron, where electricity cannot supersede gases or fuels. Here, a strong demand for hydrogen can clearly be expected. **Underground gas storage** facilities such as the salt domes used today are available technology to store hydrogen in very large quantities and over seasonal time-scales. It is more complex for the case of pore storages, such as that studied in [5], where biological methanation reactions could be used to upgrade the heating value of a stored gas mixture. If this option is drawn in the future, regions that lack salt deposits suitable for hydrogen storage could benefit from pore storage potentials.

In the **mobility sector**, it appears obvious that small vehicles, especially those that make up the large fleets of private passenger cars, may be equipped with electric propulsion systems and batteries at lower cost and higher energy efficiency compared to hydrogen or other power-to-gas products. The case is different for heavy-duty transportation on the road, rail, waterways, and, obviously, aircrafts, where the high energy densities of (liquid) fuels or compressed/liquefied gases remain advantageous compared to battery energy storage.

The research field of **Energy System Analysis** delivers insights generated in the toolbox of energy system models. Researchers cope with the above-mentioned uncertainties concerning future energy mixes by parametrizing their energy system models with assumptions for the studied technologies and optimizing the energy systems within predefined boundary conditions. The latter are based on scenarios that contain economic and regulatory base assumptions, which are in turn based on scenarios. Scenario variations allow for identifying sensitivities and assess the robustness of results. While specific quantitative results are to be interpreted carefully and only against the background of all assumptions and boundary conditions, general trends that robustly emerge for various scenarios lead to resilient results.

As far as a green energy transition is a technical issue with available solutions at hand, decision makers do have the option to realize the transformation of fossil-dominated into renewable energy systems. Various disciplines of research, some of which have been mentioned here, support policy makers with technology developments and results from their simulation-based tool-boxes. The state of progress in power-to-gas energy research suggests that future research focus should be placed on technology implementation in energy systems. By addressing the details of real-world applications of power-to-gas technologies, the concepts have to prove themselves suitable for the manifold requirements and boundary conditions. By turning to applied research, future studies should follow the progressing technology readiness of power-to-gas technologies and prepare the next steps towards mass roll-out. Detailed technical studies of power-to-gas technologies in interaction and competition with existing technologies in the field are needed. In addition, technology acceptance research from the social science perspective, as well as the development of legal and regulatory frameworks and appropriate funding schemes, are needed to guide policy makers in creating investment security for decision makers in renewable energy markets worldwide.

Funding: This research received no external funding.

Acknowledgments: The author wants to thank Juri Scholten, Jörg Leicher, Manfred Lange, Frank Burmeister, Rolf Albus, Klaus Görner, and all authors of the Special Issue "Progress in Power-to-Gas Energy Systems".

Conflicts of Interest: The author declares no conflict of interest.

References

1. Zhao, Y.; McDonell, V.; Samuelsen, S. Residential Fuel Transition and Fuel Interchangeability in Current Self-Aspirating Combustion Applications: Historical Development and Future Expectations. *Energies* **2022**, *15*, 3547. [CrossRef]
2. Leicher, J.; Schaffert, J.; Cigarida, H.; Tali, E.; Burmeister, F.; Giese, A.; Albus, R.; Görner, K.; Carpentier, S.; Milin, P.; et al. The Impact of Hydrogen Admixture into Natural Gas on Residential and Commercial Gas Appliances. *Energies* **2022**, *15*, 777. [CrossRef]
3. Glanville, P.; Fridlyand, A.; Sutherland, B.; Liszka, M.; Zhao, Y.; Bingham, L.; Jorgensen, K. Impact of Hydrogen/Natural Gas Blends on Partially Premixed Combustion Equipment: NO_x Emission and Operational Performance. *Energies* **2022**, *15*, 1706. [CrossRef]
4. Tetzlaff, D.; Alagarasan, V.; Simon, C.; Siegmund, D.; Puring, K.J.; Marschall, R.; Apfel, U.-P. [NiFe]-(Oxy)Sulfides Derived from $NiFe_2O_4$ for the Alkaline Hydrogen Evolution Reaction. *Energies* **2022**, *15*, 543. [CrossRef]
5. Zauner, A.; Fazeni-Fraisl, K.; Wolf-Zoellner, P.; Veseli, A.; Holzleitner, M.-T.; Lehner, M.; Bauer, S.; Pichler, M. Multidisciplinary Assessment of a Novel Carbon Capture and Utilization Concept including Underground Sun Conversion. *Energies* **2022**, *15*, 1021. [CrossRef]
6. Breuer, J.L.; Scholten, J.; Koj, J.C.; Schorn, F.; Fiebrandt, M.; Samsun, R.C.; Albus, R.; Görner, K.; Stolten, D.; Peters, R. An Overview of Promising Alternative Fuels for Road, Rail, Air, and Inland Waterway Transport in Germany. *Energies* **2022**, *15*, 1443. [CrossRef]
7. Karjunen, H.; Inkeri, E.; Tynjälä, T. Mapping Bio-CO_2 and Wind Resources for Decarbonized Steel, E-Methanol and District Heat Production in the Bothnian Bay. *Energies* **2021**, *14*, 8518. [CrossRef]
8. Schaffert, J.; Gils, H.C.; Fette, M.; Gardian, H.; Brandstätt, C.; Pregger, T.; Brücken, N.; Tali, E.; Fiebrandt, M.; Albus, R.; et al. Integrating System and Operator Perspectives for the Evaluation of Power-to-Gas Plants in the Future German Energy System. *Energies* **2022**, *15*, 1174. [CrossRef]
9. Lu, Y.; Pesch, T.; Benigni, A. Simulation of Coupled Power and Gas Systems with Hydrogen-Enriched Natural Gas. *Energies* **2021**, *14*, 7680. [CrossRef]
10. Ringsgwandl, L.M.; Schaffert, J.; Brücken, N.; Albus, R.; Görner, K. Current Legislative Framework for Green Hydrogen Production by Electrolysis Plants in Germany. *Energies* **2022**, *15*, 1786. [CrossRef]
11. ENGIE; GERG; DGC; GAS.BE; Electrolux; BDR Thermea; GWI; DVGW-EBI; CEA. THyGA Project-Testing Hydrogen Admixture for Gas Applications (EU Project, Fuel Cells and Hydrogen 2 Joint Undertaking, Grant Agreement No. 874983). Available online: https://thyga-project.eu/ (accessed on 26 September 2022).
12. Schweitzer, J. Project THyGA: Experimental Evaluation of Hydrogen-Natural Gas Blends on Domestic and Commercial Appliances. In Proceedings of the 28th World Gas Conference, Daegu, Korea, 23 May 2022.

13. Milin, P.; Schweitzer, J.; Schaffert, J.; de Wit, K.; Blanchard, L.; Krishnaramanujam, K.; Temperato, S.; Beghi, M.; Judd, R.; Kostereva, A. THyGA Project: Paving the Way for Hydrogen and Natural Gas Blends for the Domestic and Commercial Sectors. In Proceedings of the 28th World Gas Conference, Daegu, Korea, 23 May 2022.
14. MDPI energies Special Issue "Progress in Power-to-Gas Energy Systems". Available online: https://www.mdpi.com/journal/energies/special_issues/Progress_P2GISSN1996-1073 (accessed on 13 October 2022).

energies

Article

Residential Fuel Transition and Fuel Interchangeability in Current Self-Aspirating Combustion Applications: Historical Development and Future Expectations

Yan Zhao *, Vince McDonell * and Scott Samuelsen

UCI Combustion Laboratory, University of California, Irvine, Irvine, CA 92697-3550, USA; gss@apep.uci.edu
* Correspondence: yanz33@uci.edu (Y.Z.); mcdonell@ucicl.uci.edu (V.M.)

Abstract: To reduce greenhouse gases and air pollutants, new technologies are emerging to reduce fossil fuel usage and to adopt more renewable energy sources. As the major aspects of fuel consumption, power generation, transportation, and industrial applications have been given significant attention. The past few decades witnessed astonishing technological advancement in these energy sectors. In contrast, the residential sector has had relatively little attention despite its significant utilization of fuels for a much longer period. However, almost every energy transition in human history was initiated by the residential sector. For example, the transition from fuelwood to cheap coal in the 1700s first took place in residential houses due to urbanization and industrialization. The present review demonstrates the energy transitions in the residential sector during the past two centuries while portending an upcoming energy transition and future energy structure for the residential sector. The feasibility of the 100% electrification of residential buildings is discussed based on current residential appliance adoption, and the analysis indicates a hybrid residential energy structure is preferred over depending on a single energy source. Technical considerations and suggestions are given to help incorporate more renewable energy into the residential fuel supply system. Finally, it is observed that, compared to the numerous regulations on large energy-consumption aspects, standards for residential appliances are scarce. Therefore, it is concluded that establishing appropriate testing methods is a critical enabling step to facilitate the adoption of renewable fuels in future appliances.

Keywords: energy transition; residential appliances; renewable energy adoption; fuel interchangeability; hydrogen; combustion performance

Citation: Zhao, Y.; McDonell, V.; Samuelsen, S. Residential Fuel Transition and Fuel Interchangeability in Current Self-Aspirating Combustion Applications: Historical Development and Future Expectations. *Energies* **2022**, *15*, 3547. https://doi.org/10.3390/en15103547

Academic Editors: Johannes Schaffert and Tapas Mallick

Received: 12 February 2022
Accepted: 24 March 2022
Published: 12 May 2022

1. Introduction

When fossil fuel consumption and its impact on climate change is discussed, power generation, transportation, and industrial applications receive significant attention due to their large market share and the advanced technologies being developed and adopted by these sectors. An aspect that generally receives little attention is the residential energy sector.

The urbanization and industrialization of England in the 1700s forced London residents to abandon the use of suddenly expensive fuelwood and turn to cheaper coal. Since then, the residential sector has always been in the pioneer position relative to energy transition in the world. The large use of fossil fuels during and after the Industrial Revolution emerged after residential markets had created a thriving trade [1,2].

The past few decades have witnessed astonishing combustion technology advancement in gas turbines, internal combustion engines, and other industrial applications. Moreover, the strict emission regulations on high-energy-consuming devices greatly boosted their combustion technology advancement. Take heavy-duty gas turbines, for example: water was injected into the combustor to reduce NO_X emissions in the 1970s. As emission regulations are becoming more stringent, numerous technologies have been developed to reduce NO_X emissions, including rich burn-quick mix-lean burn (RQL) technology [3], dry low

NO_X (DLN) lean premixed combustion [4], mild combustion [5], etc. Post-combustion pollutant-elimination technologies, such as Selective Catalytic Reduction (SCR), are also being adopted to further decrease gas turbine NO_X emissions down to the single-digit ppm level [6]. Although the emission levels of the high-energy-consumption units are already very low, a lot of investment and effort is still being invested to further decrease or possibly eliminate the emissions from those devices altogether. In contrast, the emission level of some appliances can stay at around a couple hundred ppm without attracting much attention from policymakers. Moreover, the combustion technologies used in gas appliances are the same as 100 years ago, which were mainly adopted from Bunsen flames [7,8].

As more renewable energy technologies are being developed, the idea of incorporating solar panels, wind turbines, batteries, fuel cell units, etc., into residential homes is also promising [9,10]. Discussions have occurred regarding the 100% electrification of residential homes to eliminate both the greenhouse gas emissions and air pollution from the residential sector and shift them to power-generation units, which might make pollutantssrelatively easy to be regulated and controlled [11,12]. However, very few studies considered the current status of gas and electric appliances' market share and the potential for an increase in carbon emissions due to the 100% electrification of residential appliances.

Renewable gases such as biogas and renewable hydrogen are becoming available at a much lower price than most non-combustion renewable energy sources and can also be generated from multiple renewable sources. For example, biogas can be produced from wood and agricultural products, landfill waste, sewage, etc. [13]. Injecting purified biogas into existing natural gas pipelines has already been adopted and has gained much success in reducing carbon emissions [14]. Besides biogas, hydrogen is also a competitive and promising fuel for the future due to its high energy density on the mass base and the many ways it can be generated. The carbon-free property of hydrogen inherently reduces greenhouse gas emissions when generated renewably. Renewable sources for hydrogen include biomass, solar energy, wind power, and water electrolysis using grid power [15,16]. The adoption of these renewable gases in existing natural gas pipelines should be easier and cheaper for end users without much capital investment and maintenance. Biogas or hydrogen injection into the existing natural gas pipeline needs technical evaluation and fuel-interchangeability studies. However, gas appliance manufacturers and gas utility companies are still using the flame indices from the 1920s to guide their fuel-interchangeability investigations. Therefore, more comprehensive considerations should be given to guide appliance performance upgrades while adopting renewable fuels.

As many future fuel choices and energy-consumption structures are being predicted, this paper summarizes the historical energy transitions in the residential sector and their interactions with other sectors. The past experience allows us to glimpse the future energy structure. The feasibility of electrifying residential homes by 100% is discussed based on the comparison between the existing electric/gas appliances and their energy consumption. Furthermore, the technical considerations of fuel interchangeability in residential appliances are summarized and discussed, including combustion technologies, fuel properties, and flame indices. The lack of international standard testing methods and regulations on residential appliances and the representative regional regulations on appliance performance are also summarized.

2. Residential Fuel Consumption Historical Variation

The fuel consumption shares among different energy sources vary along with human society's development. Figure 1 shows the energy consumption variation for the past two centuries of the U.S.. Before the 19th century, wood was the dominant fuel. With the start of the Industrial Revolution, wood could no longer meet the massive energy demand due to its low energy density and the supply scarcity caused by fast forest destruction. Meanwhile, coal consumption significantly increased with its worldwide discovery and the advancement of coal-mining technologies. Coal consumption surpassed that of wood in the late 19th century and played an essential role in the Industrial Revolution. The oil and gas industry started to rise around the 1900s, and their market share surpassed coal consumption in the 1950s. Till today, oil, natural gas, and coal still dominate the energy sources in the world.

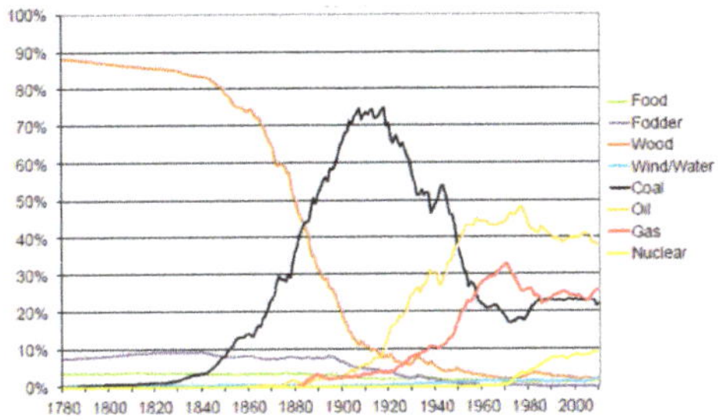

Figure 1. Transition of energy consumption in the U.S. 1780–2010 [17].

The energy transitions of the U.K., France, Netherland, Russia, Japan, China, Nigeria, etc., were also investigated [18–20]. Although the timeline or specific numbers differ, a similar trend of energy transition was found worldwide: from biomass (crop residuals, fuelwood, etc.) to coal, then oil and gas, and afterward from fossil fuel to nuclear or other renewable powers.

Although power generation, transportation, and industrial applications reflect the majority of fuel consumption, the residential sector is also an important and often the most price-sensitive sector relative to fuel utilization. Yet, this sector has significantly fewer technological advancements and research development, likely due to the low profit margins and relatively low cost associated with appliances. However, it was noted that domestic energy consumption and needs shaped the use of energy sources, especially at the initial stage of their rise [21–27]. Therefore, investigating the residential section of energy consumption is of great importance, especially in considering future fuel consumption scenarios. The following sections summarize the development of residential fuel consumption.

2.1. Before the 1900s: Transition from Wood to Coal

The transition from wood to coal in the residential sector was accompanied by population growth and the rapid development of industry. This combined effect created a large increase in demand for fuelwood. As fuelwood became more limited, people sought alternative fuels to replace what potentially looked like increasingly more expensive fuelwood. This residential fuel transition, on a massive scale, first occurred in England.

The population in England increased dramatically by more than 80%, from 3.02 million in 1541 to 5.47 million in 1656 [28]. Despite this, the population explosion did not exert a significant influence on the fuelwood supply for residential houses. Residential fuelwood could still be secured in a lot of regions by planting trees in backyards, on roadsides, on otherwise infertile slope land, or in fuelwood groves to supply nearby farms or villages [18]. In England, the fuelwood prices were actually fairly stable between 1550 and 1650 [29]. Wilson [30] also noted that, from 1450 to 1650, while timber prices in England increased, they did not increase as much as the cost of other agricultural products. In fact, timber became cheaper year by year relative to inflation, even with increasing demand from the residential sector. Hence, the population growth did not exert a high demand for wood supply and was not the major reason for the fuelwood price increase. It was the rise of industry that exhausted the fuelwood supply and drove up the price significantly. It forced the residential sector to seek alternative fuels other than fuelwood.

Before the 18th century, industry applications also used fuelwood as their energy source. Surprisingly, even though coal possesses more than 33% more energy density than fuelwood and at a lower price, the industry did not switch their fuel from fuelwood to coal very quickly. The smoky and sulfurous nature of coal made its adoption difficult for industries that required smokeless or odorless fuels [31]. Further, the slow transition from fuelwood to coal in industry was not because of a lack of clean coal-burning technology, but more because the industry was still profiting a lot from burning fuelwood. The energy density benefit of replacing fuelwood with coal could not balance out the capital investment of adjusting the combustion devices and coal transportation. Similar to the current situation, momentum in the industry and conservative tendencies usually slow new technology adoption. Fuel interchangeability studies from fuelwood to coal in industrial applications were already conducted in the Elizabethan Age. In 1590, John Thornbrough, the Dean of York, was granted a seven-year project to remove the "piercing and acrimonious spirits" for beer making and processing alum. By 1610, there were already patent applications using coal as fuel in the baking of malt, bread, bricks, tiles, and smelting of bell metal copper, brass, iron, lead, and glass [32,33]. Although the possibility of utilizing coal was realized and numerous studies were conducted to adapt combustion devices from wood to coal, coal remained an insignificant industry fuel until the 17th century.

With the development of industrial technologies, fuelwood could not meet the energy demand anymore due to its low energy density and limited supply caused by the slow recovery of forests. According to the study results on historical energy consumption, 20 kg of charcoal could only produce 1 kg of iron (600 MJ/kg) in England during the Middle Ages [18]. The glass industry was also wood-intensive. To produce 1 kg of glass, 2.4 tons of wood needed to be consumed for both heating and to obtain potassium (90 MJ/kg). Salt production is also a major wood-consuming sector, which demands as much as 500–600 MJ wood consumption per kilogram of salt produced [34]. This large amount of fuelwood requirement in the industrial sector caused severe forest destruction, and eventually, fuelwood prices increased. Therefore, residents sought cheaper alternative fuels. In the latter half of the 17th century, the coal price in England was less than half of fuelwood per unit of gross energy [35]. This fuel price gap between fuelwood and coal was more significant in densely populated London than the England average. In Table 1 [19], the fuel prices and the general living cost level in London are normalized to 100 from the year 1451 to 1500. It can be seen that coal saw a steady increasing trend with the living cost over the next 190 years. Fuelwood prices started to increase significantly by the end of the 1500s. By 1642, the cost of fuelwood was more than double the price of coal.

Table 1. Indices of estimated price movements in London, 1451–1642.

Commodity	1451–1500	1531–1540	1551–1560	1583–1592	1603–1612	1613–1622	1623–1632	1633–1642
General	100	105	132	198	251	257	282	291
Fuelwood	100	94	163	277	366	457	677	780
Coal	100	89	147	186	295	371	442	321

As the residents in London increasingly adopted coal as the substitute for fuelwood, imports of coal to London increased significantly in the last decade of the 16th century. Table 2 [36] shows the coal imports records from the late 16th century to the late 17th century. As can be seen, the coal consumption in London increased significantly from 10,785 tons to 361,189 tons in a century.

It should be noted that this coal consumption increase was mainly caused by the residential sector's needs, since the technology for using coal in industry was still not widely adopted in the 17th century. The evidence for this can be found in the iron industry. In the 17th century, iron smelting was a voracious consumer of fuelwood in London. The dependence of smelting on the increasingly scarce and high-priced fuelwood supplies in the 17th century led to a situation in which the conversion of iron ore into metal in England was temporarily checked. However, this situation did not make the iron industry turn to coal combustion. Instead, a policy was released by the government to encourage imports of intermediate product pig iron instead of raw iron ore. Therefore, the rapid growth of coal production from the mid-16th century to the late 17th century largely represents the substitution of coal for wood as a household fuel [19].

Table 2. Imports of coal into London, 1580s–1680s.

Year	Period	Tons	Notes
1580	12 March–18 September	10,785	
1585–1586	Michaelmas–Michaelmas	23,867	
1591–1592	Michaelmas–Michaelmas	34,757	
1605–1606	Christmas–Christmas	73,984	
1614–1615		91,599	One week missing
1637–1638		142,579	Two weeks missing; a year of bad trade
1667–1668	Midsummer–Midsummer	264,212	
1680–1681	Michaelmas–Michaelmas	361,189	

The earliest coal shipped to London was from Newcastle as ballast, which was known as "sea-coal" [37]. The burning of sea-coal was reported to cause smoke with a pungent odor [38]. Therefore, presumably, sea-coal was of low quality and might have had a relatively high sulfur content. Sea-coal was first adopted by the poor residents in London due to its cheap price. To adapt the wood-burning fireplace to sea-coal, chimneys were built or modified to vent the smoke out of the house. Therefore, the number of chimneys of residential houses in London increased greatly around the 1550s. The acceptance of sea-coal by the upper classes and nobility followed the steps of lower-class residents with a time lag. The use of sea-coal as a domestic heating source in royal house was a slow process since Queen Elizabeth I was "greatly grieved and annoyed with the taste and smoke of sea-coal". After her death in 1603, James VI of Scotland became James I of England, during whose governing period the popularity of using coal in residential houses was gained. Scotland was short of wood but had less indigenous sulfurous coal, which led to coal being used in houses of Scottish nobles much earlier than in England. Therefore, the new king continued his old habit of using coal for domestic heating in England. Such behavior aided the adoption of coal as a domestic fuel for wealthy London households [32].

It was the high demand for coal consumption in residential households that spurred the early development of coal mining and transportation. Coal provided the power source for and encouraged the Industrial Revolution. Although a bit later, the U.S. went along a very similar pathway in the fuel transition from fuelwood to coal. In the U.S., coal surpassed fuelwood's energy supply in the 19th century, which was around 200 years later than the energy transition of Britain [39]. The early records in the U.S. of coal trading were in Pennsylvania, and the major market for the first decade of the anthracite coal trade in Pennsylvania was in American homes [24,40]. Therefore, when the early coal boosters tried to quantify the potential market in New York, they counted the numbers of homes, not the numbers of factories or steam engines [21].

Similar to the situation of Britain, the early coal utilization in the U.S. was hindered by the poor transportation situation around the coalfields on the East Coast. Even if Pennsylvania's anthracite coalfields were only a hundred miles away from the eastern seaboard, transporting coal to Philadelphia and New York was still extremely expensive in the early 19th century. In the year 1810, it cost more to ship coal fewer than a hundred miles to Philadelphia than it did to deliver comparable shipments three thousand miles from England [41]. However, the War of 1812 tripled the price of imported coal from Britain. During this time, emphasis on energy security resulted in local anthracite coal in Philadelphia gaining a higher market share on the East Coast [42]. The real booster of the U.S. domestic coal consumption on the East Coast was the construction of the Schuylkill Canal in the 1820s. The Canal decreased coal's transportation cost significantly. In the 1810s, one ton of anthracite coal cost around 20 U.S. dollars. However, this price kept dropping to 8.4 dollars in 1820 and then 6.5 dollars in 1830. In the 1840s, the coal price on the East Coast of the U.S. decreased to around 4 dollars per ton [43]. Due to the availability of cheap coal, fuel-interchangeability studies on fuelwood to coal in residential appliances were conducted by scientists and appliance manufacturers. Between the years 1815 and 1839, 329 patents for coal-combustion stoves were issued in the U.S. [44]. In the early adoption stage, a coal-burning stove cost around 30 dollars, while the average laborer only earned 2 dollars a day [45]. However, by the early 1830s, a domestic coal stove cost less than 10 dollars [21].

The Industrial Revolution further sped up the energy transition from fuelwood to coal, especially when steam engines became widely adopted in the late 18th and early 19th centuries. However, as noted above, this energy transition was first initiated by the residential sector. After coal was widely adopted in residential cooking and heating, the accessibility of cheap coal paved its way to more applications in industry, thus fostering the Industrial Revolution.

2.2. From the 19th Century to the 1950s: Coal to Manufactured Gases

Besides the transition to coal utilization, the residential market also played an essential role in the development of the petroleum industry and trade, especially in its initial development stage in the latter half of the 19th century. The combination of urbanization and industrialization in both America and Europe led to a great need for lighting. Therefore, petroleum was first refined to obtain kerosene, which was a lighting source to replace diminishing and increasingly expensive whale oil [21,27,46,47]. The early development era of the petroleum industry is also called "The Age of Illumination" [48].

At the early age of the oil industry, instead of being a major petroleum product as it is today, gasoline was only a surplus byproduct that was usually burned at refineries, converted to gaseous fuel for gas lights, or just dumped into the atmosphere due to its high volatility. Being a waste product, cheap gasoline stimulated the development of engines in the late 19th and early 20th centuries. For example, when Mercedes entered the U.S. market in the early 20th century, the price per car was USD 12,450 (~USD 374 k in 2021 values), but the gasoline price was only 7 cents/gallon (around USD 2/gallon in 2021 values) [47]. With the accessibility of cheap and abundant gasoline, the auto industry invested more money into automobile technologies, and the prices dropped significantly. For example, when Ford first introduced its Model T to the market in 1908, the price was USD 950, which dropped down to USD 269 by 1923 [49]. Before electric lighting and automobiles gained popularity in the first decade of the 20th century, kerosene was the major focus of the petroleum industry for providing light. Since the beginning of the petroleum industry, techniques were focused on maximizing the production of kerosene until the second decade of the 20th century. As shown in Figure 2, although most crude oil is refined to gasoline and diesel today, in the 1870s, more than 80% of a barrel of crude petroleum was transformed into illuminating oil [17,48].

Figure 2. Shares of U.S. oil refinery output from 1874 to 1919 [17].

As petroleum (kerosene) found its use in domestic lighting applications in the late 19th century and early 20th century, the energy transition in domestic cooking and heating changed from coal to different kinds of manufactured gases. Although it is hard to track who was the first to produce manufactured gas, Becker and Clayton are believed to be the early pioneers who invented methods to produce combustible gases from coal. In 1681, Johann Becker from Germany discovered that combustible gases could be generated by heating coal in the absence of air. Three years later, John Clayton from England collected combustible gas using a similar coal treatment method and called this gas the "Spirit of Coals". The commercialization of manufactured gas in residential and industrial applications was initiated by the foundation of London and Westminster Gas Light and Coke Company in 1812. In the same year, this company built the world's first commercialized gas networks in Great Peter Street, Westminster, London, by laying wooden pipes and illuminated Westminster Bridge with gas lamps on New Year's Eve in 1813 [50]. Due to its clean combustion performance and ease of transporting through pipelines compared to coal, manufactured gas quickly gained popularity in the early 19th century. The prosperity of the gas industry was accompanied by emerging inventions of gas-combustion appliances. In 1826, the world's first gas stove was designed in England by James Sharp [51]. In the 1850s, Robert Bunsen invented the aspirated burner (Bunsen burner), which significantly influenced gas applications in residential and industrial applications till today [52]. In the 1870s, Dr. Carl Auer von Welsbach invented the incandescent gas light mantle, which solved the inefficient combustion of open-flame burners in streetlamps [53]. Moreover, water heaters, room heaters, and many other appliances, such as soldering irons and hair-curling tongs, appeared on the scene in the mid- to late 1800s [50]. To meet the increasing customer base, more gas networks were built in Britain along with manufactured gas production sites, which are shown in Figure 3. Due to the widespread application of manufactured gas in the cities of Britain, manufactured gas was also called "town gas" in the U.K. [54].

Figure 3. Known and suspected locations of manufactured gas sites in the British Isles [55].

The United States started the manufactured gas era at the same time as Britain. In 1816, the first manufactured gas company in America—the Gas Light Company of Baltimore—was founded. On 7 February 1817, the first manufactured gas streetlamp was lit in Baltimore. This was followed by the foundation of the Boston Gas Light Company in 1822 and the New York Gas Light Company in 1825 [50].

It should be noted that the early manufactured gas companies generated their gases from coal, except for the Gas Light Company of Baltimore, which produced its gases from the distillation of pine tar. However, the company switched to coal gas in 1822 due to the maturity of the coal gas production technologies and more satisfactory gas qualities [50]. Manufactured gas is a general term for multiple man-made gases, including coal gas, producer gas, water gas, carbureted water gas, oil (petroleum) gas, etc. The name of a particular manufactured gas depends on both the source and the gas production procedure. Overlaps in the naming system for manufactured gases are apparent. For example, producer gas can also be called coal gas if it is generated by coal. Water gas is produced in a gasification process in a carbon fuel bed with steam, and coal is decomposed into hydrogen and carbon monoxide [56]. Carbureted water gas is water gas with added carbon contents to increase the heating value of the gas, which is achieved by adding oil to hot gas in the presence of steam and is then thermally cracked to gaseous constituents [57]. Oil (petroleum) gas is produced in a gasification process where oil is thermally cracked in the presence of steam to produce a fuel gas. Reference [58] introduced the simplified processes for the production procedures of different manufactured gases.

Due to the variety of sources and production procedures, manufactured gases vary in composition and heating values. Table 3 shows the species percentages in selected gaseous fuels. The major combustible fuels include hydrogen, carbon monoxide, methane, and ethane. The diluents are mainly nitrogen and carbon dioxide. It should be noted a range exists for given specific gas species in different manufactured gases or natural gas. Table 3 shows a few representative values from [58] and [59].

Table 3. Properties of selected manufactured gases and natural gas.

Contents	Producer Gas	Water Gas	Carbureted Water Gas	Coke Oven Gas	Blast Furnace Gas	Oil Gas	Natural Gas
H_2	15%	49.7%	40.5%	57.0%	3.7%	50.0%	0.0%
CO	24.7%	39.8%	34%	5.9%	26.3%	10.2%	0.0%
CH_4	2.3%	1.3%	8.9%	29.7%	0.0%	27.6%	94.5%
C_2H_6	0.0%	0.1%		1.1%	0.0%	-	0.5%
CO_2	4.8%	3.4%	-	1.5%	12.9%	2.6%	0.2%
N_2	52.2%	5.5%	-	0.7%	57.1%	5.1%	4.0%
O_2	0.2%	0.2	-	0.0%	0.0%	0.2%	0.3%
Lower Heating Value (MJ/m^3)	5.8	11.4	20.1	21.5	3.9	19.7	35.7

Compared to wood and coal burning, gaseous fuel combustion offers numerous advantages, including the absence of ash, cleanliness, ease of end-user control, flexible combustion performance, etc. Moreover, the pipeline construction decreased the price of gaseous fuel transport dramatically compared to solid fuels. The calorific value of manufactured gases can also be altered by mixing various gases based on the end user's needs.

However, the variety of gaseous species and the lack of fuel species regulation also brought numerous problems. The major threats brought by manufactured gases were explosion accidents due to the wide flammable range of carbon monoxide and hydrogen, along with carbon monoxide poisoning. For example, the Fire Research Station of the U.K. reported 1007 town gas explosion accidents between 1957 and 1968, which is a time period approaching the end stage of the town gas era [60]. The safety awareness in the 1950s and 1960s period was relatively high, and the technologies of handling manufactured gas were also mature, but explosion tragedies still occasionally occurred. Therefore, it is believed that more accidents happened at the early age of the manufactured gas era in the 19th century without detailed records.

Carbon monoxide poisoning was another significant issue with manufactured gas use. The leakage of manufactured gas from the pipeline into residential houses can be detrimental. As shown in Table 3, the carbon monoxide percentage in manufactured gases ranges from around 6% to 40%, and it can result in human death at the ppm level in minutes [61]. Studies show that carbon monoxide poisoning can be more harmful to the elderly compared to younger people. Chalke et al. [62] conducted experiments and pointed out that people over 65 years old are more prone to carbon monoxide poisoning due to loss of the sense of smell.

Moreover, the ease of access to carbon monoxide through the residential pipeline also tended to increase the suicide rate [63]. Figure 4 shows the relationship between CO percentage in the residential gas supply and the suicide rates in England and Wales. In the 1940s, the CO percentage in town gas of U.K. was between 10% and 20%. In the 1950s, a method of manufacturing gas from oil products and naphtha began to be adopted, which decreased the CO percentage in this noncoal-based gas down to around 1%. Meanwhile, natural gas was discovered worldwide, and it quickly gained popularity in the U.K. due to its high heating value, CO-free, clean combustion performance, and continuously decreasing price due to its abundant supply. By 1971, around 69% of the gas in the U.K. domestic gas supply system was natural gas. As can be seen in Figure 4a, CO was almost depleted from the U.K. residential gas supply system by 1975. Figure 4b shows the correlated suicide rates of England and Wales, which clearly notes that the rate of CO poisoning suicide significantly decreased, as there was less access to CO from pipelines in homes.

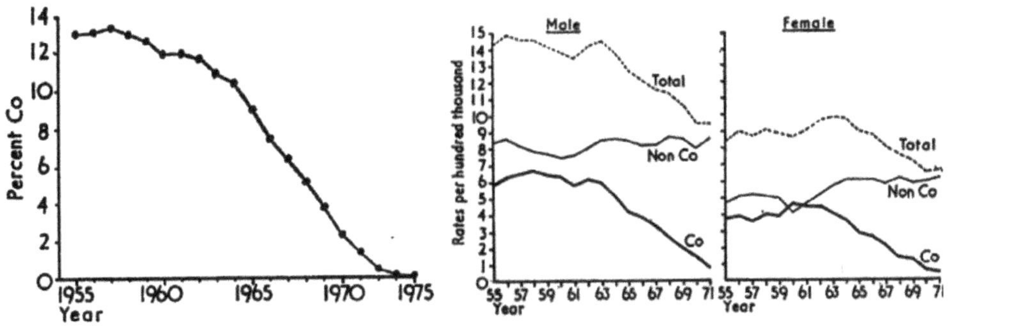

(a) (b)

Figure 4. Relationship between residential gas species and suicide rates in U.K.: (**a**) CO percentage in the domestic gas of U.K. (**b**) England and Wales: suicide rates by mode of death [64].

2.3. *From the 1950s to the 1990s: Manufactured Gas to Natural Gas and LNG*

The nation-wide organized transition from manufactured gas to natural gas occurred in the U.K., which is a good example of a government-assisted energy transition. In 1948, the U.K. Parliament passed the Gas Act and nationalized the gas industry through amalgamation. At that time, 1050 gas works existed, supplying a total of 2119 million therms to 11.3 million consumers [50]. The Gas Act also established the British Gas Council with 12 Area Gas Boards, which were responsible for arranging their own supplies and finances. The British Gas Council consisted of the 12 chairmen of the Area Boards, a Deputy Chairman, and a Chairman, whose role was to advise the government of the policies in the gas industry and coordinate different Area Boards.

A series of policies promoted by the British Gas Council significantly changed the British gas industry. For example, after the Gas Council was founded, 622 old gasworks were closed, some of the larger works were extended and linked together, and 21,200 miles of new gas mains were laid. Therefore, by 1962, the town gas production units decreased from 1050 in 1949 down to 341, of which 74 units produced around 73% of the nation's town gas [65]. With the development of the industrial sector, the gas consumption in industry rose from 639 million therms (in 1953) to 819 million therms (in 1960).

Contrary to the wide adoption of town gas in industry, the town gas consumption in residential houses decreased from 1366 million therms (in 1953) to 1268 million therms (in 1960). This was mainly due to the increase in the price of town gas, which led the residential sector to look for alternative fuels. From 1950 to 1960, the town gas price increased by 65%; however, the electricity price only increased by 25% over the same time period. The major reason for the town gas price increase was the increasing cost and difficulty of sustaining suitable supplies of coking coal, for which the steel industry was also a major customer [66]. Facing these difficulties, the British Gas Council proposed three solutions:

1. The gasification of alternative, lower-grade coal.
2. Gas production from petroleum instead of coal.
3. The introduction of natural gas to enrich manufactured gas.

Considering the cost and technology availabilities, the third solution became the final answer to solve the dilemma of the British gas industry. In 1953, the British Council started to work with the BP Exploration Company to find natural gas reserves all over the British Isles. The discovery of natural gas fields around the world also stimulated the British Gas Council to explore the overseas market. In the late 1950s, British Gas Council started a plan to explore the North Sea and also started to import LNG from the American Gulf Coast, Algeria, etc. [65].

In this manufactured-gas-to-natural-gas transition, combustion devices' adaptation to natural gas posed a significant challenge. Natural gas has a volumetric heating value around twice that of town gas. Other fuel property differences, such as density, flame speed, etc., also introduced difficulties to existing combustion devices in switching from town gas to natural gas without modification. At first, it was proposed to blend other gaseous fuels into natural gas to make a fuel mixture that possessed similar burning properties to town gas while still keeping the heating value of this new fuel higher than that of town gas. This type of gas was once called "GS gas". However, experiments showed that the old town gas appliance burners still needed modification, even with the specially made "GS gas". Therefore, with the abundant supply and the decreasing price of natural gas, the British Council determined to take a one-step action, which was to replace the old residential appliance burners with new ones that were compatible with natural gas, instead of adopting the more expensive two-step method of exchanging both the fuel and appliances.

Starting from 1966, the appliance-conversion act took almost 10 years, with 40 million appliances from 14 million end users. By 1972, appliances from 6 million families were converted, and in the same year, the British Gas Council changed its name to the British Gas Corporation with more concentrated power authorized by the government after the 1972 Gas Act [66–68]. By the end of 1978, around 13.4 million town gas end-users converted to natural gas [50]. Table 4 [50] shows the progress of the British gas end-users converting from town gas to natural gas.

This town-gas-to-natural-gas transition in residential homes was a huge task in British energy development history. More than 13.5 million sites had to be visited with 35 million appliances (200 million burners) and more than 8000 residential appliance models to convert. Because conversion always brings inconvenience to the residential gas end-users, the British Gas Council (British Gas Corporation after 1972) decided to provide advantageous terms for a new appliance purchase, or even offer it free of charge, when the old appliances could not be converted successfully. The British Gas Council also decided to supply the replacement parts to residential houses, so the conversion could be completed on-site. This was more efficient and cheaper compared to shipping the appliances to specific places and conducting conversion there. Figure 5 shows an on-site town gas purging using natural gas in England. To achieve this fuel conversion, not only were the training/education programs provided to technicians but the marketing strategies/advertisement of natural gas to the public were also equally, if not more, emphasized. For example, the first demonstration fuel conversion project was completed on Canvey Island, which gained public confidence in the fuel conversion. To assist the transition, the Conversion Executive was founded to coordinate among the British Gas Council, Area Boards, and the Society of British Industries, which represented the benefit of appliance manufacturers and contractors. At the same time, public relation strategies were adopted, including providing conversion handbooks/manuals, developing a better understanding of local conditions, and approaching gas end-users of different social classes/income levels using varying methods [69,70].

Table 4. Progress of the customer conversion to natural gas in the U.K. from the 1960s to the 1970s.

Year (April to March)	Total Customers (000s)	Annual Number Converted to NG (000s)	Cumulative Number Converted to NG (000s)
1967/68	13,210	51	51
1968/69	13,265	418	469
1969/70	13,347	1093	1562
1970/71	13,372	2029	3591
1971/72	13,390	2407	5998
1972/73	13,506	2100	8098
1973/74	13,559	2108	10,206
1974/75	13,682	1674	11,880
1975/76	13,925	1131	13,011
1976/77	14,200	329	13,340
1977/78	14,516	98	13,438

Figure 5. Town-gas-to-natural-gas transition field work in a Midlands town in England [50].

The successful conversion in Central London of Buckingham Palace, Parliament, the Bank of England, and Westminster Abbey acquired symbolic status in fuel conversion of the British upper class. However, different problems were encountered in other places of Britain. For example, cultural and linguistic issues became a hindering stone in the North East/West, East Midlands, etc. Specific lobby teams were developed to reach out to Asian immigrant communities, where women especially expressed greater concern for the work of fuel conversion [65]. In May 1968, a town gas explosion occurred at Ronan Point of London, which resulted in the death of four people and the injury of seventeen. The tragic accident further shook the public confidence in town gas during its campaign with natural gas. This also provided a window of opportunity for the natural gas supporters to reduce the public concern about converting from town gas to natural gas. For example, the government and the British Gas Council commissioned a report by Prof. Frank Morton of Manchester

University addressing the concerns with natural gas and delivering the advantages of the fuel transition, such as reducing explosion accidents and gas poisoning [65,71].

The final cost of the fuel transition from town gas to natural gas in the U.K. was GBP 577 million in 1978 (~GBP 2.8 billion in 2021), which was less than the estimated GBP 400 million in 1966, considering inflation. However, other costs, such as writing off obsolete plants, brought the total bill to GBP 1027 million (~GBP 5.3 billion in 2021). In the April 1977/March 1978 Annual Report, British Gas Acclaimed: "The cost of the entire conversion programme was met without external subsidy of any kind" [50].

The United States experienced a different procedure of adopting natural gas in residential homes compared to the British way. Before the gas transition, the U.K. had already built a relatively mature town gas industry and gas transporting system beginning in the 1800s and then experienced an organized state-led town-gas-to-natural-gas transition by an almost monopolistic strategy developed by the British Gas Council and the British Government. However, being the world's largest bastion of free market and entrepreneurship, the U.S. natural gas industry was developed in parallel with its manufactured gas industry due to the abundant and cheap supply of natural gas from international trades. From the late 1800s to the early 1900s, U.S. gas companies provided both manufactured gas and natural gas separately to customers depending on their specific needs. Figure 6 shows a gas company that used a wagon to advertise domestic gas usage in the early 1900s. In the U.S., the sale of natural gas exceeded that of manufactured gas in 1935. In the late 1940s, manufactured gas began to be phased out and ultimately replaced by natural gas in the 1950s [50]. However, at that time, the U.K. had not yet started the conversion project.

Figure 6. Advertisement for gas cooking by Atlanta Gas Light Company in the 1900s [50].

The first recorded usage of natural gas in the U.S. was from Fredonia, New York, in 1821. Local residents used logs and later lead pipes to transport the gas to nearby houses for illumination [72]. This first known natural gas practical application in New York was four years before the foundation of the New York Gas Light Company. However, in the United States, from the discovery of natural gas in 1821 to its sale exceeding manufactured gas in 1935, enormous volumes of natural gas were flared due to the lack of local usage or in hopes that oil might be lying under the gas. It was estimated that around 76 trillion cubic feet of natural gas were wasted from the early days of the oil industry to the late 1940s [50]. In the late 1800s, the natural gas industry started to thrive due to the development of pipeline transportation technology and the discovery of natural gas on U.S. land. In 1883, the Penn Fuel Company started to distribute natural gas by pipeline to customers in Pittsburgh's East Liberty and Lawrenceville neighborhoods from Haymarket Well [73]. The first long-distance pipeline (217-mile/14- to 18-inch) transporting natural gas was completed in 1925 by the Mongolia Gas Company of Texas from northern Louisiana to Beaumont, Texas [74]. In 1947, the original oil-transporting pipelines (the "Big Inch" and the "Little Big Inch") connecting Texas and the U.S. East Coast converted the fuel to natural gas after they were sold to Texas Eastern Transmission Corporation after the Second World War [75]. By 1966, natural gas pipelines were lying in every one of the lower 48 states. The natural gas consumption by American residential end-users also increased dramatically, as shown in Table 5 [50].

Table 5. U.S. domestic natural gas consumption from 1945 to 1977.

Year	Total Customers (Millions)	Total Consumption (Trillion Btu)	Average Consumption (Million Btu/Customer)
1945	18.6	775	41.6
1950	22.1	1384	62.6
1955	26.3	2239	85.2
1960	30.4	3188	104.8
1965	34.3	3999	116.5
1970	38.1	4924	129.2
1975	40.9	4991	121.9
1977	41.7	4946	118.7

The natural gas consumed in the U.S. was not only from domestic sites but had also been also imported from Canada, Mexico, or Algeria in liquid form since the 1950s. The natural gases on the energy market possess distinctive properties depending on their sources.

Table 6 [76] summarizes the components and their relative abundance in investigated natural gases on the U.S. market. As shown, the major functional specie of natural gas is methane, together with some higher hydrocarbons. The diluents are usually nitrogen, carbon dioxide, or a trace of oxygen. It should be noted that the contents of the natural gas in Table 6 are only the ones investigated. Natural gas from non-investigated resources might extend the percentage of these species into a larger range.

Table 6. Variation in composition of natural gas in the U.S. market.

Component	Volume (%)			
	Mean	Standard Deviation	Minimum	Maximum
CH_4	93.0	5.5	73	99
C_2H_6	3.0	2.6	0	13
C_3H_8	1.0	1.4	0	8
C_4H_{10}	0.5	1.0	0	7
C_5H_{12}	0.1	0.3	0	3
C_6H_{14}	0.1	0.1	0	1
N_2	1.5	2.9	0	17
CO_2	0.5	0.5	0	2

With natural gas species varying from different sources, the natural gas either mined or purchased by the U.S. gas utility companies also varied significantly. Lacking the unified national natural gas quality regulation, gas companies in the U.S. developed their own standards for the natural gas species, such as the heating value and sulfur contents. Table 7 [77,78] shows the natural gas quality regulations of some gas utility companies in the United States. Some of these regulations were modified every few years. For example, in 2019, SoCalGas decreased their natural gas minimum heating value from 990 Btu/scf to 970 Btu/scf [77].

Fuel-interchangeability studies in the United States started from the manufactured gas era and are still underway for natural gas from different sources. Although lacking a government-led uniform gas-transition strategy like the U.K., early records of fuel-interchangeability studies being conducted on manufactured gases and natural gas in the late 1800s and early 1900s from some U.S. institutes are noted.

Table 7. Natural gas quality standards adopted by gas companies of the United States.

Gas Company	Heating Value (Btu/scf)		Water Content	Various Inerts			Hydrogen Sulfide (H_2S)
	Min	Max	Lbs/MMscf	CO_2	O_2	Total Inerts	(Grain/100 scf)
SoCalGas	970	1150	7	3%	0.2%	4%	0.25
Dominion Transmission	967	1100	7	3%	0.2%	5%	0.25
Equitrans LP	970	-	7	3%	0.2%	4%	0.3
Florida Gas Transmission Co.	1000	1110	7	1%	0.25%	3%	0.25
Colorado Intrastate Gas Co.	968	1235	7	3%	0.001%	-	0.25
Questar Pipeline Co.	950	1150	5	2%	0.1%	3%	0.25
Gas Transmission Northwest Co.	995	-	4	2%	0.4%	-	0.25

In the late 19th century, the large cost increase in gas-making materials resulted in frequent manufactured gas price increases in the United States. In some localities, the gas price increase compelled a decrease in standards of quality. At that time, natural gas production was not as abundant as today. Therefore, a variety of gases were available on the market, including natural gas, manufactured gas, or artificially lower-standard manufactured gas. As a result, the investigation of residential burner mechanisms operating on multiple gases was conducted. The American Gas Association started to work on the project of improving the gas-utilization efficiency, which was one of its earliest tasks after its foundation in 1918. Meanwhile, the engineering section of the U.S. Bureau of Standards started to investigate the design and operation of atmospheric burners, which were, and still are, commonly adopted in residential appliances. Through these studies in the early 1900s, a better understanding of residential appliance burner working principles was obtained [79]. Later in 1932, other than the burner efficiency, reducing emissions from combustion such as carbon monoxide also became a consideration in residential burner design [80]. Therefore, more detailed residential burner optimization started to be conducted considering efficiency, flame characteristics, carbon monoxide emission, etc. [81]. The test setup for residential burners is shown in Figure 7.

Figure 7. Apparatus setup for studying burner performance at the Bureau of Standards of the U.S. in the 1920s [79,81].

One of the most successful fuel-interchangeability studies on residential burners was conducted by the American Gas Association started in 1927. This six-year-long project was called Mixed Gas Research, which conducted around 175,000 individual tests involving more than 250 gases. In 1936, AGA released their general formula, the "AGA Index C", to represent the gas interchangeability, where C = the index of change in performance of appliances. The AGA Index C $= \mathrm{H}/\sqrt{D}$, where H is the gas heating value and D is the gas-specific gravity [82]. This index is nowadays referred to as the Wobbe Index, which is widely adopted around the world in gaseous-fuel interchangeability studies. Not soon after, researchers realized that a single Wobbe Index was not enough to predict all the combustion behaviors. Therefore, more indices regarding flame blow-off, flashback, yellow tip, and so on were developed to examine the performance of residential burners. From the perspective of technical development, these preliminary fuel-interchangeability studies in the U.S. since the 1900s assisted the transition from manufactured gas to natural gas. The research results also benefited the historical energy conversion of the U.K. in the 1950s.

2.4. Since the 1980s: Renewable Gases

With the depletion of natural gas reserves and the urgency to reduce carbon emissions to combat climate change, replacing natural gas with renewable gases in pipelines has been of interest not only to fossil-fuel-dependent countries but also to all of human society. Although many complexities remain to be sorted out, reducing carbon emissions by replacing fossil fuels with renewable energy is a strategy that has international consensus, as evidenced by the endorsement of the United Nations Framework Convention on Climate Change in 1992 [83]. For the residential sector, one of the most effective ways of adopting renewable energy is to replace pipeline natural gas with renewable gases [84], as shown in Figure 8. The benefits of adopting renewable fuels in natural gas pipeline infrastructure include:

1. Reducing carbon emissions, thus combating climate change.
2. Extending the gaseous fuel from singular fossil fuel sources to more renewable sources, such as biogas or hydrogen.
3. Obtaining energy security by alleviating the influence of natural gas price fluctuation.

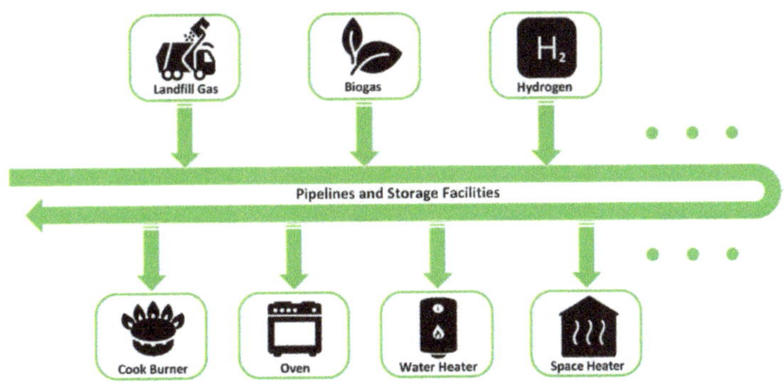

Figure 8. Blending renewable gases into existing natural gas infrastructure.

2.4.1. Biogas

Biogas can be produced from various sources, including wood and agricultural products [85–87], landfill waste [88–90], and sewage [91–93]. Although biogas production sources may vary, the functional part in different sources is mainly biomass. The biogas sources influence the contents of biogas significantly, and Table 8 [94] shows typical biogas contents.

Table 8. Typical biogas contents.

Species	Units	Biogas		
		Sewage Gas	Agricultural Gas	Landfill Gas
CH_4	%	66–75	45–75	45–55
CO_2	%	20–35	25–55	20–30
CO	%	<0.2	<0.2	<0.2
N_2	%	3.4	0.01–5	10–25
O_2	%	0.5	0.01–2	1–5
H_2	%	trace	0.5	0.0
H_2S	$mg/N{\cdot}m^3$	<8000	10–30	<8000
NH_3	$mg/N{\cdot}m^3$	Trace	0.01–2.5	trace
Siloxanes	$mg/N{\cdot}m^3$	<0.1–5.0	trace	<0.1–5.0

The variation of biogas species is not only related to biomass sources but also to gasification technology and procedures. The two major species in biogas are methane and carbon dioxide. Other common fuel species such as carbon monoxide and hydrogen are present at much lower levels. Other species in raw biogas include nitrogen, oxygen, ammonia, hydrogen sulfide, etc. However, some of these trace gases might be detrimental to pipeline or combustion devices through direct contact or combustion. Therefore, before blending biogas into a pipeline or supplying it to combustion devices, the gas needs to be purified or processed to be "upgraded" to meet certain standards.

The major functional species in biogas is methane, which is usually called biomethane due to its production source. The first known large-scale project for blending biomethane into the natural gas grid was in Moenchengladbach, Germany. The sewage gas plant had a raw gas capacity of 400 m^3/day and injected around 20 million m^3 of purified biogas (biomethane) into the local gas grid from 1982 to 1996 [95]. This practice was later followed by the other countries in Europe [96], as well as the U.S. [97], Canada [98], Japan [99], etc. [100,101].

2.4.2. Renewable Hydrogen

Hydrogen has long been touted as a long-term solution for carbon reduction, and hydrogen from renewable sources has increasingly gained attention [15,102,103]. Hydrogen is the most abundant element in the universe and is also a carbon-free combustible fuel. Besides combustion, hydrogen can also be readily used in other applications, such as fuel cells, industrial refineries, or ammonia production for agricultural fertilizers. Moreover, being the lightest element, hydrogen has a much higher heating value on the mass base compared to fossil fuels, which also makes it a competitive energy carrier in addition to being a fuel. Therefore, storing renewable energy (solar, wind, etc.) into hydrogen, transporting hydrogen through pipelines, or directly blending hydrogen into the existing natural gas infrastructure can be an effective way to achieve renewable-energy-adoption and carbon-reduction goals.

Renewable hydrogen as a fuel supply and an energy carrier to power the world is far from being a new idea. The notion of using hydrogen from electrolization dates back 100 years. In 1923, the famous British scientist J.B.S. Haldane published the book *Science and the future* [104], which noted, "The country (U.K.) will be covered with rows of metallic windmills working electric motors which ... will be used for the electrolytic decomposition of water into oxygen and hydrogen. These gasses will be liquefied, and stored in vast vacuum jacketed reservoirs, probably sunk in the ground". Considering hydrogen as a highly condensed energy carrier, Haldane also mentioned that hydrogen "will enable wind-energy to be stored, so that it can be expended for industry, transportation, heating, and lighting, as desired". Compared to coal and petroleum, which were fuels widely used in the U.K. in the 1920s, Haldane also mentioned the environmental benefit of using hydrogen as "no smoke or ash will be produced". Despite the promising utilization of hydrogen, Haldane expressed his concerns regarding the capital investment by saying "the initial costs will be very considerable". As natural gas storage and transporting technology have developed, the current cost of utilizing hydrogen in existing natural gas systems should be much lower than in 1923, when Haldane made this prediction. Today, hydrogen has already found a wide range of applications in industry and transportation and is being considered for residential heating.

Beyond Haldane's suggestion of utilizing wind power, technologies are now available to generate renewable hydrogen from other sources, including renewable feedstocks [105], solar energy [106], nuclear power [107], or directly from water electrolysis by renewable grid power (P2G, Power to Gas) [108]. Hydrogen is categorized by color coding according to various generation sources. Black or brown hydrogen uses black (bituminous) or brown (lignite) coal in the hydrogen-production process, which is the most environmentally damaging method as both the carbon dioxide and carbon monoxide generated during the process are not recaptured. Gray hydrogen is the most common form used currently and is generated from methane or natural gas through steam reformation. Hydrogen is labeled blue when the carbon generated from steam reforming is captured and stored (CSS). Green hydrogen is produced from renewable energy sources, such as solar or wind power, using water electrolysis. Pink hydrogen is created through the electrolysis of water by nuclear energy. Hydrogen has already been integrated into pipeline infrastructure, as previously discussed. We can recall that Britain once had a high percentage of hydrogen (around 50 vol%) in their town gas for over half a century. Therefore, blending hydrogen into our existing pipeline should not be as difficult as the very early adoption of hydrogen in the manufactured gas era. However, since current pipeline infrastructures around the world are mainly designed for natural gas, blending hydrogen into the natural gas system warrants some technical considerations regarding the fuel property differences between natural gas and hydrogen.

Hydrogen has a much lower density compared to natural gas, and its high flame speed and wide flammability range limit its percentage in the current natural gas infrastructure. Therefore, the hydrogen utilization in the natural gas grid is still in the demonstration stage, and the hydrogen-blending percentage is still on the lower end (generally under 10–20 vol%). Reference [109] lists the demonstration projects of hydrogen injection into existing natural gas pipeline infrastructure.

Before trying to increase the percentage of hydrogen in residential pipelines, multiple sectors need to be considered, including pipeline material, sealing, and compression system adjustment. From a combustion perspective, the ability of end uses, such as residential appliances being able to safely operate on hydrogen/natural gas mixtures, is important to assess. Further, what adjustments should be made to existing appliances to increase their tolerance level of hydrogen? Ironically, the process may look like a reversal of the retrofit program undertaken in the U.K. to adapt appliances to switch from town gas to natural gas.

Some preliminary studies have been conducted and are still ongoing in several countries. In the 1970s, Pangborn et al. [110] projected technical considerations on a typical residential atmospheric burner (designed for natural gas) operating on hydrogen. In Europe, the NATURALHY project was initiated in 2004, which aims to prepare for the hydrogen economy. After the start of this project, De Vries et al. conducted both theoretical and experimental research to investigate the feasibility of using hydrogen in our current natural gas appliances. However, these studies mainly stayed on either atheoretical level or only tested on fundamental burners [111–114].

In 2018, Jones et al. conducted tests on a commercial cooktop burner designed for natural gas and successfully injected 34.7% hydrogen into the cooktop burner without inducing flashback [115]. Around the same time, the California Energy Commission (U.S.) sponsored the University of California, Irvine, to conduct a more comprehensive study on existing natural gas appliances operating on natural gas/biogas and natural gas/hydrogen mixtures. Using both experimental and numerical methods, this study investigated the combustion performance of multiple commercial and residential appliances, including cooktop burners [116–118], oven burners [119], water heaters [120,121], room furnaces [122], etc. The investigated combustion performance included flashback/blow-off limits, ignition behavior, burner temperature, emissions (CO, UHC, NO, NO_2, N_2O, NH_3), combustion noise, efficiency, etc. In 2019, the University of Zaragoza worked with the BSH Home Appliances Group in Spain on reaction mechanism simplification to generate numerical models to predict combustion behaviors of appliances operating on hydrogen-rich fuel [123]. Additionally, in 2019, an appliance test project was initiated by E.ON, which is one of the largest companies in Europe working on P2G programs. E.ON planned to raise the hydrogen percentage in the natural gas pipeline grid to as much as 20% in Schopsdorf (Germany). To help achieve this goal, 400 commercial and residential appliances were selected for tolerance tests [124]. These activities are summarized in Table 9.

Table 9. Tests on residential appliances operating on hydrogen-enriched natural gas.

Organization	Starting Time	Appliances/Burners	Description	References
Gas Technology Institute, U.S.	1977	Atmospheric burner	Provided technical considerations of appliances operating on hydrogen-rich natural gas.	[110]
DNV GL Oil & Gas/University of Groningen, Netherlands	2004	Fundamental aspirating burner	Conducted fundamental analysis on natural gas/hydrogen mixture properties and conducted experiments on a Bunsen burner.	[111,114]
Swansea University, U.K.; King Saud University, Saudi Arabia	2018	Cooktop burner	Conducted experiments on a cooktop burner operating on over 30 vol% hydrogen.	[115]
University of California, Irvine, U.S.	2018	Cooktop burner, oven burner, room furnace, water heaters, etc.	Conducted experiments on multiple appliances and tested multiple performances (ignition, emissions, burner temperature, etc.) of various appliances operating on hydrogen/natural gas mixtures.	[116,122]
University of Zaragoza, BSH Home Appliances Group, Spain	2019	Cooktop burner	Developed simplified reaction mechanism to simulate the combustion performance of a cooktop burner.	[123]
E.ON/DVGW, Germany	2019	Various appliances	Around 400 residential appliances were tested while operating on hydrogen/natural gas mixtures.	[124]

As mentioned, hydrogen was already integrated into gas pipeline systems during the manufactured gas era. Replacing part of pipeline natural gas with renewable hydrogen is thus plausible. Technical considerations regarding hydrogen compression, leakage, pipeline material, etc., can draw lessons from the manufactured gas era. To increase the hydrogen tolerance level of current natural gas appliances, insight can also be obtained from the designs of manufactured gas burners. However, it should be noted that, since the manufactured gas era, modern residential burner designs have more requirements. Besides safety concerns, other requirements such as low emissions, high efficiency, ease of operability, longer lifetime period, and visual appeal must be considered in modern residential appliances design. As a result, while "reversing" the transition from manufactured gas to natural gas can give guidance, shifting from natural gas to natural gas/hydrogen mixtures and perhaps ultimately 100% renewable hydrogen will require more effort.

2.5. Summary

When it comes to fuel consumption, technology advancement, or emission control, the power generation, industry, and transportation sectors are usually first in line. The residential sector, which is most related to our daily lives, is usually overlooked. However, as discussed above, nearly every energy transition in human history was either initiated by or highly influenced by the residential sector. From fuelwood to coal, then to oil and different gaseous fuels, industry consumed large amounts of resources, which led to higher prices being less affordable to residents. The old fuel depletion due to the industrial sector drives the residential sector to seek less-expensive alternative fuels. For large fuel-consumption sectors such as industry, when the benefit of adopting the new fuel can balance out with the cost of throwing out their old fuel-consumption devices, the fuel transition will occur in these high-energy-consuming sectors. Therefore, in the energy transition, industry usually has a time delay in adopting the new fuel due to the capital investment required for their devices to adapt to the new energy source. However, the residential sector is more flexible and price-sensitive due to the relatively low capital investment in appliances and more flexibility in consumer adoption/replacement tendencies. Examples besides appliances include electric vehicles, rooftop solar, IOT devices in the home, etc. This places the residential sector in the pioneering position when energy transition occurs. Therefore, investigating the energy transition in the residential sector is not only beneficial to the residential sector study itself but can also help foresee future fuel transitions in other energy-consuming sectors.

Fuel transitions have also been assisted or promoted by other factors, such as government policies and stricter emission regulations. For example, after the Great Smog of London in 1952, the necessity of replacing coal with cleaner fuels was finally widely accepted by the U.K. society. The Clean Air Act 1956 passed by the Parliament of the U.K. was principally in response to the smog, which accelerated the transition of the U.K. from coal consumption to gas adoption. This successful fuel transition is a great example of how rational policy can play a leading role in energy transitions and emission control.

Yet, it is clear that air pollution problems in London were evident well before the 1950s. People were suffering from respiratory deceases due to coal combustion, and church roofs were being damaged by the high sulfur content of coal ashes in the smog hundreds of years ago before the London Smog in 1952. Numerous policies were put in place before the Clean Air Act 1956, but they all failed in some manner. The successful adoption of the Clean Air Act of 1956 is usually attributed to the human reflection and awakening after the deadly tragedy of the 1952 London Smog, which finally ended the coal era in London. Under this government-led energy transition, another critical factor that helped facilitate the transition was the large amount of natural gas discovered in the North Sea in the 1950s. While this discovery did not have the public impact of the London Smog episodes, it enabled Europe to use cheaper and cleaner natural gas than coal. As a result, one must ask: Did the discovery of cheap natural gas or the Clean Air Act 1956 end the smog tragedy? If the Clean Air Act were proposed in the Victorian era, would it be as successful as the one in 1956, or would it have been another failed policy like its predecessors? It is apparent that transitions require many factors to properly align in order for action to occur.

The discussion above touches on general trends in fuel transition. Of course, not all countries in the world were participating in this transition or experiencing the same processes. Even in the U.K. and the U.S., not every city was participating in the energy transition. For example, while the U.S. transitioned into the natural gas era more than 50 years ago, around 10% of U.S. residential houses still use fuelwood or coal for residential heating. Other countries may not follow this procedure or may skip several steps. For example, some developing countries in Africa or Asia continue to use biomass combustion for residential heating, including dry bushes, tree leaves, and residuals of crops, which corresponds to a fuel-consumption stage even preceding the fuelwood era in the developed world. However, these later-developed countries may be able to skip some energy-transition steps and directly enter the renewable gas era by adopting biomass gasification technologies if assisted by the developed world. Certainly, the world must adopt renewable resources in the near future to help combat climate change.

3. Domestic Gas Appliances vs. Electric Appliances

3.1. Lighting Market

A current debate is ongoing relative to transitioning from gas fuel combustion appliances to those that run on electricity [125,126]. Such debates between residential devices operating on electricity vs. gas fuel have been taking place since the introduction of technologies that enabled widespread availability of the electric infrastructure in the 1870s. The first battlefield between the gas and electricity industries was in the lighting market.

Before the 1870s, the lighting market was mainly dominated by kerosene and gas lighting. However, as shown in Table 10, electrical systems started to be introduced at this time, offering a possible alternative for lighting. In 1877, the first electric arc lamp was tested on the street of Paris, which triggered interest in adopting electric lighting in Europe. In 1878, London started to perform tests on electric streetlamps; however, this project was ended a few months later due to the high cost of relatively scarce electricity. While the electric lighting systems were being tested in the 1870s, the gas industry invested in high-efficiency gas lamps to compete with electric lamps. In 1877, Dr. Carl Auer von Welsbach opened the first factory in London to produce incandescent gas light mantles, but the gas mantles did not gain much popularity until the 1890s. From the 1870s to the 1880s, electric lighting technologies advanced rapidly, including the invention of the light bulb by Thomas Edison. The U.K. government passed the Electric Lighting Act in 1882 to promote electric lighting adoption. Ironically, the market share of electric lighting remained limited. In fact, the gas mantle lighting system started to take market share back from electric lighting due to electricity price fluctuations.

Table 10. Lighting system development path since the 1870s.

Year	Event	References
Early 1870s	Introduction of the electric dynamo posed a competitive threat to kerosene and gas lighting.	[127]
1877	Carbon arc lamp was first used in Paris in May 1877, which triggered the interest of London.	[128]
	Dr. Carl Auer von Welsbach of Vienna invented the incandescent gas light mantle. Welsbach established the first factory to make gas mantles in London in 1877.	[53,129]
1878–1879	On 14 December 1878, the Holborn Viaduct electric lighting was switched on in London. Afterward, four circuits, each with four lamps, were powered by a 20-horse-power Robey steam engine that drove a Gramme alternator and exciter. However, the experimental lighting was shut down on 9 May 1879. The city engineer reported that the cost was 3.75 times that of gas.	[128]
1879	Edison invented the long-life light bulb and applied for a patent.	[130,131]
1882	Electric Lighting Act (U.K.) led to the establishment of electricity undertakings on a statutory basis, but the high cost of electricity hindered the wide application of electric lighting.	[132,133]
1890s	Gas mantle started to be widely used for streetlights.	[133]
1900s–1910s	By the early 1900s, the cost of electric lighting was becoming closer to that of gas. However, only the wealthy could afford to wire their houses until the 1910s.	[21]
After 1910s	World War I resulted in coal shortage and the coal gas quality fluctuation, which gave a chance for electric lighting popularization. After World War II, the power-generation technologies and electric grid were rapidly developed, which brought down the electric lighting price, and electric lighting found wide applications not only in street lighting but also in residential houses.	[134]

In the time around World War I, electric lighting started to gain favor due to shortages in coal gas and related price fluctuations. As the technologies for electricity generation and transport matured after World War II, electricity prices dropped significantly, and electric lighting dominated all market sectors as it still does today.

3.2. Cooking, Air/Water Heating, and Other Applications

The majority of energy consumption in residential homes is associated with lighting and heating. Heating includes cooking, air (space) heating, water heating, and other small applications such as irons, dryers, and kettles.

It is believed that the first successful commercial cooking range was from the United States. In 1915, George A. Hughes from Iowa applied for a patent on his invention—an electric range with simple heating element wires set in clay bricks [135]. In 1918, Hughes merged his company with General Electric. In the 1920s, electric appliances started to appear in residential homes in both the U.S. and Europe due to the thrust of appliance manufacturers such as General Electric, Westinghouse, and Siemens [136–138]. Figure 9 shows the evolution of the appearance of electric cooking ranges. Not much change is evident even over 100 years. When it was first developed (as shown in Figure 9a), the electric wires were in parallel; then, they were optimized into coil shapes to promote better contact and support to the cooking utensils. Figure 9b shows an image from the TV show "The French Chef" with Julia Child cooking with electric stove tops in the early 1960s. Even today, the basic electric stove tops remain similar, as shown in Figure 9c.

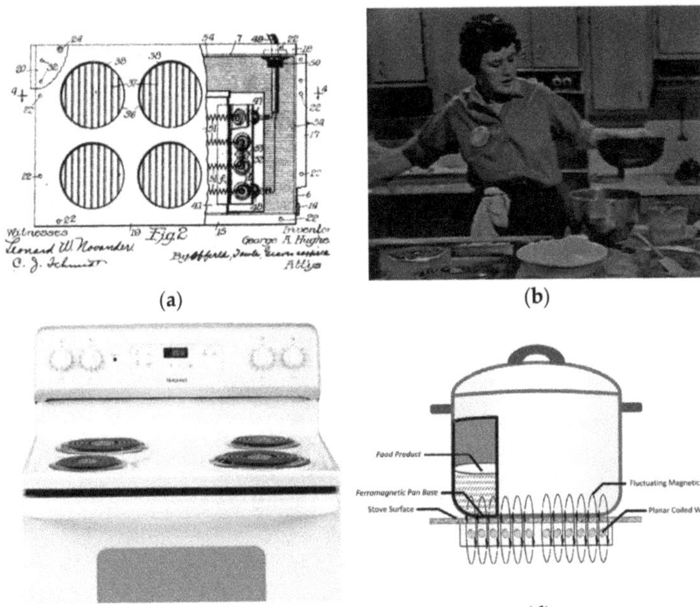

(a) (b) (c) (d)

Figure 9. Historic appearance of electric cooking range: (**a**) Cooking range invented by Hughes, 1915 [135]. (**b**) Julia Child cooking with an electric range, 1963 [139]. (**c**) Electric cooking range on market, 2022. (**d**) Induction cooking schematic [140].

In recent years, another type of electric cooking technology, induction, has gained popularity. Induction cooking technology is based on the principle of static magnetic fields. Alternating current flowing through an induction coil generates a magnetic field with changing polarities. This alternating field "induces" eddy currents that are trapped within the steel material that forms the base of the pan and transfers heat to the food it contains. The induction cooking surface provides a "clean" appearance and higher efficiency compared to traditional electric stoves. However, it is generally more expensive and also requires specific cooking pots and pans that are made of magnetizable materials.

Compared to gas cooking appliances, electric cooking appliances usually have higher operating efficiencies. In the U.S., the average efficiency of gas stoves is around 40%, which is significantly lower than the 74% efficiency of their electric competitors [141]. In Europe, electric cooking appliances have similar efficiencies as that of the U.S., but gas cooking appliances have a higher range of efficiency, from 52% up to 69% [140].

However, despite their early appearance in the market and evolution over 100 years, electric appliances do not dominate the current residential market. The high price and maintenance fee of electric appliances and electricity itself is part of the reason. Additionally, when considering the "source to pot" efficiency, it is evident that many losses occur.

As shown in Figure 10, the total system efficiency of utilizing electric appliances not only includes the appliance efficiency itself but also includes the efficiency of electricity production and transmission process. Currently, most of the electricity in the world is generated from fossil-fuel-based power plants, whose highest efficiency is around 60% [142]. Further, according to the U.S. EPA, in 2019, the average efficiency of fossil-fueled power plants in the United States was only 36% [143]. Figure 10b compares the efficiencies of cooking appliances using multiple energy sources. As shown, even if the end-use efficiency of electric cooking appliances is assumed to be relatively high at 80%, the total efficiency is only around 18%, which is much lower than the natural gas appliance efficiency of 45%.

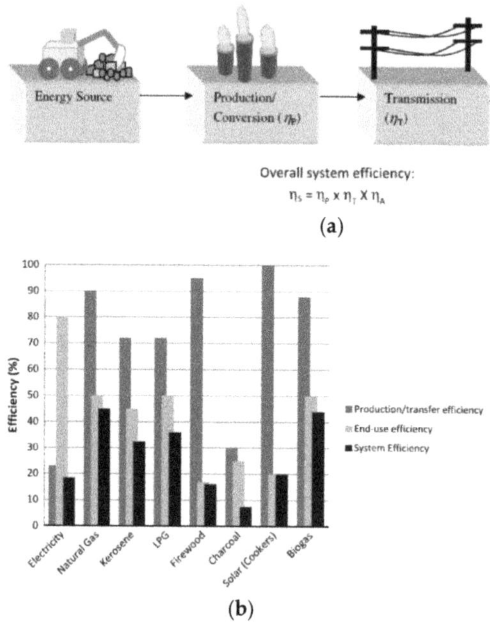

Figure 10. Efficiency of residential cooking appliances using different energy sources: (**a**) System efficiency considerations of utilizing electric appliances. (**b**) Efficiency comparison among cooking appliances [140].

Figure 10b shows an efficiency comparison of different types of residential cooking appliances. Cooking applications exhibit smaller variations in system efficiency due to the large energy loss in the cooking process. However, when it comes to air-heating and water-heating appliances, the gas appliance total system efficiency is higher than the electric competitors. The operating efficiency of residential gas-fired air furnaces is typically greater than 80% [122,144]. Tankless water heater efficiencies are typically greater than 90% [120]. With the large gap between electric appliances and gas appliances in the water-heating and space-heating applications, it does not seem likely that energy savings will result from electrifying residential homes.

One possible advantage of electrifying residential houses is shifting the emissions of greenhouse gases and other pollutants such as CO/NO_X to the electricity generating sector, which is mainly power plants. It is relatively easy to control the emissions from a few power plants rather than from millions of households. However, as electrification increases, more demand for electricity will occur, which will likely drive prices up. Further, with the higher penetration of clean but intermittent renewable energy sources, such as solar and wind, electrifying residential homes may add extra dispatchable fossil-fuel-based power and/or expensive energy-storage devices. This might increase the total greenhouse gas emissions, which is quite the opposite of what is intended. Combined with the decrease in electric appliance efficiency compared to gas-fueled appliances, an even greater impact on the electrical system results again, foreshadowing higher electricity prices.

Figure 11a shows the U.S. natural gas consumption in the residential sector from 1949 to 2020. After the 1970s, natural gas consumption reached a plateau. Starting in the late 1980s, natural gas consumption showed a slowly increasing trend, with a small surge before the COVID-19 pandemic. As shown in Figure 11b, in 2019, about 16.1% of natural gas was consumed in the residential sector, which is the third-largest percentage after power generation and the industrial sector. In the near term, if renewable power supplies are not available, electrification of the residential sector will require this 16.1% to be provided by power generation. As shown in Figure 11c, this electrification will increase the total natural gas consumption by 25%, and the natural gas consumption percentage in the electricity generation sector will increase from 36.3% to 59%, if not assisted by renewable electricity supply. The total U.S. natural gas consumption will also increase by 16% up to more than 36 M cubic feet per year. This assumption is relatively conservative since the efficiency difference of the cooking appliances in Figure 10b is used to predict the system efficiency decrease in the residential sector. If considering the larger system efficiency gap of water heaters and air furnaces between using gas and electricity, even more natural gas would be consumed in the electric-power-generation sector.

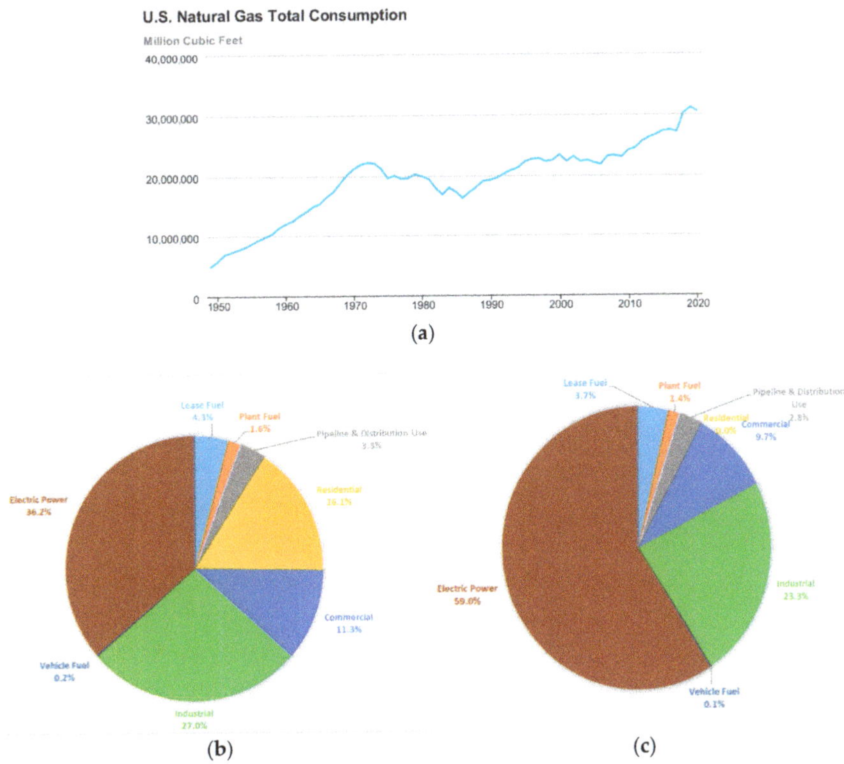

Figure 11. Natural gas consumption situation in the United States and future prediction: (**a**) U.S. natural gas residential consumption from 1949 to 2020 [145]. (**b**) U.S. NG consumption in 2019 (before pandemic) [145]. (**c**) U.S. NG consumption prediction.

The challenge for this transition is significant. The State of California in the United States is one of the most ambitious regions in the world, promoting renewable energy and reducing carbon emissions. In September 2018, California passed Senate Bill 100, setting a world-leading precedent by committing to 100% renewable electricity by 2045 [146]. Together with the renewable electricity goal, California also aims to electrify home appliances [11,147–149]. Yet, even with these actions, a significant time lag in available renewable electricity resources is likely, which warrants an approach with diverse sources in the near term. Figure 12 shows the natural gas consumption by different appliances in residential homes of California (U.S.), which reveals that water heating, air (space) heating, and cooking dominate the use of natural gas [150]. The column charts show the energy sources for these three sectors in large representative cities of California. More than 60% of the households in California use natural gas to cook, and more than 80% of the households use natural gas to heat water and space [151]. Based on the current status, electrifying Californian residential homes will face great challenges. The increased demand for renewable electricity will likely increase utility costs for consumersm, which may hinder the progress of electrifying residential homes. It is estimated by the California Building Industry Association (CBIA) that it could cost (1) USD 7200 per household to upgrade wiring and electrical panels to purchase new appliances, (2) USD 877 higher annual utility costs per household, and (3) an overall cost increase of USD 4.3 to USD 6.1 billion per year in California [152]. It should be noted that when the U.K. was undergoing the transition, the cost of replacing appliances across the country was covered by the government-owned British Gas Council. For countries with a free market today, the cost of converting gas appliances to more expensive electric appliances is less likely to be covered 100% by electricity utility companies.

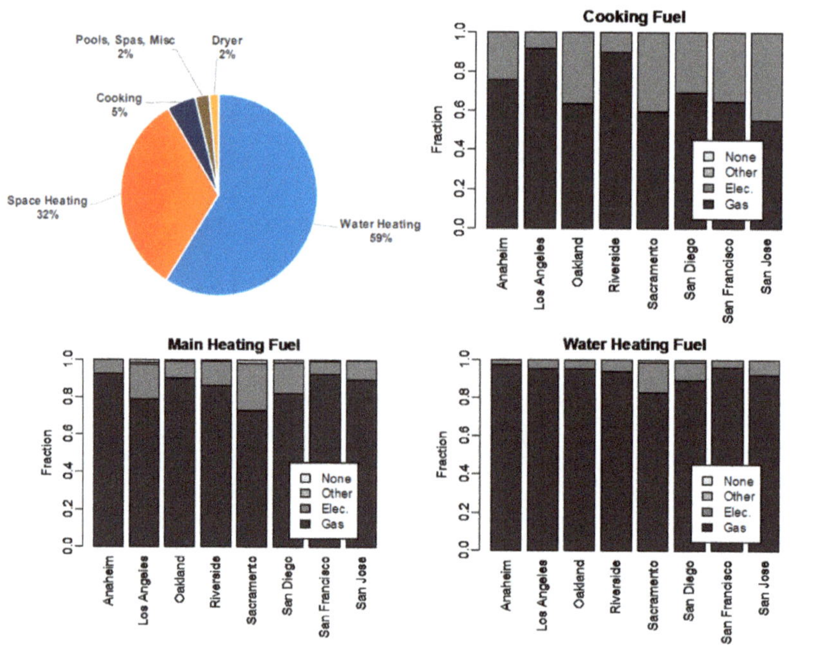

Figure 12. Energy-consumption status by residential appliances in California (U.S.) [150,151].

It is apparent that electrifying residential homes must be assisted by renewable energy sources. This is exacerbated by the fact that current renewable electricity availability is far less than that needed to replace the fossil fuel power supply. Governments around the world are trying to establish policies to increase the percentage of renewable electricity generation.

The discussions above focused on energy consumption and cost concerns. However, another important challenge is people's acceptance of using electric over gas appliances. If not considering possible higher electricity bills, people may have less concern for adopting electric appliances for heating air and water because they cannot feel the difference between gas and electric appliances from the heated air or water. However, when it comes to cooking, a clear preference of gas over electricity is evident. In 2014, a survey was conducted among 100 professional chefs across the United States, and 96 reported they prefer using gas burners. The reported advantages of using gas stoves include fast response speed and controllability, high adaptability of cookware, simple cleaning, and inexpensive maintenance [153]. Advantages of gas over electric cooking appliances have been highlighted since the introduction of electric cookstoves 100 years ago. Given the limited advances in either gas or electric cooking appliances in past decades, sentiment regarding the preferred fuel has not changed. Even though Julia Child used electric stoves on the TV show in the 1960s, it still did not popularize them enough to displace gas ranges. It is also interesting to note that while Julia Child used electric appliances on TV, she actually had a gas stove in her own house [154]. The reason she used the electric stove on TV was more of a sponsor issue. When WGBH planned to shoot the scene at the demonstration kitchen of the Boston Gas Company, they learned that the kitchen was dismantled in 1962. As a result, the Cambridge Electric Light Company offered to provide their electric demonstration kitchen for the "The French Chef" [155]. Moreover, General Electric was the co-sponsor for this show and provided the electric stove. Besides her personal use of a gas range, in published letters to her book editor and friend Avis Devoto, it was also clear that Julia Child strongly preferred cooking with gas [156]. It should be noted that the promotion of any specific energy form might have been motivated by the pursuit of profits.

3.3. Summary

It took less than 40 years for electricity to take over residential lighting since the first arc lamp was lighted in 1877 on a street in Paris. However, more than 100 years have passed since electric appliances for cooking and heating were first introduced, and they still remain a small portion of the heating appliances market. This might be caused by the intrinsic disadvantage of electric appliances for cooking and heating. In reality, people do not necessarily need electricity to heat food, air, and water.

For lighting systems, electricity has the intrinsic advantage of converting to light with high efficiency. However, combustion converts most of the energy in the fuel into heat, which is what is needed for cooking and domestic heating. If heat is required to first be converted into electricity in power plants before the output electricity is transported into residential homes, it needs to be converted to heat again. The energy loss in this process is significant.

Currently, the major consideration of electrifying residential homes is to reduce emissions from the residential sector and allocate the emissions into the power-generation sector. As discussed above, due to the efficiency loss in this appliance-conversion process, there might be an increase in greenhouse gas emissions if this conversion is not assisted by the high uptake of renewable electricity from solar, wind, or hydro power. This renewable electricity adoption might be practical for cities that have access to renewable energy sources. California, for example, has implemented policies to reduce carbon emissions by requiring solar panels to be added to any new residence constructed from 2020 onward [157]. However, this policy might not be practical for cities such as London or Seattle, where solar energy is much less abundant than in California. Even if London could utilize the electricity from offshore wind farms, energy loss in the electricity transport through the grid would occur.

Therefore, instead of electrifying all residential homes, another strategy has already been in commission: replacing pipeline natural gas with renewable gases, including biogas and hydrogen. This method of adopting renewable energy requires much less investment for replacing appliances and wiring adaptation for electric appliances and takes advantage of existing infrastructure widely prevalent in most developed regions.

Table 11 compares various aspects of electrification vs. using renewable fuels.

Table 11. Comparison between electrifying residential homes and adopting renewable gases in residential pipelines.

	100% Electrifying Residential Homes	Adopting Renewable Gases in Pipelines
System Efficiency	System efficiency decreases due to the energy loss in the electricity-generation process.	The efficiency should not change significantly.
Carbon Reduction	Shifting the carbon emissions to power plants may increase total carbon emissions if not assisted by renewable energy sources.	Carbon emissions will drop in proportion to the amount of biomethane or renewable hydrogen adopted.
Combustion Pollutants	Shifts pressure to power plants to reduce emissions.	Pollutants do not change much [116].
Appliance upgrade	Need to replace all the gas appliances with electric appliances, potentially together with upgrading the house's wiring system.	Do not need appliance upgrades for biomethane adoption. Need studies on hydrogen fuel interchangeability.
Cost	Electric appliances are usually of a higher price, and electricity price is more expensive than gas.	Gas price might increase due to limitation of renewable gas source availability.
Technology Availability	Technology available.	Technology available.
Energy Security	Might be vulnerable when blackout occurs due to a single type of energy source.	Energy system is more diverse.
People's Acceptance Level	Gas cooking appliances are usually preferred.	Appliances' performances should be consistent or improved.

A full energy transition from gas to electricity in the residential sector remains a current area of debate. Regardless, if it is ultimately achieved, residential gas pipelines should be and already are adopting renewable fuels.

4. Technical Considerations of Adopting Renewable Fuels in Residential Burners

Adopting renewable fuels in residential homes requires the technical evaluation of fuel-interchangeability studies on residential burners, including burner design principles and properties of renewable fuels.

4.1. Working Principles of Residential Burners

Residential burners are usually mounted in domestic appliances. Compared to higher heating load gas burners designed for turbines or industrial applications, residential burners usually adopt simple designs that are easily manufactured, maintained, or replaced if damaged. Lacking complicated fuel/air flow control systems, most residential burners adopt self-adjusting fuel/air mixing technologies. The most common, if not the only, burner type in residential appliances is self-aspirating, also called an aspirated burner, whose flame type is usually partially premixed. This combustion technology was first introduced in the 1850s by the German chemist Robert Wilhelm Bunsen, who invented the later Bunsen burner and influenced the combustion applications and the gas industry significantly [52].

Figure 13a shows a model of the Bunsen burner. It is an atmospheric burner with a jet from a gas-aspirating part, and the air can be entrained through adjustable shutters into a fuel/air mixing tube before reaching the burner port for combustion. Due to the pre-mixing of fuel and air before combustion, the soot formation and carbon monoxide emissions are significantly diminished, and the flame length or position is more controllable by adjusting the air shutter. Therefore, the Bunsen burner gained popularity not soon after it was introduced to the public and was modified into various applications. Figure 13b shows a domestic aspirating burner for cooking and water heating in the late 19th century. As shown, the fuel is supplied to the burner through a valve at the bottom and then entrains some air through the small circular openings on the burner surface before reaching the flame zone. Till today, self-aspirating burners are still dominant in residential appliances. Figure 13c shows a representative cooktop burner that is available on the market today. As shown, fuel is injected into the burner head while entraining a small amount of surrounding air. Part of the mixture flows out of the ignition ports and arrives at the ignitor through the ignition tube. After ignition, the flame propagates to the burner head, and a full flame is established. Figure 13d shows that most of the current representative appliances on the market, including cooking, air-heating, and water-heating applications, use self-aspirating technology. As can be seen, most of the flames appear blue in color due to their relatively low equivalence ratio at operating conditions, which helps reduce soot, unburned hydrocarbons, and carbon monoxide emissions. However, due to different heating purposes, some appliances might glow yellow/red in color. For example, outdoor grillers usually have yellow tips to enhance heat transfer in grilling. A gas fireplace's flame appearance is close to a diffusion flame because it needs the soot radiation to heat the surrounding space.

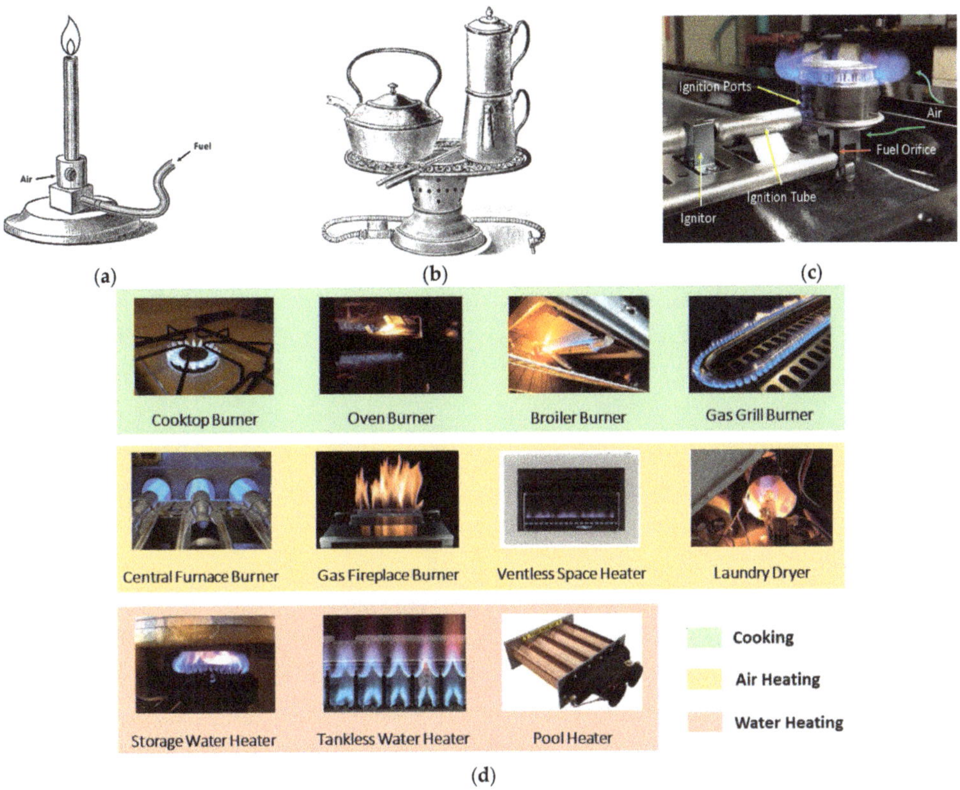

Figure 13. Domestic burner technology and appearance. (**a**) Bunsen burner. (**b**) Domestic burner from the 19th century [50]. (**c**) Current cooktop burner. (**d**) Representative residential appliance burners on market.

Due to the simple burner design and outstanding combustion performance, self-aspirating combustion is not only the most common combustion technique in residential homes but also in laboratories and the light industry.

Although these residential burners adopt the same combustion technology, their appearances might still be distinct from each other. Figure 14 shows the most common self-aspirating burner types in residential burners. Figure 14a shows the configuration of a flat self-aspirating burner, which usually has the fuel and primary air inlet at the burner bottom and flame at the top. The primary air is usually not adjustable for this simple design, and it is hard to replace the designated fuel with alternative fuels. This burner type is widely adopted in cookstoves and storage water heaters. Figure 14b is a tube burner, which is usually equipped with an air shutter at the fuel-injection location. This burner is used in ovens, broilers, outdoor grillers, etc. Figure 14c is a Venturi jet burner, which has the primary air suction openings at the smallest diameter of the burner. The Venturi effect produces a low pressure at the primary air opening area, thus helping aspirate more primary air into the burner. This burner type is usually used for higher-heating-load residential appliances, such as laundry dryers and room furnaces.

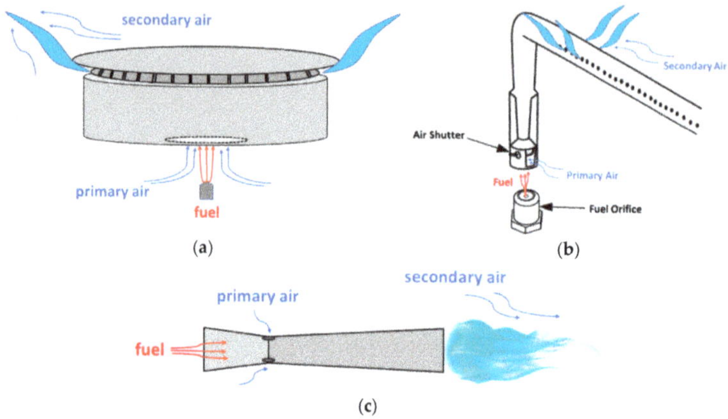

Figure 14. Self-aspirating or aspirated burner types in residential appliances. (**a**) Flat/surface burner. (**b**) Tube burner. (**c**) Venturi burner.

4.2. Fuel Properties

As discussed in the earlier chapters, the most used fuel in residential pipelines is natural gas, whose major content is methane. In the manufactured gas era, the major contents were carbon monoxide, hydrogen, and some higher hydrocarbons. Due to the toxicity of carbon monoxide, it is considered to be a dangerous substance in the residential pipeline and has already been eliminated in most residential pipeline systems around the world. However, due to its carbon-free property and ease of production from renewable energy sources, hydrogen might play an important role in future residential pipeline systems. Therefore, Table 12 lists the key physical and chemical properties of representative residential gaseous fuels. The major fuel properties are from [158,159], with exceptions noted.

Table 12. Property comparison among common residential fuels (at 298.15 K and 1 atm).

Fuel Properties	Unit	Propane (C_3H_8)	Ethane (C_2H_6)	Methane (CH_4)	Carbon Monoxide (CO)	Hydrogen (H_2)
Density	kg/m^3	1.808	1.219	0.648	1.131	0.0813
Viscosity	10^{-5} Pa·s	0.82	0.94	1.11	1.8	0.89
Laminar Flame Speed (ϕ =1)	m/s	0.46	0.45	0.43	0.17 [160]	2.1
Low Flammability	ϕ	0.51	0.50	0.46	0.34	0.1 *
	vol%	2	3	5	12.5	4
High Flammability	ϕ	2.83	2.72	1.64	6.76	7.2 *
	vol%	11	14	15	74	75
Ignition Energy (ϕ =1)	10^{-5} J	30.5	42	33	-	2
Quenching Distance (ϕ =1)	mm	2.0	2.3	2.5	-	0.64
Adiabatic Flame Temperature (ϕ =1)	K	2267	2259	2226	2400	2318
Lower Heating Value	MJ/m^3	83.9	57.9	32.4	11.5	9.8
	MJ/kg	46.4	47.5	50.0	10.2	120.1
Higher Heating Value	MJ/m^3	91.1	63.3	36.0	11.5	11.6
	MJ/kg	50.4	51.9	55.5	10.2	142.1
Wobbe Index	MJ/m^3	73.7	62.4	48.6	11.8	44.3

* It should be noted that the flammability range in [158] has errors. The correction is made through personal communication with Prof. Stephen Turns, and a corrigendum has been published under [117].

Lower hydrocarbons have smaller sizes of molecules, which result in a lower density and ease of leakage. Currently, the leakage of natural gas into the atmosphere from residential buildings is becoming a concern for climate change [161]. Additionally, if pipeline gas contains carbon monoxide, the leakage of this gas can be detrimental due to the density similarity between carbon monoxide and the air, which makes it harder for it to dissipate out of residential houses. Hydrogen has the smallest molecule and thus the lowest density among all the combustible fuels; therefore, hydrogen can more easily leak from pipelines. However, some studies also show that the leakage rate of hydrogen is at the same level as natural gas in low-pressure infrastructures such as a pipeline in residential buildings [162]. Due to the high reactivity of hydrogen, the leakage danger can be a big challenge in its residential applications. Some researchers are also arguing that the leakage of hydrogen might be less dangerous compared to other fuels because the low density of hydrogen increases its ventilation rate. Both numerical and experimental studies were conducted to evaluate the danger level of hydrogen leakage in residential houses, and results show that the ventilation rate of leaked hydrogen is highly dependent on the leakage rate and location [163,164]. There is no definite conclusion yet regarding the danger level of natural gas and hydrogen leakage.

Another important fuel property is the flame speed of the fuel. Table 12 compares the laminar flame speed of different fuels at stoichiometric conditions. As shown, listed hydrocarbons have a very similar flame speed, which is between 0.4 and 0.5 m/s. It is important to note that the laminar flame speed of hydrogen is around five times that of methane. This might increase the danger level of flashback when replacing the pipeline's natural gas with hydrogen.

Another property difference is the fuel flammability range in air. Methane, which is the major content in natural gas, has a relatively small range of 5–15%. This property makes it a "safe" gas to be used in residential buildings. Ethane and propane have similar flammability properties. However, carbon monoxide and hydrogen have much wider flammable ranges, especially hydrogen. Hydrogen can be ignited when its volume percentage in the air is between 4% and 75%. This property brings extra challenges to fuel/air flow rate control and burner design.

The high reactivity of hydrogen is also reflected in its low ignition energy and smaller quenching distance. However, these properties also have positive influences on combustion performance. For example, the ease of hydrogen ignition can reduce the ignition time, which decreases the fuel leakage in the ignition process. Moreover, the high reactivity can also increase the flame stability to allow combustion to take place at a fuel-lean condition, which helps decrease thermal NO_X emissions by deviating the reaction from the stoichiometric condition. On the other hand, the adiabatic flame temperature of hydrogen is slightly higher than that of methane (around 4% higher in stoichiometric conditions). Therefore, in stoichiometric conditions, if the heat-release rate is kept constant, the theoretical thermal NO_X of hydrogen flames is slightly higher than that of methane.

An important fuel property for combustion is the heat-release rate, which is closely related to the heating value and the Wobbe Index. The heating value is either interpreted on a volume base or on a mass base. A higher heating value is usually used to describe the total heat release from the fuel, including the latent heat from the water in the exhaust. In contrast, a lower heating value does not consider the latent heat. It should be noted that natural gas is usually sold from utilities to customers with a higher heating value; however, most residential appliances cannot recover the latent heat. This means that we pay the gas bill for the higher heating value of the fuel, but we only use the lower heating value of the gas. It is important to note that some modern high-efficiency appliances, when equipped with water-condensing heat exchangers, can recover the latent heat from the exhaust. These appliances are usually of higher heating loads, such as room furnaces or water heaters. Appliances with water-condensing systems are usually labeled as having an energy efficiency higher than 90%.

As shown in Table 13, the volumetric heating value of a gaseous fuel decreases as the fuel reaches lower hydrocarbons or carbon/hydrogen-free. However, on a mass base, the heating values of hydrocarbon fuels are at a similar level. Although the volumetric lower heating value of hydrogen is at 9.8 MJ/m^3, which is around 30% of methane, the heating value of hydrogen on the mass base is more than twice that of natural gas. The Wobbe Index is usually correlated to the heat-release rate in combustion applications. It should be noted that although methane and hydrogen have distinct heating values, their Wobbe Index values are very similar, which is a positive sign for fuel interchangeability.

Table 13. AGA flame code classifications.

Code	Flame Description
+5	Flames lifting from ports with no flame on 25% or more of the ports.
+4	Flames tend to lift from ports but become stable after short period of operation.
+3	Short inner cone; flames may be noisy.
+2	Inner cones distinct and pointed.
+1	Inner cones and tips distinct.
0	Inner cones rounded; soft tips.
−1	Inner cones visible; very soft tips.
−2	Faint inner cones.
−3	Inner cones broken at top; lazy wavering flames
−4	Slight yellow streaming in the outer mantles or yellow fringes on tops of inner cones. Flames deposit no soot on impingement.
−5	Distinct yellow in outer mantles or large volumes of luminous yellow tips on inner cones. Flames deposit soot on impingement.

4.3. Interchangeability Considerations

Although all of the fuel properties can influence the combustion performance, there are some essential parameters that need to be considered ahead of other ones when it comes to fuel interchangeability. For example, the heating value and the Wobbe Index are two priorities regarding the heat-release properties of the fuels.

4.3.1. Heating Value and Wobbe Index

The major function of fuel is to convert its chemical energy into heat—this potential is presented as the heating value. Due to the similarity between natural gas and methane properties, methane is used to conduct analysis. The higher heating value and Wobbe Index of methane and other fuel mixtures are plotted in Figure 15.

Figure 15a shows the heating value curves of methane and other fuel mixtures. The volumetric heating value of methane is 36.0 MJ/m^3 (298 K, 1 atm). With alternative fuels mixed with methane, the higher heating value of fuel mixtures shows a linear change. When methane is mixed with fuels of a higher heating value, such as propane and ethane, the volumetric heating value shows an increasing trend. However, when methane is mixed with hydrogen and carbon monoxide, the heating value of the mixture shows a decreasing trend. Due to the similarity of the higher heating value between hydrogen and carbon monoxide, the two lines overlap.

Differing from the fuel mixture heating value plot, the Wobbe Index plot does not show a linear plot due to the definition of the Wobbe Index.

$$\text{Wobbe Index} = \frac{\text{Higher Heating Value}}{\sqrt{\text{Fuel Specific Gravity}}} \tag{1}$$

While the Wobbe Index is widely adopted today, the concept of using the heating value over the square root of fuel density as a fuel-interchangeability parameter originates from an American Gas Association (AGA) study in the 1920s, as mentioned in the previous discussion. In this study, a "C-index" of change in performance of appliances was established as a critical parameter, which is the predecessor of the widely used Wobbe Index today.

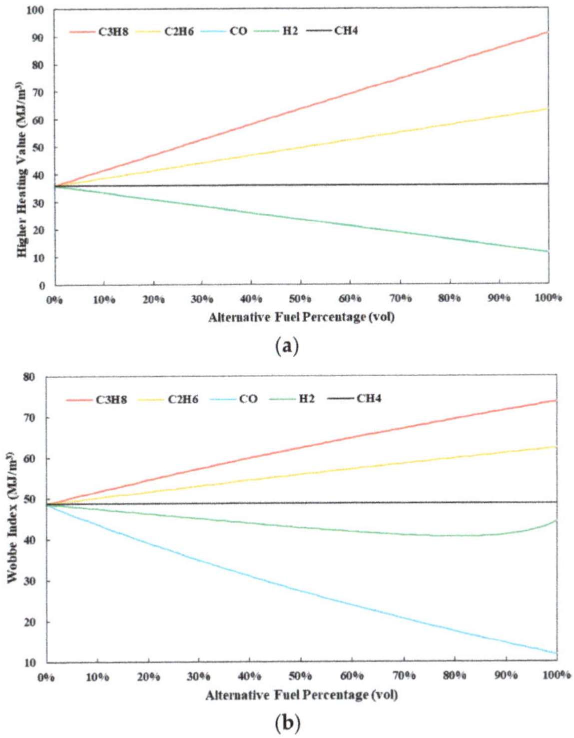

Figure 15. HHV and WI of methane/alternative fuel mixtures. (**a**) Higher heating value of methane/alternative fuel mixtures. (**b**) Wobbe Index of methane/alternative fuel mixtures.

To illustrate why the Wobbe Index can be used to predict the interchangeability of the heat-release rate, the Bernoulli equation can be used as a starting point, and a few assumptions can be made that are applicable to typical operating conditions of appliances. By assuming the fuel flow is at a steady state and the flow is incompressible (constant fluid density) and inviscid, the Bernoulli equation is reduced to Equation (2).

$$p_1 + \frac{1}{2}\rho_f V_1^2 = p_2 + \frac{1}{2}\rho_f V_2^2 \tag{2}$$

In the above equation, p_1 is the natural gas pressure in a household pipeline, which is at a low level, around 2000 Pa. For most appliances, combustion usually takes place in open-air conditions; therefore, p_2 is the atmospheric pressure. Equation (2) can be rearranged to solve for the volumetric flow of the gas exiting the fuel port:

$$\dot{V}_f = A_2 \sqrt{\frac{2\Delta p}{\rho_f\left(1 - \left(\frac{A_2}{A_1}\right)^2\right)}} \tag{3}$$

As shown, the volume flow rate of the fuel is a function of the fuel density, fuel flow areas, and pressure drop from the household pipeline to the combustion device. However, due to the relatively small orifice area for the combustion device compared to the gas feed cross-section, $(A_2/A_1)^2$ can usually be ignored, which leads to Equation (4).

$$\dot{V}_f = A_2\sqrt{\frac{2\Delta p}{\rho_f}} \tag{4}$$

Therefore, the heat output of a combustion device can be presented as

$$\dot{q} = \dot{V}_f HHV = \sqrt{\frac{2\Delta p}{\rho_{air}}}\frac{HHV}{\sqrt{sg_f}} \tag{5}$$

Based on the assumption of constant density and neglecting the viscous effect, the heat output of the combustion device is only a function of the heating value over the square root of the fuel-specific density.

$$\text{Wobbe Index} = \frac{HHV}{\sqrt{sg_f}} \sim \dot{q} \tag{6}$$

The reason why the Wobbe Index is good at predicting the heat output of different combustion devices is due to the validity of the assumptions in practice. The Mach number of the gas flow in the household pipeline is much smaller than 0.3. Therefore, it can be assumed that the gas is incompressible in the pipeline. When the gas flows through the orifice of a combustion device, the gas pressure drop can accelerate the flow, which will result in friction loss. However, the gauge pressure in the pipeline is only at around 2000 Pa, which is less than 2% of the atmospheric pressure. Therefore, the pressure loss is not significant.

The Wobbe Index of methane and alternative fuel mixtures (assuming alternative fuel takes $x\%$ of the volume in the fuel mixture) are given by Equation (7).

$$WI_{Mix} = \frac{(1-x\%)HHV_{CH4} + x\%HHV_{Al}}{\sqrt{\frac{(1-x\%)\rho_{CH4}+x\%\rho_{Al}}{\rho_{Air}}}} \tag{7}$$

As shown in Figure 15b, although the heating values of hydrogen and methane are significantly different, they have similar Wobbe Index values. This observation leads to the consideration that part of the pipeline natural gas can be replaced with hydrogen without influencing the heat output in current combustion devices. In other words, a 5 kW burner will retain a nearly 5 kW rating when operating on natural gas/hydrogen mixtures.

4.3.2. AGA Indices

Even though the heating value and Wobbe Index are essential indicators for fuel interchangeabilities, they cannot predict all the other combustion performances, such as flame characteristics, flashback limits, and emissions. For example, due to the nonlinear character of the Wobbe Index definition, a 37% hydrogen/63% methane mixture has the same Wobbe Index as pure hydrogen. These two different fuel classes might have the same heat-release rate in a combustion device; however, their combustion performances should differ from each other significantly. Therefore, there are also other flame indices that can help predict the combustion performances besides the heat-release rate, such as flame lifting, flash-back, and a yellow tip.

The most widely applied flame indices are AGA flame indices regarding flame lifting (I_L), flame flashback (I_F), and yellow tip (I_Y) [82,165,166]. It should be noted here that in [82], the flame-lifting index (I_L) was wrong.

$$I_L = \frac{K_a}{\frac{f_a a_s}{f_s a_a}\left\{K_s - \log\left(\frac{f_a}{f_s}\right)\right\}} \tag{8}$$

$$I_F = \frac{K_s f_s}{K_a f_a}\sqrt{\frac{H_s}{39934}} \quad (9)$$

$$I_Y = \frac{f_s a_a Y_a}{f_a a_s Y_s} \quad (10)$$

- K: lifting limit constant; a: volume of air theoretically required for complete combustion; f: primary air factor; H: a higher heating value of the fuel (unit: kJ/N·m^3); Y: yellow tip coefficient.
- Subscripts a and s: designating adjustment and substitute gases, respectively.

AGA flame codes are listed in Table 13. When the code stays at low absolute values, the flame is predicted to be acceptable. Gas utility companies usually have their own AGA index regulations. For example, the SoCalGas Company in the U.S. has the following requirement: $I_L \leq 1.06$, $I_F \leq 1.2$, $I_Y \geq 0.8$.

4.3.3. Weaver Indices

Another series of the interchangeability index is the Weaver index series [82]. This series included the considerations of all AGA indices and three other additional indices. The additional indices are the consideration of the Wobbe Index, primary air entrainment, and incomplete combustion prediction.

The heating load index (J_H) is the ratio of the substitute gas Wobbe Index over the original gas Wobbe Index. As J_H becomes close to 1, it becomes easier to keep the heating load constant while replacing the original fuel with the substitute gas.

$$J_H = \frac{H_s\sqrt{D_a}}{H_a\sqrt{D_s}} = \frac{WI_s}{WI_a} \quad (11)$$

- D: specific density of the fuel

The primary air index (J_A) shows the change in the primary air that accompanies a change of gas. It is indicated by Weaver that this index not only provides an accurate measure of primary air to burn two gases but an almost accurate measure of secondary air as well. The ideal value of J_A is also 1.

$$J_A = \frac{a_s\sqrt{D_a}}{a_a\sqrt{D_s}} \quad (12)$$

- a: volume of air theoretically required for complete combustion

The lifting index (J_L) notes the relative tendency for flames of the two gases to lift from the burner ports. When J_L = 1, the two gases are interchangeable regarding the flame position.

$$J_L = J_A \frac{S_s}{S_a}\frac{100 - Q_s}{100 - Q_a} \quad (13)$$

- S: flame speed of the fuel in the air.
- Q: percentage of oxygen in the gas.

The flashback index (J_F) shows the relative tendency for flames of the two gases to flash back into the burners. J_F = 0 indicates there is no difference between the gases in this respect.

$$J_F = \frac{S_s}{S_a} - 1.4J_A + 0.4 \quad (14)$$

The yellow tip index (J_Y) indicates the tendency of fuel-producing soot, therefore yellow tips, in flames. As shown in the definition, the higher content of carbon in a fuel, the higher the index can be. Therefore, the ideal value for J_Y is 0.

$$J_Y = J_A + \frac{N_s - N_a}{110} - 1 \quad (15)$$

- N: the number, in 100 molecules of gas, of the total carbon atoms minus one, except for methane.

The last index in this series is the incomplete combustion index, which is used to represent the tendency of a fuel to produce unburned hydrocarbons or carbon monoxide in the exhaust. When the two gases are equally likely to liberate carbon monoxide during combustion, $J_I = 0$.

$$J_I = J_A - 0.366\frac{R_s}{R_a} - 0.634 \quad (16)$$

- R: the ratio of hydrogen to carbon atoms contained in the fuel.

4.3.4. Other Indices

The AGA indices and Weaver indices are the two most widely applied flame index series regarding fuel interchangeability. Although the Weaver index series (six total) is more comprehensive in covering flame characteristics, the three AGA indices are more commonly adopted by gas utilities around the world. This is because gas utilities usually have their own regulations on fuel species, the heating value, and the Wobbe Index. These basic regulations from utility companies ensure that the combustion performance of the substitute fuel does not deviate from the original fuel too significantly. After the substitute fuel meets all requirements, the application of the AGA Index can help further test the flame lifting, flashback, and yellow tip performance.

Besides the AGA and Weaver indices, there are also other indices developed around the world to predict interchangeability among gaseous fuels. They are usually considered or adopted in scientific research because they are either too focused on one or two combustion performances or too complicated to be applied in industrial applications. The other indices are listed in Table 14.

Table 14. Flame indices besides AGA and Weaver indices.

Name	Description	Reference
Knoy index	Similar to Wobbe Index, which predicts heat-release rate interchangeability.	[167]
Dutton indices	This series of indices mainly considers the incomplete combustion behavior or near blow-off condition for higher hydrocarbons. It consists of incomplete combustion index, lift index, and soot index.	[168]
Delbourg indices	This series is composed of two indices: yellow tip index and soot-formation index.	[168–170]

5. Residential Burner Performance Evaluation

5.1. Efficiency

Efficiency, which is directly related to the utility bill, is one of the primary considerations for a residential householder choosing an appliance. Therefore, improving efficiency has always been a motive for appliance manufacturers to upgrade their products. To guide the residential sector to a higher energy-efficiency level, regulations are also being improved to promote this process worldwide [171]. Among all the energy-efficiency-promoting programs, the most influential one might be the ENERGY STAR program executed by the U.S. Environmental Protection Agency and the U.S. Department of Energy in 1992 [172]. Over the past few years, this program has been adopted worldwide, including in the European Union, Japan, and Australia.

For gas appliances, the efficiency of an appliance is closely related but not equal to the burner's combustion efficiency. The combustion efficiency of a burner is the ratio of the total heat release over the total chemical energy in the fuel. Therefore, the major combustion efficiency loss is from the unburned hydrocarbons or carbon monoxide emissions. In modern residential appliances, these emissions are mostly at the ppm level, which makes the combustion efficiency close to 100%. Therefore, the appliance efficiency loss is most likely from aspects regarding the appliance's inner structure design, including the burner location, heat-exchanger arrangement, etc.

It should also be noted that a high appliance efficiency can also be coupled with a low combustion efficiency, which might increase the emission levels. For example, in a cooktop burner support rack optimization process, it was found that the lower rack could shorten the distance between the flame and the cooking utensil, which helps reduce radiation loss to the environment, thus increasing efficiency. However, this also results in flame impingement onto the cooking utensil surface and can increase the carbon monoxide emission level and also degrade the cooking utensil's lifespan [173]. Therefore, improving appliance efficiency is a task that involves numerous aspects; sometimes, a compromised efficiency level should be adopted to optimize other aspects of the appliance's performance.

Among various residential appliances, water heating and air heating efficiency are relatively easy to measure using the energy output/input method. Most of the existing room furnaces can easily achieve an energy efficiency of 80%, and water heater efficiency can be higher than 90% with a latent-heat-recovering (water-condensing) system [120,122].

Compared to the mature efficiency testing method for air/water-heating appliances, there is yet to be a universal efficiency testing standard for cooking appliances. In contrast to air/water heating, heating food to a certain temperature is not the only purpose of cooking, so it is relatively hard to define a perfect cooking process. However, efforts are still being made around the world to quantify cooking efficiency to help improve cooking appliances' performance. As shown in Table 15, the most common ways to evaluate cooking efficiency are food cooking, water boiling, and object heating. Cooking a specific food is an intuitive way to evaluate cooking performance. For example, Oberascher [174] used boiling eggs and cooking potatoes to evaluate the cooking performance of different cooking appliances. To evaluate the cooking results, the egg yolk status was scaled to five levels: very liquid, liquid, a firm outer edge, can be separated from the egg white, firm, and dry. The egg yolk status was also coupled with the yolk color judgment to distinguish the cooking efficiency of different appliances. Cheng [175] conducted experiments on meat and used the meat moisture loss rate as one of the cooking efficiency measurement methods. The food cooking method is effective in evaluating the cooking performance at one location and time, but it is relatively hard to repeat the results for other testing locations or even the same location at different times due to the tested food quality variation. Moreover, it is also hard to generate a universal quantified standard to evaluate the cooked food characteristics related to cooking efficiency.

Table 15. Cooking appliance efficiency evaluation methods.

Name	Method	Reference
Food cooking	Cooking a specific food (egg, etc.) and evaluating the food appearance to interpret efficiency	[174,175]
Water boiling	Boiling water and calculating the ratio of heat absorbed by water over the total heat-release rate	[116,117,176–180]
Object heating	Heating some materials (usually metal rod/disk) and measure the temperature change of the material to quantify the efficiency	[181,182]

Therefore, the water boiling test is more widely adopted to evaluate cooking appliances' efficiency. To avoid the influence of cooking utensils on cooking efficiency, the American National Standards Institute (ANSI) regulated the dimension of the water boiling pot and testing method [176]. Although boiling water deviates from the actual cooking process, it is relatively easy to quantify the cooking efficiency using Equation (17). Efficiency is defined as the heat absorbed by water (including the latent heat from the escaped water vapor) over the total heat released from the fuel. Most cooking appliances cannot recover heat from the water vapor in the exhaust; therefore, a lower heating value is usually adopted to calculate the heat released from the fuel.

$$\eta_{cooking} = \frac{c_{p,\,water} * m_{water} * \Delta T + h_{vapor} * \Delta m_{vapor}}{V_{fuel} * LHV} \tag{17}$$

Heating an intermediate object instead of food or water is also a method to evaluate the cooking efficiency by measuring the object's temperature variance over time. The material should be able to resist high temperatures and have high thermal conductivity; therefore, metals such as steel or aluminum are usually adopted [181,182]. This method requires the temperature of the intermediate object being relatively uniform over time, so it is more commonly adopted in oven testing.

It should be emphasized that the cooking speed and cooking efficiency are different from each other when evaluating the cooking performance of different fuel classes. If the heat-release rate is constant for two testing conditions, a faster cooking time can be regarded as higher efficiency. However, when the fuel classes have different heating values or Wobbe Index values, less cooking time can also result from a higher heating load of the burner. In this case, faster cooking, which might result in overcooking, may become a non-ideal situation. For example, Zhao et al. [116] evaluated the cooking efficiency of a cooktop burner operating on natural gas and biogas. The results showed that the cooking time increased as natural gas was replaced by biogas due to the heating load decreasing. However, the cooking efficiency stayed at a constant level. The cooking performance of a cooktop burner operating on natural gas/hydrogen mixtures was also tested. It was found that both the cooking time and efficiency did not change much up to 50 vol% hydrogen added into natural gas [117].

5.2. Emissions

Besides increasing efficiency, reducing emissions from appliances is also one of the major incentives for appliance manufacturers due to the stricter emission regulation worldwide. In the past few decades, combustion technologies have been advanced to reduce emissions, mainly in power generation, transportation, and industry. However, the emissions from the residential sector did not draw as much attention as other energy-utilization sectors. Residential appliances are closely related to our daily life, so the emissions from them not only harm the environment but might also threaten our health. Therefore, the existing emission regulations on appliances and the current emission levels of existing appliances should be understood as projecting the future.

5.2.1. Residential Appliance Emission Regulations

Currently, the emission regulations on power generation, transportation, and industrial applications are frequently updated and widely adopted worldwide. These regulations also play a significant role in motivating combustion technology advancement in power generation. It is believed that strict emission regulations for residential appliances can also stipulate the development of residential combustion technologies.

One of the most active agencies promoting appliances standards is the American National Standard Institute (ANSI) and the Canadian Standards Association (CSA). They update the ANSI-CSA standards on appliance performance every few years. Currently, the ANSI-CSA standards are not only adopted in North America but are also being learned and adopted by more countries worldwide. However, the ANSI-CAS standards only cover carbon monoxide emissions for safety reasons while lacking regulation on other emissions, such as nitrogen oxides. Therefore, the ANSI-CSA standards have to be coupled with other standards to form a more comprehensive emission-regulation series, especially standards regulating nitrogen oxides emissions. One of the most pro-active regions regulating NO_X emissions and promoting new combustion–pollutants control technologies is California

(U.S.). To deal with the Los Angeles smog hazard caused by NO_X-O_3 photochemical reactions, several environmental-protection and emission-regulation-promoting agencies were founded after the U.S. Congress passed the Clean Air Act in 1963. These agencies include the U.S. Environmental Protection Agency (EPA, 1970) promoting federal regulations, the California Air Resources Board (CARB, 1967) regulating vehicle emissions, and the South Coast Air Quality Management District (SCAQMD, 1976) issuing emission standards on standard sources of air pollution. Among these agencies, the SCAQMD issues emission regulations for multiple sectors, including industrial and residential applications.

Table 16 lists the existing emission regulations for residential appliances in North America, which is also considered as one of the strictest regulation series in the world. It should be noted that most of these regulations are only adopted in very few places around the world. For example, some of the NO_X emission regulations from SCAQMD are not even extensively adopted in California itself. However, these regulations should indicate the future of appliance emissions levels worldwide.

Table 16. Representative emission regulations for residential appliances in North America.

	Regulation Title	Agency	Adopted Time	Major Contents
Water-heating appliances	ANSI Z21.10.1 [183] ANSI Z21.10.3 [184]	ANSI-CSA	2017	CO < 800 ppm (heating load ≤ 75,000 Btu/h). CO < 400 ppm (heating load > 75,000 Btu/h).
	Rule 69.5.1 [185]	SDCAPCD	2017	NO_X < 10 ng/J (calculated as NO_2) or NO_X < 15 ppm (@ 3% O_2, dry).
	Rule 1121 [186]	SCAQMD	2004	NO_X < 10 ng/J (calculated as NO_2) or NO_X < 15 ppm (@ 3% O_2, dry).
Space-heating appliances	HSC-1988 [187]	California Law	1997	No person shall sell, or offer for sale, any new or used unvented heater that is designed to be used inside any dwelling house or unit, with the exception of an electric heater or decorative gas logs for use in a vented fireplace.
	ANSI Z21.86 [188]	ANSI-CSA	2016	CO: less than 200 ppm in the air free sample.
	Rule 4905 [189]	SJVAPCD	2018	NO_X: 14 ng of oxides of nitrogen (calculated as NO_2) per joule of useful heat delivered to the heated space.
	Rule 1111 [190]	SCAQMD	2018	NO_X: 14 ng of oxides of nitrogen (calculated as NO_2) per joule of useful heat delivered to the heated space.
Cooking appliances	Rule 1153 [191]	SCAQMD	1995	VOC of commercial bakery ovens (≥2 million Btu/h) should be less than 50 pounds/day.
	Rule 1138 [192]	SCAQMD	1997	Chain-driven charbroiler must be equipped with catalytic oxidizer reducing PM and VOC.
	Rule 1131 [193]	SCAQMD	2003	The VOC content of each solvent used ≤120 g per liter of material.
	Rule 1153.1 [194]	SCAQMD	2014	Commercial ovens: CO: 800 ppm (@ 3% O_2). NO_X: 40 ppm (@ 3% O_2)-500 °F-60 ppm (@ 3% O_2).
	ANSI Z21.1 [176]	ANSI-CSA	2016	CO: cooking appliances less than 800 ppm.
	Rule 4692 [195]	SJVUAPCD	2018	In lieu of SCAQMD-Rule 1138. The catalytic oxidizer shall have a control efficiency ≥ 83% for PM-10 emissions and a control efficiency ≥ 86% for VOC emissions.

For residential water-heating appliances, the ANSI-CSA regulates that CO should be lower than 400 ppm or 800 ppm in an undiluted exhaust sample for water-heating appliances of different heating loads [183,184]. The most widely adopted NO_X standard in the U.S. is the SCAQMD-14 ng/J emission regulation [186]. The water heaters that adopt this regulation are usually labeled low-NO_X water heaters on the market. The ultra-low-NO_X 10 ng/J standard is from the San Diego County Air Pollution Control District (SDCAPCD) [185], which is being adopted by more and more residential water-heater manufacturers.

Space heating is also a major emission contributor from residential houses. In 1997, California banned the sale of ventless space heaters in residential homes. Currently, vented heaters being sold or installed in California, the CO emission should be lower than 200 ppm [188]. Additionally, the maximum NO_X emission of residential water heaters dropped from 40 ng/J to 14 ng/J. If the residential water heaters on the market cannot meet this standard by the year 2021, a penalty will be paid by manufacturers for each water heater unit sold in California [122,189,190].

Currently, the major emission regulations on cooking appliances are mainly for VOC or particle emissions from the food, which requires ventilation in the kitchen. The threatening emissions to human health from kitchens are mainly particle emissions instead of emissions from kitchen flames [196,197]. However, residential houses with bad ventilation might have CO and NO_X accumulation in the living space, which creates health issues. Even if the direct influence of residential flame emissions on human health is negligible compared to particle emission from food itself, reducing kitchen flame emissions should also help reduce the total emissions released into the atmosphere. Due to the uncertainness of kitchen burners' operating conditions, ANSI-CSA imposes a relatively "generous" restriction on CO emission (400 ppm/800 ppm) compared to water and space heaters. Although there are no direct NO_X emission regulations on residential cooking appliances, SCAQMD restricts the NO_X emission of commercial ovens to lower than 40 ppm at an operating temperature lower than 500 °F. With the oven temperature exceeding 500 °F, the NO_X emission limit is extended to 60 ppm.

It should be noted that the emission regulations in Table 16 are mainly for gaseous fuels. There are still a lot of developing countries and even some areas in developed countries that use wood, coal, or biomass as residential heating/cooking energy sources. For these areas, particle emissions might be a larger or more direct threat to human health.

5.2.2. Emission Evaluation Methods' Conclusion

As can be seen in Table 16, there are multiple ways to interpret emission levels of an appliance: for instance, the volume percentage in an undiluted exhaust sample (ppm), volume percentage at a corrected oxygen concentration (ppm @ 3% O_2), and emission level per unit energy output (ng/J). The lack of universal standards is reflected in the different ways to interpret emission levels. In fact, not only the emission level interpretations of residential appliances are hard to achieve a consensus on, but there are also various sorts of emission-testing methods.

As shown in Figure 16a, secured exhaust sampling is an ideal way to sample the emissions from an appliance's exhaust-venting location. This is an effective method to avoid air dilution in the exhaust. However, this method is only practical for confined combustion appliances such as room furnaces or water heaters. Before all the exhaust is vented into the atmosphere, it will first be directed into a pipe, which the sampling probe can be located at.

However, a large number of residential burners have open-air flames, which means combustion takes place in an open space and dilution from the surrounding air is inevitable. Therefore, quartz enclosure and hood sampling methods are usually adopted, as shown in Figure 16b,c, respectively. The quartz enclosure sampling method is more suitable for small-sized burners. When adopting this method, it is necessary to make sure that the sampled mixture is homogeneous. If not, multiple locations within the quartz enclosure should be tested over a certain time period, and the average should be calculated to present the emission level of a burner. The exhaust hood sampling method is more commonly adopted for kitchen burners. The exhaust hood collects all the emissions before venting them into the atmosphere. Before the sample is taken from the hood, the exhaust mixture should be homogeneous. The distance between the hood and the burner should be especially emphasized. The hood should be close enough to the burner so that all the exhaust can be collected; otherwise, the tested emission level will be lower than the actual level. However, if the hood is too close to the flame, it might cut off the source of the oxidants (air) to the flame, which results in a carbon monoxide increase in the exhaust. ANSI Z21.1 recommends this gap should be around 5 inches (12.7 cm).

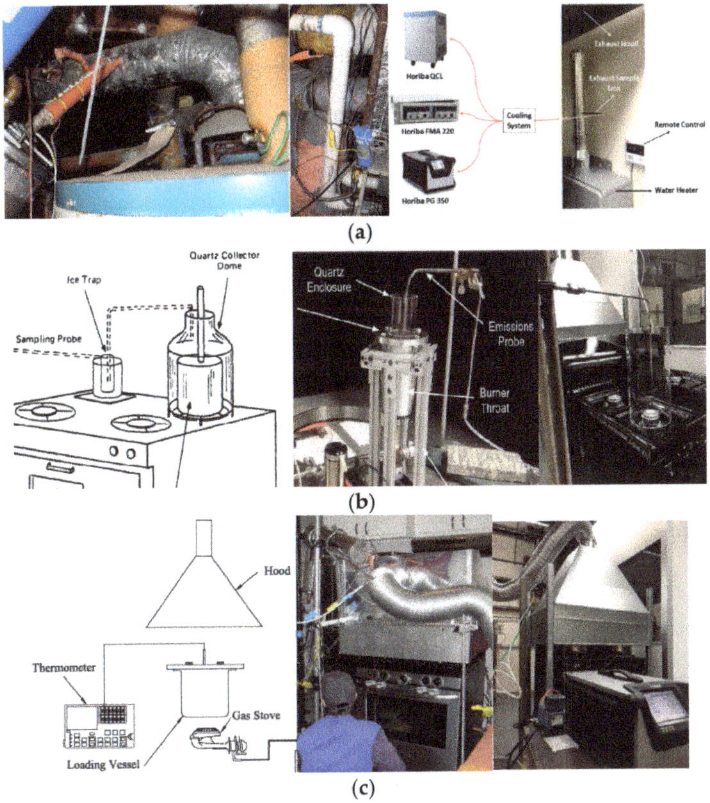

Figure 16. Emission testing methods of residential appliances. (**a**) Secured sampling [120,198]. (**b**) Quartz enclosure sampling [116,199,200]. (**c**) Exhaust hood sampling [117,201].

As shown above, the emission tests on residential burners are sometimes naturally diluted. Even for confined combustion, such as water heaters and room furnace burners, the secondary air from atmosphere also dilutes the emission to a lower absolute level. Therefore, the emissions are usually corrected under a certain reference level so they can be compared with each other.

A common emission correction method is to covert the absolute value of the emission to a certain oxygen level in the exhaust. As shown in Equation (18), the emission is corrected at 3% O_2.

$$[X]_{3\%\ O_2,\ ppm} = [X]_{abs,\ ppm}\frac{20.9 - 3}{20.9 - [O_2]_{abs,\%}} \tag{18}$$

The oxygen correction method has already been verified to be valid in industry, and so it is applied to the residential sector. Zhao et al. [119] compared the influence of the air dilution on an oven's emission and found that the oxygen correction method could interpret the emission level without bias. As shown in Figure 17a,b, two exhaust methods were adopted and compared with each other: flue collar sampling without dilution and hood sampling with significant dilution. As shown in Figure 17c, the emissions have periodic readings due to the oven burner's automatic on/off switching to keep the temperature constant. As can be seen in the figure, by the end of the first period, the flue collar sampling method has a NO reading of 35 ppm with 15% O_2 in the exhaust. When the exhaust hood is adopted, the NO emission level is diluted down to around 10 ppm with 19% O_2 in the exhaust. However, when the emissions readings of these two methods are both corrected to 3% O_2, they show a very similar emission reading in spite of the significant difference in dilution.

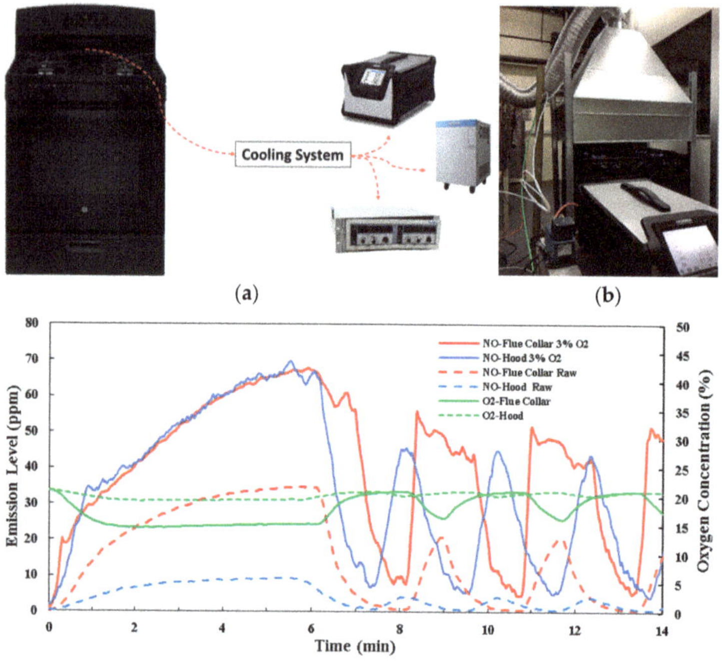

Figure 17. Validation of emission correction method using an oven burner [119]. (**a**) Flue collar sampling, (**b**) Exhaust hood sampling, (**c**) oven burner emission plot.

Another common reference level is 12% CO_2. This is because the carbon dioxide percentage in the dry exhaust for natural gas oxidized in the air in stoichiometric conditions is close to 12% [202]. This correction method is sometimes called the "air-free" method. The 12% CO_2 correction is shown in Equation (19). However, it should be noted that this correction method might increase the corrected emission level when carbon-free fuel (such as H_2) is added to the original fuel. This is due to the intrinsic property of Equation (19): the CO_2 percentage of the exhaust is in the denominator, which decreases as the carbon-free fuel percentage increases. For example, Zhao et al. [122] tested a room furnace burner operating on natural gas/hydrogen mixtures with the hydrogen percentage increasing from 0 to 40%. Although the 3% O_2 correction method shows a flat emission curve, the 12% CO_2 gives an obvious increasing trend for all emissions.

$$[X]_{12\%\ \mathrm{CO_2,\ ppm}} = [X]_{\mathrm{abs,\ ppm}} \frac{12}{[\mathrm{CO_2}]_{\mathrm{abs},\%} - [\mathrm{CO_2}]_{\mathrm{air},\%}} \tag{19}$$

Equation (20) presents the calorific correction, which is also considered as a reliable emission correction method due to its independence from the exhaust species. In theory, the representation provided by calorific correction should be the most robust because it is independent of both O_2 and CO_2 in the clarification. However, it requires more information to be known in addition to the adjusted measured emissions levels, which makes it inherently more difficult to implement.

$$[X]_{ng/J} = \frac{0.1 \cdot [X]_{abs,\ ppm}}{[CO_2]_{abs,\%} - [CO_2]_{air,\%}} \frac{mol\ CO_2}{MJ\ Fuel} M_{X,\ g/mol} \tag{20}$$

5.3. Flame Characteristics

As shown in Table 13, most flames of residential appliances should have a relatively small absolute code value. The flame cones are usually blue in color and have laminar flame characteristics. However, there are also exceptions. Since stable laminar flames are more likely to have a low flow rate, they are more for lower-heating-load appliances. Higher-heating-load appliances, such as room furnaces and laundry dryers, tend to have turbulent flames. Moreover, not all flames are supposed to have blue color. Gas fireplaces need to generate soot to promote radiation; therefore, the flames are usually yellowish or orangish. Although the yellow tip index was developed to guide fuel selection, avoiding yellow tips, outdoor grillers are allowed to have yellow tips, which help increase the heat-transfer rate to the food.

When replacing natural gas with renewable gases, the flame characteristics should be re-evaluated. For example, a hydrogen flame is invisible under sunlight, which might limit its use in existing gas fireplaces or outdoor grillers. The light flame color of hydrogen-rich fuel might also cause danger if an existing flame cannot be identified. However, researchers around the world also find a random reddish glow in hydrogen flames. It is believed that the color is from a contaminant in hydrogen, which might be related to the hydrogen embrittlement effect on metals [118].

5.4. Ignition Performance

Appliance ignition is also a crucial aspect of evaluating the performance of an appliance. Failure of ignition can result in natural gas leakage into residential homes, which endangers the safety of residents. Moreover, a long ignition time also results in methane leakage into the atmosphere for natural gas appliances. The greenhouse gas effect of methane is about 20 times that of carbon dioxide; therefore, the methane emissions from a delayed ignition also contribute to the greenhouse effect [161,203]. Some regulations set an upper limit for appliance ignition time. For example, ANSI Z21.1 [176] requires that residential cooking appliances achieve ignition within four seconds. Higher-heating-load appliances need an even stricter ignition time limit. ANSI Z21.86 [188] regulates that room furnace ignition should be completed within 0.8 s.

Ignition is an unsteady process, which might become a limiting factor for fuel-interchangeability studies. For example, Zhao et al. [117,204,205] replaced more than 75% of natural gas with hydrogen without having flashback in a cooktop burner under steady operating conditions. However, ignition flashback only occurs at 20% hydrogen addition. This is because, in steady operating conditions, the mixture within the burner is usually fuel-rich and out of the flammability range. In ignition conditions, when the fuel starts to fill the burner, the existing air within the burner makes the mixture fall into the flammability range. This causes ignition flashback.

Therefore, ignition performance should draw special attention to the burner design or residential appliance regulations.

5.5. Other Aspects

Other performance metrics are also considered in residential appliance evaluations, such as the burner temperature and combustion noise [117,119,122,206,207]. Modern appliances not only require high durability and working performance, but they are also expected to occupy a small space and have a neat design or even esthetic appearance.

6. Conclusions and Future Prospects

This paper summarizes the historical development of the residential fuel transition and appliances. Future energy transition and energy structure in the residential sector are projected. Following are the major conclusions and future prospects:

Almost every energy transition in human history was initiated in the residential sector. This residential energy source transition is either forced by the higher price of traditional fuels or a result of new needs. For example, the urbanization and industrialization in Europe first caused scarcity of fuelwood and forced the residential sector to turn to coal combustion. As coal needs developed in residential heating, mining technology was advanced, and coal's transportation costs decreased. When the coal price became lower than fuelwood, the industry sector started to switch its energy source to coal, which boosted coal utilization worldwide. The oil industry was first initiated by residential lighting needs. Kerosene was extracted from crude oil to extend the lighting duration and quality. By that time, gasoline was just an abandoned by-product due to its high volatility, which made it difficult to store and use safely in the residential sector. However, as transportation needs increased, the cheap price of gasoline witnessed the booming of the transport engine industry. Natural gas was first regarded as a waste product in the oil industry and was usually vented into the atmosphere. This cheap or almost free gaseous fuel was first transported and utilized in nearby residential houses. Now, natural gas is one of the major fuel sources in the world. Thus, the residential sector fuel choice is interactive with large energy consumption sectors, such as industry and transportation, and on many occasions in history, it was the lead end use for exploring future fuel sources.

The power generation, transportation, and industrial sectors have received attention for centuries due to high energy consumption and demand and the adoption of novel technologies. This is in sharp contrast with the simple light and heating needed by the residential sector. This leaves sectors with a large energy consumption with time lag for energy transition due to their high demand for capital investment for equipment upgrades when facing an energy transition. In contrast, residential appliances are more focused on fuel itself and therefore are quicker to respond and to adapt during an energy transition. The inertia of power generation, transportation, and industrial applications transitioning to renewable energy systems inhibits steps that could be taken against climate change. The more flexible residential sector is, therefore, in the leading position again for transitioning into a renewable future.

The challenge of electrifying the residential sector by 100% is significant, perhaps not practical, based on the current power-generation technologies. Most of the residential needs, such as cooking, air heating, and water heating, come from a low grade of energy: heat. By electrifying the residential sector, the energy loss in electricity generation, transport, and conversion from electricity to heat is significant. Currently, more than 70% of the power in the world is still from combustion. Because the renewable energy sources are far from meeting the current energy needs in the world, electrifying the residential sector might increase the burden on power generation using combustion technologies and might result in more greenhouse gas emissions due to the efficiency loss in the electrification process.

The energy transition of the 21st century might be different from previous transitions. Most of the previous transitions were among existing fossil fuels, and the technology was mainly combustion. However, due to the scarcity and environmental impact of fossil fuels, future energy sources must involve various energy forms and a large number of new technologies. This energy transition in the residential sector will not only be led by the fuel market, but it also highly involves the policymaking by governments considering carbon emission reduction, air pollution control, and energy security issues.

Even if new energy sources are being invented in great numbers, it is very unlikely that the old energy sources will be completely abandoned. For example, even in developed countries, large numbers of people still use traditional energy sources such as burning crop residuals or coal. Renewable gaseous fuels such as biogas and renewable hydrogen are becoming strong competitors serving as an energy source for the future. These renewable gases can be generated by biomass or P2G technologies from solar, wind, hydropower, etc. Compared to the high capital investment and expensive operation and maintenance of fuel cells and solar panels for residential houses, replacing current pipelines' natural gas with renewable gases can be an easier method for the residential energy transition. For a long period of time, maybe hundreds of years to come, combustion technology will coexist with other technologies. As pipelines' natural gas can be replaced by renewable gases and renewable gases are energy carriers for different renewable technologies, combustion might be able to sustain its market share in residential applications, if not increase its share.

Incorporating renewable gases in residential houses requires a re-evaluation of the interchangeability criteria. Technical terms such as fuel mixture properties and flame indices need to be considered collectively. The flame indices being used by the residential sector are from the 1930s, by which time greenhouse gas emissions or air pollutants such as NO_X were not yet considered or understood at all. Therefore, for future fuel-interchangeability studies in residential appliances, new flame indices might need to be developed.

Compared to the frequently updated regulations on gas turbines, car engines, or industrial combustion devices, little attention has been paid to residential appliances. The emission regulations on appliances are mostly local rules, and the residential sector also lacks international testing standards. In the future, more efforts will be needed in policymaking as much as the technology advancement for appliances.

In conclusion, the energy transition in the residential sector in the developed world experienced several stages: from biomass (agricultural residuals) to fuelwood, to coal (heating)/oil (lighting), to manufactured gas, to natural gas/electricity, and now, to the renewable hybrid energy sources.

It should also be noted that we live in a very unbalanced world today, and energy utilization significantly differs in different countries and even in different regions in the same country. Some regions might stay in one energy form for a long period of time, and some might skip several transition steps. For example, in the past few years, some residential sectors in east Asia, such as China, transitioned from coal combustion for residential heating directly into natural gas consumption, skipping the manufactured gas era. Moreover, the Clinton Foundation developed and donated solar stoves to Africa and helped a lot of residential homes there transfer directly from burning crop residuals to renewable energy cooking.

Although renewable energy forms are promising, a single renewable energy form is not likely to be dominant again on earth like fossil fuels. For example, newly built southern California residential homes are required to install solar panels on their rooftops. However, this policy might become unreasonable for residents in places with rainy weather, such as London or Seattle. Direct renewable energy adoption in residential houses requires a large amount of capital investment; therefore, intermediate energy carriers are more likely to be adopted, at least in the near future. However, the competition between renewable electricity and renewable gases could last for a long period of time in the residential sector.

Author Contributions: Conceptualization, Y.Z., V.M. and S.S.; methodology, Y.Z. and V.M.; formal analysis, Y.Z.; investigation, Y.Z. and V.M.; resources, Y.Z.; writing—original draft preparation, Y.Z.; writing—review and editing, V.M.; visualization, Y.Z.; supervision, V.M. and S.S.; project administration, V.M.; funding acquisition, V.M. All authors have read and agreed to the published version of the manuscript.

Funding: This research was funded by the California Energy Commission (Contract PIR-16-017).

Institutional Review Board Statement: Not applicable.

Informed Consent Statement: Not applicable.

Data Availability Statement: Not applicable.

Conflicts of Interest: The authors declare no conflict of interest or any personal circumstances or interest that may be perceived as inappropriately influencing the representation or interpretation of the reported research results.

References

1. Fouquet, R. The slow search for solutions: Lessons from historical energy transitions by sector and service. *Energy Policy* **2010**, *38*, 6586–6596. [CrossRef]
2. Fouquet, R.; Pearson, P.J. Past and prospective energy transitions: Insights from history. *Energy Policy* **2012**, *50*, 1–7. [CrossRef]
3. Samuelsen, S. Rich burn, quick-mix, lean burn (RQL) combustor. In *Gas Turbine Handbook*; 2006; Chapter 3.2.1.3, pp. 227–233. Available online: https://www.netl.doe.gov/sites/default/files/gas-turbine-handbook/3-2-1-3.pdf (accessed on 20 March 2022).
4. Dunn-Rankin, D. (Ed.) *Lean Combustion: Technology and Control*; Academic Press: Cambridge, MA, USA, 2011.
5. Cavaliere, A.; de Joannon, M. Mild combustion. *Prog. Energy Combust. Sci.* **2004**, *30*, 329–366. [CrossRef]
6. Forzatti, P. Present status and perspectives in de-NOx SCR catalysis. *Appl. Catal. A Gen.* **2001**, *222*, 221–236. [CrossRef]
7. Jones, H.R. *The Application of Combustion Principles to Domestic Gas Burner Design*; Taylor & Francis: London, UK, 1989.
8. Singer, B.C. *Natural Gas Variability in California: Environmental Impacts and Device Performance: Literature Review and Evaluation for Residential Appliances: PIER Final Project Report*; California Energy Commission: Sacramento, CA, USA, 2007.
9. Krishnamurti, T.; Schwartz, D.; Davis, A.; Fischhoff, B.; DE Bruin, W.B.; Lave, L.; Wang, J. Preparing for smart grid technologies: A behavioral decision research approach to understanding consumer expectations about smart meters. *Energy Policy* **2011**, *41*, 790–797. [CrossRef]
10. Arabul, F.K.; Arabul, A.Y.; Kumru, C.F.; Boynuegri, A.R. Providing energy management of a fuel cell–battery–wind turbine–solar panel hybrid off grid smart home system. *Int. J. Hydrogen Energy* **2017**, *42*, 26906–26913. [CrossRef]
11. Wei, M.; Nelson, J.H.; Greenblatt, J.B.; Mileva, A.; Johnston, J.; Ting, M.; Yang, C.; Jones, C.; E McMahon, J.; Kammen, D.M. Deep carbon reductions in California require electrification and integration across economic sectors. *Environ. Res. Lett.* **2013**, *8*, 014038. [CrossRef]
12. Fowler, R.; Elmhirst, O.; Richards, J. Electrification in the United Kingdom: A Case Study Based on Future Energy Scenarios. *IEEE Power Energy Mag.* **2018**, *16*, 48–57. [CrossRef]
13. Weiland, P. Biogas production: Current state and perspectives. *Appl. Microbiol. Biotechnol.* **2009**, *85*, 849–860. [CrossRef]
14. Urban, W. Biomethane injection into natural gas networks. In *The Biogas Handbook*; Woodhead Publishing: Wolfgang, Germany, 2013; pp. 378–403.
15. Turner, J.; Sverdrup, G.; Mann, M.K.; Maness, P.C.; Kroposki, B.; Ghirardi, M.; Evans, R.J.; Blake, D. Renewable hydrogen production. *Int. J. Energy Res.* **2008**, *32*, 379–407. [CrossRef]
16. Hosseini, S.E.; Wahid, M.A. Hydrogen production from renewable and sustainable energy resources: Promising green energy carrier for clean development. *Renew. Sustain. Energy Rev.* **2016**, *57*, 850–866. [CrossRef]
17. O'Connor, P.; Cleveland, C. US energy transitions 1780–2010. *Energies* **2014**, *7*, 7955–7993. [CrossRef]
18. Smil, V. *Energy Transitions: History, Requirements, Prospects*; ABC-CLIO: Santa Barbara, CA, USA, 2010; Available online: https://www.abc-clio.com/products/a2598c/ (accessed on 20 March 2022).
19. Humphrey, W.S.; Stanislaw, J. Economic growth and energy consumption in the UK, 1700–1975. *Energy Policy* **1979**, *7*, 29–42. [CrossRef]
20. Edomah, N.; Foulds, C.; Jones, A. Energy Transitions in Nigeria: The Evolution of Energy Infrastructure Provision (1800–2015). *Energies* **2016**, *9*, 484. [CrossRef]
21. Jones, C. The Carbon-Consuming Home: Residential Markets and Energy Transitions. *Enterp. Soc.* **2011**, *12*, 790–823. [CrossRef] [PubMed]
22. Nef, J.U. An Early Energy Crisis and its Consequences. *Sci. Am.* **1977**, *237*, 140–151. [CrossRef]
23. Kernot, C. *The Coal Industry*; Elsevier: Cambridge, UK, 2000.
24. Lewis, W.D.; Binder, F.M. Coal Age Empire: Pennsylvania Coal and Its Utilization to 1860. *Technol. Cult.* **1976**, *17*, 140. [CrossRef]
25. Bowers, B.; Anastas, P. *Lengthening the Day: A History of Lighting Technology*; Oxford University Press: Oxford, UK, 1998.
26. Rose, M.H. *Cities of Light and Heat: Domesticating Gas and Electricity in Urban America*; Penn State Press: University Park, TX, USA, 1995.
27. Gibbs, L.; Anderson, B.; Barnes, K.; Engeler, G.; Freel, J.; Horn, J.; Ingham, M.; Kohler, D.; Lesnini, D.; MacArthur, R.; et al. *Motor Gasolines Technical Review*; Chevron Products Company: San Ramon, CA, USA, 2009.
28. Outhwaite, R.B.; Houston, R.A. The Population History of Britain and Ireland, 1500–1750. *Econ. Hist. Rev.* **1993**, *46*, 821. [CrossRef]
29. Fouquet, R. *Heat, Power and Light: Revolutions in Energy Services*; Edward Elgar Publishing: Cheltenham, UK, 2008.
30. Wilson, C. *England's Apprenticeship, 1603–1763*; Longman Group: London, UK, 1967.
31. Evelyn, J. *Fumifugium*; Rota: Exeter, UK, 1772.
32. Brimblecombe, P. *The Big Smoke (Routledge Revivals): A History of Air Pollution in London since Medieval Times*; Routledge: London, UK, 2012.
33. Price, W.H. *The English Patents of Monopoly*; Harvard University Press: Cambridge, MA, USA, 1920.
34. Peter, S.R. *The Subterranean Forest*; The White Horse Press: Cambridge, UK, 2001.
35. Hatcher, J. *The History of the British Coal Industry*; Clarendon Press: Oxford, USA, 1993.
36. Nef, J.U. *The Rise of the British Coal Industry*; Routledge: London, UK, 2013.
37. Brimblecombe, P. Attitudes and Responses towards Air Pollution in Medieval England. *J. Air Pollut. Control Assoc.* **1976**, *26*, 941–945. [CrossRef]
38. Jenner, M. The politics of London air John Evelyn's Fumifugium and the Restoration. *Hist. J.* **1995**, *38*, 535–551. [CrossRef]

39. O'Connor Peter, A. Aspects of Energy Transitions: History and Determinants. Ph.D. Thesis, Boston University, Boston, MA, USA, 2014.
40. Höök, M.; Aleklett, K. Historical trends in American coal production and a possible future outlook. *Int. J. Coal Geol.* **2009**, *78*, 201–216. [CrossRef]
41. Taylor, G.R. *The Transportation Revolution, 1815–1860*; Routledge: London, UK, 2015.
42. Powell, H.B. *Philadelphia's First Fuel Crisis: Jacob Cist and the Developing Market for Pennsylvania Anthracite*; Pennsylvania State University Press: State College, PA, USA, 1978.
43. Taylor, R.C. *Statistics of Coal*; JW Moore: Archbald, PA, USA, 1848.
44. Brewer, P.J. From Fireplace to Cookstove: Technology and the Domestic Ideal in America. *J. Early Repub.* **2001**, *21*, 566. [CrossRef]
45. Adams, S.P. Warming the poor and growing consumers: Fuel philanthropy in the early republic's urban north. *J. Am. Hist.* **2008**, *95*, 69–94. [CrossRef]
46. Auzanneau, M. *Oil, Power, and War: A Dark History*; Chelsea Green Publishing: White River Junction, VT, USA, 2018.
47. Van Vactor, S.A. *Historical Perspective on Energy Transitions (May 14, 2018)*; USAEE Working Paper; SSRN: Online, 2018.
48. Williamson, H.F.; Daum, A.R. *The American Petroleum Industry: 1859–1899. The Age of Illumination*; Northwestern University Press: Evanston, IL, USA, 1959.
49. Gordon, R.J. *The Rise and Fall of American Growth: The US Standard of Living since the Civil War*; Princeton University Press: Princeton, NJ, USA, 2017.
50. Peebles, M.W. *Evolution of the Gas Industry*; Macmillan International Higher Education: London, UK, 1980.
51. Clendinning, A. Gas Cooker. *Vic. Rev.* **2008**, *34*, 56–62. [CrossRef]
52. Jensen, W.B. The Origin of the Bunsen Burner. *J. Chem. Educ.* **2005**, *82*, 518. [CrossRef]
53. Baumgartner, E. Carl Auer von Welsbach: A pioneer in the industrial application of rare earths. In *Episodes from the History of the Rare Earth Elements*; Springer: Dordrecht, The Netherlands, 1996; pp. 113–129.
54. Webber, W.H. *Town Gas and Its Uses for the Production of Light, Heat, and Motive Power*; A. Constable & Company, Limited: London, UK, 1907.
55. Hatheway, A.W.; Doyle, B.C. Technical History of the Town Gas Plants of the British Isles. In Proceedings of the International Association of Engineering Geologists Conference, 2006; Volume 564. Available online: http://citeseerx.ist.psu.edu/viewdoc/download?doi=10.1.1.472.7729&rep=rep1&type=pdf (accessed on 20 March 2022).
56. Newsome, D.S. The water-gas shift reaction. *Catal. Rev. Sci. Eng.* **1980**, *21*, 275–318. [CrossRef]
57. Murphy, B.L.; Sparacio, T.; Shields, W.J. Manufactured gas plants—Processes, historical development, and key issues in insur-ance coverage disputes. *Environ. Forensics* **2005**, *6*, 161–173. [CrossRef]
58. Hamper, M.J. Manufactured Gas History and Processes. *Environ. Forensics* **2006**, *7*, 55–64. [CrossRef]
59. Ray, H.S. *Energy in Minerals and Metallurgical Industries*; Allied Publishers: New Delhi, India, 2005.
60. Fry, J.F. Town gas explosions in dwellings. *Fire Saf. Sci.* **1970**, *813*, 1.
61. Ernst, A.; Zibrak, J.D. Carbon monoxide poisoning. *N. Engl. J. Med.* **1998**, *339*, 1603–1608. [CrossRef]
62. Chalke, H.D.; Dewhurst, J.R.; Ward, C.W. Loss of Sense of Smell in Old People. A Possible Contributory Factor in Accidental Poisoning from Town Gas. *Public Health* **1958**, *72*, 223–230. [CrossRef]
63. Clarke, R.V.; Mayhew, P. The British Gas Suicide Story and Its Criminological Implications. *Crime Justice* **1988**, *10*, 79–116. [CrossRef]
64. Kreitman, N. The coal gas story. United Kingdom suicide rates, 1960–1971. *J. Epidemiol. Community Health* **1976**, *30*, 86–93. [CrossRef] [PubMed]
65. Arapostathis, S.; Carlsson-Hyslop, A.; Pearson, P.J.G.; Thornton, J.; Gradillas, M.; Laczay, S.; Wallis, S. Governing transitions: Cases and insights from two periods in the history of the UK gas industry. *Energy Policy* **2013**, *52*, 25–44. [CrossRef]
66. Williams, M. *A History of the British Gas Industry*; Oxford University Press: Oxford, UK, 1981.
67. Tiratsoo, E.N. *Natural Gas*; Scientific Press: Beaconsfield, UK, 1972.
68. Elliott, C. *The History of Natural Gas Conversion in Great Britain*; Cambridge Information and Research Services: Cambridge, UK, 1980.
69. Falkus, M.E. *Always under Pressure: A History of North Thames Gas since 1949*; Palgrave Macmillan: London, UK, 1988.
70. Falkus, M. The Canvey Experiment and the Beginnings of Conversion. In *Always under Pressure*; Palgrave Macmillan: London, UK, 1988; pp. 89–122.
71. Morton, F. *Report of the Inquiry into the Safety of Natural Gas as a Fuel*; HM Stationery Office: London, UK, 1970.
72. Lash, G.G.; Lash, E.P. Early History of the Natural Gas Industry, Fredonia, New York. In Proceedings of the AAPG Annual Conventional and Exhibition, Houston, TX, USA, 6–9 April 2014.
73. Tarr, J.A.; Clay, K.; Clay, J.A.T. Boom and Bust in Pittsburgh Natural Gas History: Development, Policy, and Environmental Effects, 1878–1920. *Pa. Mag. Hist. Biogr.* **2015**, *139*, 323. [CrossRef]
74. Troxel, C.E. Long-distance natural gas pipe lines. *J. Land Public Util. Econ.* **1936**, *12*, 344–354. [CrossRef]
75. Liu, H. *Pipeline Engineering*; CRC Press: Boca Raton, FL, USA, 2003.
76. Rao, A. *Sustainable Energy Conversion for Electricity and Coproducts: Principles, Technologies, and Equipment*; John Wiley & Sons: Hoboken, NJ, USA, 2015.

77. Southern California Gas Company. *Rule No. 30: Transportation of Customer-Owned Gas*; Southern California Gas Company: Los Angeles, CA, USA, 2019.
78. American Gas Association. *Report #4A Natural Gas Contract Measurement and Quality Clauses*; American Gas Association: Washington, DC, USA, 2009.
79. Berry, W.M.; Brumbaugh, I.V.; Moulton, G.F.; Shawn, G.B. *Design of Atmospheric Gas Burners*; US Government Printing Office: Washington, DC, USA, 1921.
80. Brumbaugh, I.V.; Jones, G.W. *Carbon Monoxide in the Products of Combustion from Natural Gas Burners*; US Government Printing Office: Washington, DC, USA, 1922.
81. Eiserman, J.; Weaver, E.; Smith, F. A method for determining the most favorable design of gas burners. *Bur. Stand. J. Res.* **1932**, *8*, 669. [CrossRef]
82. Weaver, E. Formulas and graphs for representing the interchangeability of fuel gases. *J. Res. Natl. Inst. Stand. Technol.* **1951**, *46*, 213. [CrossRef]
83. Protocol, K. *United Nations Framework Convention on Climate Change*; Kyoto Protocol: Kyoto, Japan, 1997.
84. DNV. *Hydrogen and Other Routes to Decarbonization of the Gas Network*; FCH2 Conference: Birmingham, UK, 2018.
85. Hobson, P.N. Production of biogas from agricultural wastes. In *Advances in Agricultural Microbiology*; Butterworth Scientific: London, UK, 1982; pp. 523–550.
86. Osman, G.A.; El-Tinay, A.H.; Mohamed, E.F. Biogas production from agricultural wastes. *J. Food Technol.* **2006**, *4*, 37–39.
87. Mackuľak, T.; Prousek, J.; Švorc, Ľ.; Drtil, M. Increase of biogas production from pretreated hay and leaves using wood-rotting fungi. *Chem. Pap.* **2012**, *66*, 649–653. [CrossRef]
88. Gardner, N.; Probert, S. Forecasting landfill-gas yields. *Appl. Energy* **1993**, *44*, 131–163. [CrossRef]
89. Bove, R.; Lunghi, P. Electric power generation from landfill gas using traditional and innovative technologies. *Energy Convers. Manag.* **2006**, *47*, 1391–1401. [CrossRef]
90. Kamalan, H.; Sabour, M.; Shariatmadari, N. A review on available landfill gas models. *J. Environ. Sci. Technol.* **2011**, *4*, 79–92. [CrossRef]
91. Midilli, A.; Dogru, M.; Howarth, C.R.; Ling, M.J.; Ayhan, T. Combustible gas production from sewage sludge with a downdraft gasifier. *Energy Convers. Manag.* **2001**, *42*, 157–172. [CrossRef]
92. Elango, D.; Pulikesi, M.; Baskaralingam, P.; Ramamurthi, V.; Sivanesan, S. Production of biogas from municipal solid waste with domestic sewage. *J. Hazard. Mater.* **2007**, *141*, 301–304. [CrossRef] [PubMed]
93. Acelas, N.Y.; López, D.P.; Brilman, D.W.; Kersten, S.R.; Kootstra, A.M. Supercritical water gasification of sewage sludge: Gas production and phosphorus recovery. *Bioresour. Technol.* **2014**, *174*, 167–175. [CrossRef]
94. Deublein, D.; Steinhauser, A. *Biogas from Waste and Renewable Resources: An Introduction*; John Wiley & Sons: Hoboken, NJ, USA, 2011.
95. Wellinger, A.; Murphy, J.D.; Baxter, D. (Eds.) *The Biogas Handbook: Science, Production and Applications*; Woodhead Publishing: Sawston, UK, 2013.
96. Scarlat, N.; Dallemand, J.-F.; Fahl, F. Biogas: Developments and perspectives in Europe. *Renew. Energy* **2018**, *129*, 457–472. [CrossRef]
97. Gamba, S.; Pellegrini, L.A. Biogas upgrading: Analysis and comparison between water and chemical scrubbings. *Chem. Eng.* **2013**, *32*, 1273–1278.
98. Morin, P.; Marcos, B.; Moresoli, C.; Laflamme, C.B. Economic and environmental assessment on the energetic valorization of organic material for a municipality in Quebec, Canada. *Appl. Energy* **2010**, *87*, 275–283. [CrossRef]
99. Fujino, J.; Morita, A.; Matsuoka, Y.; Sawayama, S. Vision for utilization of livestock residue as bioenergy resource in Japan. *Biomass-Bioenergy* **2005**, *29*, 367–374. [CrossRef]
100. Persson, M.; Jönsson, O.; Wellinger, A. Biogas upgrading to vehicle fuel standards and grid injection. In *IEA Bioenergy Task*; IEA: Paris, France, 2006; Volume 37, pp. 1–34.
101. Available online: www.iea-biogas.net (accessed on 20 March 2022).
102. Kim, S.H.; Kumar, G.; Chen, W.H.; Khanal, S.K. Renewable hydrogen production from biomass and wastes (ReBioH2-2020). *Bioresour. Technol.* **2021**, *331*, 125024. [CrossRef] [PubMed]
103. Chaubey, R.; Sahu, S.; James, O.O.; Maity, S. A review on development of industrial processes and emerging techniques for production of hydrogen from renewable and sustainable sources. *Renew. Sustain. Energy Rev.* **2013**, *23*, 443–462. [CrossRef]
104. Haldane, J.B.; Daedalus, O. *Science and the Future. A Paper Read to the Heretics, Cambridge, on February 4th. 1923*; Kegan Paul: Whitelackington, UK, 1924.
105. Das, D.; Veziroğlu, T.N. Hydrogen production by biological processes: A survey of literature. *Int. J. Hydrogen Energy* **2001**, *26*, 13–28. [CrossRef]
106. Muradov, N.Z.; Veziroğlu, T.N. "Green" path from fossil-based to hydrogen economy: An overview of carbon-neutral technologies. *Int. J. Hydrogen Energy* **2008**, *33*, 6804–6839. [CrossRef]
107. Sorgulu, F.; Dincer, I. Cost evaluation of two potential nuclear power plants for hydrogen production. *Int. J. Hydrogen Energy* **2018**, *43*, 10522–10529. [CrossRef]
108. Götz, M.; Lefebvre, J.; Mörs, F.; McDaniel Koch, A.; Graf, F.; Bajohr, S.; Reimert, R.; Kolb, T. Renewable Power-to-Gas: A technological and economic review. *Renew. Energy* **2016**, *85*, 1371–1390. [CrossRef]

109. Quarton, C.J.; Samsatli, S. Power-to-gas for injection into the gas grid: What can we learn from real-life projects, economic assessments and systems modelling? *Renew. Sustain. Energy Rev.* **2018**, *98*, 302–316. [CrossRef]
110. Pangborn, J.; Scott, M.; Sharer, J. Technical prospects for commercial and residential distribution and utilization of hydrogen. *Int. J. Hydrogen Energy* **1977**, *2*, 431–445. [CrossRef]
111. De Vries, H.; Florisson, O.; Tiekstra, G.C. Safe Operation of Natural Gas Appliances Fueled with Hydrogen/Natural Gas Mixtures (Progress Obtained in the Naturalhy-Project). In Proceedings of the International Conference on Hydrogen Safety, San Sebastian, Spain, 11–13 September 2007.
112. Van Essen, V.M.; De Vries, H.; Levinsky, H.B. Possibilities for admixing gasification gases: Combustion aspects in domestic natural gas appliances in The Netherlands. In Proceedings of the International Gas Union Research Conference, Seoul, Korea, 19–21 October 2011.
113. De Vries, H.; Mokhov, A.V.; Levinsky, H.B. The impact of natural gas/hydrogen mixtures on the performance of end-use equipment: Interchangeability analysis for domestic appliances. *Appl. Energy* **2017**, *208*, 1007–1019. [CrossRef]
114. De Vries, H.; Levinsky, H.B. Flashback, burning velocities and hydrogen admixture: Domestic appliance approval, gas regulation and appliance development. *Appl. Energy* **2019**, *259*, 114116. [CrossRef]
115. Jones, D.R.; Al-Masry, W.A.; Dunnill, C.W. Hydrogen-enriched natural gas as a domestic fuel: An analysis based on flash-back and blow-off limits for domestic natural gas appliances within the UK. *Sustain. Energy Fuels* **2018**, *2*, 710–723. [CrossRef]
116. Zhao, Y.; Choudhury, S.; McDonell, V. Influence of Renewable Gas Addition to Natural Gas on the Combustion Performance of Cooktop Burners. In Proceedings of the ASME 2018 International Mechanical Engineering Congress and Exposition, Pittsburgh, PA, USA, 9–15 November 2018.
117. Zhao, Y.; McDonell, V.; Samuelsen, S. Influence of hydrogen addition to pipeline natural gas on the combustion performance of a cooktop burner. *Int. J. Hydrogen Energy* **2019**, *44*, 12239–12253. [CrossRef]
118. Zhao, Y.; Leytan, K.N.S.; McDonell, V.; Samuelsen, S. Investigation of visible light emission from hydrogen-air research flames. *Int. J. Hydrogen Energy* **2019**, *44*, 22347–22354. [CrossRef]
119. Zhao, Y.; McDonell, V.; Samuelsen, S. Experimental assessment of the combustion performance of an oven burner operated on pipeline natural gas mixed with hydrogen. *Int. J. Hydrogen Energy* **2019**, *44*, 26049–26062. [CrossRef]
120. Zhao, Y.; Morales, D.; McDonell, V. Influence of Blending Hydrogen and Biogas Into Natural Gas on the Combustion Performance of a Tankless Water Heater. In Proceedings of the ASME 2019 International Mechanical Engineering Congress and Exposition, Salt Lake City, UT, USA, 11–14 November 2019.
121. Choudhury, S.; McDonell, V.G.; Samuelsen, S. Combustion performance of low-NO_X and conventional storage water heaters operated on hydrogen enriched natural gas. *Int. J. Hydrogen Energy* **2019**, *45*, 2405–2417. [CrossRef]
122. Zhao, Y.; McDonell, V.; Samuelsen, S. Assessment of the combustion performance of a room furnace operating on pipeline natural gas mixed with simulated biogas or hydrogen. *Int. J. Hydrogen Energy* **2020**, *45*, 11368–11379. [CrossRef]
123. Gimeno-Escobedo, E.; Cubero, A.; Ochoa, J.S.; Fueyo, N. A reduced mechanism for the prediction of methane-hydrogen flames in cooktop burners. *Int. J. Hydrogen Energy* **2019**, *44*, 27123–27140. [CrossRef]
124. E.ON. *Hydrogen Levels in German Gas Distribution System to be Raised to 20 Percent for the First Time*; E.ON: Essen, Germany, 2019.
125. Lewis, J. Short-run and long-run effects of household electrification. In *Economic History Workshop*; Queen's University: Kingston, ON, Canada, 2014.
126. Deason, J.; Borgeson, M. Electrification of Buildings: Potential, Challenges, and Outlook. *Curr. Sustain./Renew. Energy Rep.* **2019**, *6*, 131–139. [CrossRef]
127. Brittain, J.E. The International Diffusion of Electrical Power Technology, 1870–1920. *J. Econ. Hist.* **1974**, *34*, 108–121. [CrossRef]
128. Bourne, R. The beginnings of electric street lighting in the City of London. *Eng. Sci. Educ. J.* **1996**, *5*, 81–88. [CrossRef]
129. Gutmann, V. More light: A short historical sketch of Carl Auer von Welsbach. *J. Chem. Educ.* **1970**, *47*, 209. [CrossRef]
130. Moran, M.E. The light bulb, cystoscopy, and Thomas Alva Edison. *J. Endourol.* **2010**, *24*, 1395–1397. [CrossRef] [PubMed]
131. Edison, T.A. Electric Lamp. U.S. Patent 223,898, 27 January 1880.
132. Hughes, T.P. British Electrical Industry Lag: 1882–1888. *Technol. Cult.* **1962**, *3*, 27. [CrossRef]
133. Shiman, D.R. Explaining the collapse of the British electrical supply industry in the 1880s: Gas versus electric lighting prices. *Bus. Econ. Hist.* **1993**, *21*, 318–327.
134. Fouquet, R.; Pearson, P.J.G. Seven Centuries of Energy Services: The Price and Use of Light in the United Kingdom (1300–2000). *Energy J.* **2006**, *27*, 139–177. [CrossRef]
135. Hughes, G.A. Electrical Stove. U.S. Patent 1,169,827, 1 February 1916.
136. Bowden, S.; Offer, A. Household appliances and the use of time: The United States and Britain since the 1920s. *Econ. Hist. Rev.* **1994**, 725–748. [CrossRef]
137. Ronald, W. *The Electrical Workers: A History of Labor at General Electric and Westinghouse, 1923–1960*; University of Illinois Press: Champaign, IL, USA, 1987.
138. Morton, D.L., Jr. Reviewing the history of electric power and electrification. *Endeavour* **2002**, *26*, 60–63. [CrossRef]
139. WGBH. *Picture Clipped form The French Chef by Julia Child—The Potato Show*; WGBH: Boston, MA, USA, 1963.
140. Hager, T.J.; Morawicki, R. Energy consumption during cooking in the residential sector of developed nations: A review. *Food Policy* **2013**, *40*, 54–63. [CrossRef]

141. DOE. *U.S. Department of Energy: Office of Energy Efficiency and Renewable Energy. Technical Support Document: Energy Efficiency Program for Consumer Products and Commercial and Industrial Equipment: Residential Dishwashers, Dehumidifiers, and Cooking Products, and Commercial Clothes Washers (Chapters 4–6; Appendix 6A)*; DOE: Washington, DC, USA, 2008.
142. Breeze, P. *Gas-Turbine Power Generation, Chapter 7: Combined Cycle Power Plants*; Academic Press: Cambridge, MA, USA, 2016.
143. United States Environmental Protection Agency. Combined Heat and Power Partnership. Available online: https://www.epa.gov/chp/chp-benefits (accessed on 20 March 2022).
144. Han, H.J.; Jeon, Y.I.; Lim, S.H.; Kim, W.W.; Chen, K. New developments in illumination, heating and cooling technologies for energy-efficient buildings. *Energy* **2010**, *35*, 2647–2653. [CrossRef]
145. EIA. *EIA NG Consumption in 2018*; EIA: Washington, DC, USA, 2018. Available online: https://www.eia.gov/dnav/ng/ng_cons_sum_dcu_nus_a.htm (accessed on 20 March 2022).
146. California Legislative Information; Senate Bill No. 100. 2018. Available online: https://leginfo.legislature.ca.gov/faces/billTextClient.xhtml?bill_id=201720180SB100 (accessed on 20 March 2022).
147. Mahone, A.; Subin, Z.; Orans, R.; Miller, M.; Regan, L.; Calviou, M.; Saenz, M.; Bacalao, N. On the path to decarbonization: Electrification and renewables in California and the Northeast United States. *IEEE Power Energy Mag.* **2018**, *16*, 58–68. Available online: https://ieeexplore.ieee.org/abstract/document/8386887 (accessed on 20 March 2022). [CrossRef]
148. Raghavan, S.V.; Wei, M.; Kammen, D.M. Scenarios to decarbonize residential water heating in California. *Energy Policy* **2017**, *109*, 441–451. [CrossRef]
149. Ebrahimi, S.; Mac Kinnon, M.; Brouwer, J. California end-use electrification impacts on carbon neutrality and clean air. *Appl. Energy* **2018**, *213*, 435–449. [CrossRef]
150. DNV GL Energy Insights USA, Inc. *2019 California Residential Appliance Saturation Study*; California Energy Commission: Sacramento, CA, USA, 2020.
151. Fischer, M.L.; Chan, W.; Jeong, S.; Zhu, Z.; Lawrence Berkeley National Laboratory. *Natural Gas Methane Emissions from California Homes*; California Energy Commission: Sacramento, CA, USA, 2018; Publication Number: CEC-500-2018-021.
152. Nemec, R. California Reports Show Homeowners Prefer NatGas over Electrification. Natural Gas Intelligence. 2018. Available online: https://www.naturalgasintel.com/california-reports-show-homeowners-prefer-natgas-over-electrification/ (accessed on 20 March 2022).
153. Kellner, S. 5 Reasons Top Chefs Prefer Gas Cooktops. The Daily Meal. 2014. Available online: https://www.thedailymeal.com/cook/5-reasons-top-chefs-prefer-gas-cooktops (accessed on 20 March 2022).
154. Child, J.; Prud'homme, A. *My Life in France*; Knopf Doubleday Publishing Group: New York, NY, USA, 2006.
155. Fitch, N.R. *Appetite for Life: The Biography of Julia Child*; Knopf Doubleday Publishing Group: New York, NY, USA, 2012.
156. Reardon, J. (Ed.) *As Always, Julia: The Letters of Julia Child and Avis DeVoto*; Houghton Mifflin Harcourt: Boston, MA, USA, 2010.
157. California Energy Commission. 2019 Building Energy Efficiency Standards. Available online: https://www.energy.ca.gov/programs-and-topics/programs/building-energy-efficiency-standards/2019-building-energy-efficiency (accessed on 20 March 2022).
158. Turns, S.R. *Introduction to Combustion*; McGraw-Hill Companies: New York, NY, USA, 1996.
159. Glassman, I.; Yetter, R.A.; Glumac, N.G. *Combustion*; Academic Press: Burlington, MA, USA, 2014.
160. Dong, C.; Zhou, Q.; Zhao, Q.; Zhang, Y.; Xu, T.; Hui, S. Experimental study on the laminar flame speed of hydrogen/carbon mon-oxide/air mixtures. *Fuel* **2009**, *88*, 1858–1863. [CrossRef]
161. Fischer, M.L.; Chan, W.; Delp, W.; Jeong, S.; Rapp, V.; Zhu, Z. An Estimate of Natural Gas Methane Emissions from California Homes. *Environ. Sci. Technol.* **2018**, *52*, 10205–10213. [CrossRef] [PubMed]
162. Mejia, A.H.; Brouwer, J.; Mac Kinnon, M. Hydrogen leaks at the same rate as natural gas in typical low-pressure gas infrastructure. *Int. J. Hydrogen Energy* **2020**, *45*, 8810–8826. [CrossRef]
163. Barley, C.D.; Gawlik, K. Buoyancy-driven ventilation of hydrogen from buildings: Laboratory test and model validation. *Int. J. Hydrogen Energy* **2009**, *34*, 5592–5603. [CrossRef]
164. Hajji, Y.; Jouini, B.; Bouteraa, M.; Elcafsi, A.; Belghith, A.; Bournot, P. Numerical study of hydrogen release accidents in a residen-tial garage. *Int. J. Hydrogen Energy* **2015**, *40*, 9747–9759. [CrossRef]
165. Interchangeability of Other Fuel Gases with Natural Gas. In *American Gas Association Research Bulletin No. 36*; American Gas Association: Washington, DC, USA, 1946.
166. Interchangeability of Various Fuel Gases with Manufactured Gas. In *American Gas Association Research Bulletin No. 60*; American Gas Association: Washington, DC, USA, 1950.
167. Briggs, T. The Combustion and Interchangeability of Natural Gas on Domestic Burners. *Combustion* **2014**, *4*, 67–87.
168. Honus, S.; Kumagai, S.; Yoshioka, T. Replacing conventional fuels in USA, Europe, and UK with plastic pyrolysis gases—Part II: Multi-index interchangeability methods. *Energy Convers. Manag.* **2016**, *126*, 1128–1145. [CrossRef]
169. Visser, P. The testing of cookstoves: Data of water-boiling tests as a basis to calculate fuel consumption. *Energy Sustain. Dev.* **2005**, *9*, 16–24. [CrossRef]
170. Grimsby, L.; Rajabu, H.; Treiber, M. Multiple biomass fuels and improved cook stoves from Tanzania assessed with the water boiling test. *Sustain. Energy Technol. Assess.* **2016**, *14*, 63–73. [CrossRef]
171. Turiel, I. Present status of residential appliance energy efficiency standards—An international review. *Energy Build.* **1997**, *26*, 5–15. [CrossRef]

172. Brown, R.; Webber, C.; Koomey, J. Status and future directions of the Energy Star program. *Energy* **2002**, *27*, 505–520. [CrossRef]
173. Ashman, P.J.; Junus, R.; Stubington, J.F.; Sergeant, G.D. The effects of load height on the emissions from a natural gas-fired domestic cooktop burner. *Combust. Sci. Technol.* **1994**, *103*, 283–298. [CrossRef]
174. Oberascher, C.; Stamminger, R.; Pakula, C. Energy efficiency in daily food preparation. *Int. J. Consum. Stud.* **2011**, *35*, 201–211. [CrossRef]
175. Cheng, Q.; Sun, D.-W.; Scannell, A.G. Feasibility of water cooking for pork ham processing as compared with traditional dry and wet air cooking methods. *J. Food Eng.* **2005**, *67*, 427–433. [CrossRef]
176. *ANSI Z21.1-2016/CSA 1.1-2016*; Household Cooking Gas Appliances; ANSI: New York, NY, USA, 2016.
177. Decker, T.; Baumgardner, M.; Prapas, J.; Bradley, T. A mixed computational and experimental approach to improved biogas burner flame port design. *Energy Sustain. Dev.* **2018**, *44*, 37–46. [CrossRef]
178. Natarajan, R.; Karthikeyan, N.; Agarwaal, A.; Sathiyanarayanan, K. Use of vegetable oil as fuel to improve the efficiency of cooking stove. *Renew. Energy* **2008**, *33*, 2423–2427. [CrossRef]
179. Karunanithy, C.; Shafer, K. Heat transfer characteristics and cooking efficiency of different sauce pans on various cooktops. *Appl. Therm. Eng.* **2016**, *93*, 1202–1215. [CrossRef]
180. Anozie, A.; Bakare, A.; Sonibare, J.; Oyebisi, T. Evaluation of cooking energy cost, efficiency, impact on air pollution and policy in Nigeria. *Energy* **2007**, *32*, 1283–1290. [CrossRef]
181. Probert, D.; Newborough, M. Designs, thermal performances and other factors concerning cooking equipment and associated facilities. *Appl. Energy* **1985**, *21*, 81–222. [CrossRef]
182. Cernela, J.; Heyd, B.; Broyart, B. Evaluation of heating performances and associated variability of domestic cooking appliances (oven-baking and pan-frying). *Appl. Therm. Eng.* **2014**, *62*, 758–765. [CrossRef]
183. *ANSI Z21.10.1-2017/CSA 4.1-2017*; Gas Water Heaters, Volume I, Storage Water Heaters with Input Ratings of 75,000 btu per Hour or Less; ANSI: New York, NY, USA, 2017.
184. *ANSI Z21.10.3-2017/CSA 4.3-2017*; Gas-Fired Water Heaters, Volume III, Storage Water Heaters with Input Ratings above 75,000 btu per Hour, Circulating and Instantaneous; ANSI: New York, NY, USA, 2017.
185. San Diego County Air Pollution Control District. *Rule 69.5.1 Natural Gas-Fired Water Heaters*; San Diego County Air Pollution Control District: San Diego, CA, USA, 2015.
186. South Coast Air Quality Management District. *Rule 1121—Control of Nitrogen Oxides from Residentialtype, Natural Gas-Fired Water Heaters*; South Coast Air Quality Management District: Diamond Bar, CA, USA, 2004.
187. California Legislative Information. Health and Safety Code–HSC-Division 13. Housing-Part 3. Miscellaneous-Chapter 12: Heating Appliances and Installations. Available online: https://law.justia.com/codes/california/2013/code-hsc/division-13/part-3/chapter-12/ (accessed on 20 March 2022).
188. *ANSI Z21.86-2016/CSA 2.32-2016*; Vented Gas-Fired Space Heating Appliances; ANSI: New York, NY, USA, 2016.
189. San Joaquin Valley Air Pollution Control District. *Rule 4905: Natural Gas-Fired, Fan-Type Central Furnaces*; San Joaquin Valley Air Pollution Control District: Fresno, CA, USA, 2020.
190. South Coast Air Quality Management District. *Rule 1111: Reduction of NOX Emissions from Natural Gas-Fired, Fan-Type Central Furnaces*; South Coast Air Quality Management District: Diamond Bar, CA, USA, 2018.
191. South Coast Air Quality Management District. *Rule 1153: Commercial Bakery Ovens*; South Coast Air Quality Management District: Diamond Bar, CA, USA, 1995.
192. South Coast Air Quality Management District. *Rule 1138: Control of Emissions from Restaurant Operations*; South Coast Air Quality Management District: Diamond Bar, CA, USA, 1997.
193. South Coast Air Quality Management District. *Rule 1131: Food Product Manufacturing and Processing*; South Coast Air Quality Management District: Diamond Bar, CA, USA, 2003.
194. South Coast Air Quality Management District. *Rule 1153.1: Emissions of Oxides of Nitrogen from Commercial Food Ovens*; South Coast Air Quality Management District: Diamond Bar, CA, USA, 2014.
195. San Joaquin Valley Air Pollution Control District. *Rule 4692: Commercial Charboiling*; San Joaquin Valley Air Pollution Control District: Fresno, CA, USA, 2018.
196. Schwela, D. Cooking smoke: A silent killer. *People Planet* **1997**, *6*, 24. [PubMed]
197. Wang, L.; Xiang, Z.; Stevanovic, S.; Ristovski, Z.; Salimi, F.; Gao, J.; Wang, H.; Li, L. Role of Chinese cooking emissions on ambient air quality and human health. *Sci. Total Environ.* **2017**, *589*, 173–181. [CrossRef]
198. Singer, B.C.; Apte, M.G.; Black, D.R.; Hotchi, T.; Lucas, D.; Lunden, M.M.; Mirer, A.G.; Spears, M.; Sullivan, D.P. *Natural Gas Variability in California: Environmental Impacts and Device Performance Experimental Evaluation of Pollutant Emissions from Residential Appliances*; Lawrence Berkeley National Lab. (LBNL): Berkeley, CA, USA, 2009.
199. Moschandreas, D.J.; Relwani, S.M.; Billick, I.H.; Macriss, R.A. Emission rates from range-top burners—Assessment of measurement methods. *Atmos. Environ.* **1987**, *21*, 285–289. [CrossRef]
200. Therkelsen, P.; Cheng, R.; Sholes, D.; California Energy Commission. *Research and Development of Natural Draft Ultra-Low Emissions Burners for Gas Appliances*; Lawrence Berkeley National Lab. (LBNL): Berkeley, CA, USA, 2015; Publication number: CEC-500-2016-054.
201. Hou, S.-S.; Lee, C.-Y.; Lin, T.-H. Efficiency and emissions of a new domestic gas burner with a swirling flame. *Energy Convers. Manag.* **2007**, *48*, 1401–1410. [CrossRef]

202. Jahnke, J.A. *Continuous Emission Monitoring*; John Wiley & Sons: Hoboken, NJ, USA, 2000.
203. Blake, D.R.; Rowland, F.S. Continuing Worldwide Increase in Tropospheric Methane, 1978 to 1987. *Science* **1988**, *239*, 1129–1131. [CrossRef]
204. Zhao, Y.; McDonell, V.; Samuelsen, S. Corrigendum to "Influence of hydrogen addition to pipeline natural gas on the combustion performance of a cooktop burner"[Int J Hydrogen Energy 44 (2019) 12239–12253] & An introduction to combustion: Concepts and applications [McGraw-Hill (2012)]. *Int. J. Hydrogen Energy* **2021**, *46*, 10586–10588. [CrossRef]
205. Zhao, Y. Impact of Increased Renewable Gases in Natural Gas on Combustion Performance of Self-Aspirating Flames. Ph.D. Dissertation, University of California, Irvine, CA, USA, 2020.
206. Glanville, P.; Fridlyand, A.; Sutherland, B.; Liszka, M.; Zhao, Y.; Bingham, L.; Jorgensen, K. Impact of Hydrogen/Natural Gas Blends on Partially Premixed Combustion Equipment: NO_x Emission and Operational Performance. *Energies* **2022**, *15*, 1706. [CrossRef]
207. Zhao, Y.; Hickey, B.; Srivastava, S.; Smirnov, V.; McDonell, V. Decarbonized combustion performance of a radiant mesh burner operating on pipeline natural gas mixed with hydrogen. *Int. J. Hydrogen Energy* **2022**, *in press*. [CrossRef]

energies

Article

The Impact of Hydrogen Admixture into Natural Gas on Residential and Commercial Gas Appliances

Jörg Leicher [1,*], Johannes Schaffert [1], Hristina Cigarida [1], Eren Tali [1], Frank Burmeister [1], Anne Giese [1], Rolf Albus [1], Klaus Görner [1], Stéphane Carpentier [2], Patrick Milin [2] and Jean Schweitzer [3]

1 Gas- und Wärme-Institut Essen e.V. (GWI), Hafenstrasse 101, 45356 Essen, Germany; johannes.schaffert@gwi-essen.de (J.S.); hristina.cigarida@gwi-essen.de (H.C.); eren.tali@gwi-essen.de (E.T.); frank.burmeister@gwi-essen.de (F.B.); anne.giese@gwi-essen.de (A.G.); rolf.albus@gwi-essen.de (R.A.); klaus.goerner@uni-due.de (K.G.)
2 ENGIE Lab CRIGEN, 93240 Stains, France; stephane.carpentier@engie.com (S.C.); patrick.milin@engie.com (P.M.)
3 Danish Gas Technology Center (DGC), 2950 Hørsholm, Denmark; jsc@dgc.dk
* Correspondence: joerg.leicher@gwi-essen.de; Tel.: +49-201-3618-278

Abstract: Hydrogen as a carbon-free fuel is commonly expected to play a major role in future energy supply, e.g., as an admixture gas in natural gas grids. Which impacts on residential and commercial gas appliances can be expected due to the significantly different physical and chemical properties of hydrogen-enriched natural gas? This paper analyses and discusses blends of hydrogen and natural gas from the perspective of combustion science. The admixture of hydrogen into natural gas changes the properties of the fuel gas. Depending on the combustion system, burner design and other boundary conditions, these changes may cause higher combustion temperatures and laminar combustion velocities, while changing flame positions and shapes are also to be expected. For appliances that are designed for natural gas, these effects may cause risk of flashback, reduced operational safety, material deterioration, higher nitrogen oxides emissions (NOx), and efficiency losses. Theoretical considerations and first measurements indicate that the effects of hydrogen admixture on combustion temperatures and the laminar combustion velocities are often largely mitigated by a shift towards higher air excess ratios in the absence of combustion control systems, but also that common combustion control technologies may be unable to react properly to the presence of hydrogen in the fuel.

Keywords: hydrogen; combustion; admixture; blend; H2NG; power-to-gas; emissions; decarbonisation; pollutants; appliance technology

Citation: Leicher, J.; Schaffert, J.; Cigarida, H.; Tali, E.; Burmeister, F.; Giese, A.; Albus, R.; Görner, K.; Carpentier, S.; Milin, P.; et al. The Impact of Hydrogen Admixture into Natural Gas on Residential and Commercial Gas Appliances. *Energies* **2022**, *15*, 777. https://doi.org/10.3390/en15030777

Academic Editor: Andrzej L. Wasiak

Received: 18 November 2021
Accepted: 6 January 2022
Published: 21 January 2022

1. Introduction

Climate change and the resulting need to reduce the emission of greenhouse gases (GHG) while still providing energy for a growing world population is one of the major challenges of the 21st century, affecting all sectors of society and economy. While the widespread use of electricity from renewable sources is one option to reduce energy-related greenhouse gas emissions, the use of hydrogen as a carbon-free fuel is also considered a promising decarbonisation option, particularly in hard-to-abate applications, e.g., aviation, heavy duty road and ship transport or some industrial high temperature processes.

In Europe, natural gas is the second most important primary energy source (after oil) today [1]. The European gas industry considers hydrogen (H_2) to be essential for the decarbonisation of their business model. They support the creation of dedicated hydrogen infrastructures supplying hydrogen to end-users [2], but also prepare for the injection of hydrogen into existing natural gas pipelines in order to reduce CO_2 emissions quickly and ramp up demand for hydrogen. In Germany, for example, the German association for gas and water (DVGW) plans to increase permissible hydrogen concentrations in natural gas from just below 10 vol% to 20 vol% in the near future [3]. The situation is similar in other European nations [4,5] and in the EU itself [6].

In the European Union, natural gas consumption is distributed relatively evenly across the residential and commercial, industrial, and power generation sectors, while vehicles only play a negligible role [7]. The residential and commercial sector is the biggest, both in terms of gas consumption and the number of installed appliances. It is estimated that the stock of installed appliances accounts for more than 200 million residential and commercial appliances within the European Union [8]. This includes heating systems, appliances for warm water production, cooking and catering devices and micro-CHP appliances (CHP: combined heat and power), but also other applications such as decorative fires.

Within the framework of the Horizon 2020 project "THyGA–Testing Hydrogen for Gas Applications" [9], nine EU-based research organizations and companies investigate how the admixture of hydrogen in natural gas can affect appliances in the residential and commercial sector, looking at natural gas blends (H2NG) with up to 60 vol% H_2. Measurements are being carried out for up to 100 appliances of different types and technologies. The measurement campaigns are accompanied by a market analysis [8], theoretical investigations into the impact of hydrogen on combustion processes in these appliances [10], a literature review [11], and analyses to assess if and how materials in pipes and fittings may be affected by H_2 [12]. Additional investigations study how the potential negative effects of H_2 admixture might be mitigated, and how certification and standardization processes may have to be adapted for high hydrogen admixture levels.

In this paper, hydrogen admixture to natural gas is analysed from the perspective of combustion theory. The effects on relevant fuel characteristics such as Wobbe Index, calorific values, air requirements, and laminar combustion velocities are discussed, and the potential impact on typical end-use equipment in the field are deduced and discussed.

2. Materials and Methods

Calculations were done using the COSILAB software suite [13]. As a reaction model, an adiabatic chemical equilibrium was chosen to determine adiabatic combustion temperatures while a freely propagating one-dimensional premixed flame model was used to determine the laminar combustion velocities using the reaction mechanism GRI 3.0 [14], which includes 53 species and 325 reaction equations.

All values are given in the ISO reference system of 15 °C/15 °C, with a reference pressure of 1.01325 bar (1 atm), which is used in the European gas quality standard EN 16726 [15].

3. Results and Discussion

3.1. Natural Gas and Hydrogen/Natural Gas Blends (H2NG)

In the residential and commercial sector, natural gas is exclusively used as a fuel to provide low-temperature heat, which is then used for space heating, food preparation or to produce warm water, to name the most common applications. With the exception of fuel cell CHP appliances, gas is burned directly with burners to produce a hot flue gas. Therefore, the changing fuel properties due to the admixture of hydrogen into natural gas must be considered when assessing how residential and commercial gas appliances may respond to higher levels of H_2 in natural gas.

Natural gas (which mostly consists of methane, CH_4) and hydrogen differ significantly in their physical properties. Hence, in many ways, the question of which level of hydrogen in natural gas is acceptable to both legacy and new appliances is a question of gas quality.

Gas quality and its impact on gas-fired appliances and equipment in different sectors have been investigated by both the gas industry and equipment manufacturers and operators of equipment for quite some time (e.g., [16–19]). There are regulations in place in many countries which specify a number of criteria which a gas must comply with so that it may be injected into public gas grids. Common gas quality criteria are the relative density d, the (volumetric) gross calorific value (GCV) and the Wobbe Index (WI), a criterion for fuel gas interchangeability.

If two gases have the same Wobbe Index and are burned with the same burner nozzle and with the same nozzle pressure, they will release the same amount of heat [20]. This means that an appliance can fulfil its purpose, i.e., satisfy a given heat demand with both gases without the need to physically modify the hardware.

While this is a very reduced way to tackle a complex topic, it is convenient to quantify fuel interchangeability in this manner, at least for chemically similar fuels.

Figure 1 shows how relative density, which is the ratio of the standard density of the fuel and the standard density of air, gross calorific value and Wobbe Index change when hydrogen is blended with methane, representing natural gas in this consideration. While both d and GCV decline linearly with higher levels of H_2, the reduction of the Wobbe Index is far less pronounced, and also non-linear. For example, pure methane and pure hydrogen differ by about 70% in terms of the GCV, but only by about 10% in terms of the Wobbe Index.

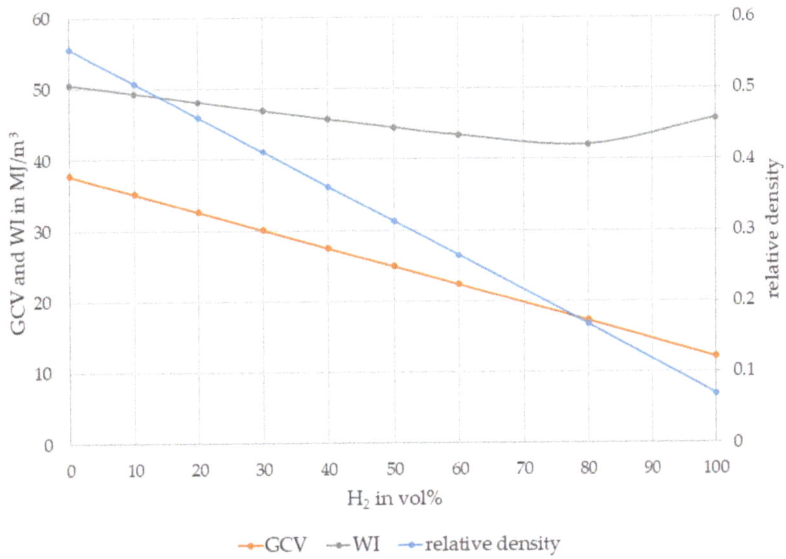

Figure 1. Relative densities, gross calorific values and Wobbe Indices for CH_4/H_2 blends. All values given in the ISO reference system 15 °C/15 °C.

So far, only a binary mixture of methane and hydrogen was considered. Natural gas, however, consists not only of methane, but also contains higher hydrocarbons (e.g., ethane and propane) or inert species such as carbon dioxide or nitrogen. Gas compositions vary depending on where the gas was extracted, how it was processed, and whether it was mixed with other natural gases in the gas infrastructure. At any given location within a gas network, local gas composition can change over time.

It is therefore important for all market partners to specify the gas that is being transported and used. As it is impractical to prescribe gas compositions for grid operations, it is common practice to specify the gas quality using a small number of relevant criteria. In the European gas quality standard EN 16726 [15], for example, a range for the relative density is given as well as a minimum Methane Number, while the EASEE-gas Common Business Practice from 2005 [21], a voluntary agreement within the European gas industry to facilitate cross border gas trading in the EU, also specifies a range of permissible Wobbe Indices for H-Gases.

Figure 2 shows three typical natural gases (Russian H-gas and North Sea H-gas, the two most important H-gas qualities in the EU as well as CH_4 as a reference) in a gas quality diagram along with the limits imposed on relative density and Wobbe Index by EN 16726 and the EASEE-gas Common Business Practice, respectively. It is obvious that the most restrictive limit to hydrogen admixture, at least from a regulatory perspective, is the density criterion, and that the Wobbe Index range is far less critical in this context. Hydrogen admixture above 30 vol% was not considered in this diagram, as most public discussions about H_2 admixture into natural gas grids focus on concentrations up to 20–30 vol%. The diagram also underlines that permissible hydrogen limits must consider the quality and composition of the natural gas that the hydrogen is being admixed to.

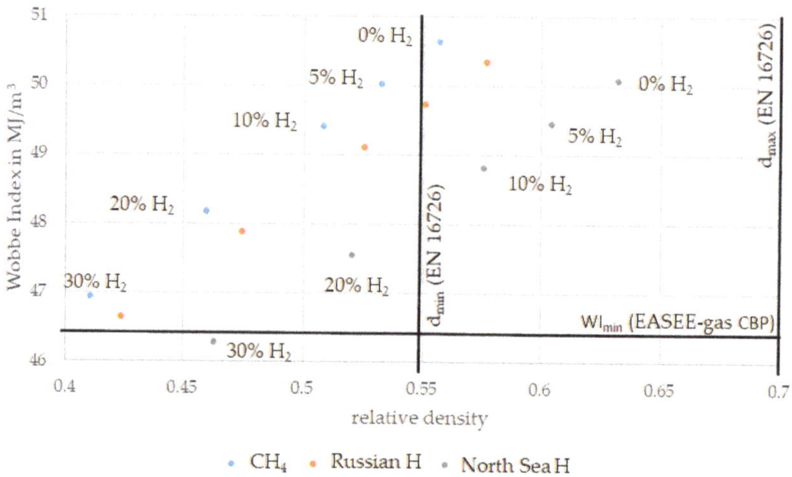

Figure 2. Hydrogen admixture to natural gases in a gas quality diagram. All values given in the ISO reference system 15 °C/15 °C.

Other combustion-related aspects should be considered as well, however. One of the main concerns in the context of H_2 admixture into natural gas and its impact on end-use equipment relates to expected higher combustion temperatures. With higher levels of hydrogen, the adiabatic combustion temperature of the fuel blend increases (cf. Figure 3), as long as other operational parameters like the air excess ratio λ remain constant. Temperatures are important since they affect many different aspects of a combustion process. They may cause local overheating of components, but they can also lead to increased emissions of nitrogen oxides (NO_X).

Another issue to consider is an increase in the laminar combustion velocity S_L. Combustion velocities are crucial for flame stabilization in premixed burners. Most residential and commercial appliances use premixed (heating appliances) or partially premixed burners (gas hobs and ovens), in contrast to industrial burner systems where non-premixed systems are more common [22]. As combustion processes in residential appliances are usually laminar [23], the laminar combustion velocity is the relevant property for this application. In a premixed laminar burner, the flame will stabilize where there is an equilibrium between local laminar combustion velocities and the local flow speed.

Figure 4 shows S_L plotted over the equivalence ratio φ (=1/λ), calculated for atmospheric pressure p = 1.01325 bar. A freely propagating one-dimensional flame model was used in combination with the GRI 3.0 reaction mechanism [14] to calculate laminar combustion velocities, and the values agree well with data from both simulations and measurements found in the literature (see [24], for example).

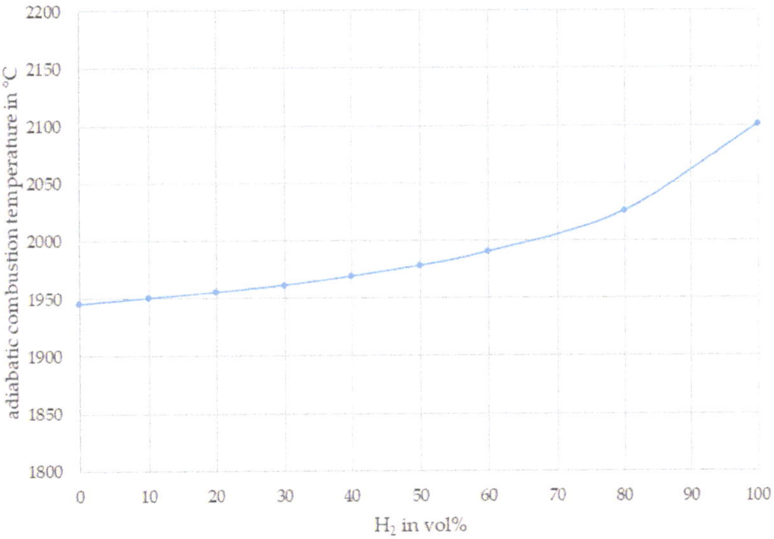

Figure 3. Adiabatic combustion temperature of CH_4/H_2 blends at stoichiometric conditions (p = 1.01325 bar).

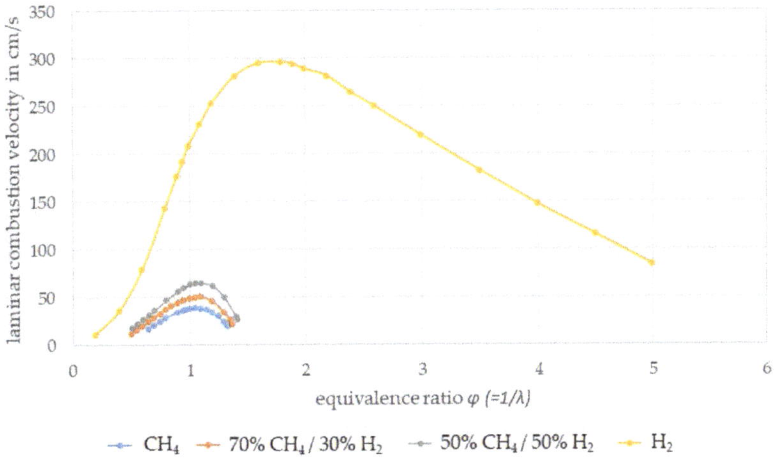

Figure 4. Laminar combustion velocity of CH_4, CH_4/H_2 blends and H_2 as a function of the equivalence ratio (=1/λ).

It can be seen that S_L increases significantly once H_2 is admixed to CH_4. As a consequence, there are concerns that higher levels of H_2 in natural gas may cause flashbacks in appliances that are not designed for it, especially at partial load when flow speeds are lower anyway. In a flashback, the flame moves upstream into the burner itself because the local combustion velocity is higher than the local flow speed, leading to a safety shutdown or, in the worst case, to damage in the burner. Given the strong impact of H_2 admixture on the laminar combustion velocities of a natural gas/hydrogen blend and the safety-related implications, this is obviously an aspect to consider.

As previously stated, the Wobbe Index is often used as the primary criterion to assess the impact of varying fuel gas compositions on combustion equipment, particularly for residential and commercial appliances, or to specify permissible gas qualities.

Table 1 highlights why looking only at the Wobbe Index is insufficient when discussing the impact of hydrogen admixture on end-use equipment. In this table, fuel properties for pure methane (CH_4, representing natural gas (H-gas)), pure hydrogen, and two blends of CH_4 with an inert (nitrogen (N_2) and carbon dioxide (CO_2), respectively) are compared. The methane blends were chosen in such a way that they have almost identical Wobbe Indices as pure hydrogen. It can be seen that, despite near identical Wobbe Indices, all other given fuel properties are very different when comparing H_2 with the blends. Thus, while the Wobbe Index is a useful fuel gas interchangeability criterion as long as certain assumptions are met (which generally is the case for residential and commercial applications, less so in industrial equipment [25,26]), it becomes far less meaningful when discussing chemically very different fuel gases or more complex combustion applications.

Table 1. Fuel properties of CH_4, two CH_4/inert blends and 100% H_2 [1].

	Unit	100% CH_4	94% CH_4/ 6% CO_2	92% CH_4/ 8% N_2	100% H_2
WI	MJ/m^3	50.64	45.28	45.27	45.78
GCV	MJ/m^3	37.80	35.53	34.78	12.10
d	-	0.5571	0.6157	0.5901	0.0698
T_{ad} (λ = 1)	°C	1982	1971	1974	2096
S_L (λ = 1)	cm/s	38.57	36.79	37.52	209

[1] All concentrations are given in vol%, ISO reference system 15 °C/15 °C.

It is important to realise that the changes in fuel properties due to the admixture of hydrogen in natural gas are only one aspect when assessing the impact of hydrogen on both legacy and new appliances. The actual technological implementation of the combustion process in a given appliance is just as important and has a profound impact on how the appliance will respond to changes in fuel. Two combustion systems may respond very differently, despite being confronted with the same change in fuel gas composition.

For this reason, extensive measurements of representative equipment are essential when discussing hydrogen admixture and its impact on appliances in the residential and commercial sector, as well as in other end-use sectors.

3.2. The Air Excess Ratio and the Impact of Combustion Control Systems

The air excess ratio λ is a crucial operational parameter for all kinds of combustion processes. Changes in the air excess ratio can impact temperatures, efficiency, heat transfer and pollutant formation, but also affect safety-related aspects such as flame stability. Residential and commercial appliances are usually adjusted on-site [26,27] to an air excess ratio specified by the manufacturer (based on prescribed O_2 or CO_2 concentrations in the flue gas) with the locally distributed gas at the time of adjustment. If the fuel gas composition changes, the actual air excess ratio of the system can also change. This would be the case in uncontrolled appliances. Modern appliances are often equipped with a combustion control system which adapts the air supply to the combustion process, based on an input signal. In this manner, these appliances always operate at the intended air excess ratio, even if the fuel gas composition changes [28]. There are, however, still many appliances in the field which have no such combustion control [8,29].

One consequence of the admixture of hydrogen to natural gas is that the minimum air requirement Air_{min}, i.e., the minimum amount of air that is necessary to achieve complete combustion, is reduced. In an appliance with combustion control, this is, in theory, counteracted by reducing the volume flow of air accordingly, but in an uncontrolled system where the volume flow of air remains constant, an increased H_2 concentration will lead to an increase of the air excess ratio λ.

This shift in the air excess ratio can be estimated by the following equation:

$$\frac{\lambda_2}{\lambda_1} = \frac{Air_{min,1}}{Air_{min,2}} \cdot \sqrt{\frac{d_2}{d_1}} = \frac{CARI_1}{CARI_2} \approx \frac{W_{S,1}}{W_{s,2}} \tag{1}$$

where λ is the air excess ratio, Air_{min} the minimum air requirement of a fuel gas (in volumetric terms), d the relative density and W_S the superior Wobbe Index of the fuel. *CARI* stands for the Combustion Air Requirement Index which is closely correlated to the Wobbe Index. This equation is valid for combustion systems with constant nozzle diameters and nozzle pressures, which is generally the case with appliances in the residential and commercial sector.

Similar to the (superior) Wobbe Index, which can be derived from Bernoulli's equation as

$$W_s = \frac{H_S}{\sqrt{d}}, \tag{2}$$

CARI is defined as

$$CARI = \frac{Air_{min}}{\sqrt{d}}. \tag{3}$$

Thus, if an appliance was adjusted to a gas with a given Wobbe Index and is then supplied with a fuel gas with a lower WI (e.g., due to hydrogen admixture), the air excess ratio will increase and vice versa. This means that if an uncontrolled appliance was originally adjusted with natural gas and is then supplied with a natural gas/hydrogen blend, it will operate at a higher air excess ratio and is thus even less likely to produce carbon monoxide (CO). However, the inverse is also true: if an appliance were to be adjusted in the field with a hydrogen/natural gas blend and the local fuel gas composition changes to lower hydrogen concentrations, the appliance's air excess ratio will be reduced, potentially leading to increased CO emissions. Given that today, the vast majority of gas appliances is adjusted in the field to an unknown local gas quality [26,27], common installation and commissioning practices may have to be re-considered if the widespread injection of hydrogen into natural gas grids is to take place in the near future.

Most burners in residential and commercial appliances are fully premixed. Therefore, any shift in the air excess ratio in a burner system will have a direct impact on the chemical processes in the flame front during combustion. This can have profound consequences, e.g., in the context of flame stabilization. Figure 4 shows that the laminar combustion velocity will increase with higher levels of hydrogen in natural gas, as long as the air excess ratio λ remains constant. The equivalence ratio φ (=1/λ) is used on the x-axis in this diagram here for better visibility.

In an uncontrolled system, this increase in S_L due to the presence of hydrogen will be counteracted by the shift of λ, so that the net change of S_L (and thus the propensity for flashback) is significantly reduced if the appliance is operated with air excess ratios higher than unity (or, correspondingly, equivalence ratios below 1). For most residential appliances, this is common practice: residential heating appliances are usually adjusted for λ values between 1.2 and 1.4 [30] to minimize carbon monoxide emissions. Gas hobs or other cooking devices may be an exception here, since they are often designed with partially premixed burner systems where regions with a sub-stoichiometric fuel-air mixture can exist, although the combustion process as a whole will be safely super-stoichiometric. Such systems are therefore more sensitive to flashback due to hydrogen admixture since in this case the change of the combustion velocity due to the shifting air excess ratio and the change in fuel composition will stack up, leading to a significant increase of the actual combustion velocity.

Combustion temperatures in uncontrolled appliances are also affected by the shifting air excess ratio: although hydrogen admixture leads to higher combustion temperatures of the fuel blend, this will be largely compensated if the air excess ratio is not actively controlled. Therefore, NO_X emissions in premixed uncontrolled appliances tend to decline as they are very much dependent on local temperatures.

These considerations also indicate that a combustion control system enforcing a constant air excess ratio may not be beneficial when it comes to hydrogen admixture, at least not for appliances in the residential and commercial sector, where premixed combustion is common. This is in contrast to non-premixed burners where combustion control generally helps reduce the increased NO_X formation due to hydrogen [31,32].

The different behaviour between these two forms of combustion can also be explained by the air excess ratio λ. In a premixed burner in which fuel and oxidizer are thoroughly mixed prior to injection into the combustion chamber, the air excess ratio is homogenously distributed in the reaction zone, there are no local differences in λ. This means that the actual combustion process will occur at the λ set point of the burner, and any change in the air excess ratio (e.g., due to hydrogen admixture into natural gas and the lack of a control system) will directly affect the chemical processes in the flame front.

In a non-premixed burner, however, fuel and oxidizer are injected into the combustion chamber separately, and the flows mix downstream of the burner, so that there is a non-uniform distribution of local λ inside the combustion chamber. A non-premixed flame will stabilise where the local λ equals unity, and most of the heat release and chemical conversion processes will occur there, always under roughly stoichiometric conditions. Thus, any change in fuel composition due to hydrogen admixture (and correspondingly, flame temperature) will directly affect the main combustion processes and also NO_X formation in such a burner system, while these effects are largely mitigated in an uncontrolled premixed burner system since the local λ shifts as well.

A premixed burner in an appliance with combustion control will behave similarly to a non-premixed burner in this regard, albeit at the chosen air excess ratio, not unity.

These effects are visualised in Figures 5 and 6 for combustion systems without, and with, air excess ratio control, respectively, where the composition of the supplied fuel gas switches from pure methane to a blend of CH_4 and 30 vol% H_2, and have been corroborated, e.g., in [33].

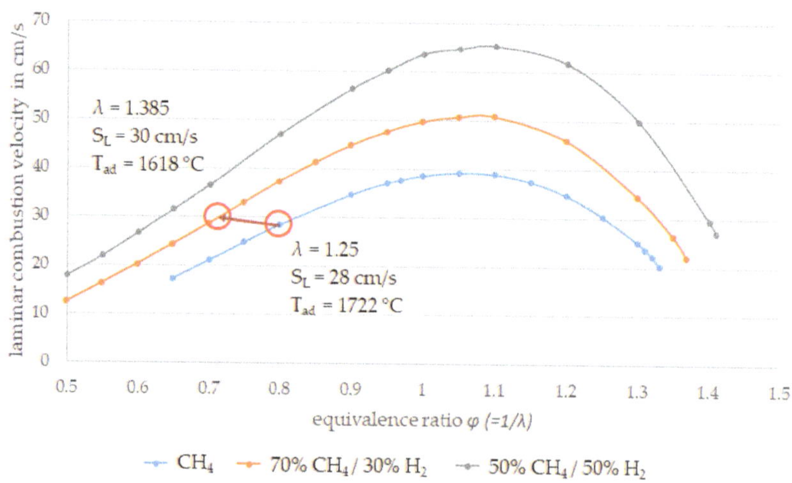

Figure 5. Effects of 30 vol.% H_2 admixture on laminar combustion velocity and adiabatic combustion temperature for an appliance without combustion control.

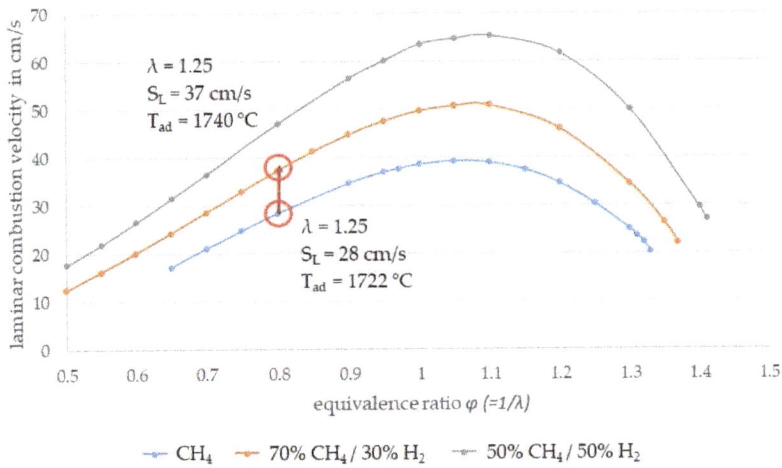

Figure 6. Effects of 30 vol.% H_2 admixture on laminar combustion velocity and adiabatic combustion temperature for an appliance with combustion control.

Another question in the context of combustion control is whether or not control systems that were originally developed to compensate for different natural gas compositions will also work reliably with hydrogen/natural gas blends. The primary purpose of a combustion control system in a residential appliance is to maintain a setpoint λ value, independent of the fuel gas that the appliance is supplied with and its original adjustment. It is, for the most part, a safety feature to prevent excessive CO formation.

Many control systems in the residential and commercial sector are based on measurements of the flame ionization current. This current will have a maximum at stoichiometric conditions, and the control system can use this information to re-adjust an appliance if the gas composition (and hence the minimum air requirement of the fuel) changes. However, measurements carried out within the THyGA project show that this approach can be unsuited for natural gas/hydrogen blends, as is visualized in Figure 7. This diagram shows measurements of how an appliance with combustion control responds to changing fuel gas compositions, both for minimum (Q_{min}) and maximum load (Q_{max}). For both loads, the hydrogen concentration was increased stepwise from 0 to 40 vol.%, the rest being methane (CH_4). The volume flows of fuel gas and the resulting air excess ratios (calculated from the measured O_2 concentration in the flue gas) are also shown. The plot shows that the control system is able to maintain a constant air excess ratio at minimum load, but fails to do so for maximum load, resulting in higher air excess ratio with higher H_2 concentrations.

Nevertheless, the control system has at least some effect. For example, in a completely uncontrolled system, a hydrogen concentration of 40 vol.% should have shifted the air excess ratio to a value of about 1.7. Instead, it was found to stabilise at 1.6 in the experiment.

The failure of the control system is probably due to the fact that hydrogen admixture does not only change the chemical processes during combustion, but also the shape and length of a flame, particularly in premixed burners. The flame ionisation current signal, however, is dependent both on the processes within the flame front, in particular the concentration of certain ions in the reaction zone, and on the relative position of the electrodes to the flame. If the flame position and shape change (e.g., due to a change in the fuel, or the thermal load of the appliance), this can impact the ionisation signal [34] and hence lead to an inappropriate response of the control system, as the effects of both the relative change of the flame position and the chemical effects in the flame front are superimposed.

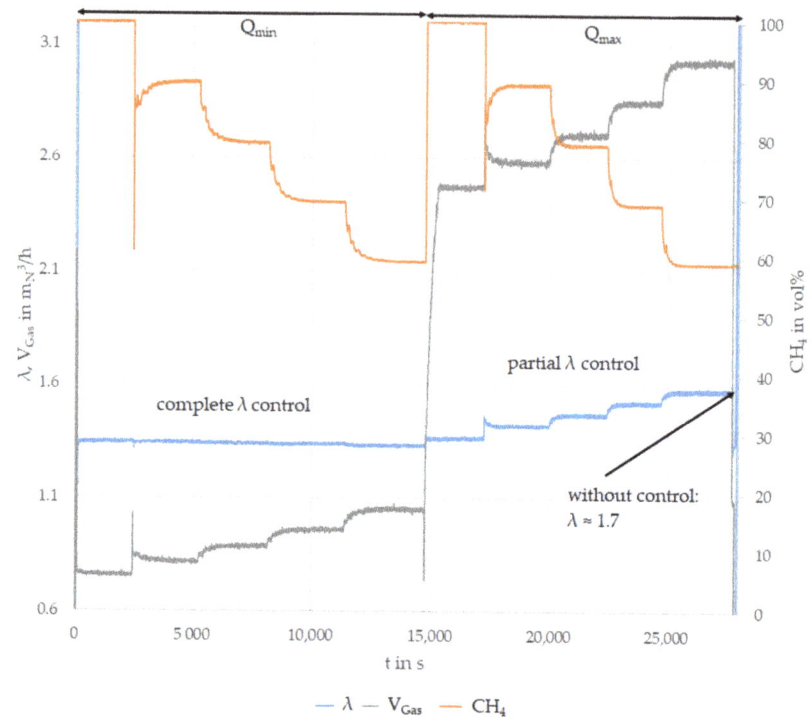

Figure 7. Response of a combustion-controlled appliance to various levels of hydrogen in methane at minimum and full load.

This effect is also shown in Figure 8, taken from [28], where several appliances with combustion control were investigated with fuel gases with different Wobbe Indices. While the appliances were able to maintain almost constant air excess ratios despite varying Wobbe Indices, this changed once the change in the Wobbe Index was caused by the admixture of hydrogen (highlighted data points). Again, the systems responded by shifting towards higher air excess ratios, indicating that the measurement and control systems were unable to detect the presence of hydrogen and react appropriately.

It is worth pointing out that shifting towards higher air excess ratios is generally not safety-relevant since a higher λ value usually leads to reduced CO emissions, unless the air excess ratio is extremely high. However, based on the first measurements both within the THyGA project and other investigations (e.g., [28]), hydrogen admixture can severely reduce the effectiveness of combustion control systems in residential and commercial appliances, at least for systems working with flame ionisation measurements. Control systems based on flue gas component measurement, e.g., by measuring the O_2 content in the exhaust gas, should perform better in this regard. This has already been demonstrated with industrial burner systems [31,35] where this control approach is very common.

The main issue is that if the air excess ratio is actually maintained at a set point value despite varying levels of H_2 in natural gas, the mitigating effects of the λ shift on combustion velocities and temperatures (and thus also NO_X formation, which, in gas combustion, is primarily dependent on temperatures) cannot be exploited.

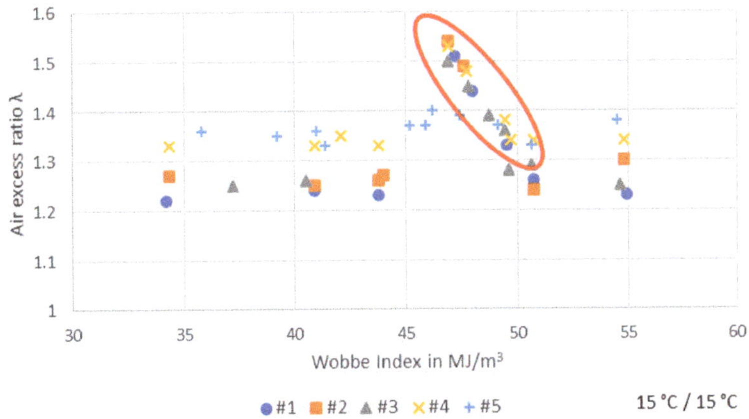

Figure 8. Impact of changing Wobbe Indices on the air excess ratio in gas appliances with combustion control. The highlighted data points show where the Wobbe Index change was due to H_2 admixture. The different colors indicate different combustion-controlled appliances. Recreated from [28].

4. Conclusions

It is likely that hydrogen will play a major part in future energy systems, for large-scale energy storage and transmission, and as a means to help decarbonise hard-to-abate applications, e.g., in the mobility and industrial sectors. There are also plans to promote the direct injection of hydrogen into the existing natural gas infrastructure. As a consequence, appliance and equipment populations in the EU could be supplied with hydrogen/natural gas blends in the near future, across all end-use sectors.

Given the significant differences between the physical and chemical properties of natural gas and hydrogen, switching from natural gas to hydrogen/natural gas blends or even pure hydrogen can affect combustion processes in residential and commercial appliances in terms of performance, but also in terms of safety. It is obvious that the consequences will become more pronounced with higher levels of hydrogen in the fuel gas. In many ways, the question of how appliances and equipment respond to higher levels of hydrogen in natural gas is a gas quality issue. Hydrogen admixture has an impact on gas quality criteria such as relative densities, calorific values or Wobbe Indices, but also on other combustion aspects such as adiabatic combustion temperatures and laminar combustion velocities. It can be shown that the Wobbe Index alone is not well-suited to assess the impact of the presence of hydrogen in a fuel on an appliance.

It is, however, important to not only look at the changing fuel properties, but also at how combustion processes are implemented in appliances and equipment across all end-use sectors.

Different combustion technologies will behave quite differently when supplied with hydrogen/natural gas blends. As most gas appliances in operation today were never designed with hydrogen in mind, it is therefore important to identify potential issues due to hydrogen admixture into natural gas, determine acceptable H_2 concentration limits and develop mitigation options where required.

Within the framework of the European project "THyGA", such investigations are being carried out for appliances in the residential and commercial sector, the biggest end-use sector for natural gas in the EU, both in terms of gas consumption and in terms of the number of installed devices.

Theoretical considerations and first measurements indicate that the effects of hydrogen admixture on combustion temperatures (relevant for potential thermal overheating of components and NO_X emissions) and the laminar combustion velocities (important for flame stabilisation) are often largely mitigated by a shift towards higher air excess ratios, at least in residential premixed gas appliances. This shift occurs when a combustion process was adjusted for a fuel gas and is then supplied with another fuel gas with a lower Wobbe Index and is inevitable in an appliance without combustion control (barring manual re-adjustment) but can also occur in controlled systems.

Current measurement technology installed in residential appliances is often unable to properly detect the changes caused by hydrogen admixture, so that the control systems fail to respond adequately. There is, however, the question of whether or not maintaining a constant air excess ratio is actually beneficial in this case. Partially premixed appliances (e.g., gas hobs or ovens) are likely to be more sensitive to the presence of hydrogen in natural gas than fully premixed systems (e.g., boilers and heating appliances) since partially premixed burners are at a greater risk to experience a flashback.

The mostly theoretical investigations presented here are only a first step in the THyGA project and will be followed up by extensive measurement campaigns for a variety of different combustion technologies and gas appliances typical for residential and commercial gas utilization. They give a first indication, however, that many existing appliance types can safely be operated with higher levels of hydrogen. These investigations also point to further relevant aspects, e.g., the performance of combustion control systems and the question of how to properly adjust appliances in a future where higher and fluctuating levels of hydrogen may be found in the gas grids.

Author Contributions: Conceptualization, J.L., S.C., J.S. (Johannes Schaffert), J.S. (Jean Schweitzer), P.M., A.G.; methodology, J.L.; software, J.L., E.T.; validation, all; formal analysis, J.L.; investigation, all; resources, J.L., A.G., F.B., R.A.; data curation, J.L., H.C., E.T.; writing—original draft preparation, J.L., J.S. (Johannes Schaffert), S.C., J.S. (Jean Schweitzer); writing—review and editing, all; visualization, J.L., J.S. (Johannes Schaffert), S.C.; supervision, P.M., K.G.; project administration, P.M., J.S. (Jean Schweitzer), J.S. (Johannes Schaffert); funding acquisition, P.M., J.S. (Jean Schweitzer), J.S. (Johannes Schaffert). All authors have read and agreed to the published version of the manuscript.

Funding: This research was funded by Fuel Cells and Hydrogen Joint Undertaking under grant agreement No. 874983. This Joint Undertaking receives support from the European Union's Horizon 2020 research and innovation programme, Hydrogen Europe and Hydrogen Europe research.

Institutional Review Board Statement: Not applicable.

Informed Consent Statement: Not applicable.

Acknowledgments: The authors want to thank all partners of the THyGA project consortium, especially the following persons for discussions and/or administrative help during the THyGA project: Kris de Wit, Andrea Manini, Maurizio Beghi, Philipp Fischer, Manfred Lange, Lisa Blanchard, Laurent Briottet, Krishnaveni Krishnaramanujam, Jörg Endisch, Regis Anghilante, Robert Judd, Alexandra Kostereva and Alberto J. García Hombrados.

Conflicts of Interest: The authors declare no conflict of interest. The funders had no role in the design of the study; in the collection, analyses, or interpretation of data; in the writing of the manuscript, or in the decision to publish the results.

References

1. BP Plc. *Statistical Review of World Energy 2020*; BP Plc: London, UK, 2020.
2. Jens, J.; Wang, A.; van der Leun, K.; Peters, D.; Buseman, M. *Extending the European Hydrogen Backbone—A European Hydrogen Infrastructure Vision Covering 21 Countries*; Guidehouse: Utrecht, The Netherlands, 2021.
3. DVGW. *Wasserstoff—Schlüssel für das Gelingen der Energiewende in Allen Sektoren*; Deutscher Verein des Gas und Wasserfaches e.V. (DVGW): Bonn, Germany, 2019.
4. Department for Business, Energy and Industrial Strategy. *UK Hydrogen Strategy*; Department for Business, Energy and Industrial Strategy: London, UK, 2021.
5. French Government. *Stratégie Nationale Pour Le Développement de l'Hydrogéne Décarboné En France*; French Government: Paris, France, 2020.
6. European Commission. *A Hydrogen Strategy for a Climate-Neutral Europe*; European Commission: Brussels, Belgium, 2020.
7. Eurogas. *EUROGAS Statistical Report 2015*; Eurogas: Brussels, Belgium, 2016.

8. Flayyih, M.; Schaffert, J.; Burmeister, F.; Albus, R.; Görner, K.; Milin, P.; Carpentier, S.; Krishnaramanujam, K.; Endisch, J.; de Wit, K.; et al. *Market Segmentation of Domestic and Commercial Natural Gas Appliances*; Testing Hydrogen admixture for Gas Applications; GWI, ENGIE, EBI, gas.be, DGC: Essen, Germany, 2020.
9. THyGA—Testing Hydrogen Admixture for Gas Applications. Available online: Thyga-project.eu (accessed on 15 November 2021).
10. Leicher, J.; Schaffert, J.; Carpentier, S.; Albus, R.; Görner, K. *Impact of Hydrogen Admixture on Combustion Processes—Part I: Theory*; Testing Hydrogen Admixture for Gas Applications; Gas- und Wärme-Institut Essen e.V., ENGIE S.A.: Essen, Germany, 2020.
11. Schaffert, J.; Fischer, P.; Leicher, J.; Burmeister, F.; Flayyih, M.; Cigarida, H.; Albus, R.; Görner, K.; Milin, P.; Carpentier, S.; et al. *Impact of Hydrogen Admixture on Combustion Processes—Part II: Practice*; Testing Hydrogen Admixture for Gas Applications; GWI, ENGIE, EBI, gas.be, DGC: Essen, Germany, 2020.
12. Blanchard, L.; Briottet, L. *Non-Combustion Related Impact of Hydrogen Admixture—Material Compatibility*; Testing Hydrogen Admixture for Gas Applications; CEA: Grenoble, France, 2020.
13. The COSILAB Code. Available online: www.rotexo.com (accessed on 15 November 2021).
14. Smith, G.P.; Golden, D.M.; Frenklach, M.; Moriarty, N.W.; Eiteneer, B.; Goldenberg, M.; Bowman, C.T.; Hanson, R.K.; Song, S.; Gardiner, W.C., Jr.; et al. GRI-Mech 3.0. Available online: www.me.berkeley.edu/gri_mech (accessed on 15 November 2021).
15. Comité Européen de Normalisation. *EN 16726:2019-11—Gas Infrastructure—Quality of Gas—Group H*; Comité Européen de Normalisation: Brussels, Belgium, 2019.
16. National Gas Council. *White Paper on Natural Gas Interchangeability and Non-Combustion End Use*; National Gas Council: Washington, DC, USA, 2005.
17. Leicher, J. Effects of Natural Gas Quality Changes on Industrial Combustion Processes. In Proceedings of the TOTeM 42 "Industrial heating: Furnaces, Process Heaters, Kilns—Design of Safe, Fuel and Environmentally Efficient Thermal Equipment", IJmuiden, The Netherlands, 24–25 June 2014.
18. Meuzelaar, D.J. Gas Quality: The Orphan of the Gas Industry? *Energy Delta Inst. Q.* **2012**, *4*, 9–11.
19. Lantoine, L.; Ourliac, M.; Buchet, P. Wobbe Index Measurement and Control for Industry: A Mature Technology Facing New Challenges. In Proceedings of the International Gas Union Research Conference (IGRC), Rio de Janeiro, Brazil, 24–26 May 2017.
20. Wobbe, G. La Definizione Della Qualita Del Gas. *d'Industria Gas Degli Acquedotti* **1926**, *XV*, 165–172.
21. EASEE-Gas. *Common Business Practice: Harmonisation of Natural Gas Quality*; European Association for the Streamlining of Energy Exchange-Gas (EASEE-Gas): Paris, France, 2009.
22. Dreizler, A.; Pitsch, H.; Scherer, V.; Schulz, C.; Janicka, J. The Role of Combustion Science and Technology in Low and Zero Impact Energy Transformation Processes. *Appl. Energy Combust. Sci.* **2021**, *7*, 100040. [CrossRef]
23. Jones, H.R.N. *The Application of Combustion Principles to Domestic Burner Design*; Taylor & Francis: Abingdon, UK, 1989.
24. Hermanns, R.T.E. *Laminar Burning Velocities of Methane-Hydrogen-Air Mixtures*; Technische Universiteit Eindhoven: Eindhoven, The Netherlands, 2007.
25. Ourliac, M. Deal with Gas Quality Variations and Melt Glass with Syngas from Gasification. In Proceedings of the TOTeM 44: "Gaseous Fuels in Industry and Power Generation: Challenges and Opportunities", Essen, Germany, 14–15 March 2017.
26. Leicher, J.; Giese, A.; Görner, K.; Werschy, M.; Krause, H.; Dörr, H. Natural Gas Quality Fluctuations—Surveys and Statistics on the Situation in Germany. *Energy Procedia* **2017**, *120*, 165–172. [CrossRef]
27. Ruillard, R. L'ajustement Sur Site Par Les Artisans Chauffagistes. In Proceedings of the Colloque d'AFG sur la Qualité du Gaz, Paris, France, 19 March 2012.
28. Carpentier, S.; Milin, P.; Mostefoui, N.; Nitschke-Kowsky, P.; Schweitzer, J.; Sadegh, N.; Thibaut, O. Self-Regulated Gas Boilers Able to Cope with Gas Quality Variation—State of the Art and Performances. Project Report. 2018. Available online: https://www.dgc.dk/sites/default/files/filer/publikationer/R1804_self_regulated_boilers.pdf (accessed on 15 November 2021).
29. Kemna, R.; van Elburg, M.; Corso, A. *Space and Combination Heaters—Ecodesign and Energy Labelling—Task 2: Market Analysis*; VHK, BRG Building Solutions: Delft, The Netherlands, 2019.
30. Dzubiella, M.; Hack, S.; Gleim, E.; Hesse, W.; Vogt, A.; Brämer, R. *Entwicklungsstand Gasadaptiver Verbrennungsregelungssysteme für den Bereich der Gebäudebeheizung*; Deutscher Flammentag: Darmstadt, Germany, 2017.
31. Leicher, J.; Nowakowski, T.; Giese, A.; Görner, K. Power-to-Gas and Hydrogen Admixture into the Natural Gas Grids: Impact on Industrial Firing Systems. In Proceedings of the IFRF Members' Conference, Sheffield, UK, 30–31 May 2018.
32. Nowakowski, T. *Schlussbericht: Untersuchung Der Auswirkung von Wasserstoff-Zumischung Ins Erdgasnetz auf industrielle Feuerungsprozesse in Thermoprozesstechnischen Anlagen*; Gas- und Wärme-Institut Essen e.V. (GWI): Essen, Germany, 2017.
33. Levinsky, H.B. Why Can't We Just Burn Hydrogen? *Challenges When Changing Fuels in an Existing Infrastructure. Prog. Energy Combust. Sci.* **2021**, *84*, 100907.
34. Ding, Y.; Durox, D.; Darabiha, N.; Schuller, T. Combustion State Monitoring of Premixed Heating Appliances with Flame Ionization Current and Chemiluminiscence. *Combust. Sci. Technol.* **2019**, *192*, 382–401. [CrossRef]
35. Hemmann, P. Predictive Compensation of Fluctuating Gas Quality. In Proceedings of the Conference on Glass Problems, Columbus, OH, USA, 7–10 November 2016.

energies

Article

Impact of Hydrogen/Natural Gas Blends on Partially Premixed Combustion Equipment: NO_x Emission and Operational Performance

Paul Glanville *, Alex Fridlyand, Brian Sutherland, Miroslaw Liszka, Yan Zhao, Luke Bingham and Kris Jorgensen

Gas Technology Institute, 1700 S Mount Prospect Rd, Des Plaines, IL 60018-1804, USA; afridlyand@gti.energy (A.F.); bsutherland@gti.energy (B.S.); mliszka@gti.energy (M.L.); yzhao@gti.energy (Y.Z.); lbingham@gti.energy (L.B.); kjorgensen@gti.energy (K.J.)
* Correspondence: pglanville@gti.energy

Abstract: Several North American utilities are planning to blend hydrogen into gas grids, as a short-term way of addressing the scalable demand for hydrogen and as a long-term decarbonization strategy for 'difficult-to-electrify' end uses. This study documents the impact of 0–30% hydrogen blends by volume on the performance, emissions, and safety of unadjusted equipment in a simulated use environment, focusing on prevalent partially premixed combustion designs. Following a thorough literature review, the authors describe three sets of results: operating standard and "ultra-low NO_x" burners from common heating equipment in "simulators" with hydrogen/methane blends up to 30% by volume, in situ testing of the same heating equipment, and field sampling of a wider range of equipment with 0–10% hydrogen/natural gas blends at a utility-owned training facility. The equipment was successfully operated with up to 30% hydrogen-blended fuels, with limited visual changes to flames, and key trends emerged: (a) a decrease in the input rate from 0 to 30% H_2 up to 11%, often in excess of the Wobbe Index-based predictions; (b) NO_x and CO emissions are flat or decline (air-free or energy-adjusted basis) with increasing hydrogen blending; and (c) a minor decrease (1.2%) or increase (0.9%) in efficiency from 0 to 30% hydrogen blends for standard versus ultra-low NO_x-type water heaters, respectively.

Keywords: hydrogen; natural gas; combustion; partially premixed; water heater; furnace; appliances; NO_x emissions; hythane; hydrogen-blended gas

Citation: Glanville, P.; Fridlyand, A.; Sutherland, B.; Liszka, M.; Zhao, Y.; Bingham, L.; Jorgensen, K. Impact of Hydrogen/Natural Gas Blends on Partially Premixed Combustion Equipment: NO_x Emission and Operational Performance. *Energies* **2022**, *15*, 1706. https://doi.org/10.3390/en15051706

Academic Editor: Johannes Schaffert

Received: 4 January 2022
Accepted: 21 February 2022
Published: 24 February 2022

1. Introduction

The interest in hydrogen in North America on the part of the energy industry is growing rapidly, as a means of supporting climate change mitigation goals with this flexible low-carbon energy carrier. As an energy vector, not unlike electricity, low-carbon hydrogen can be generated in multiple ways, as a means of storing renewable energy ("green" H_2) or decarbonizing fossil natural gas with integrated carbon capture ("blue", "turquoise" H_2). This flexibility has driven a rapid scale up in investment and interest, from numerous utilities initiating programs to inject hydrogen into natural gas networks to Canada's national hydrogen strategy and the U.S. Dept. of Energy's Earthshot program to reach a goal of USD 1/kg H_2 [1,2].

The scale of the decarbonization challenge is not trivial, with a combined U.S./Canadian natural gas network of 5.4 million km serving 85 million homes and businesses, where natural gas combustion in U.S. and Canadian buildings and industry are responsible for a combined 1077 Mt CO_2e/year [3–6]. However, with significant potential as a decarbonized energy vector, blending hydrogen into gas grids serving buildings and industry can serve as both an important short-term way of addressing the scalable demand for hydrogen, driving down costs of generation, storage, and distribution and an important long-term strategy to decarbonize 'difficult-to-electrify' end uses, including those with significant thermal demands, in older buildings, and in cold climates [7–9].

Hydrogen utilization represents one of several important and emerging shifts in the energy industry towards broad decarbonization. While this concept is not new, with North American development of the "green hydrogen" concept going back to the 1970s [10], only recently are multiple large-scale pilots and demonstrations underway, as summarized in Table 1. These efforts build on a prior coordinated shift from one piped gas to another, the transition from manufactured gas to natural gas in the early 20th century. This prior transition is covered well by Tarr in a comprehensive historical account, highlighting that industry-wide, the full transition took 30 to 40 years to accomplish, with the greatest effort concerning the conversion of end-use equipment [11].

Table 1. Selected North American hydrogen/natural gas blending demonstrations.

Location	Details of Demonstration *
Canada—Alberta [12]	ATCO Gas will inject 5% of H_2 by volume starting in late 2022, in a section of its customer network serving approximately 2000 customers.
Canada—Ontario [13]	Enbridge Gas will inject 2% H_2 by volume in a network serving approximately 3600 customers in the Toronto metropolitan area in 2022.
US—California [14]	A joint effort of San Diego Gas & Electric and SoCalGas to perform multiple demonstrations of blending initially from 1 to 5% H_2 by volume up to 20%, in multiple portions of their networks, from 2021 to 2026.
US—Utah [15]	Beginning with 5% H_2 injection at a training facility in the Salt Lake City region, Dominion Energy may expand to customer networks starting in 2022.
US—Oregon [16]	Testing at training facility at 5% H_2 blended, NW Natural may also expand into customer networks into 2022–2023.
US—Hawaii [17]	Not a blending demonstration per se, but Hawaii Gas has long operated a distribution network on Oahu delivering a manufactured gas containing 10–15% H_2 by volume serving approximately 30,000 customers.

* Information current as of 2021.

Concerning these risks of blending hydrogen into the existing natural gas networks in the U.S., several excellent overviews were performed with focus on infrastructure concerns, including a National Renewable Energy Laboratory (NREL) and Gas Technology Institute (GTI) technical review focused on pipeline distribution concerns [18], followed by two comprehensive industry reviews, prepared jointly for the American and Canadian Gas Associations and the Pipeline Research Council International, respectively. Additionally, the heating, ventilation, air-conditioning, and refrigeration (HVAC/R) industry commissioned its own review with a focus on end use equipment [19]. Studies largely point to the European "NaturalHy" Project [20], from 2004 to 2009 that concluded that minor adjustments to equipment in Europe could accommodate fuel blends with up to 20% hydrogen by volume, though given variations in equipment in the U.S. versus Europe, the 2013 study pointed to 5–15% as a range that would "appear to be feasible with very few modifications to existing pipeline systems and end-use appliances" [18]. More recent industry reviews agreed that up to a 20% limit was generally suitable, though the HVAC/R industry's detailed failure analysis approach concluded that only currently (as of 2021) produced equipment should be safe to operate with up to a 20% hydrogen blend, provided that no adjustments are made regarding the reduction in heating capacity, a conclusion largely based on an attempt to certify one piece of North American equipment in Europe [13]. Citing the efficiency benefit of newer products, the study also recommended that existing equipment be replaced and did not specify a hydrogen blend tolerance for equipment currently in operation [19].

Up until recently, most investigations of blended hydrogen's impacts on building equipment, including the highly cited "NaturalHy" project [20], were performed in Europe, and the equipment evaluated differed from that in use in the U.S. and Canada. However, with the growing interest in hydrogen's role as a low-carbon energy carrier, in addition to the continued research in Europe [21], there is a renewed interest in North America, with recent laboratory investigations by UC Irvine [22] and Appliance Engineering [23], in addition to the work described in this paper.

Using a laboratory and field-based approach, the authors investigated the impact of hydrogen-blended natural gas on conventional unadjusted fuel-fired equipment frequently found in North American buildings, specifically the impact on this equipment's performance, emissions, and safety in a simulated use environment. In these buildings, where natural gas (>95% methane) remains the predominant fuel for heating, this study focuses on space and water heating equipment, which consumes 95–97% of natural gas in these applications [24,25], with simplified diagrams of these burners shown in Figure 1. Miscellaneous appliances such as hearth products and cooking equipment, in addition to water heaters and furnaces, were also examined in a field environment, focusing on partially premixed equipment and the resulting NO_x emissions. While residential-sized equipment is evaluated in this study, note that the designs examined are often simply scaled-up in size for commercial building applications, where variants of burners shown in Figure 1 are applied in residential-sized and commercial-sized equipment alike.

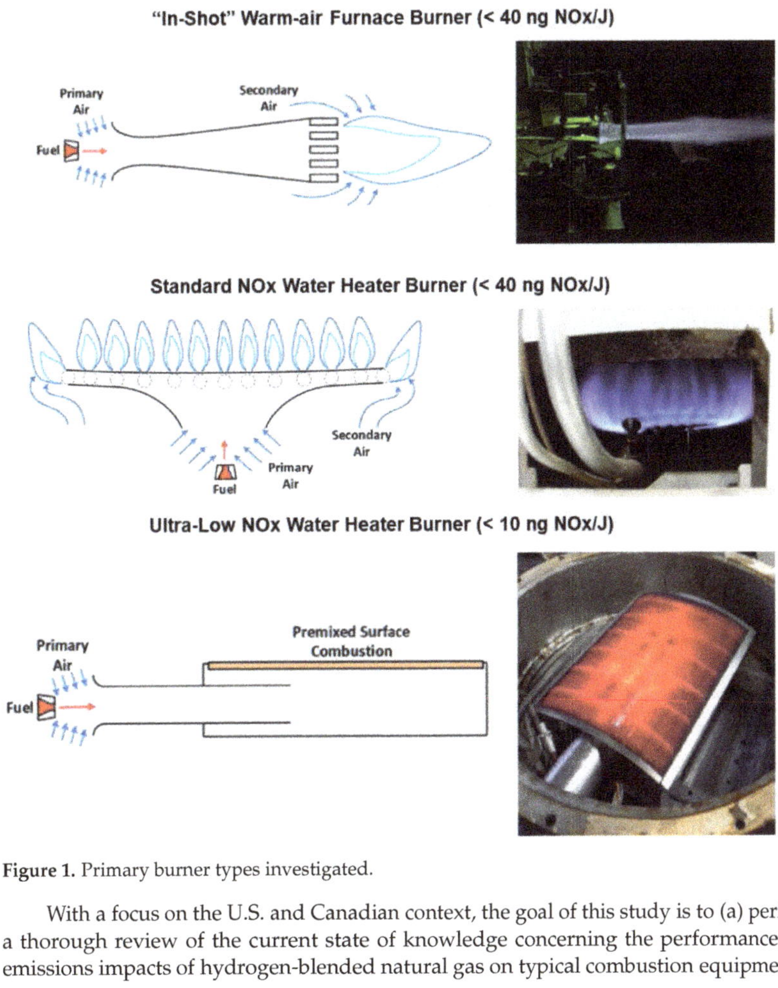

Figure 1. Primary burner types investigated.

With a focus on the U.S. and Canadian context, the goal of this study is to (a) perform a thorough review of the current state of knowledge concerning the performance and emissions impacts of hydrogen-blended natural gas on typical combustion equipment in homes and businesses, (b) expand these datasets with laboratory and field-based sampling of partially premixed type burners and combustion equipment operating with up to 30% hydrogen blends, with a focus on operational performance and NO_x emission impacts, and (c) draw distinctions between common variations, including natural versus induced draft, high versus standard efficiency, and standard versus "ultra-low NO_x" designs.

2. Background

For combustion equipment designed to operate with standard gaseous fuels (natural gas, liquified propane, and manufactured gas), hydrogen presents numerous challenges as a fuel when blended, including its faster flame speed, increased flame temperature, reduced volumetric density, wider flammability range, reduced flame luminosity, and other factors [21–23]. Appendix A provides an overview of the fuel gas quality impacts of blending hydrogen into natural gas.

As apparent in the aforementioned reviews, predicting the tolerance of blended hydrogen in the wide array of combustion-based end-use equipment is challenging, in large part due to the limited datasets available. For earlier studies, hydrogen tolerance of existing end use equipment was based largely on a small number of older European studies [18], while more recent assessments analyzed an expanded dataset for residential and commercial-sized equipment; however, they remain limited (<30 pieces of equipment) [19]. This presents a challenge to utilities when considering the injection of hydrogen into existing natural gas networks, as the short-term and long-term impacts on the wide diversity of combustion equipment downstream remain uncertain.

$$\text{Combustion Air Requirement} = \frac{(\text{Air to Fuel Ratio})_{\text{stoichiometric}}}{\sqrt{SG_{\text{fuel}}}}$$

2.1. *Equipment Testing Data*

Despite noted challenges with predicting equipment impacts on gas quality alone, for fuel-fired heating and cooking equipment in North America considered in this study, general trends do apply to the major combustion system types with hydrogen blending. As with this study, the following applies to "moderate" levels of hydrogen blending into natural gas at or less than 30% by volume, though equipment-specific impacts can vary:

1. All unadjusted equipment will see reductions in heating output with increased hydrogen added. For steady-state (i.e., on/off) equipment, manual adjustments are possible, but may not be necessary. For equipment meeting a thermal demand, equipment may be manually or automatically adjusted to compensate, and unadjusted equipment will compensate with longer runtimes.
2. Partially premixed combustion systems will likely see an increase in primary aeration, resulting in the potential for concerns with flame stability and temperature, leading to flashback and increased thermal NO_x emissions, respectively. However, the available test data show that for moderate ranges of blending (<30%), flame stability is generally not an issue and NO_x emissions are stable or decline [22,23], as will also be shown later in this paper. As a class, these are the most common combustion system types in North America, due to low cost and high reliability, including most furnaces, water heaters, boilers, cooking equipment, and hearth products.
3. Premixed combustion system impacts will vary by the control of fuel/air mixing, as the impact of hydrogen addition varies accordingly. For common pneumatically controlled fuel/air mixing, the air flow remains approximately constant as hydrogen is added, and thus combustion shifts to being leaner (λ increases), which can counteract the impact hydrogen has on flame temperature, speed, and stability. For electronically (or "digitally") controlled fuel/air mixing, often a constant-λ approach is employed, the equipment automatically compensates for the change in fuel properties with added hydrogen, requiring additional compensation to avoid flame stability issues. Premixed systems are commonly used in high-efficiency equipment where the precise control

and modulation can be valued, and pressurization of the combustion chamber(s) is needed to overcome heat exchanger pressure losses. Examples of common equipment classes that utilize premixed combustion include tankless water heaters, combi boilers, fuel-fired heat pumps, micro-combined heat and power, and equipment required to meet ultra-low emission requirements (<14 ng NO_x/J).

4. Non-premixed (diffusion) combustion systems have a greater tendency towards flame lift, though these have been observed to be minor in practice at moderate ranges of blending (up to 30%). While there are many examples of non-premixed combustion in daily life, from candle flames to wood fires, these are not common with gaseous fuels due to the poor combustion control. Examples are limited to decorative flames (e.g., gas lights), log lighters, and individual pilot lights.

While published datasets of equipment testing are scarce, an excellent review provided additional insights, largely based on the testing of European-style premixed combustion systems (e.g., hot water domestic boilers) [21]. Broadly, with increasing hydrogen blending, efficiency impacts are generally small (<2%) or within measurement error. Flame ionization sensors showed measurable declines in the control signal requiring further investigation; however, this impact did not warrant overall safety concerns. Similarly, an impact on ignition was not observed for hydrogen addition. The impact on flame temperature was mixed, though generally studies showing the region near the flame did increase in temperature but combustion chamber temperatures declined due to the increase in excess air levels. Regarding emissions, generally CO and NO_x emissions are shown to decrease or remain the same with added hydrogen. Regarding flame stability issues such as flashback, this is observed in some studies with higher hydrogen blend ratios, at or above 20% for fuel-rich combustion and at or above 40–50% for standard combustion. These stability issues are not well characterized in the literature, some appearing to be random, and cannot be explained by hydrogen addition alone [21].

For the North American context, these findings do not always translate to the prevalent partially premixed combustion-type water heaters, furnaces, and cooking equipment. Two recent datasets provide insights on North American appliances, though studies differed in the equipment tested, the operating conditions, the test method and instrumentation employed, and the analytical approaches. Additionally, one study included the operation of all equipment at 5 and 15% hydrogen blended with methane, while the second study varied the blending ratio into natural gas by appliance, depending on observations and experimental limitations. With these disclaimers noted, the following consistent results between datasets concerning furnaces, boilers, and water heaters that primarily use partially premixed combustion system designs can be observed [22,23,26–28]:

1. Equipment de-rating was a consistent result, wherein hydrogen blending decreases the input rate of equipment that the shift in Wobbe Index generally underpredicts, where more than a 3.5% de-rate is observed at 15% H_2 in most instances.
2. The impact on CO and NO_x emissions from unadjusted equipment with hydrogen blending is inherently complex and it is a common misconception, particularly for NO_x, that hydrogen blending rates are proportional to rates of emissions. In principle, unadjusted partially premixed equipment will experience competing factors towards CO and NO_x emission increases owing to the shifts in the gas quality and availability and distribution of combustion air. In most cases, for the furnaces, boilers, and water heaters, the 15% H_2 case had CO emissions within $\pm$10 ppm air-free (AF) from baseline and NO_x emissions $\pm$5 ppm AF from baseline, though some boilers saw significant decreases from the baseline of both. In all cases, the overall fuel/air ratio shifted lower as predicted, as observed with stack O_2 and CO_2 measured.

The current study seeks to both (a) expand the dataset for a broader range of equipment types, with variation within categories (high vs. low efficiency) and (b) quantify these impacts through steady and dynamic experiments, simulated use and extreme scenarios, from 0 to 30% hydrogen blends.

2.2. Partially Premixed Burner Typologies

Figure 2 schematically illustrates the three types of burners tested in the laboratory as part of this study. These burners belong to a broader category of "self-aspirating" or "inspirating" burners, whereby some or all the air required for combustion is entrained into the burner body by an expanding fuel gas jet through momentum transfer. Most commonly, these types of burners are implemented as "partially premixed" burners, where less than 100% of the air required for complete combustion is injected as "primary air". "Secondary air" is then required to complete combustion outside the burner body. Flames from these types of burners exhibit a distinct "double flame" structure, where a bright inner-cone of a rich-premixed flame is visible, surrounded by a duller outer cone diffusion flame. Gas manifold pressures of 3.5–12 mbar are commonly used with these types of burners to inject gas into the body of the burner. Both the "pancake" water heater and the "in-shot" burners illustrated in Figure 2a,b are examples of partial premix systems. Other types of appliances where these styles of burners are common include gas ranges, clothes dryers, decorative fireplaces, space heaters, older boilers, grills, commercial ovens and fryers, among others.

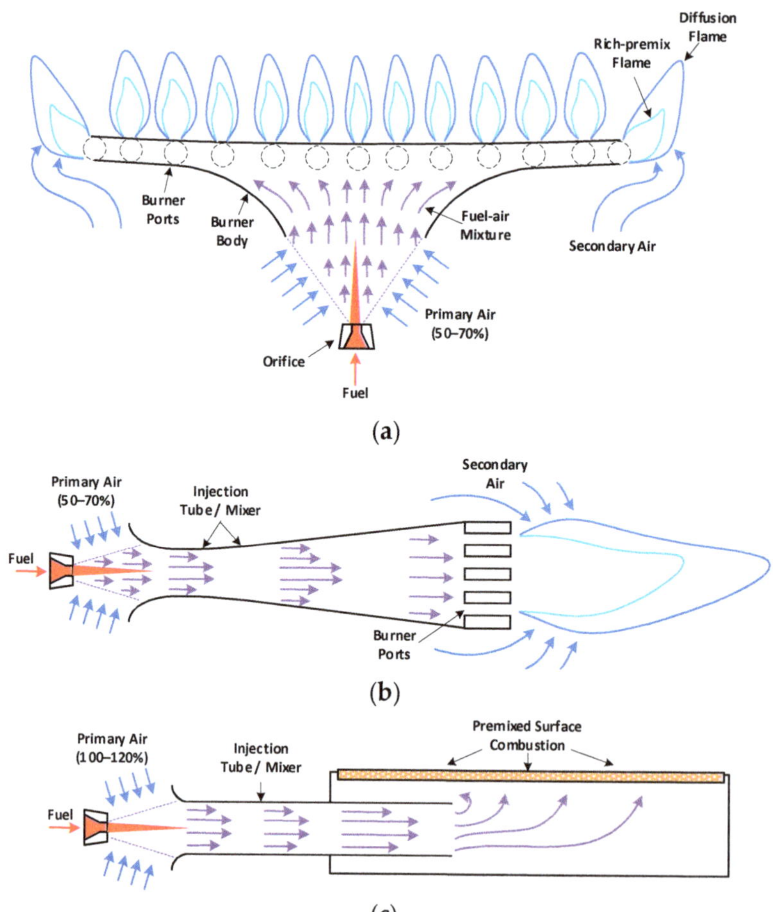

Figure 2. Schematic illustrations of typical North American atmospheric burners, including: (**a**) a "pancake" partially premixed burner from a storage water heater; (**b**) an "in-shot" partially premixed burner from a modern residential furnace; (**c**) an premixed ultra-low NO_x burner from a storage water heater.

The prevalence of partially premixed burners in North America can be in the large part attributed to their low cost (often stamped steel or cast iron), simplicity (can be unpowered), as well as stable and efficient operation [29]. The amount of primary air injected is typically 50–70% [30], leading to a rich-premix flame with a laminar speed of less than 50% of the maximum [31]. The reduced flame speed makes the burner more resistant to flashback, while consuming most of the fuel. To complete combustion, secondary air is entrained into a diffusion flame either through a naturally induced draft (pancake) or a forced draft (in-shot). In the latter case, induced draft is required to allow the burner to operate in a horizontal orientation without allowing the flame to impinge on the heat exchanger (common in North American furnaces).

Additionally illustrated in Figure 2c is a unique fully premixed self-aspirating burner. In North America, these types of burners are used for ultra-low NO_X water heaters, as required by local laws in both California and Utah. To achieve ultra-low NO_X levels of emissions, these burners rely on a radiant screen (metal wire, perforated plate, or ceramic) to absorb some of the heat of combustion and radiate it back out along the surface. This phenomenon has the effect of reducing the gaseous flame temperature and stabilizes the flame near the surface with an overall smaller reaction volume [29,30]. This in turn reduces the formation of NO_X [32]. What makes the burner in Figure 2c unique is how it achieves fully premixed operation. Instead of using a blower and a pressurized combustion system, the burner in Figure 2c relies on self-aspiration to inject nearly 100% of the air required for complete combustion (by means of a large port area [30]). To get up to 115–120% of stoichiometric air for complete combustion, this burner relies on a natural draft established inside the water heater flue to draw additional air through the burner inlet, which is positioned outside the combustion chamber (i.e., the burner outlet is at a negative pressure relative to the burner inlet). Regardless of whether they are partially or fully premixed, the types of burners depicted Figure 2 have operating characteristics (firing rate, fraction primary air, and port loading) that are sensitive to the geometry, operating conditions, as well as the gas properties. While a self-aspirating burner can be designed to operate using any type of gaseous fuel [30], if the fuel properties suddenly change, the same burner may become susceptible to flashback, flame lift, or other instabilities.

3. Methods

In this study, a comprehensive approach was used to characterize the impacts of hydrogen blended with natural gas on common North American fuel-fired equipment. First, the authors built and operated two partially premixed combustion system "simulators" to represent a storage-type water heater and warm-air furnace combustion chambers. These simulators were used to evaluate common burners and their controls, while permitting imaging and direct observation of qualitative impacts on flame appearance, stability, and other factors. Second, the authors identified and acquired five appliances, including conventional (standard NO_X) and ultra-low-NO_X versions, and designed flexible test stands to evaluate each appliance with natural gas mixtures with increasing hydrogen content, with a focus on mass-market products. For these laboratory tests, the burners and equipment were operated with pipeline natural gas and mixtures of methane/hydrogen ranging from 0 to 30% hydrogen by volume. Finally, the authors travelled to a North American utility-owned training facility, consisting of a collection of small buildings, to perform field sampling of emissions from fifteen (15) appliances that included water heaters, furnaces, ranges, ovens, dryers, and a fireplace. In the field, emission measurements were taken for all appliances operated with 100% natural gas and a blend of approximately 5% hydrogen and 95% natural gas, with one water heater also tested at a 10% hydrogen blend. While many aspects of equipment operation were examined, the primary focus concerned the measured emissions of NO_X. Figure 3 below highlights this progression of testing for water heaters, from the simulator, to in situ testing in the laboratory, and then the field.

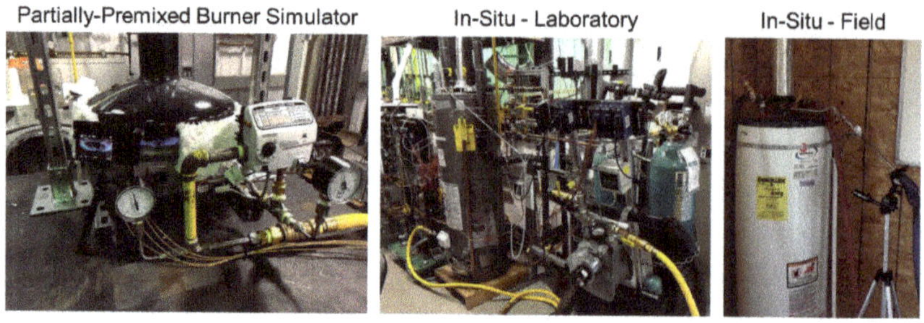

(a) (b) (c)

Figure 3. Example of approach with: (**a**) water heater simulator testing; (**b**) laboratory testing; (**c**) field testing.

3.1. Laboratory Testing

Two sets of tests were performed with a focus on the two primary categories of equipment in North American homes, the fuel-fired storage-type water heater and the warm-air furnace, one or both used in more than half of North American homes [24]. Custom fueling rigs were built for open air and in situ testing, which fed the experimental equipment with natural gas, 100% methane, and hydrogen/methane mixtures from 5 to 30% hydrogen with increments of 5%. An upper limit of 30% was selected a priori due to the anticipated suitability with most equipment, based on the literature review and discussions with manufacturers; however, this is not suggestive as an upper limit of hydrogen tolerance for any equipment tested. Fuel mixtures were supplied from cylinders with a simplified process, and the instrumentation diagram of the fueling rig is shown in Figure 4, while the instrumentation, analyzers, and other equipment used are listed in Table 2 below. The semi-portable fueling rig comprised a cylinder cart and an instrumentation cart which allowed the rig to be easily moved around to various test locations. No filters were used with either the gas from the cylinder or from the building supply. The gas supply from the high-pressure cylinders was allowed to expand naturally, resulting in a ~6 °C degree drop in some instances. An improvement for the future will be to better control the gas supply temperature, which varied in the present study by 5–6 °C.

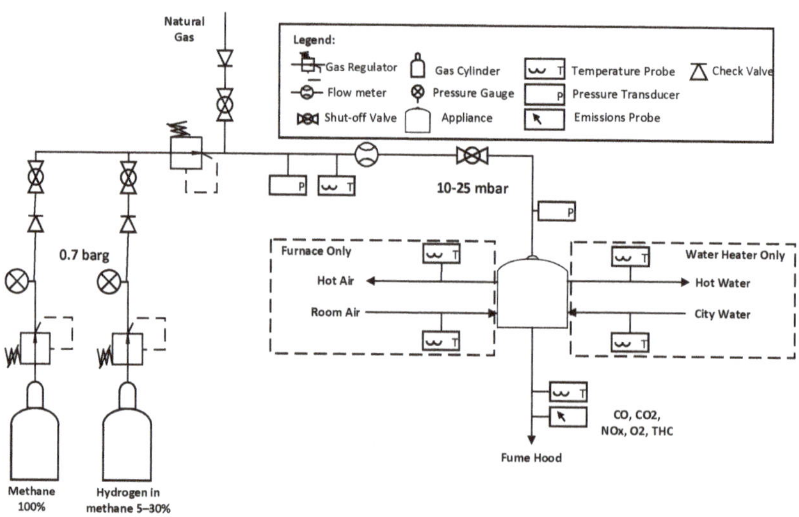

Figure 4. Simplified Process and Instrumentation Diagram (P&ID) of experimental fueling rig.

Table 2. Summary of instrumentation used in simulator and in situ laboratory testing.

Measurement	Instrument/Analyzer Used	Calibration Range/ Instrument Accuracy
Total Hydrocarbons (THC)	Rosemount Analytical 400A	800 ppm
NO_x (NO and NO_2)	Ecophysics CLD 700EL	80 ppm NO
O_2	Rosemount Analytical X-Stream	8% O_2
CO/CO_2	Rosemount Analytical X-Stream	400 ppm CO, 18% CO_2
Fuel Pressure	Dwyer ISDP-008	±31.1 Pa
Fuel Flow	Elster DTM-200A Gas Meter	±1% of reading (prec.: 17.7 pulses per L)
Gas/Air Temperatures	T-type/K-type thermocouples	±0.75% of reading
Atmospheric Pressure	Traceable Excursion-Trac Barometer	±406 Pa
Water pressure	Ashcroft G2 (0–6.9 bar)	±1.0% full scale
Water Temperatures	Omega P-M-1/10-1/8-6-0-P-3 RTDs	1/10 DIN (less than ± 0.08 °C at 60 °C, less than ± 0.04 °F at 10 °C)
Supply Water Flow	Dwyer MFS2-3	±1% of reading

Regarding the natural gas supplied during testing, house gas analysis of the natural gas supply to the lab showed it to contain 93.5% methane by volume and have a higher heating value of 38.8 MJ/m^3 (1042.4 Btu/scf). For the water heater burners, the ignition process was controlled and the fuel supplied, using an unmodified storage water heater gas valve in simulator testing with a 10.0 mbar manifold pressure. Supply pressure was kept at or slightly above 17.0 mbar. For the furnace simulator, the "high" and "low" firing rates were controlled by manifold pressures of 8.7 and 3.7 mbar, respectively. All in situ testing was initially calibrated to manufacturer requirements with pipeline natural gas conditions and then held constant for methane and hydrogen/methane blends. All tests were performed at an altitude of 196 m above sea level.

3.1.1. Simulator Testing Details

The primary goal with testing using "simulators" of water heaters and furnaces was to gather quantitative and qualitative data on the short-term operation of the appliance burners alone, while providing physical and visual access not afforded by testing equipment (burners in situ). Representing most installed gas-fired water heaters, both the standard and ultra-low NO_x type, and warm-air furnaces in North America, four burners were evaluated using simulators. Burners were removed from appliances, installed in simulated operating environments, and operated with standard controls and boundary conditions (e.g., fuel pressure). Exhaust properties were measured, including temperature and composition, in addition to fuel inlet conditions (temperature, pressure, and flow) and burner surface temperatures at multiple locations. Visually, photography and video were used to capture the dynamic impacts of hydrogen addition on flame ignition, start-up, and steady operation, provided that successful start-up was demonstrated, using a digital single-lens reflex (DSLR) camera.

Test durations were up to 10 min, until loss of flame, or until flame instability and/or an unsafe combustion condition was observed (e.g., >400 ppm CO air-free). If the measured surface temperatures were observed to climb for the duration of the 10 min test period, then the test period was extended such that no appreciable trend of increasing burner surface temperatures was observed for at least 5 min. The purpose of these tests was to observe and record the ignition process, and to determine whether stable combustion was achieved and could be maintained afterwards.

Each burner was first tested with natural gas, then 100% methane, followed by hydrogen–methane mixtures from 5 to 30% hydrogen by volume in increments of 5%. Natural gas served as the baseline for the burner operation and adjustments, with properties such as gas pressures, orifice sizes, and simulator controls held constant for subsequent fuels. Both "cold" and "hot" starts were performed, where the latter represented cycling, operating the burner after a loss of flame/re-light operation. As the blended hydrogen fraction increased, the original objective was to terminate the test at a hydrogen level where instabilities or inconsistent operation were observed. However, in all cases, tests were performed up to the a priori limit of 30%. The following issues of concern were monitored with increasing hydrogen blending:

1. Uneven flame distribution and hot spots: particularly for ultra-low NO_x burners. These burners have a larger burner port/flame holder surface area to decrease the flame temperature for NOx control.
2. Overheating of burner material: this could occur in any of the burners, but most readily in the "pancake" and "in-shot" burners because of the higher port loading compared to the other burners.
3. Flashback and/or formation of a diffusion flame at the burner orifice: The higher flame speed and wider flammability range of hydrogen makes it possible for a flame to occur where it would otherwise not be possible with methane or natural gas. The simple flow-through design of the "in-shot" furnace burners makes them particularly susceptible to this.

The water heater simulator (Figure 3) approximated the operating environment of a water heater while providing ease of visual access to the burners. The simulator was based on a combustion chamber and flue segments from an unassembled water heater. The burners were tested in the bottom portion of a water heater combustion chamber, and a storage water heater flue piece was suspended above the combustion chamber with a ~2.5 cm gap between the top of the combustion chamber and the flue section. The gap between the bottom portion of the combustion chamber and the fuel input assembly was partially covered with ceramic fiber insulation to minimize flue gas dilution and to establish a draft for the burners. A small opening was left to serve as an observation port and to allow for ease of recording unobstructed videos of the ignition and combustion process. In addition to measuring fuel properties (temperature, pressure) and exhaust gas analysis, burner surface temperatures were measured during simulator testing. The diagrams in Figure 5 show the location and naming of welded surface thermocouples for the ultra-low NO_x (ULN) burner designs #1 and #2. Given the similarities in geometry, the circular Standard NO_x "pancake burner" (Figure 1) had similar locations and identical naming to those thermocouples shown for ULN burner #2.

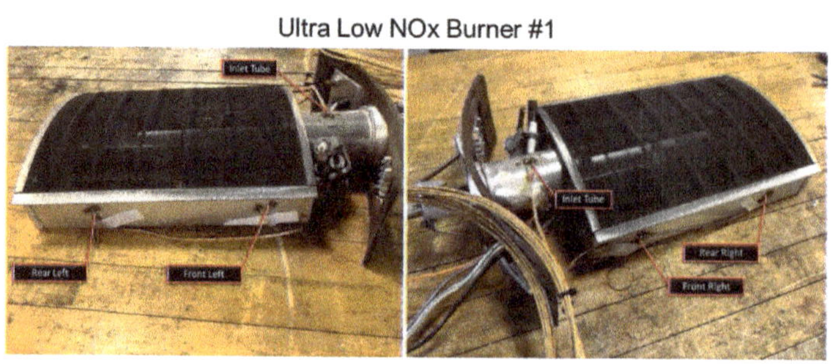

(a)

Figure 5. *Cont.*

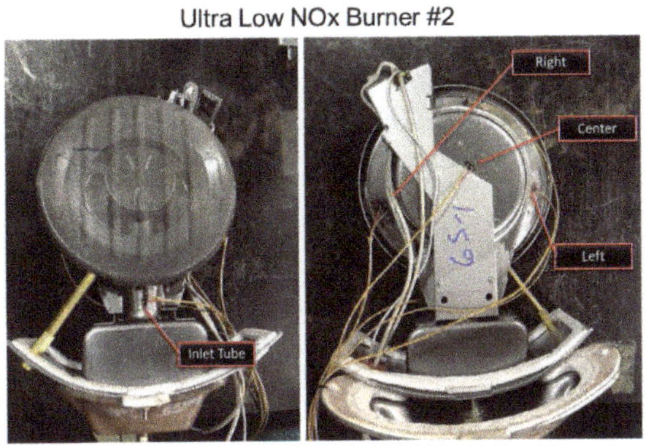

(b)

Figure 5. Ultra-low NO_x water heater burners with positions of thermocouples: (a) #1 and; (b) #2.

The induced draft warm-air furnace simulator was constructed to approximate the operating environment and for visual access to the "in-shot" burner, with an example pictured in Figure 6. As the typical furnace flame is a "loose" flame, flame stability and structure were a key component of the testing, and unique methodologies of qualifying these were provided using the simulator. The simulator comprised a solid steel metal U-pip with a burner and borosilicate glass tube at the inlet, a water-cooled heat exchange loop, and an induced-draft blower at the outlet. The system was completely sealed outside of the inlet and outlet portions with insulation provided on the body of the loop, as shown in Figure 6, covered with fiberglass insulation for safety.

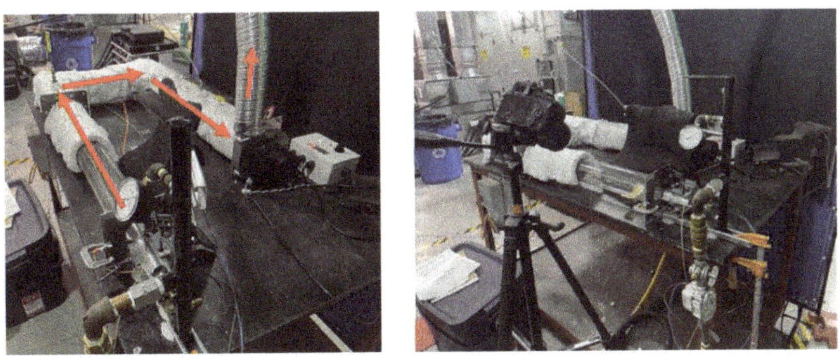

(a) (b)

Figure 6. Photos of furnace simulator: (a) flow path; (b) placement of camera.

The "in-shot" burner used was a typical design found in residential forced-air furnaces, made of stamped sheet metal and carry-over flamelets on either side of the main burner outlet for ignition propagation, with an orifice size 2.08 mm providing the fuel. The burner was situated 3.2 cm from an inlet plate that has an inlet hole 3.2 cm in diameter with the burner being concentric with the inlet hole—the placement and dimensions were taken from a non-condensing residential furnace on site later used for in situ testing.

During tests, the visible flame resided within the borosilicate glass tube, immediately after the inlet plate, and the glass tube was 46 cm long, 9 cm in diameter, and 0.3 cm thick. The upstream end of the tube was compressed into glass fiber insulation that covers the inlet plate, and the downstream end of the tube was inserted 5 cm into the metal framing of the simulator body, with insulation compressed radially at the interface of the glass and body. The "U"-shaped steel flow path downstream of the glass and viewing tube and temperature measurement point contained water-cooled copper tubing loops. An inducer fan was installed at the outlet of the U-tube, modulated by a custom-built controller providing a pulse width modulation signal with digital tachometer output. Measurements included flue gas composition and the following temperatures: (1) that of a thermocouple welded to the burner body itself, approximately 1 cm from the burner face, (2) that of a thermocouple placed 20 cm downstream of the glass tube-to-metal housing transition, with the bead in the centerline, recessed 0.6 cm into the body of a shielding tube to limit radiant heat from the flame, and (3) that of a thermocouple placed 10 cm upstream of inducer blower in order to prevent damage to the impeller. Emission measurements were performed 15 cm upstream of the inducer, immediately after the copper cooling coils.

Two firing rates were used based on a common two-stage reference residential furnace, with a 5.9 kW "High" fire and a 4.1 kW "Low" fire, set with natural gas as a reference fuel based on manifold pressures noted previously. The inducer blower speed was adjusted such that that the horizontal flame did not impinge on the glass observation tube and to minimize exhaust temperatures to protect the inducer. The nominal stack O_2 concentration was 14% by volume (dry), set with natural gas. The inducer speed was then kept constant for the remaining tests. For ignition control, a standard control module was used with a self-grounding spark ignitor and separate flame sense rod. The spark ignitor was placed 1 cm downstream from the burner face and the flame sense rod placed 1.3 cm from the burner face, both placed in the center of the primary burner face. The fuel flow was controlled via an unmodified gas valve which in turn was controlled via a signal from the ignition module, operated with a manual switch.

3.1.2. In Situ Testing Details

In situ equipment testing was performed with three water heaters and two furnaces. The appliances were operated with simulated loads, with imposed draws on the water heater and simulated thermostat calls for the furnace. Each appliance was first adjusted using natural gas (supply pressure and flue installation) and then tested with hydrogen blended with methane in the 0–30% range by volume, in 5% increments. The appliances were installed and operated in a manner consistent with manufacturer requirements (aside from fuel mixture). The three water heaters selected used the three open air burners as shipped, the standard NO_x "pancake" burner, ULN burner #1, and ULN burner #2, respectively. As is common in industry, the furnaces both used variations on the "in-shot" burner design. While four manufacturers were represented amongst the five products selected, where the "pancake" burner and ULN #1 burner-type water heaters were from the same OEM, an important note is that *results should not be viewed as manufacturer-specific, but reflecting the authors' operation of these appliances as per this test plan.* Testing and analysis of the results was not performed in consultation with equipment manufacturers and, due to the nature of laboratory testing, may not reflect the impacts observed in a field environment.

Table 3 summarizes the equipment tested, noting that the efficiency of water heaters and furnaces are shown in terms of the uniform energy factor (UEF) or annual fuel utilization efficiency (AFUE), common in North America [33,34]. For the water heaters, measurements of inlet and outlet water temperatures and flow rates facilitated an energy balance. The furnace testing used an air handler unit (AHU) for ducting, which was instrumented for air-side temperature and pressure measurements, allowing for an air-side energy balance to be completed. Fuel flow was measured using the previously described fueling rig.

Each steady-state operating point was deemed complete based on observation of the burner surface and exhaust temperatures. For water heater tests, the test was concluded once the aquastat setpoint temperature was reached and the ignition controller turned off the burner. The ignition process used existing hardware, either a pilot light or electronic ignition, and for each fuel the tests were conducted from a "cold" start with the furnaces at room temperature (~20 °C) and with water heaters after they had been flushed with cold water such that the inlet and outlet temperature from the water heater were within 2.8 °C of the incoming water temperature at the time of testing, and a re-ignition attempt immediately after the previous test to represent a "hot" start. Throughout all test points, emission measurements (CO, CO_2, NO_X, O_2, THC on a dry basis) and stack temperature were measured to determine the appliance performance and combustion efficiency.

Table 3. Equipment used in in situ laboratory testing.

Equipment Name	Burner Type	Description and Key Features
Standard Water Heater	"Pancake" Burner	Standard NO_X, 0.62 UEF, 189 L, 11.7 kW input
ULN Water Heater #1	ULN Burner #1	Ultra-low NO_X, 0.64 UEF, 151 L, 11.7 kW r input
ULN Water Heater #2	ULN Burner #2	Ultra-low NO_X, 0.58 UEF, 151 L, 11.1 kW input
Non-Condensing Furnace	"In-Shot" Burners	Standard NO_X, 80% AFUE, High Fire = 23.4 kW, Low Fire = 18.8 kW
Condensing Furnace	"In-Shot" Burners	Standard NO_X, 95% AFUE, Input (Single Stage) = 16.4 kW

Specific to the water heaters, a cycling test and the first draw with full recovery was completed for the standard water heater and ULN water heater #1, with only methane and the fuel blend containing 30% hydrogen to measure the recovery efficiency as defined by the U.S. standard [33]. Each water heater also underwent a "slug test" where the water heater was operated continuously by imposing a constant draw with the fuel mixture containing 5% hydrogen initially. The fuel supply then was switched to the mixture containing 30% hydrogen. The water heater was further operated for at least 5 min, and then the fuel was switched again to the mixture containing 5% hydrogen. This test was conducted to see if rapid changes in fuel composition would negatively impact the stability of the burner.

3.2. Field Equipment Sampling

To supplement the laboratory tests, field sampling was performed at a natural gas utility training facility, wherein a wide range of common fuel-fired equipment is operated in a simulated residential environment, with 5–7 groups of equipment installed in 14 mock homes. Sampling occurred over the span of a week, where hydrogen blending was performed on-site with pipeline natural gas into the network serving this facility, which is located in a high-altitude region (>1 km above sea level).

The appliances were tested as installed using unmodified controls. Emission measurements were taken at locations that were most convenient to minimize the alteration of appliance operation. In some instances, the exhaust flue was partially removed from the appliance to provide a location for measuring emissions, but no other modifications were made that could have provided better access to measurement locations. Table 4 highlights the residential equipment sampled, which excludes dryers due to challenges with drawing an accurate sample from a highly diluted exhaust stream without significant modification to the appliance. Relevant to the equipment emission sampling periods, the actual hydrogen blending and fuel heating values are shown in Table 5.

For all stack measurements, a Bacharach model PCA400 was used for the emission tests. The emissions analyzer was calibrated prior to travel by the manufacturer, with a National Institute of Standards and Technology (NIST) traceable certificate of calibration available upon request. The measurement ranges and accuracies for each of the reported measurements and derived values are listed below in Table 6. Except for dryers, most appliances were able to be run in a continuous fashion to allow enough time for the sensors to become fully saturated and meet response time requirements.

Table 4. Equipment used in field sampling.

Location	Equipment Name	Burner Type	Description and Key Features
A	Water Heater #1	"Pancake" Burner	Standard NO_X, 0.59 UEF, 151 L, 11.7 kW input
B	Water Heater #2	"Pancake" Burner	Standard NO_X, 0.59 UEF, 151 L, 11.7 kW input
D	Water Heater #3	ULN Burner #2	Ultra-low NO_X, 0.62 UEF, 144 L, 10.6 kW input
E	Water Heater #4	"Pancake" Burner	Standard NO_X, 0.59 UEF, 151 L, 10.6 kW input
D	Furnace #1	"In-shot" Burners	Standard NO_X, 80% AFUE, Input (Single Stage) = 25.8 kW
E	Furnace #2	"In-shot" Burners	Standard NO_X, 80% AFUE, High Fire = 14.7 kW, Low Fire = 10.3 kW
B	Wall Furnace #1	"In-shot" Burners	66% AFUE, Input = 14.7 kW
G	Wall Furnace #2	"Ribbon" Burners	Input = 23.4 kW
C	Fireplace #1	Perforated Burner	Input = 8.8 kW
A	Range/Oven #1	Standard Range Burner	Max. Input = 15.5 kW
E	Range/Oven #2	Standard Range Burner	Max. Input = 19.9 kW
F	Range/Oven #3	Standard Range Burner	Max. Input = 19.2 kW

Table 5. Gas quality details during field sampling.

Sampling Day	Hydrogen Blend (Actual %)	Heating Value (Average, MJ/m^3)
1 *	4.49	37.63
2	5.18 ± 0.39	37.56 ± 0.15
3	0.00	38.90 ± 0.26
4 **	10.02 ± 0.43	35.95 ± 0.11

* Applies to Location D only. ** Applies to Water Heater #3 only.

Table 6. Field emission sampling analyzer details.

Qty.	Range	Resolution	Accuracy	Response Time
O_2	0 to 20.9%	0.1%	±0.3 %	T90 < 20 s
CO	0 to 10,000 ppm	1 ppm	±10 ppm (0 to 200) ±5% reading (201 to 2000)	T90 < 40 s
NO	0 to 3000 ppm	1 ppm	±3 ppm (0 to 50) ±5% reading (51 to 2000)	T90 < 30 s
NO_2	0 to 500 ppm	1 ppm	±3 ppm (0 to 50) ±5% reading (51 to 500)	T90 < 40 s
T_{flue}	−20 °C to 1200 °C	0.05 °C	±0.5 °C	T90 < 70 s

Most of the tests used sample periods of 10 min with data collected at 5 s intervals. The emissions analyzer was purged outdoors in fresh air before each 10 min sample run. The emissions analyzer also underwent CO auto-zeroing during each startup. As cooling would occur in the corrugated stainless-steel tubing (CSST) where used, a sheathed Type K thermocouple was added to the ambient temperature port of the side of the Bacharach PCA400 to give an indication of the flue gas temperature in addition to the sample gas temperature. This helped to provide a rough determination of appliance operation. The sample lines water trap was adjusted to the vertical position to ensure proper sampling.

The filter in the water trap assembly was checked daily to ensure no water had condensed in the probe/CSST and bypassed the water trap to soak the filter. Examples of appliance exhaust sampling methods are shown in Figure 7.

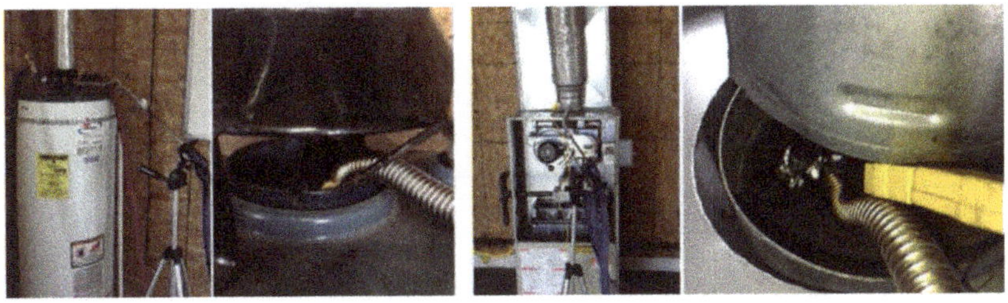

(a) (b)

Figure 7. Field-testing exhaust sampling examples for: (**a**) water heaters; (**b**) furnaces.

4. Results

4.1. Laboratory Testing

With the partially premixed combustion simulators built and commissioned, recreating storage-type water heaters and warm-air furnaces, four common burners in use in North America were installed and tested with natural gas and 0–30% hydrogen blended with methane, using 5% increments. ULN burner #1 was operated "as-shipped" and with a common orifice to the "pancake" burner to match nameplate input rates, the latter noted as "orifice". For in situ testing, three water heaters were installed and operated with natural gas and hydrogen/methane blends ranging from 0 to 30%, with cold starts (cold tank), hot starts, "slug tests" varying fuel mixtures dynamically, and a recovery efficiency test for two of the three water heaters. Of the three water heaters used, each contained one of the burner types tested in simulator testing. Two furnaces were installed and tested with the same range of fuel mixtures, a non-condensing dual-stage furnace and a condensing single-stage furnace. All in situ water heaters and furnaces were tested "as-shipped", without adjustments to the burner or its operating settings.

4.1.1. Simulator Test Results

All water heater burners and their pilot lights were consistently able to operate with natural gas and 0 to 30% mixtures of hydrogen blended into methane. Ignition was not an issue with all mixtures and flashback was not observed. ULN burner #2 had a localized and slightly lifted flame in a region on the flame holder which became more pronounced with greater hydrogen fractions, though this did not present an observable operational issue.

For furnace "in-shot" burners, hot and cold starts were performed for both high and low firing cases. Within less than 0.3 s from the initiation of the ignitor, the flame typically reaches the full length of the viewing tube, though it required 2–5 s to reach an appearance of a steady flame. While the steady state flame structure appeared to be impacted by hydrogen blending, with a shortened flame with up to 30% hydrogen blended in, ignitions were not observed to differ in duration or nature with increasing hydrogen. No sustained issues were observed with startup over the range of firing rates and mixtures, both for cold and hot starts, though an intermittent stability issue determined to be an artifact of the test setup is discussed later.

Still images of the burners from all tests are shown in Figure 8, where for "blue" flames (pancake and in-shot) it is difficult to draw conclusions, though it is important to note that the shift in color between more teal and blue flames is unexplained and is likely influenced by lighting/camera factors. For radiant burners, ULN #1 and #2 water heater burners, the cooling of the burner is apparent with increasing hydrogen, as is the portion of ULN #2, which has a slightly lifted flame in the rear, closer to the ignitor, both attributed to the reduced heat output.

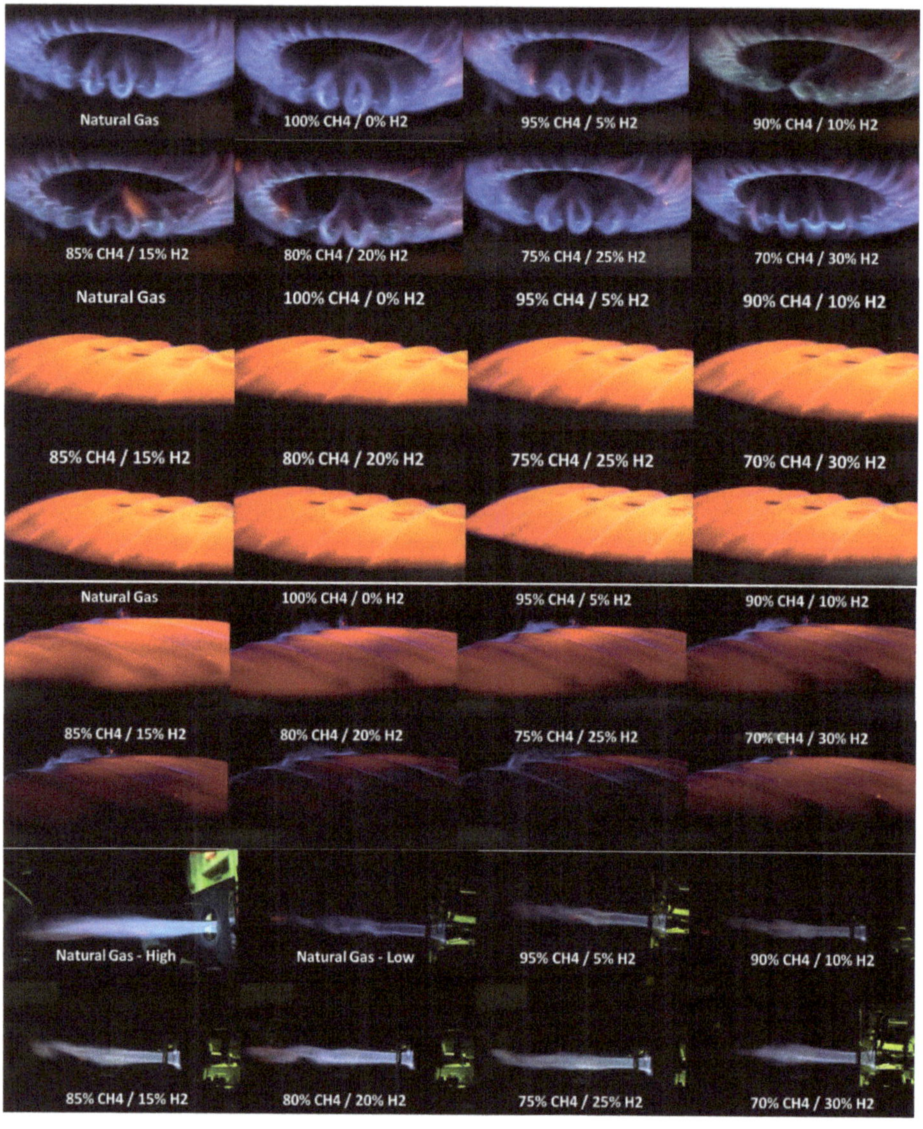

Figure 8. Still images of water heater and furnace burners during simulator testing.

For all water heater burners and most furnace burner test cases, there were no adverse combustion characteristics observed when the hydrogen content in the fuel was increased during the ignition of pilot/main burners. A typical ignition of the in-shot burner at high-fire and 30% H_2 mixture is shown in Figure 9. The exception was one stability issue intermittently observed, namely flashback with a 20 and 25% H_2 mixture, though this was determined to be an artifact of the simulator test setup. Initially when observed, subsequent efforts to re-create this flashback via rapid cycling of the burner were successful, but only every 5–10 on-cycles.

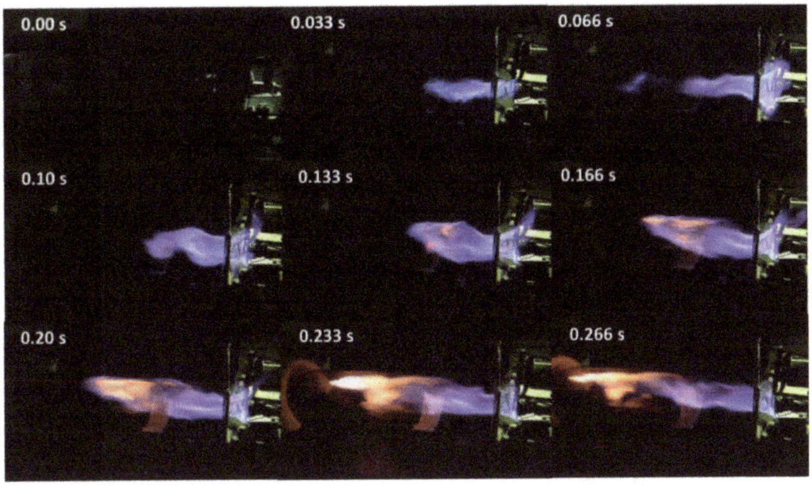

Figure 9. High-fire "hot" ignition of in-shot burner with 30% H_2 mixture.

The authors sought to recreate this flashback event systematically and isolate the impact of (a) duration between ignition calls when short-cycling and (b) the fuel mixture. A series of tests were repeated for natural gas, 10% hydrogen, 20% hydrogen, and 30% hydrogen, operating 10 sequential ignition cycles at high fire with a 15 s on-cycle and a range of delays between cycles of 2, 5, and 10 s. The inducer fan remained at the same setting as with prior testing. Finally, for the 30% hydrogen mixture only, the delay to energize the ignitor was varied between the default setting of 0.5 s up to 1.5 s. Over all conditions, flashback was not recreated, which suggests that other factors outside of burner operation, potentially including environmental factors, were responsible for the seemingly random flashback events. Despite the inability to systematically recreate flashback, initial observations confirm that the simple flow-through design of the "in-shot" furnace burners makes is more susceptible to flashback.

As expected, input rates were observed to decline with increasing hydrogen blends, with Figure 10 highlighting the decline for each water heater burner, with a scaled comparison of the calculated shift in WI. As with parallel studies, actual de-rating differs from the WI-based prediction, likely due to non-idealized hydrodynamics relative to orifice sizing practices. For the pancake burner and ULN burner #2, both radial burners with the orifice assembly within the combustion chamber, the WI-based prediction underestimates the de-rating suggesting the addition of hydrogen has a non-linear impact on burner fuel and air flow. By contrast, for the ULN #1 burner with its ample and rectilinear flow path and orifice assembly external to the combustion chamber, the WI-based prediction overestimates de-rating, suggesting that these dynamics are less important. Furnace de-rating from simulator tests does not show consistent declines, due to the constant inducer fan and artifacts of the simulator design, with high-fire and low-fire input rates only decreasing between 0 and 30% H_2 by 1.4 and 1.9%, respectively. The heat input rate was obtained by taking the average of the stabilized fuel flowrate measurements. Therefore, the experiment error was mainly from the Elster DTM-200A Gas Meter, which was $\pm 1\%$ of the heat input rate readings.

Regarding the NO_x and CO emissions in Figures 11 and 12, water heater and furnace burners show moderate shifts in CO emissions and declines in NO_x emissions with increasing hydrogen blended, owing in the large part to the excess air dilution impacts. In terms of magnitude with the data shown below for the three burners, the one exception is ULN #2 with CO emissions, showing slightly greater than ± 1 10 ppm air free (AF) CO, which may be due to observed localized flame lifting. Furnace in-shot burners show similar increases in CO emissions from 25 to 30% H_2. Measured NO_x emission levels of the ULN burners were under 50 ppm; therefore, the accuracy was ±3 ppm of the analyzer readings. The pancake burner and the in-shot burner NO_x emission levels were above 50 ppm. Therefore, their emission levels were ±5% of the readings. ULN #2 burner had the highest CO emissions, which had an accuracy range of ±5% of the readings. The measurement accuracy of the rest burners was ±10 ppm. It should be noted that the emission analyzer and sampling system was not optimized for ultra-low NO_x measurements, and so unaccounted for uncertainties for ULN #1 and ULN #2 water heaters may still be present.

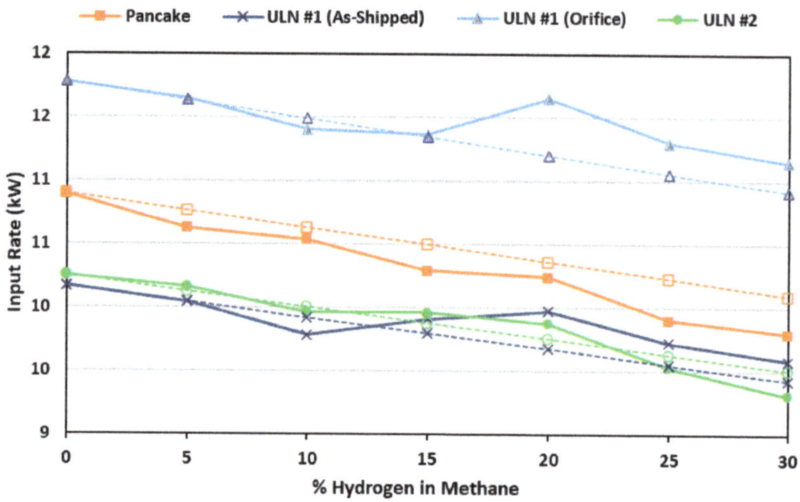

Figure 10. De-rating of water heater burners as measured (solid lines) and compared to Wobbe Index shift calculation (dashed lines).

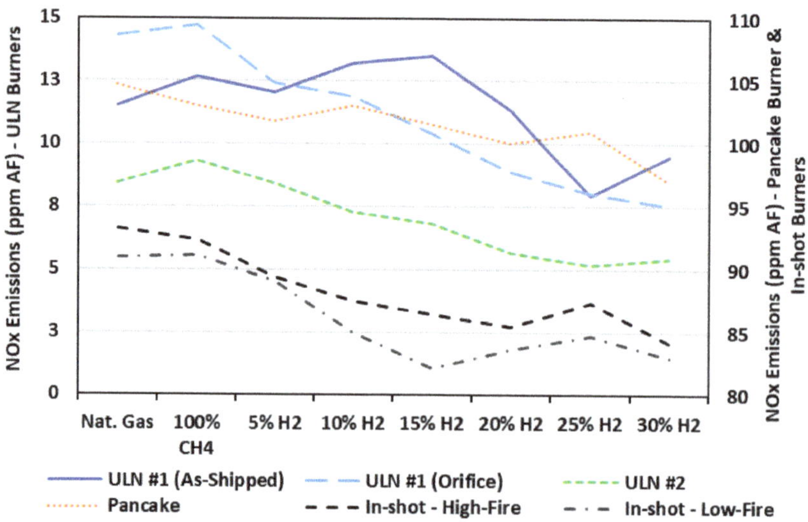

Figure 11. NO_x emissions (air-free) for simulator tests.

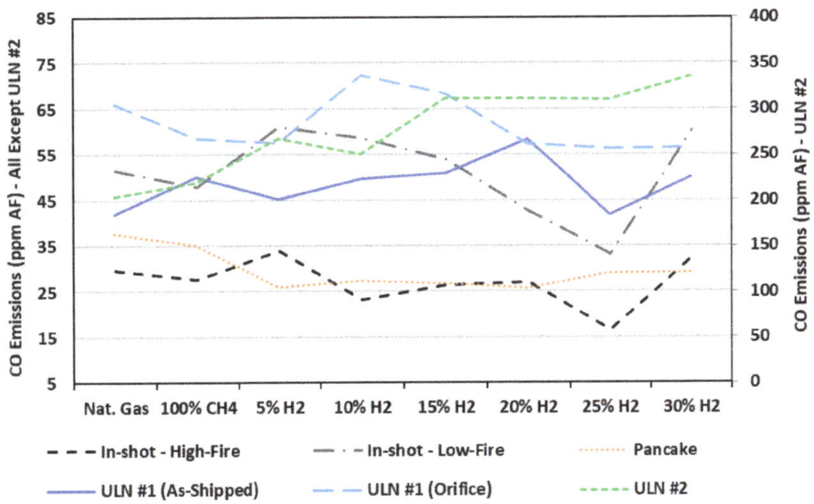

Figure 12. CO emissions (air-free) for simulator tests.

4.1.2. In Situ Test Results

For water heater testing, as with open air testing, there were no observable issues with ignition from a cold start or hot cycling for the three water heaters from 0 to 30% hydrogen, with pilot lights functional over the full range as well. When observed in the open air, with examples in Figure 13, ULN water heater #2 did not have the same noticeable lifted flame portion towards the burner rear while operating within the water heater seen in the figure below, suggesting this was largely an artifact of the simulator operation itself (e.g., insufficient draft through the burner inlet).

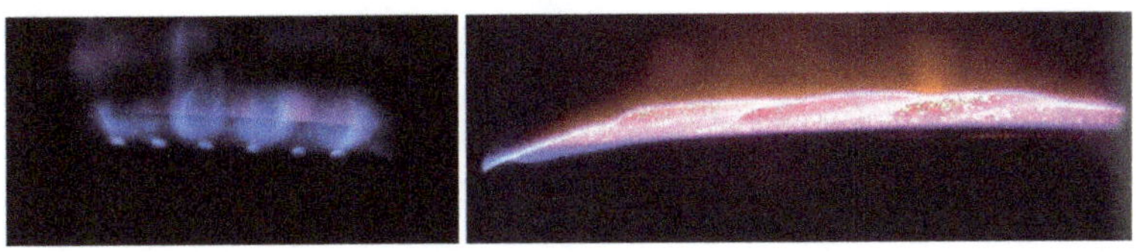

Figure 13. Standard water heater (**Left**, Steady State) and ULN Water Heater #2 (**Right**, 30 s After Ignition) operating in situ with 30% H_2.

For both furnaces over all conditions tested, the cold and hot startups were successful except for one instance of flashback, due to an operational error. Ignitions were successful and there was no discernable difference between the natural gas, 100% methane, and 30% hydrogen blended fuels, although the visual access was much more limited than simulator testing, as seen in Figure 14. According to the images below, the condensing furnace combustion was not noticeably different across the range of fuels used, for example. In the instance of flashback, the system was operating with natural gas, shut down, then switched to a 5% hydrogen mixture. Typically, there would be a purge time between switching fuels of ~1 min, followed by operating the system. During the purge time between natural gas and 5% hydrogen/methane, the flame receded to the orifice for ~20 s, after which the flame returned to stabilize at the flame holder at the end of the burner. There were abnormalities with recorded measurements, such as inlet pressure, suggesting that there was a test rig malfunction, and despite repeated attempts to recreate this flashback via transition from natural gas to 5% hydrogen blended fuel, this flashback event was not repeated. While inconclusive, this suggests that furnaces may be more sensitive to rapid shifts in hydrogen content than water heaters, which is worthy of subsequent investigation.

Further investigation into the combustion stability of in-shot burners is necessary, but was beyond the scope of this study.

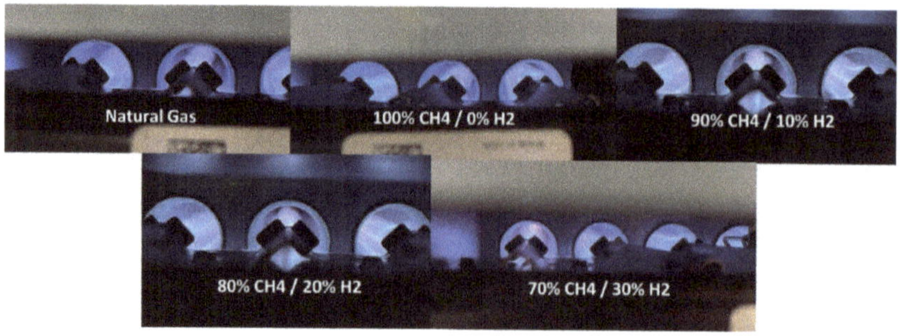

Figure 14. Visualization of condensing furnace burners at a steady state.

Concerning ignition, most furnaces use a "rolling ignition", wherein one burner is ignited and the flame "rolls" across the other burners to compete ignition. This sequence is clear from the images below in Figure 15 for the non-condensing furnace with a 30% hydrogen blend. Comparing across high vs. low fire, cold vs. hot start, and a range of fuel compositions, the timing of this rolling ignition was quantified, where cold starts, higher hydrogen blends, and high firing rates all tend to delay ignition across the four burners, with a maximum increase of 233 ms from 5% H_2 to 30% H_2 observed for the right-most burner. Visual access prevented a similar analysis of the condensing furnace, in addition to its use of multiple ignition points.

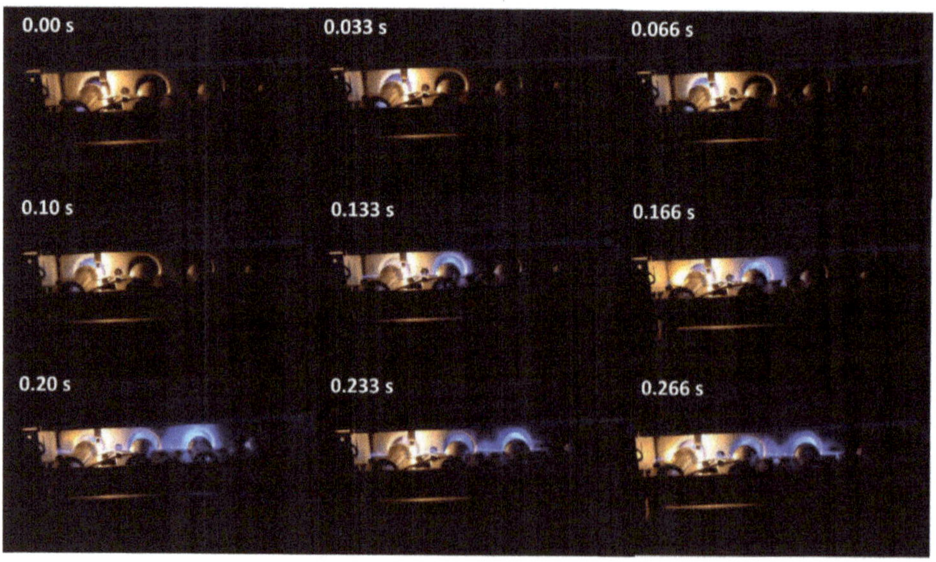

Figure 15. Rolling ignition of a non-condensing furnace at a 30% H_2 blend.

As anticipated, de-rating was observed for all water heaters from 0 to 30% hydrogen blends, ranging from 7.4% (ULN #1) to 9.1% (ULN #2) and 11.2% (Pancake). When comparing these values in Figure 16, the deviation from WI-based de-rate tightens. In the case of ULN #1, it shows a near perfect prediction of de-rating with the WI. As with the individual burners being tested, the experimental error was from the Elster DTM-200A Gas Meter, which was ±1% of the heat input rate readings. Both the pancake and ULN #2 type water heaters show slight underprediction of de-rating by the WI decline, though the difference is smaller than that of open-air burners. A key distinction between the ULN #1 burner and the other burners in this study is that the orifice ejecting fuel is positioned outside of the combustion chamber, a feature clear from the burner photos in Figure 5. By contrast, the orifice for the "pancake" burner is wholly within the chamber and for the ULN #2 burner is exposed to the ingested primary air. This points to a difference in the static pressure inside the combustion chamber and at the gas orifice, which explain the differing observations. The static pressure inside the water heater combustion chamber was not measured, so this hypothesis remains to be confirmed. Capacity decline with hydrogen blending for furnaces was not steady, though an overall decline of 7.8% for the condensing furnace and only 2.3% for the non-condensing furnace, from 0 to 30% H_2, respectively, was observed. A subsequent investigation is needed to study the impact of furnace operation as a function of inducer fan settings for single-stage, multi-stage, and modulating furnaces.

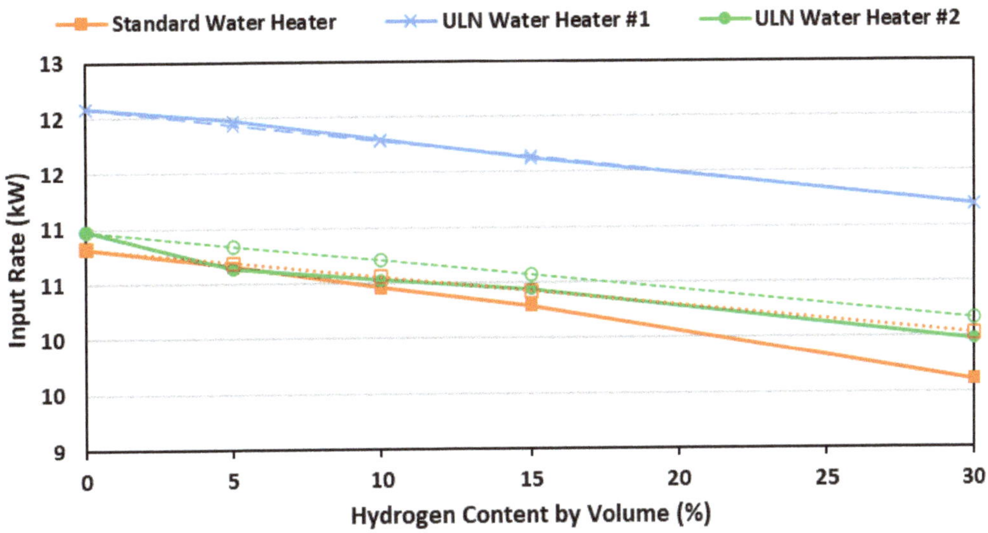

Figure 16. De-rating of water heaters as-measured (solid) and comparing to Wobbe Index shift calculation (dashed).

Regarding emissions, CO declined or remained flat for the pancake and ULN burner #1 type water heaters within narrow ranges. For the ULN burner #2 type water heater, a steady increase in CO emissions from 30 to 63 ppm AF was observed, though emissions were well below the allowable 400 ppm AF for certification via ANSI Z21.10.1. For the furnaces, the O_2 and CO_2 showed increasing dilution, from 0 to 30% H_2 blends, with CO_2 from 7.5 to 5.9% (condensing) and 6.5 to 3.7% (non-condensing high-fire). For both furnaces, a moderate increase in CO emissions was observed, at 50 ppm AF (condensing) and 10 ppm AF (non-condensing high-fire). For NO_x emissions, a consistent decline was observed with all water heaters, reducing both ULN burner NO_x emissions by approximately half, while furnaces showed a similar but less pronounced decline, as plotted in Figure 17. The CO/NO_x emission measurement accuracy was consistent with the individual burner testing. For the condensing furnace, a moderate increase (~5 °C) in burner surface temperatures up to the 30% hydrogen blend was observed, while a small decrease (~2.5 °C) was seen with the non-condensing furnace at high-fire over the same range. The temperature measurement accuracy range was ±0.5 °C. More significant shifts in surface temperatures were observed with the natural draft water heaters, as shown in Figure 18, where steady declines were observed for the radiant ULN #1 burner, while the standard pancake burner showed increases.

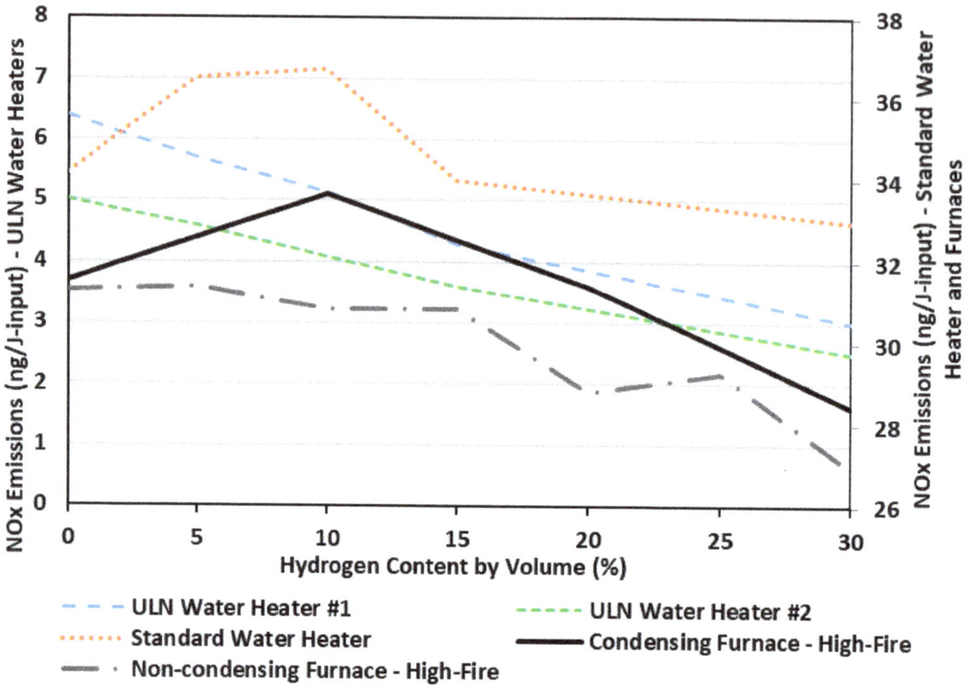

Figure 17. NO_x emissions (ng/J-input) for in situ tests.

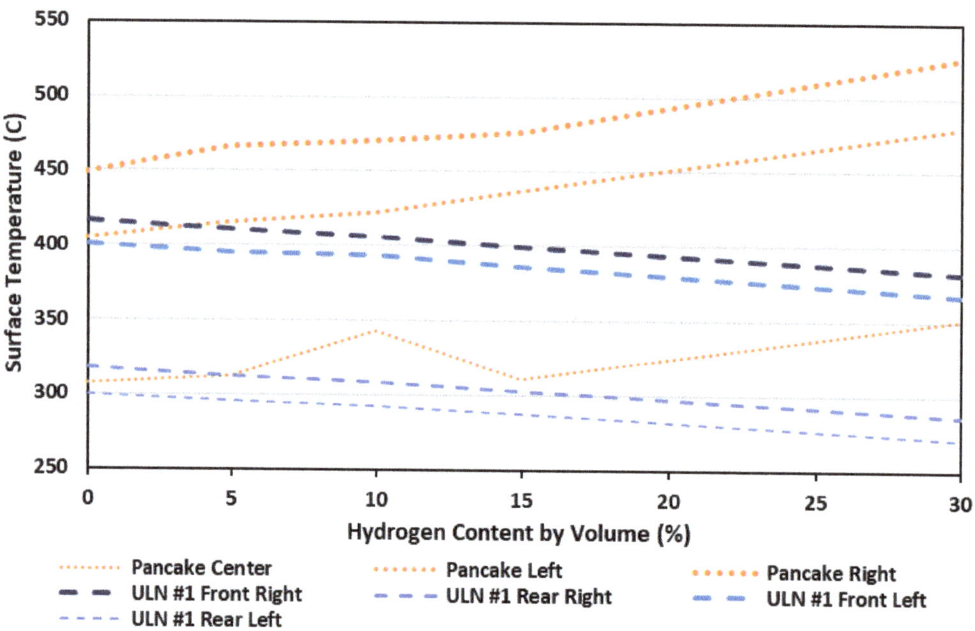

Figure 18. Burner surface temperatures during water heater in situ tests.

For the standard water heater and ULN water heater #1, the recovery efficiency test was performed as defined in the U.S. Dept. of Energy test method, which calculates an equivalent steady-state efficiency during storage tank reheating as the recovery efficiency as summarized in Figure 19. This efficiency is used in place of the true steady-state efficiency in the calculation of the uniform energy factor (UEF). Additionally, during this test, exhaust temperature and flue-gas composition were used to estimate the excess aeration. Note that for the standard water heater, the recovery duration for the 30% hydrogen blend increased by 10.4%, while the ULN water heater #1 ran for 20.7% longer (an artifact of the recovery efficiency procedure). Shifting from 100% methane to 30% hydrogen/70% methane increased the measured total excess air and decreased the flue gas temperatures. For the standard water heater, the impact of dilution was apparent as a minor efficiency penalty counteracting the expected improvement in efficiency due to de-rating. For the ULN #1 burner, a radiant burner in contrast to the "pancake" burner, the radiant heat transfer may be improved with the 30% hydrogen case in addition to the overall reduced excess aeration. Additionally, the greater increase in recovery time for the ULN water may also play a role in progressing toward steady-state operation. It should be noted here that the temperature measurement accuracy was ±0.5 °C, and the excess air level error was under ±3% (contributed by oxygen measurement). Therefore, the efficiency variation from 100% methane to 30% hydrogen/70% methane was within the measurement accuracy range.

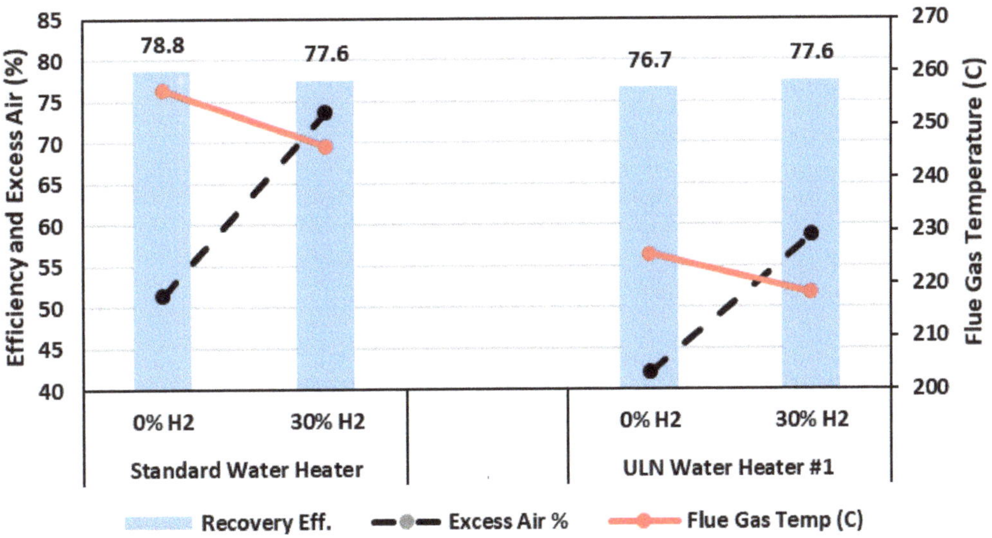

Figure 19. Impact of hydrogen blending on water heater efficiency, excess air, and flue gas temperature.

For the "slug test", the standard water heater and ULN water heater #1 were operated with a 5% hydrogen mixture, then rapidly shifted to a 30% mixture, repeating this cycle while the unit was operating. The plots in Figures 20 and 21 show the results for this test, with the sharp rise and decline in O_2 (inverse for CO_2) as the hydrogen concentration was shifted, with a tandem fall and rise in flue gas temperatures. For the "pancake" burner, the CO and NO_x emissions were not significantly impacted in this shift, while they were for the ULN water heater. Throughout the "slugs" shifting from 5 to 30%, there was no noticeable impact on the stability of the flame and equipment operation.

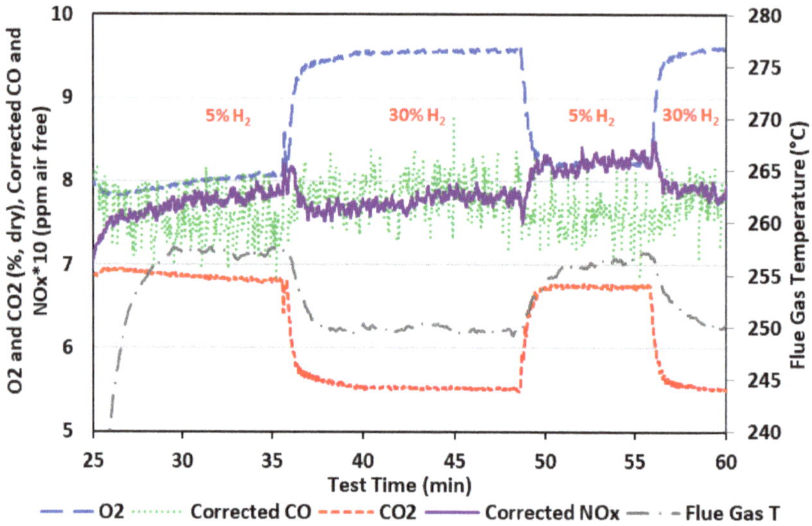

Figure 20. Slug test results for standard water heater.

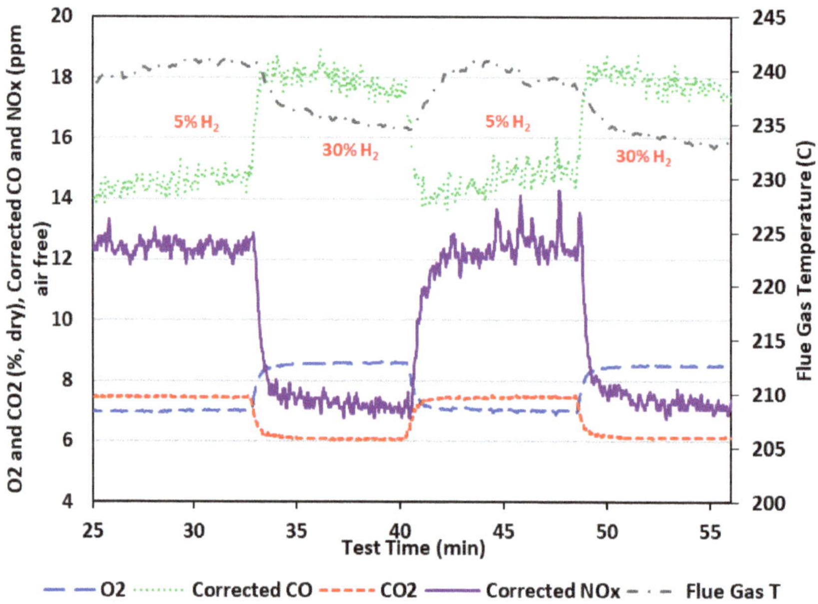

Figure 21. Slug test results for ULN water heater #1.

4.2. Field Equipment Sampling

Within the limitations of the field measurements, no significant difference in NO_x emissions was observed between natural gas and hydrogen mixtures. There were other factors outside of the control of this study, such as ambient temperature, humidity, and other weather conditions that may have affected the results.

Figure 22 shows the NO_x emissions of representative appliances operating using natural gas and natural gas/hydrogen mixtures. Generally, the 5% hydrogen addition to natural gas did not influence the NO_x emissions for these appliances.

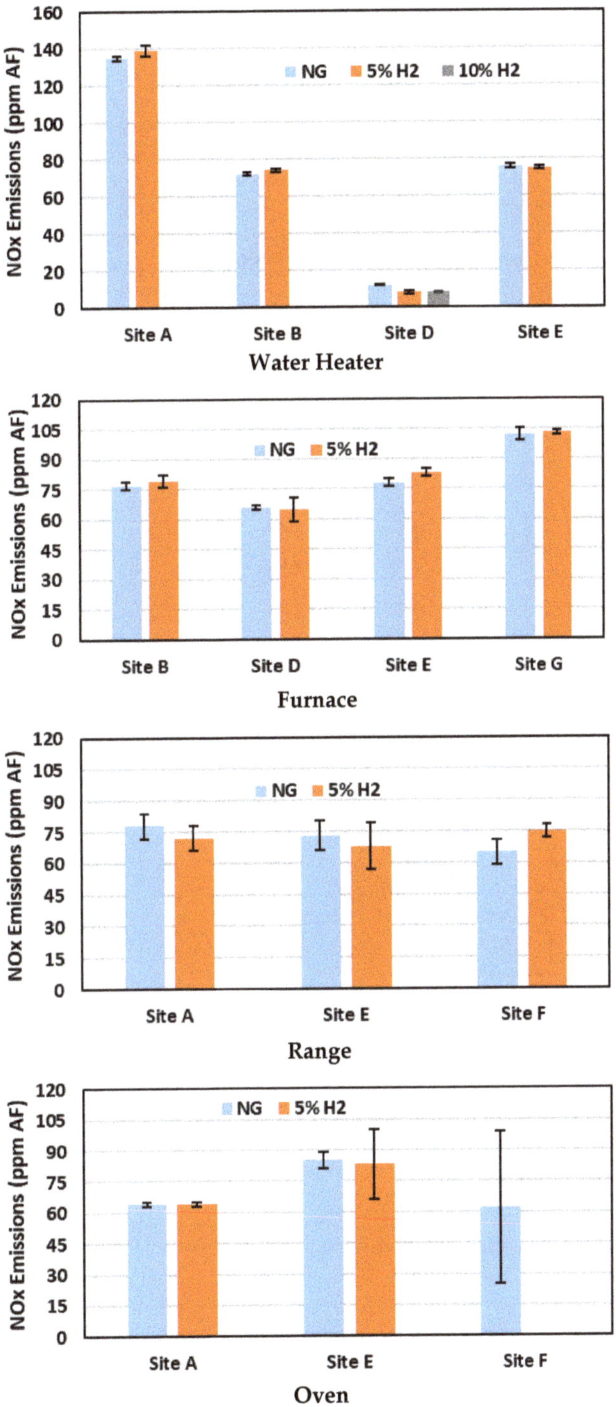

Figure 22. NO_x emissions of field-sampled equipment.

For the water heater at Site A, which was a manufactured housing-type water heater, the NO_x emissions were the highest, reaching 140 ppm in air-free conditions. The water heaters at Site B and E were conventional storage water heaters, and the NO_x emissions of these water heaters were around 75 ppm. The water heater at Site D was an ultra-low NO_x variety, which had the lowest NO_x emissions level (<10 ppm) among all appliances tested.

Warm-air furnaces at four testing locations had relatively similar NO_x emissions. The furnaces at Site B, D and E had NO_x emissions under 80 ppm. The highest NO_x emissions of furnaces were from Site G, which was a natural draft wall furnace.

Range cooktop and oven emissions were also collected. The NO_x emissions of these burners were similar among various testing sites. It should be noted that kitchen flames are usually less enclosed compared to water heaters and furnaces; therefore, the error of the emission readings are larger.

The results from sampling dryers were not reported. Due to the high dilution rate of the exhaust (~20% O_2 in the sample), the emission readings were outside of reportable accuracy. According to visual observation, it is believed that 5% hydrogen addition did not have significant impacts on the performance of the dryers. It is suggested that future dryer emission testing requires accessing the burner assembly. The fireplace emissions measurement had a similar problem. The measured NO_x emissions of the fireplace were around 10 ppm with a high dilution, and therefore, more data need to be collected on fireplaces. No evidence was observed that shows that hydrogen addition had any significant influence on CO emissions from the appliances tested.

5. Summary

5.1. Conclusions

In this effort and following a thorough review of the current state of knowledge, the authors sought to better characterize the impacts of hydrogen-blended natural gas up to 30% by volume on common partially premixed combustion equipment, including water heaters, furnaces, and miscellaneous appliances, from a whole equipment point of view and concerning the burners used. Through laboratory testing, using purpose-built "simulators" and in situ tests, and field sampling in a simulated operating environment, a series of short-term tests were performed on these components and equipment with the aim of characterizing performance, efficiency, emissions, and other factors as a function of hydrogen blending up to 30% by volume. In general, all appliances and their burners were able to tolerate this shift in fuel composition, without notable excursions in process temperatures or emissions, and anticipated trends were confirmed and further quantified for these appliances, ranging from the de-rating of heat input, to the increase in excess aeration, and to the NO_x and CO emissions. For these unadjusted, partially premixed type combustion appliances, the dominant impact of hydrogen blending is the increase in excess air, often resulting in lower NO_x emissions, surface temperatures, and other parameters.

The authors emphasize that these findings, if generalized, only apply to natural-gas appliances from mainland United States and Canada, and are as follows:

1. The combustion stability of the burners and appliances tested was not impacted by up to 30% of H_2 by volume, as evidenced by lack of flashback, flame lift, and CO emissions above 400 ppm AF.
2. While de-rating of appliances can be approximated by the WI comparison, it is not exact and higher levels of de-rate are likely to be observed in the field.
3. The efficiency of the appliances tested only varied by ~1–1.5% with 0 to 30% hydrogen-blended fuels, which is consistent with prior observations, though the changes in flue gas temperature and excess aeration did not always point to the same result. The overall efficiency ratings should be investigated in more detail; the decarbonization benefit of H_2 blending can be decreased or increased by changes in appliance efficiency.
4. Within the limits of the instrumentation and procedures used, NO_x emissions from laboratory and field measurements point to either no change or a decrease with increasing blends of hydrogen up to 30% by volume.

5. The most sensitive burners to hydrogen blending were of the "in-shot" variety, used by warm-air furnaces, tested in the laboratory. Flashback events observed were inconsistent and likely caused by either test procedures or sensitivities of the specific test stands used. Further investigation into these burners is recommended.

In broader terms, other follow-on research is recommended concerning the nuances amongst equipment and between blending levels, including (a) the atypical de-rating behavior of the non-condensing furnace, (b) disaggregating the impact of hydrogen on radiant burner output, (c) characterizing the nature and impact of ignition timing, (d) generalizing the impact of hydrogen on specific burner design features (e.g., key dimensions), (e) examining the errant flashback events observed during testing, and (f) establishing actual blend limits for the variety of partial-premixed burners and appliances in use.

5.2. Recommendations

Concerning the decarbonization of gas grids using hydrogen, a broader issue remains that the industry's knowledge of how hydrogen-blended natural gas impacts the wide diversity of stationary combustion equipment is based on a limited dataset. For residential and commercial buildings, the authors recommend expanding investigations similar to this effort in the following ways, to:

1. Expand the dataset: further quantify the emissions, efficiency, and safety impacts on a wider range of equipment types, including a greater diversity of water and space heating equipment, cooking equipment, and other miscellaneous fuel-fired appliances. Additionally, expand the scope of testing, including higher hydrogen blends, the impact of the balance fuel (e.g., natural gas), indoor and equipment component, new versus aged equipment, emerging technologies (e.g., fuel-fired heat pumps), and explore the operating envelope (fuel pressure, over/under-firing, venting matters, environmental conditions, etc.).
2. Quantify long-term impacts: long-term impacts are even more poorly understood, ranging from hydrogen-blended natural gas impacts on equipment operating life, maintenance needs, material and component degradation, and on the infrastructure (e.g., piping, venting).
3. Gain experience in the field: true in situ testing will be valuable in the field, to verify laboratory-based findings, in addition to (a) quantifying impacts on installation, operation, and maintenance of equipment, (b) establishing best practices concerning re-commissioning and troubleshooting equipment issues, (c) implementing simple retrofit packages to enable hydrogen-blended fuel tolerance, and (d) establishing the use case(s) for enhanced sensors for equipment and building systems.
4. Modernize codes and standards: to operate the equipment in this study with a 30% hydrogen/natural gas blend is to go outside its certification for safety, performance, and possibly efficiency and emissions. Modernization of these associated codes and standards is essential in parallel to expanding these laboratory and field datasets.

Author Contributions: Conceptualization, P.G., A.F. and K.J.; methodology, P.G., A.F., K.J., B.S., L.B. and Y.Z.; testing and validation, A.F., B.S., M.L. and L.B.; formal analysis, P.G., A.F., K.J., B.S., L.B. and Y.Z.; investigation, P.G., A.F., K.J., B.S., L.B. and Y.Z.; resources, P.G., A.F., K.J., B.S., L.B. and Y.Z.; data curation, P.G., A.F. and Y.Z.; writing—original draft preparation, P.G., A.F. and Y.Z.; writing—review and editing, P.G., A.F. and Y.Z.; visualization, P.G., A.F. and Y.Z.; supervision, P.G., A.F. and K.J.; project administration, P.G., A.F. and K.J.; funding acquisition, P.G., A.F. and K.J. All authors have read and agreed to the published version of the manuscript.

Funding: This research was funded in large part by Utilization Technology Development (www.utd-co.org) (accessed on 12 December 2021), under project number 22549. An expanded project report or supporting test data are available upon request to the corresponding author.

Institutional Review Board Statement: Not applicable.

Informed Consent Statement: Not applicable.

Data Availability Statement: Not applicable.

Conflicts of Interest: The authors declare no conflict of interest. The funders had no role in the design of the study; in the collection, analyses, or interpretation of data; in the writing of the manuscript, or in the decision to publish the results.

Appendix A

This informative appendix provides an overview of the fuel gas quality impacts of hydrogen blended into natural gas. In practice, the effect of increasing the quantity of hydrogen blended into natural gas on equipment is highly equipment-specific. However, general trends from the fuel properties can be illustrative. For hydrogen blended into methane, >95% of delivered natural gas in North America [25], key gas quality metrics are shown in Figure A1 as a function of hydrogen blended by volume.

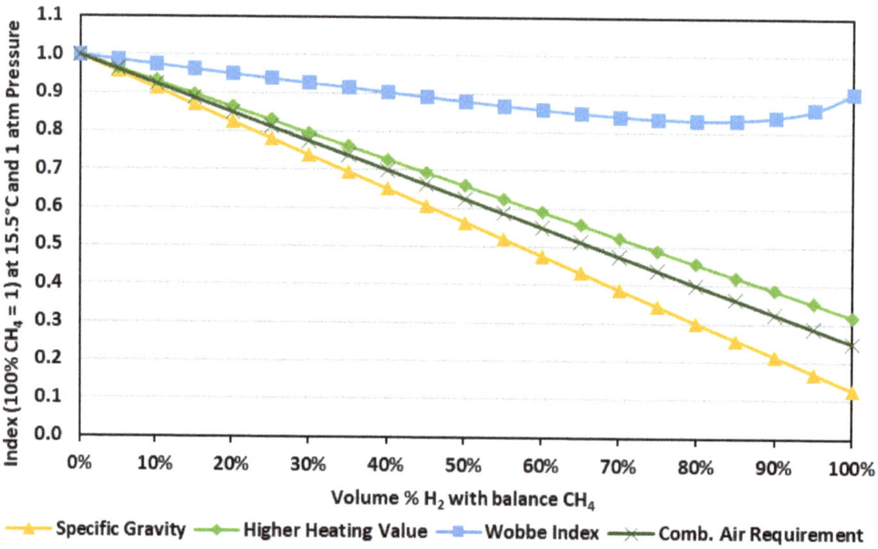

Figure A1. Impact of hydrogen blended into methane on key gas quality properties.

For equipment operating on distribution networks maintained at low pressures, typically 34 to 138 mbar delivered for typical homes and businesses, the most significant impact is the reduced volumetric density, as specific gravity (SG) is reduced by 17% for a 20% hydrogen blend and by 87% for pure hydrogen, where values for natural gas are typically 0.60 to 0.70. Similarly, the higher heating value (HHV) is reduced by 14% for a 20% hydrogen blend and 68% for pure hydrogen, driven by this volumetric density impact. A commonly used metric to judge the interchangeability of gaseous fuels, the Wobbe Index (WI), is defined in Equation (A1). As shown in Figure A1, WI is stable over the range of hydrogen addition, reducing by only 5% for a 20% hydrogen blend and by 10% for pure hydrogen. In practice, fuels with the same WI should yield the same heating rate for a given appliance with a fixed orifice pressure. However, it is imperfect for predicting the response from equipment with modern combustion controls or for certain fuel mixtures.

Analogous to the WI, the combustion air requirement (CAR), as defined in Equation (A2), posits that for appliances without active control of the air-to-fuel ratio, representing the majority of those in buildings, the actual air-to-fuel ratio is assumed to be a function of the fuel density alone for a given appliance with a fixed orifice pressure, and thus the excess aeration can be predicted for a change in fuel as the product of the air-to-fuel equivalent ratio (λ), and this index is assumed to be constant. From Figure A1, the stoichiometric combustion air-to-fuel ratio declines significantly, by 15% with a 20% hydrogen blend and by 75% for pure hydrogen, indicating the sharp reduction in combustion air necessary for hydrogen versus methane. Note that as with the WI, the CAR is imperfect in its predictive accuracy, as fuel and air mixing is a complex, turbulent process influenced by changes in fuel viscosity and other properties. The fact that hydrogen addition impacts λ is fundamental, making subsequent predictions of flame speed, emissions, and other impacts very difficult.

$$\text{Wobbe Index} = \text{HHV}/\sqrt{\text{SG}}, \tag{A1}$$

$$\text{Combustion Air Requirement} = \frac{(\text{Air to Fuel Ratio})_{\text{stoichiometric}}}{\sqrt{\text{SG}_{\text{fuel}}}}, \tag{A2}$$

It is important to emphasize the nuances of CO_2 emission reduction from blending hydrogen into natural gas, as hydrogen is primarily viewed as a decarbonization vector. In Figure A2, the following illustrative comparison is made, examining the impact of normalizing to a volume, mass, or energy basis.

1. Scaling to a mass basis is not common, as the delivered fuel is measured on a volumetric basis (ft^3, m^3). Nonetheless, hydrogen's significantly higher energy density on a mass basis (e.g., Btu/lb, MJ/kg) is shown to increase by more than 2.5 times for pure hydrogen. However, when plotted as a function of volume of hydrogen added (horizontal axis), the CO_2 emission factor on a mass basis is highly non-linear.
2. Scaling to a volumetric basis is appropriate in some circumstances, and with zero on-site CO_2 emissions from hydrogen, the emission factor declines proportionately with blending (e.g., 10% blend reduces CO_2 emissions by 10%). In practice, this is only appropriate when there is not manual or automatic compensation for the reduced heating rate (e.g., decorative gas fixtures).
3. Scaling to an energy basis is appropriate in most cases, where the fuel-fired equipment manually or automatically compensates for the reduced heating rate. For example, in a furnace operating normally as controlled by a thermostat for a given heat demand, the furnace will consume more blended fuel with longer operating times to compensate for the fuel's reduced heating value, yielding a net CO_2 reduction of 7.2% at 20% H_2 (energy basis).

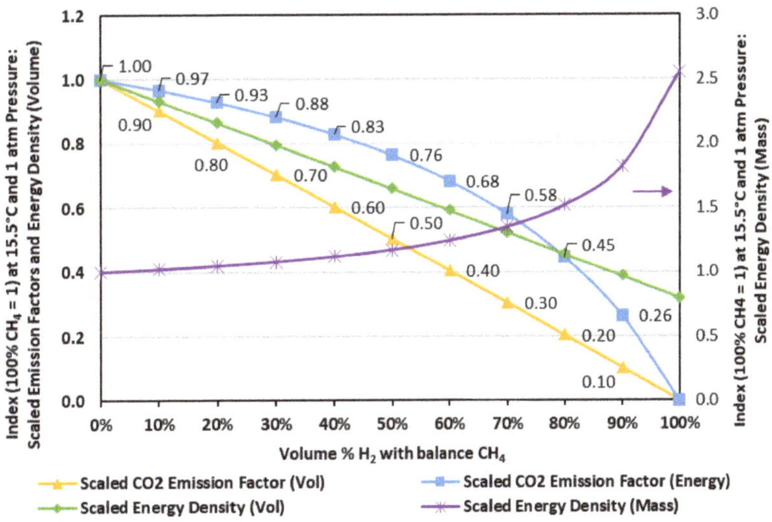

Figure A2. Impact of blended hydrogen in methane on CO_2 emission factors.

Note that this discussion neglects any upstream CO_2 emissions in hydrogen production, transmission, and distribution, in the same manner as for "site" electricity. Additionally, testing is necessary to assess the CO_2 emissions reduction from specific equipment, as the operating efficiency has been shown to be impacted by changes in the fuel mixture [22,23], as shown in this paper.

References

1. Natural Resources Canada (NRCan). Hydrogen Strategy for Canada: Seizing the Opportunities for Hydrogen. 2020. Available online: https://www.nrcan.gc.ca/sites/www.nrcan.gc.ca/files/environment/hydrogen/NRCan_Hydrogen-Strategy-Canada-na-en-v3.pdf (accessed on 12 December 2021).
2. U.S. Department of Energy (DOE). Secretary Granholm Launches Hydrogen Energy Earthshot to Accelerate Breakthroughs Toward a Net-Zero Economy. Available online: https://www.energy.gov/articles/secretary-granholm-launches-hydrogen-energy-earthshot-accelerate-breakthroughs-toward-net (accessed on 12 December 2021).
3. U.S. Energy Information Administration (EIA). Natural gas explained. Available online: https://www.eia.gov/energyexplained/natural-gas/natural-gas-pipelines.php (accessed on 12 December 2021).
4. Canadian Gas Association (CGA). Natural Gas Statistics. Available online: https://www.cga.ca/natural-gas-statistics/ (accessed on 12 December 2021).
5. Lawrence Livermore National Laboratory (LLNL). Carbon Flow Charts. Available online: https://flowcharts.llnl.gov/commodities/carbon (accessed on 12 December 2021).
6. Canadian National Energy Use Database (NEUD). NEUD Tables. Available online: https://oee.nrcan.gc.ca/corporate/statistics/neud/dpa/data_e/databases.cfm (accessed on 12 December 2021).
7. Fuel Cell and Hydrogen Energy Association (FCHEA). Road Map to a US Hydrogen Economy. 2020. Available online: https://www.fchea.org/us-hydrogen-study (accessed on 12 December 2021).
8. Fuel Cells and Hydrogen—Joint Undertaking (FCH-JU). Hydrogen Roadmap: Europe. 2019. Available online: https://www.fch.europa.eu/news/hydrogen-roadmap-europe-sustainable-pathway-european-energy-transition (accessed on 12 December 2021).
9. Saeedmanesh, A.; MacKinnon, M.; Brouwer, J. Hydrogen is Essential for Sustainability. *Curr. Opin. Electrochem.* **2018**, *12*, 166–181. [CrossRef]
10. Institute of Gas Technology (IGT). *A Hydrogen–Energy System*; Report prepared for the American Gas Association; Catalog No. L21173; Institute of Gas Technology: Des Plaines, IL, USA, 1972.
11. Tarr, J. Transforming an Energy System: The Evolution of the Manufactured Gas Industry and the Transition to Natural Gas in the United States (1807–1954). In *The Governance of Large Technical Systems*; Coutard, O., Ed.; Routledge: London, UK, 1999; pp. 19–37.
12. ATCO Gas. Fort Saskatchewan Hydrogen Blending Project. Available online: https://www.atco.com/content/dam/web/projects/projects-overview/fort-sask-hydrogen-blending-info-sheet.pdf (accessed on 12 December 2021).
13. Enbridge Gas. Clean Hydrogen Enters the Markham Energy Mix. Available online: https://www.enbridge.com/stories/2022/january/hydrogen-blending-project-enbridge-gas-cummins-operational-markham-ontario (accessed on 9 December 2021).
14. SoCalGas. Prepared Direct Testimony of Kevin Woo, David McQuilling, and Kevin Lang. Available online: https://www.socalgas.com/sites/default/files/2020-11/H2_Application-Chapter_4-Technical.pdf (accessed on 9 December 2021).
15. Dominion Energy. Hydrogen. Available online: https://www.dominionenergy.com/projects-and-facilities/hydrogen (accessed on 9 December 2021).
16. NW Natural. VISION 2050: Destination Zero. Available online: https://www.nwnatural.com/-/media/nwnatural/pdfs/vision2050bronov2021final.pdf?la=en&hash=85E02B181494D79DD7F2225778612E47 (accessed on 9 December 2021).
17. Hawaii Gas. Synthetic Natural Gas (SNG). Available online: https://www.hawaiigas.com/clean-energy/synthetic-natural-gas/ (accessed on 12 December 2021).
18. Melaina, M.; Antonia, O.; Penev, M. *Blending Hydrogen into Natural Gas Pipeline Networks: A Review of Key Issues*; Technical Report NREL/TP-5600-51995 prepared under Contract No. DE-AC36-08GO28308 for the U.S. Dept. of Energy; National Renewable Energy Laboratory: Golden, CO, USA, 2013.
19. Needley, P.; Peronski, L. *Assessment of Hydrogen Enriched Natural Gas*; Report AHRI-8024 prepared for the Air-Conditioning Heating and Refrigeration Institute; Air-Conditioning Heating and Refrigeration Institute: Arlington, VA, USA, 2021.
20. De Vries, H.; Florisson, O.; Thiekstra, G.C. Safe Operation of Natural Gas Appliances Fueled with Hydrogen/Natural Gas Mixtures (Progress Obtained in the NaturalHy-Project). In Proceedings of the International Conference on Hydrogen Safety, San Sebastian, Spain, 11–13 September 2007.
21. Schaffert, J.; Fischer, P.; Leicher, J.; Burmeister, F.; Flayyih, M.; Cigarida, H.; Albus, R.; Görner, K.; Mili, P.; Carpentier, S.; et al. Impact of Hydrogen Admixture on Combustion Processes—Part II: Practice. Deliverable D2.3 as Submitted from the THyGA Project. 2020. Available online: https://thyga-project.eu/wp-content/uploads/20201211-D2.3-Impact-of-Hydrogen-in-Practice_final.pdf (accessed on 12 December 2021).
22. McDonell, V.; Zhao, Y.; Choudhury, S. *Implications of Increased Renewable Natural Gas on Appliance Emissions and Stability*; Report CEC-500-2020-070 prepared for the California Energy Commission under Contract PIR-16-017; California Energy Commission: Sacramento, CA, USA, 2020.

23. Suchovsky, C.J.; Ericksen, L.; Williams, T.A.; Nikolic, D.J. *Appliance and Equipment Performance with Hydrogen-Enriched Natural Gases*; Report prepared for the Canadian Standards Association; Canadian Standards Association: Toronto, ON, Canada, 2021.
24. Glanville, P.; Fridlyand, A.; Keinath, C.; Garrabrant, M. Demonstration and Simulation of Gas Heat Pump-Driven Residential Combination Space and Water Heating System Performance. *ASHRAE Trans.* **2019**, *125*, 264–272.
25. Southwest Research Institute. *Compressed Natural Gas Vehicle Fuel Survey*; Report prepared for the CRC under Project No. PC-2-12; Southwest Research Institute: San Antonio, TX, USA, 2014.
26. Choudhury, S.; McDonell, V.; Samuelsen, S. Combustion performance of low-NOx and conventional storage water heaters operated on hydrogen enriched natural gas. *Int. J. Hydrog. Energy* **2020**, *45*, 2405–2417. [CrossRef]
27. Zhao, Y.; McDonell, V.; Samuelsen, S. Assessment of the combustion performance of a room furnace operating on pipeline natural gas mixed with simulated biogas or hydrogen. *Int. J. Hydrog. Energy* **2020**, *45*, 11368–11379. [CrossRef]
28. Zhao, Y.; McDonell, V.; Samuelsen, S. Influence of hydrogen addition to pipeline natural gas on the combustion performance of a cookop burner. *Int. J. Hydrog. Energy* **2019**, *44*, 12239–12253. [CrossRef]
29. Jones, H.R.N. Chaper 3—Partially aerated burners. In *The Application of Combustion Principles to Domestic Gas Burner Design*; E. & F.N.Spon: London, UK, 1990; pp. 30–75.
30. Pritchard, R.; Guy, J.J.; Connor, N.E. Chapter 7—Gas burners and burner systems. In *Handbook of Industrial Gas Utilization: Engineering Principles and Practice*; Van Nostrand Reinhold Company: New York, NY, USA, 1977; pp. 216–313.
31. Glassman, I.; Yetter, R.A. Appendix F—Laminar Flame Speeds. In *Combustion*, 4th ed.; Elsevier: Amsterdam, The Netherlands, 2008; pp. 713–720.
32. Glassman, I.; Yetter, R.A. Chapter 8—Environmental Combustion Considerations. In *Combustion*, 4th ed.; Elsevier: Amsterdam, The Netherlands, 2008; pp. 409–494.
33. U.S. Department of Energy (DOE). Appliance and Equipment Standards Rulemakings and Notices—Consumer Water Heaters. Available online: https://www1.eere.energy.gov/buildings/appliance_standards/standards.aspx?productid=32 (accessed on 12 December 2021).
34. U.S. Department of Energy (DOE). Appliance and Equipment Standards Rulemakings and Notices—Consumer Furnaces. Available online: https://www1.eere.energy.gov/buildings/appliance_standards/standards.aspx?productid=59&action=viewlive (accessed on 12 December 2021).

energies

MDPI

Article

[NiFe]-(Oxy)Sulfides Derived from $NiFe_2O_4$ for the Alkaline Hydrogen Evolution Reaction

David Tetzlaff [1,2], Vasanth Alagarasan [2], Christopher Simon [3], Daniel Siegmund [1,2], Kai junge Puring [1], Roland Marschall [3] and Ulf-Peter Apfel [1,2,*]

1 Fraunhofer Institute for Environmental, Safety and Energy Technology UMSICHT, 46047 Oberhausen, Germany; david.tetzlaff@umsicht.fraunhofer.de (D.T.); daniel.siegmund@umsicht.fraunhofer.de (D.S.); kai.junge.puring@umsicht.fraunhofer.de (K.j.P.)
2 Inorganic Chemistry I, Ruhr University Bochum, 44780 Bochum, Germany; Vasanth.Alagarasan@ruhr-uni-bochum.de
3 Physical Chemistry III, University of Bayreuth, 95447 Bayreuth, Germany; Christopher.Simon@uni-bayreuth.de (C.S.); Roland.Marschall@uni-bayreuth.de (R.M.)
* Correspondence: ulf.apfel@rub.de or ulf-peter.apfel@umsicht.fraunhofer.de

Abstract: The development of noble-metal-free electrocatalysts is regarded as a key factor for realizing industrial-scale hydrogen production powered by renewable energy sources. Inspired by nature, which uses Fe- and Ni-containing enzymes for efficient hydrogen generation, Fe/Ni-containing chalcogenides, such as oxides and sulfides, received increasing attention as promising electrocatalysts to produce hydrogen. We herein present a novel synthetic procedure for mixed Fe/Ni (oxy)sulfide materials by the controlled (partial) sulfidation of $NiFe_2O_4$ (NFO) nanoparticles in H_2S-containing atmospheres. The variation in H_2S concentration and the temperature allows for a precise control of stoichiometry and phase composition. The obtained sulfidized materials (NFS) catalyze the hydrogen evolution reaction (HER) with increased activity in comparison to NFO, up to -10 and -100 mA cm^{-2} at an overpotential of approx. 250 and 450 mV, respectively.

Keywords: hydrogen; oxysulfide; electrocatalysis; alkaline hydrogen evolution reaction

Citation: Tetzlaff, D.; Alagarasan, V.; Simon, C.; Siegmund, D.; junge Puring, K.; Marschall, R.; Apfel, U.-P. [NiFe]-(Oxy)Sulfides Derived from $NiFe_2O_4$ for the Alkaline Hydrogen Evolution Reaction. *Energies* **2022**, *15*, 543. https://doi.org/10.3390/en15020543

Academic Editor: Johannes Schaffert

Received: 29 November 2021
Accepted: 10 January 2022
Published: 13 January 2022

1. Introduction

The electrochemical hydrogen evolution reaction (HER) is a promising approach to foster hydrogen usage for replacing fossil fuels as major energy carriers [1,2]. Currently, electrocatalysts based on noble metals such as Pt are primarily used as cathode material in electrolyzers due to their outstanding performance; however, the scarcity of common noble metals impedes widespread application [3–5]. Therefore, scientists have focused on earth-abundant transition metal chalcogenides such as oxides and sulfides as cost-effective alternatives to Pt for the HER [6–8]. Transition metal oxides are characterized by their compositional and structural flexibility, which offer a high diversity in the electronic and crystal structure. Although transition metal oxides are regarded as catalytically inert for the HER, defect engineering or the introduction of dopants enabled an improved HER performance [9–11]. In contrast, transition metal sulfides can overcome the major drawbacks of metal oxides such as poor electronic conductivity, unsuitable hydrogen adsorption and limited catalytic-active sites [12–14]. To combine the properties of transition metal oxides and sulfides, researchers have focused on the synthesis of distinct transition metal oxysulfides, which contain oxygen and sulfur. Commonly, the synthesis of oxysulfides can be achieved by multiple pathways, including the sulfidation of an oxygenated phase, the oxidation of sulfides, the reduction of sulfates or by the co-insertion of oxygen and sulfur [15]. In this respect, Nelson et al. synthesized CoO_xS_y hollow nanoparticles by substituting oxide with sulfide species in CoO using ammonium sulfide in oleylamine at 100 °C [16]. The obtained electrocatalysts displayed sulfur-content-dependent HER activity, with the highest activity for $CoO_xS_{0.18}$. Another oxysulfide material was synthesized by Sarma et al. by the anodic oxidation of WS_2 sheets [17]. Here, distinct WO_xS_y materials were obtained depending on the deposition potential, which has shown the highest HER performance for the WO_xS_y material, which was deposited at 5 V.

In terms of hydrogen production, nature has established Fe- and Ni-containing enzymes (hydrogenases), which effectively perform the reversible conversion of hydrogen to protons and electrons [18–20]. Inspired by the natural evolutionary choice of transition metals, numerous Fe/Ni-containing catalysts were synthesized, which showed high catalytic activities for the HER [21–27]. Furthermore, Fe/Ni oxysulfide materials were synthesized displaying promising activities towards the HER and oxygen evolution reaction (OER) [28,29]. However, investigating the HER activity of Fe/Ni oxysulfides with different sulfur to metal (S:M) ratios and material phases has not yet been realized.

In lieu, we herein present the synthesis of various NiFe (oxy)sulfide materials by the sulfidation of $NiFe_2O_4$ with H_2S. We demonstrate a controlled sulfidation towards a nickel containing pyrite and pyrrhotite depending on the H_2S gas composition as well as on the temperature. Finally, we show the performance of the synthesized NiFe (oxy)sulfide materials towards the HER in an alkaline medium.

2. Materials and Methods

2.1. Chemicals

The sulfidizing gases H_2S/N_2 (50:50) and H_2S/H_2 (15:85) (Air Products, Hattingen, Germany), as well as KOH (Fisher Scientific, Dreieich, Germany, >85%), were purchased from commercial vendors and used without further purification.

2.2. Synthesis of Sulfidized $NiFe_2O_4$ (NFS) Materials

The synthesis of the starting material $NiFe_2O_4$ was realized according to protocols recently published in the literature [30]. For subsequent sulfidation reactions, 100 mg of $NiFe_2O_4$ was placed into a tubular furnace and purged for 10 min with H_2S/N_2 (50:50) or H_2S/N_2 (15:85). Maintaining the gas flow, the furnace was heated to $100 \leq T \leq 300$ °C and the temperature was held for 1 h. The furnace was then allowed to cool down to room temperature within approx. 20 min using pressurized air while applying a N_2 gas flow.

2.3. Characterization

2.3.1. Physical Characterization

Characterization of the investigated materials was performed by powder X-ray diffraction (PXRD) using a HUBER powder X-ray diffractometer (HUBER, Rimsting, Germany) equipped with a Mo-K_α source. The 2-Θ values were converted to values from a Cu-K_α, according to Bragg's law of diffraction.

The particle sizes of the synthesized electrocatalysts were determined using a SALD-2300 laser diffraction particle size analyzer (Shimadzu, Duisburg, Germany) equipped with a SALD-BC23 batch cell. The respective samples were prepared by dispersing approx. 10 mg of the catalyst material for 1 min in 1 mL isopropyl alcohol using an ultra-sonic bath. Subsequently, a portion of the dispersion was added to the batch cell, which was filled with isopropyl alcohol. The obtained particle sizes were calculated using the Fraunhofer approximation and the volume was chosen based on the dimensions of the particle amount.

A Gemini2 Merlin HR-FESEM (ZEISS, Oberkochen, Germany) was used for scanning electron microscopy (SEM), equipped with an OXFORD AZtecEnergy X-ray microanalysis system for energy dispersive X-ray spectroscopy (EDX). Samples were dispersed in 1 mL isopropyl alcohol and ultra-sonicated for 1 min. Afterwards, the samples were drop-casted on a flat Si Wafer for analysis. The SEM images were recorded at an acceleration voltage of 5 kV while EDX mappings were performed from 0–20 kV.

X-ray photoelectron spectroscopy (XPS) measurements were performed with a polychromatic Al-K_α X-ray source (anode operating at 14 kV and 13 mA) combined with an ultra–High-vacuum (UHV, 10^{-9} mbar) setup and a hemispherical analyzer (type CLAM2,

VG, Scientific, Thermo Fisher Scientific, Dreieich, Germany). A pass energy of 100 eV was applied to record the spectra.

2.3.2. Electrochemical Characterization

The (oxy)sulfide electrocatalysts were investigated as drop-casted materials on a glassy carbon (GC) rod electrode. For this purpose, 11.78 mg of the catalyst material was dispersed in a mixture composed of 0.30 mL water, 0.15 mL isopropanol and 0.05 mL Nafion (5 % in aliphatic alcohols) using an ultra-sonic bath for 30 min. Subsequently, 3 μL of the catalyst ink was applied on a GC electrode (d = 3 mm, 1 mg cm^{-2} catalyst), which was dried at room temperature for 30 min. Before drop-casting, the GC electrode was polished using Al_2O_3 pastes with grain sizes of 0.30 and 0.05 μm for 3 min, each followed by ultra-sonication in Milli-Q water for 5 min.

Electrochemical measurements were performed in a three-electrode setup, employing the catalyst-modified GC working electrode (WE), a Pt mesh counter electrode (CE) and a Hg/HgO (1 M KOH) reference electrode (RE) in 1 M KOH. The WE and CE were separated by utilizing an H-type electrolysis cell with both half-cells being separated by an anion exchange separator (Zirfon®, AGFA, Mortsel, The Netherlands). Electrochemical measurements were conducted using a GAMRY Reference 600 or Reference 600 + (C3-Analysentechnik, Haar, Germany) and the measured potentials were converted to the reversible hydrogen electrode (RHE) reference according to the equation: $E_{RHE} = E_{measured} + E_{Ref.} + 0.059$ pH.

For the HER experiments, the material was first electrochemically conditioned through cyclic voltammetry (CV) between 0 and −0.3 V vs. RHE at 100 mV s^{-1} until a stable voltammogram was obtained. The investigation of the electrochemical surface area (ECSA) was realized through CV measurements between −0.16 and −0.24 V vs. RHE at scan rates of 40, 80, 120, 160 and 200 mV s^{-1}, respectively. Changes in the electrochemical activity were monitored via linear sweep voltammetry (LSV) between 0 and −0.45 V vs. RHE at a scan rate of 1 mV s^{-1}. Electrochemical impedance spectroscopy (EIS) measurements were conducted at −0.4 V vs. RHE from 100 kHz to 0.10 Hz taking 7 points per decade at an amplitude of 7 mV rms. Stability tests were performed using chronopotentiometry at −10 or −100 mA cm^{-2} for at least one hour.

3. Results and Discussion

3.1. Synthesis and Physical Characterization

The synthesis of NiFe-(oxy)sulfides (NFS) was performed by utilizing $NiFe_2O_4$ (NFO) as the starting compound. The NFO precursor was treated with different H_2S gas compositions (H_2S/N_2, (50:50) and H_2S/H_2 (15:85)) at different temperatures ($100 \leq T \leq 300$ °C) to control the sulfur to metal (S:M) ratio in the materials [31]. For clarity, the synthesized materials are referred to as $NFS_T{-}N_2$ and $NFS_T{-}H_2$, where T represents the temperature in °C and N_2 and H_2 the diluting gases.

The obtained materials were analyzed by powder X-ray diffraction (PXRD) and show distinct phase formations depending on the applied temperature and H_2S gas composition (Figure 1). Treatment of $NiFe_2O_4$ in either the H_2S/N_2 or H_2S/H_2 gas mixture results in no visible change in the powder pattern compared to NFO when heated below 200 °C, indicating a conservation of the NFO phase. Starting at 200 °C, phase transformations occur, which become more prominent at higher temperatures. Using a H_2S/N_2 atmosphere, phase transformations towards nickelian pyrite ($Ni_{0.35}Fe_{0.65}S_2$) were observed, while the usage of H_2S/H_2 resulted in phase transformations towards nickelian pyrrhotite ($Ni_{0.35}Fe_{0.65}S$). Notably, the crystallinity of the synthesized materials increases with temperature and no further phase transformation occurred after heating the samples for longer than 1 h.

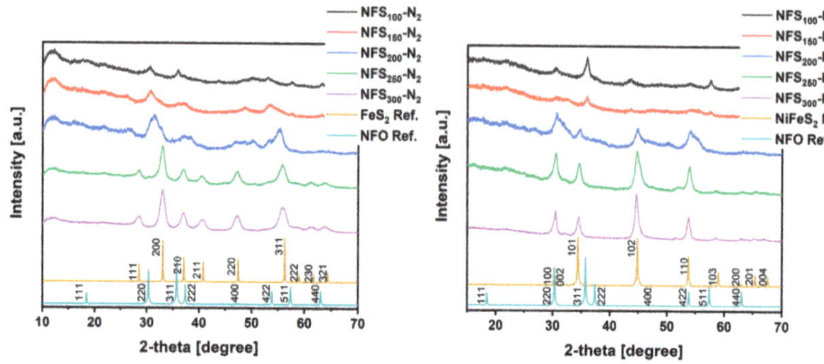

(a) (b)

Figure 1. Powder X-ray diffractograms of NFS materials synthesized at temperatures between 100 and 300 °C using (**a**) H_2S/N_2 and (**b**) H_2S/H_2 reaction atmospheres [32–34].

To further clarify the composition of the obtained NFS materials, X-ray photoelectron spectroscopy (XPS) measurements were performed (Figure S1 in Supplementary Materials). The XPS spectra show characteristic peaks of the two $p_{3/2}$ orbitals of Ni and Fe at approx. 858 and 710 eV, respectively. Furthermore, all materials show a peak in the O 1s spectrum at approx. 532 eV, which can be ascribed to either metal–oxygen bonds, low-coordinated oxygen ions at the surface or adsorbed water [29]. Interestingly, the XPS analysis reveals the presence of sulfur in each NFS sample, which indicates a sulfidation of NFO below 200 °C.

In order to quantify the S:M ratios, the synthesized materials were subjected to energy dispersive spectroscopy (EDX) (Table 1). While no phase changes in the powder patterns are observable, sulfur was detected in the $NFS_{100}-N_2$ and $NFS_{100}-H_2$ materials with S:M ratios of 1.09 ± 0.09 and 0.79 ± 0.26, respectively. For $NFS-N_2$, the S:M ratio undergoes a slight increase until 250 °C (S:M ratio = 1.49 ± 0.06) and a much steeper increase up to 300 °C with a S:M ratio of 1.98 ± 0.26. For the $NFS-H_2$ materials, the S:M ratio increases until reaching a S:M ratio of 1.20 ± 0.08 for $NFS_{200}-H_2$, which subsequently decreases to a S:M ratio of 1.15 ± 0.06 for $NFS_{300}-H_2$. According to the EDX analysis, the sum formulas of the nickelian pyrite and pyrrhotite were calculated to $Ni_{0.30}Fe_{0.71}S_{1.99}$ and $Ni_{0.30}Fe_{0.64}S_{1.06}$, at that point reaching complete sulfidation. Thus, the obtainable phases (NFO, nickelian pyrite or pyrrhotite) as well as the degree of sulfidation can be controlled by the variation of S-source gas composition and temperature.

Table 1. Energy-dispersive X-ray emission (EDX) analysis of the investigated electrocatalysts displaying the obtained Fe:Ni and the S:M ratio.

Material	Fe:Ni Ratio	S:M Ratio	Sum Formula
NFO	2.46 ± 0.10	0	$Ni_{0.30}Fe_{0.74}O_x$
$NFS_{100}-N_2$	1.93 ± 0.96	1.09 ± 0.09	$Ni_{0.3}Fe_{0.52}S_{0.87}O_x$
$NFS_{150}-N_2$	2.47 ± 0.30	1.29 ± 0.22	$Ni_{0.3}Fe_{0.74}S_{1.29}O_x$
$NFS_{200}-N_2$	2.18 ± 0.41	1.34 ± 0.10	$Ni_{0.3}Fe_{0.64}S_{1.26}O_x$
$NFS_{250}-N_2$	2.50 ± 0.33	1.49 ± 0.06	$Ni_{0.3}Fe_{0.74}S_{1.54}O_x$
$NFS_{300}-N_2$	2.33 ± 0.31	1.98 ± 0.26	$Ni_{0.3}Fe_{0.71}S_{1.99}$
$NFS_{100}-H_2$	2.43 ± 0.49	0.79 ± 0.26	$Ni_{0.3}Fe_{0.71}S_{0.80}O_x$
$NFS_{150}-H_2$	2.58 ± 0.45	1.09 ± 0.34	$Ni_{0.3}Fe_{0.76}S_{1.14}O_x$
$NFS_{200}-H_2$	2.41 ± 0.33	1.20 ± 0.08	$Ni_{0.3}Fe_{0.71}S_{1.21}O_x$
$NFS_{250}-H_2$	2.35 ± 0.37	1.11 ± 0.15	$Ni_{0.3}Fe_{0.70}S_{1.04}O_x$
$NFS_{300}-H_2$	2.18 ± 0.31	1.15 ± 0.06	$Ni_{0.3}Fe_{0.64}S_{1.07}$

Scanning electron microscopy (SEM) characterization of the obtained materials reveals comparable particle morphologies and sizes for all tested samples (Figures S2 and S3). In general, spherical shaped particles with sizes in the nanometer range are present, which form larger agglomerates with a rough surface morphology. A more detailed picture of the size distribution of the obtained NFS materials is given by laser diffraction analysis (Figure S4). In general, all materials display a broad particle size distribution. For example, the NFO precursor shows a significant number of particles in the nanometer range as well as in the micrometer range (up to 100 μm). In comparison, the particle size distribution of the NFS materials shows a shift towards larger particles of up to 400 μm, indicating agglomeration. However, a correlation of the particle sizes with the applied temperature cannot be observed and for obtaining smaller particle sizes post-synthetic milling is suggested.

3.2. Electrochemical Hydrogen Evolution Reaction

To further investigate the dependence of the alkaline HER activity on the S:M ratio of the synthesized NFS compounds and to observe trends arising from this alteration, electrodes were prepared via drop-casting and served as working electrodes in a three-electrode setup employing an H-type electrolysis cell.

For a first analysis, the electrochemical activity was determined by linear sweep voltammetry (LSV) at a scan rate of 1 mV s^{-1} (Figure 2a,d). The investigated electrocatalytic materials display distinct overpotentials vs. RHE at a current density of −10 mA cm^{-2} depending on the catalyst composition. For instance, the usage of the NFO precursor as an HER catalyst resulted in almost no catalytic activity. In contrast, all sulfidized materials show a lower overpotential compared to NFO. Using the NFS−N_2 electrocatalysts, the lowest overpotential at −10 mA cm^{-2} was observed for NFS_{100}−N_2 at 266 ± 12 mV. The overpotential increases for catalysts synthesized at higher temperatures up to 394 ± 24 mV for NFS_{250}−N_2. Surprisingly, the NFS_{300}−N_2 electrocatalyst displays an increased catalytic activity in contrast to NFS_{250}−N_2 with an overpotential of 346 ± 4 mV. In comparison, the highest HER activity for the NFS−H_2 electrocatalysts was observed for NFS_{250}−H_2 with an overpotential of 302 ± 9 mV, which increases to 408 ± 30 mV for NFS_{150}−H_2. Notably, the lowest overpotentials are observed for NFS−N_2 electrocatalysts synthesized at temperatures around 100 °C and for NFS−H_2 electrocatalysts synthesized around 250 °C, which show a S:M ratio of approx. 1:1. However, since most of the NFS−H_2 electrocatalysts display similar S:M ratios, the presence of the nickelian pyrrhotite phase seems to play a major role for increased HER activity.

To test this theory, we normalized the LSV curves by the electrochemical surface area (ECSA) to exclude particle size effects from the electrochemical activity. For this purpose, the ECSA was determined by measuring the double layer capacitance (C_{DL}) of the materials using cyclic voltammetry (Figure 2b,e). The NFS−N_2 electrocatalysts display similar C_{DL} values, with NFS_{150}−N_2 having the highest C_{DL} of 51 ± 6 mF cm^{-2}, followed by NFS_{250}−N_2 (49 ± 2 mF cm^{-2}), NFS_{200}−N_2 (45 ± 5 mF cm^{-2}) and NFS_{100}−N_2 (44 ± 3 mF cm^{-2}). Interestingly, the NFS_{300}−N_2 electrocatalyst displays a C_{DL} of 11 ± 1 mF cm^{-2}, which is several times lower than the other NFS−N_2 electrocatalysts. The NFS−H_2 electrocatalysts display the highest C_{DL} for NFS_{100}−H_2 (44 ± 3 mF cm^{-2}) and NFS_{150}−H_2 (51 ± 6 mF cm^{-2}), which decreases with increasing temperature during the synthesis down to 12 ± 2 mF cm^{-2} for NFS_{300}−H_2. The calculation of the ECSA was performed by dividing the obtained C_{DL} values by a specific capacitance C_S of 0.04 mF cm^{-2} [35]. The overpotentials from the LSV curves normalized to the ECSA were obtained at a current density of 0.02 mA $cm^{-2}{}_{ECSA}$ (Figure 2c,f). Here, the ECSA normalization for the NFS−N_2 electrocatalysts results in clearly separated overpotentials for NFS_{100}−N_2 (329 ± 13 mV) and the NFS_{150}−N_2 (353 ± 21 mV) electrocatalysts, which have shown similar low non−Normalized overpotentials. A similar trend can be observed for NFS_{250}−H_2 and NFS_{300}−H_2 catalyzing the HER with overpotentials of 300 ± 9 mV and 272 ± 9 mV, respectively. Notably, NFS_{100}−N_2 catalyzes the HER at the lowest non−Normalized overpotential, while NFS_{300}−H_2 displays the lowest normalized overpotential. Thus, consideration of particle size effects on the electrocatalytic activity is important to determine intrinsic material properties. In addition, we performed electrochemical impedance spectroscopy (EIS) measurements, which reveal similar trends compared to the non−Normalized LSV data (Figure S5). Here, $NFS_{150}-N_2$ and $NFS_{300}-H_2$ show the smallest Nyquist arcs of approx. 25 Ω, which increases to approx. 110 and 160 Ω for $NFS_{250}-N_2$ and $NFS_{150}-H_2$, respectively.

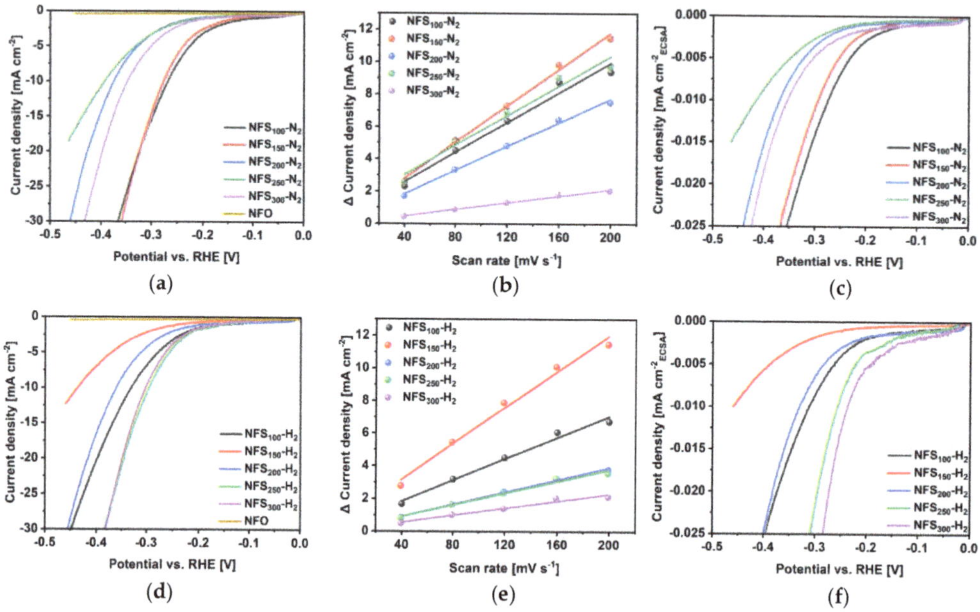

Figure 2. Electrochemical performance data of the $NFS-N_2$ electrocatalysts showing (**a**) non−Normalized LSV curves, (**b**) C_{DL} values obtained by cyclic voltammetry and (**c**) LSV curves normalized by the ECSA. Electrochemical performance data of the $NFS-H_2$ electrocatalysts showing (**d**) non−Normalized LSV curves, (**e**) C_{DL} values obtained by cyclic voltammetry and (**f**) LSV curves normalized by the ECSA.

To test the stability of the NFS electrocatalysts, we performed chronopotentiometry experiments for 1 h at a current density of −10 mA cm^{-2} (Figure 3a,b). A stable performance with a minor activation or deactivation behavior can be observed depending on the investigated electrocatalyst. For example, the NFO precursor catalyst shows the highest potential required to catalyze the HER and shows an activation behavior over the duration of the experiment. However, the overall activity after 1 h is inferior to the synthesized NFS materials. In comparison, most of the $NFS-N_2$ electrocatalysts, except for $NFS_{250}-N_2$, show a slight deactivation within 1 h. The deactivation behavior can also be observed for $NFS_{100}-H_2$ and $NFS_{200}-H_2$; however, a deactivation behavior from a particle detachment from the electrode cannot be generally excluded. Interestingly, $NFS_{250}-H_2$ and $NFS_{300}-H_2$, which catalyze the HER with the lowest potential, show a stable performance. We therefore subjected the $NFS_{300}-H_2$ electrocatalyst to an elongated electrolysis at −10 and −100 mA cm^{-2} for 10 h, respectively (Figure 3c). Here, the HER was also catalyzed with a stable performance by $NFS_{300}-H_2$.

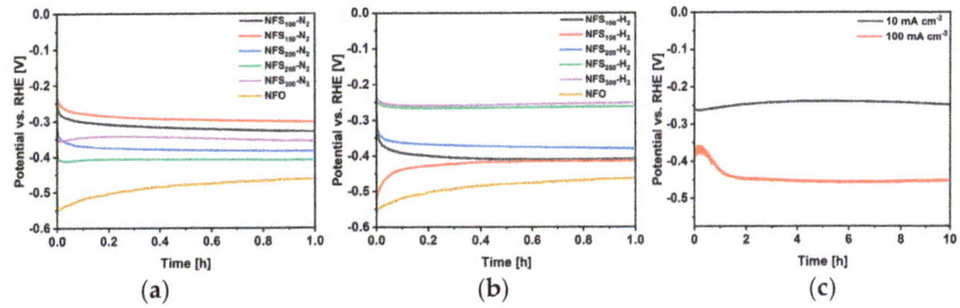

Figure 3. Chronopotentiometry data obtained from the (**a**) NFS$-N_2$ electrocatalysts at 10 mA cm^{-2}, (**b**) NFS$-H_2$ electrocatalysts at 10 mA cm^{-2} and (**c**) $NFS_{300}-H_2$ electrocatalyst at -10 and 100 mA cm^{-2} for 10 h.

It can be concluded that NiFe (oxy)sulfide materials catalyze the electrochemical HER with efficiencies depending on the S:M ratio and the materials phase. The $NFS_{100}-N_2$ and $NFS_{H2,300}$ °C show high overall activities; however, since the NFS$-N_2$ materials mostly tend to deactivate during catalysis, the usage of the fully sulfidized $NFS_{300}-H_2$ material should be prioritized, which displayed stability for 10 h at -100 mA cm^{-2}.

4. Conclusions

A series of bimetallic NiFe (oxy)sulfide materials was synthesized by heating the transition metal oxide NFO in H_2S-containing atmospheres. Depending on the choice of the H_2S gas composition and the applied reaction temperature, a control of the sulfur incorporation into NFO and the materials phase was achieved. For example, a sulfidation of NFO was observed at low temperatures of 100 °C and full conversion into the nickelian pyrite ($Ni_{0.30}Fe_{0.71}S_{1.99}$) and pyrrhotite ($Ni_{0.30}Fe_{0.64}S_{1.06}$) sulfide materials were realized at 300 °C. SEM analysis and the particle size analysis by the laser diffraction technique revealed a broad particle size distribution caused by the particle sintering of the NFS materials.

Furthermore, we assessed the electrochemical HER performance of the NFS materials in 1 M KOH. The electrochemical performance varied with the sulfur content and the materials phase. Materials with a S:M ratio of approx. 1 and/or a nickelian pyrrhotite phase catalyzed the HER with the lowest overpotentials of 266 $\pm$ 12 mV and 302 $\pm$ 9 mV vs. RHE at -10 mA cm^{-2} for $NFS_{100}-N_2$ and $NFS_{250}-H_2$, respectively. Additionally, a normalization of the geometric current density by the ECSA was performed, which revealed the lowest overpotential of 272 $\pm$ 9 mV vs. RHE at a current density of 0.02 mA $cm^{-2}{}_{ECSA}$ catalyzed by $NFS_{300}-H_2$. Here, $NFS_{100}-N_2$ and $NFS_{300}-H_2$ showed the lowest overpotential for the non$-$Normalized and the ECSA$-$Normalized LSV data, respectively. Therefore, particle size effects should be included in the consideration of the HER activity. Finally, preliminary stability measurements have revealed a rather deactivating behavior of NFS$-N_2$ materials, while the $NFS_{300}-H_2$ was able to catalyze the HER at current densities of -10 and -100 mA cm^{-2} with stable potentials for 10 h.

These results represent another step towards designing transition metal chalcogenide catalyst materials for the electrochemical HER and point the way towards the most efficient stoichiometric formulations of NiFe (oxy)sulfide-based catalysts.

Supplementary Materials: The following are available online at https://www.mdpi.com/article/10.3390/en15020543/s1, Figure S1: XPS data of the synthesized NFS materials. Figure S2: Deconvoluted XPS spectra of $NFS_{300}-H_2$. Figure S3: Deconvoluted XPS spectra of $NFS_{100}-N_2$. Figure S4: SEM images of the synthesized NFS$-N_2$ materials. Figure S5: SEM images of the synthesized NFS$-H_2$ materials. Figure S6: Representative SEM images at $NFS_{100}-H_2$ and $NFS_{100}-N_2$ at distinct magnifications. Figure S7: Particle size analysis by laser diffraction. Figure S8: Representative CV measurements for determination of C_{DL}. Figure S9: EIS spectra of the NFS materials. Figure S10: Tafel analysis of NFS materials.

Author Contributions: Conceptualization, D.T.; methodology, D.T.; validation, D.T. and V.A.; formal analysis, D.T. and V.A.; investigation, D.T. and V.A.; resources, D.T. and C.S.; data curation, D.T.; writing—original draft preparation D.T.; writing—review and editing, D.T., V.A., C.S., D.S., K.j.P., R.M. and U.-P.A.; visualization, D.T.; supervision, D.S., K.j.P. and U.-P.A.; funding acquisition, U.-P.A. All authors have read and agreed to the published version of the manuscript.

Funding: This research was funded by the German Research Foundation DFG (AP242/6–1, MA 5392/7–1). U.-P.A. is grateful for financial support from the Deutsche Forschungsgemeinschaft (under Germany's Excellence Strategy—EXC-2033—Project number 390677874), the Fraunhofer Internal Programs under Grant no. Attract 097–602175, and the Fraunhofer Cluster of Excellence CINES. Open access funding was enabled and organized by Projekt DEAL.

Institutional Review Board Statement: Not applicable.

Informed Consent Statement: Not applicable.

Data Availability Statement: Not applicable.

Acknowledgments: The authors gratefully acknowledge funding by the German Research Foundation DFG (AP242/6–1, MA 5392/7–1). U.-P.A. is grateful for financial support from the Deutsche Forschungsgemeinschaft (under Germany's Excellence Strategy—EXC-2033—Project number 390677874), the Fraunhofer Internal Programs under Grant no. Attract 097–602175, and the Fraunhofer Cluster of Excellence CINES. Open access funding was enabled and organized by Projekt DEAL. The authors thank Stephan Spöllmann (RUBION) for measuring XPS spectra.

Conflicts of Interest: The authors declare no conflict of interest.

References

1. Siegmund, D.; Metz, S.; Peinecke, V.; Warner, T.E.; Cremers, C.; Grevé, A.; Smolinka, T.; Segets, D.; Apfel, U.-P. Crossing the Valley of Death: From Fundamental to Applied Research in Electrolysis. *JACS Au* **2021**, *1*, 527–535. [CrossRef] [PubMed]
2. Chu, S.; Majumdar, A. Opportunities and challenges for a sustainable energy future. *Nature* **2012**, *488*, 294–303. [CrossRef] [PubMed]
3. Greeley, J.; Stephens, I.E.L.; Bondarenko, A.S.; Johansson, T.P.; Hansen, H.A.; Jaramillo, T.F.; Rossmeisl, J.; Chorkendorff, I.; Nørskov, J.K. Alloys of platinum and early transition metals as oxygen reduction electrocatalysts. *Nat. Chem.* **2009**, *1*, 552–556. [CrossRef] [PubMed]
4. Greeley, J.; Markovic, N.M. The road from animal electricity to green energy: Combining experiment and theory in electrocatalysis. *Energy Environ. Sci.* **2012**, *5*, 9246. [CrossRef]
5. Amer, M.S.; Ghanem, M.A.; Al-Mayouf, A.M.; Arunachalam, P.; Khdary, N.H. Low-loading of oxidized platinum nanoparticles into mesoporous titanium dioxide for effective and durable hydrogen evolution in acidic media. *Arab. J. Chem.* **2020**, *13*, 2257–2270. [CrossRef]
6. Boppella, R.; Tan, J.; Yun, J.; Manorama, S.V.; Moon, J. Anion-mediated transition metal electrocatalysts for efficient water electrolysis: Recent advances and future perspectives. *Coord. Chem.* **2021**, *427*, 213552. [CrossRef]
7. Hu, C.; Zhang, L.; Gong, J. Recent progress made in the mechanism comprehension and design of electrocatalysts for alkaline water splitting. *Energy Environ. Sci.* **2019**, *12*, 2620–2645. [CrossRef]
8. Li, A.; Sun, Y.; Yao, T.; Han, H. Earth-Abundant Transition-Metal-Based Electrocatalysts for Water Electrolysis to Produce Renewable Hydrogen. *Chem. Eur. J.* **2018**, *24*, 18334–18355. [CrossRef]
9. Li, Y.H.; Liu, P.F.; Pan, L.F.; Wang, H.F.; Yang, Z.Z.; Zheng, L.R.; Hu, P.; Zhao, H.J.; Gu, L.; Yang, H.G. Local atomic structure modulations activate metal oxide as electrocatalyst for hydrogen evolution in acidic water. *Nat. Commun.* **2015**, *6*, 8064. [CrossRef] [PubMed]
10. Li, L.; Zhang, T.; Yan, J.; Cai, X.; Liu, S.F. P Doped MoO_{3-x} Nanosheets as Efficient and Stable Electrocatalysts for Hydrogen Evolution. *Small* **2017**, *13*, 1700441. [CrossRef]
11. Zeng, H.; Chen, S.; Jin, Y.Q.; Li, J.; Song, J.; Le, Z.; Liang, G.; Zhang, H.; Xie, F.; Chen, J.; et al. Electron Density Modulation of Metallic MoO 2 by Ni Doping to Produce Excellent Hydrogen Evolution and Oxidation Activities in Acid. *ACS Energy Lett.* **2020**, *5*, 1908–1915. [CrossRef]
12. Zhu, Y.; Lin, Q.; Zhong, Y.; Tahini, H.A.; Shao, Z.; Wang, H. Metal oxide-based materials as an emerging family of hydrogen evolution electrocatalysts. *Energy Environ. Sci.* **2020**, *13*, 3361–3392. [CrossRef]
13. Guo, Y.; Park, T.; Yi, J.W.; Henzie, J.; Kim, J.; Wang, Z.; Jiang, B.; Bando, Y.; Sugahara, Y.; Tang, J.; et al. Nanoarchitectonics for Transition-Metal-Sulfide-Based Electrocatalysts for Water Splitting. *Adv. Mater.* **2019**, *31*, 1807134. [CrossRef] [PubMed]
14. Siegmund, D.; Blanc, N.; Smialkowski, M.; Tschulik, K.; Apfel, U.-P. Metal-Rich Chalcogenides for Electrocatalytic Hydrogen Evolution: Activity of Electrodes and Bulk Materials. *ChemElectroChem* **2020**, *7*, 1514–1527. [CrossRef]

15. Larquet, C.; Carenco, S. Metal Oxysulfides: From Bulk Compounds to Nanomaterials. *Front. Chem.* **2020**, *8*, 179. [CrossRef] [PubMed]
16. Nelson, A.; Fritz, K.E.; Honrao, S.; Hennig, R.G.; Robinson, R.D.; Suntivich, J. Increased activity in hydrogen evolution electrocatalysis for partial anionic substitution in cobalt oxysulfide nanoparticles. *J. Mater. Chem. A* **2016**, *4*, 2842–2848. [CrossRef]
17. Sarma, P.V.; Vineesh, T.V.; Kumar, R.; Sreepal, V.; Prasannachandran, R.; Singh, A.K.; Shaijumon, M.M. Nanostructured Tungsten Oxysulfide as an Efficient Electrocatalyst for Hydrogen Evolution Reaction. *ACS Catal.* **2020**, *10*, 6753–6762. [CrossRef]
18. Vignais, P.M.; Billoud, B. Occurrence, classification, and biological function of hydrogenases: An overview. *Chem. Rev.* **2007**, *107*, 4206–4272. [CrossRef]
19. Pandey, A.S.; Harris, T.V.; Giles, L.J.; Peters, J.W.; Szilagyi, R.K. Dithiomethylether as a ligand in the hydrogenase h-cluster. *J. Am. Chem. Soc.* **2008**, *130*, 4533–4540. [CrossRef]
20. Möller, F.; Piontek, S.; Miller, R.G.; Apfel, U.-P. From Enzymes to Functional Materials-Towards Activation of Small Molecules. *Chem. Eur. J.* **2018**, *24*, 1471–1493. [CrossRef]
21. Shang, X.; Chi, J.-Q.; Liu, Z.-Z.; Dong, B.; Yan, K.-L.; Gao, W.-K.; Zeng, J.-B.; Chai, Y.-M.; Liu, C.-G. Ternary Ni-Fe-V sulfides bundles on nickel foam as free-standing hydrogen evolution electrodes in alkaline medium. *Electrochim. Acta* **2017**, *256*, 241–251. [CrossRef]
22. McGlynn, S.E.; Kanik, I.; Russell, M.J. Peptide and RNA contributions to iron-sulphur chemical gardens as life's first inorganic compartments, catalysts, capacitors and condensers. *Phil. Trans. R. Soc. A* **2012**, *370*, 3007–3022. [CrossRef]
23. Konkena, B.; Junge Puring, K.; Sinev, I.; Piontek, S.; Khavryuchenko, O.; Dürholt, J.P.; Schmid, R.; Tüysüz, H.; Muhler, M.; Schuhmann, W.; et al. Pentlandite rocks as sustainable and stable efficient electrocatalysts for hydrogen generation. *Nat. Commun.* **2016**, *7*, 12269. [CrossRef]
24. Piontek, S.; Andronescu, C.; Zaichenko, A.; Konkena, B.; Junge Puring, K.; Marler, B.; Antoni, H.; Sinev, I.; Muhler, M.; Mollenhauer, D.; et al. Influence of the Fe:Ni Ratio and Reaction Temperature on the Efficiency of $(Fe_xNi_{1-x})_9S_8$ Electrocatalysts Applied in the Hydrogen Evolution Reaction. *ACS Catal.* **2018**, *8*, 987–996. [CrossRef]
25. Zhang, W.; Li, D.; Zhang, L.; She, X.; Yang, D. NiFe-based nanostructures on nickel foam as highly efficiently electrocatalysts for oxygen and hydrogen evolution reactions. *J. Energy Chem.* **2019**, *39*, 39–53. [CrossRef]
26. Wang, X.; Zong, X.; Liu, B.; Long, G.; Wang, A.; Xu, Z.; Song, R.; Ma, W.; Wang, H.; Li, C. Boosting Electrochemical Water Oxidation on NiFe (oxy) Hydroxides by Constructing Schottky Junction toward Water Electrolysis under Industrial Conditions. *Small* **2021**, *7*, e2105544. [CrossRef] [PubMed]
27. Zhang, Z.; Wu, Y.; Zhang, D. Potentiostatic electrodeposition of cost-effective and efficient Ni–Fe electrocatalysts on Ni foam for the alkaline hydrogen evolution reaction. *Int. J. Hydrogen Energy* **2021**, *47*, 1425–1434. [CrossRef]
28. Li, B.-Q.; Zhang, S.-Y.; Tang, C.; Cui, X.; Zhang, Q. Anionic Regulated NiFe (Oxy)Sulfide Electrocatalysts for Water Oxidation. *Small* **2017**, *13*, 1700610. [CrossRef]
29. Liu, J.; Zhu, D.; Ling, T.; Vasileff, A.; Qiao, S.Z. Anthony Vasileff; Shi-Zhang Qiao. S—NiFe2O4 ultra-small nanoparticle built nanosheets for efficient water splitting in alkaline and neutral pH. *Nano Energy* **2017**, *40*, 264–273. [CrossRef]
30. Simon, C.; Zakaria, M.B.; Kurz, H.; Tetzlaff, D.; Blösser, A.; Weiss, M.; Timm, J.; Weber, B.; Apfel, U.-P.; Marschall, R. Magnetic $NiFe_2O_4$ Nanoparticles Prepared via Non-Aqueous Microwave-Assisted Synthesis for Application in Electrocatalytic Water Oxidation. *Chem. Eur. J.* **2021**, *27*, 16990–17001. [CrossRef]
31. Bezverkhyy, I.; Danot, M.; Afanasiev, P. New Low-Temperature Preparations of Some Simple and Mixed Co and Ni Dispersed Sulfides and Their Chemical Behavior in Reducing Atmosphere. *Inorg. Chem.* **2003**, *42*, 1764–1768. [CrossRef] [PubMed]
32. Wyckoff, R.W.G. The structure of crystals. *J. Chem. Technol. Biotechnol.* **1931**, *50*, 877. [CrossRef]
33. Alsén, N. Röntgenographische Untersuchung der Kristallstrukturen von Magnetkies, Breithauptit, Pentlandit, Millerit und verwandten Verbindungen. *Geol. Fören. Stock. Förh.* **1925**, *47*, 19–72. [CrossRef]
34. Baylis, P. Crystal structure refinement of a weakly anisotropic pyrite. *Am. Mineral.* **1977**, *62*, 1168–1172.
35. McCrory, C.C.L.; Jung, S.; Ferrer, I.M.; Chatman, S.M.; Peters, J.C.; Jaramillo, T.F. Benchmarking hydrogen evolving reaction and oxygen evolving reaction electrocatalysts for solar water splitting devices. *J. Am. Chem. Soc.* **2015**, *137*, 4347–4357. [CrossRef] [PubMed]

energies

Article

Multidisciplinary Assessment of a Novel Carbon Capture and Utilization Concept including Underground Sun Conversion

Andreas Zauner [1], Karin Fazeni-Fraisl [1], Philipp Wolf-Zoellner [2], Argjenta Veseli [1,3,*], Marie-Theres Holzleitner [1], Markus Lehner [2], Stephan Bauer [4] and Markus Pichler [4]

1 Energieinstitut an der Johannes Kepler, Universität Linz, Altenberger Straße 69, 4040 Linz, Austria; zauner@energieinstitut-linz.at (A.Z.); fazeni@energieinstitut-linz.at (K.F.-F.); holzleitner@energieinstitut-linz.at (M.-T.H.)
2 Montanuniversität Leoben, Chair of Process Technology and Industrial Environmental Protection, Franz-Josef-Strasse 18, 8700 Leoben, Austria; philipp.wolf-zoellner@unileoben.ac.at (P.W.-Z.); markus.lehner@unileoben.ac.at (M.L.)
3 Johannes Kepler Institute for Energy, Johannes Kepler University Linz, Altenberger Straße 69, 4040 Linz, Austria
4 RAG Austria AG, Schwarzenbergplatz 16, 1015 Vienna, Austria; stephan.bauer@rag-austria.at (S.B.); markus.pichler@rag-austria.at (M.P.)
* Correspondence: veseli@energieinstitut-linz.at; Tel.: +43-732-2468-5655

Abstract: The current work investigates the feasibility of a novel Carbon Capture and Utilization (CCU) approach—also known as Underground Sun Conversion (USC) or geo-methanation. The overall objective of the current work is a comprehensive assessment on the technical, economic and legal aspects as well as greenhouse gas impacts to be concerned for establishing USC technology concept. This is achieved by applying multidisciplinary research approach combining process simulation, techno-economic and greenhouse gas assessment as well as legal analysis allows answering questions about technical, economic feasibility and greenhouse gas performance as well as on legal constraints related to large scale CCU using geo-methanation in depleted hydrocarbon reservoirs. CO_2 from the industry and renewable H_2 from the electrolyser are converted to geomethane in an underground gas storage and used in industry again to close the carbon cycle. Process simulation results showed the conversion rates vary due to operation mode and gas cleaning is necessary in any case to achieve natural gas grid compliant feed in quality. The geomethane production costs are found to be similar or even lower than the costs for synthetic methane from Above Ground Methanation (AGM). The GHG-assessment shows a significant saving compared to fossil natural gas and conventional power-to-gas applications. From a legal perspective the major challenge arises from a regulative gap of CCU in the ETS regime. Accordingly, a far-reaching exemption from the obligation to surrender certificates would be fraught with many legal and technical problems and uncertainties.

Keywords: power-to-gas; geo-methanation; CCU; life cycle assessment; large-scale energy storage

Citation: Zauner, A.; Fazeni-Fraisl, K.; Wolf-Zoellner, P.; Veseli, A.; Holzleitner, M.-T.; Lehner, M.; Bauer, S.; Pichler, M. Multidisciplinary Assessment of a Novel Carbon Capture and Utilization Concept including Underground Sun Conversion. *Energies* **2022**, *15*, 1021. https://doi.org/10.3390/en15031021

Academic Editor: Johannes Schaffert

Received: 29 November 2021
Accepted: 14 January 2022
Published: 29 January 2022

1. Introduction

The European Green Deal sets the scene for a climate neutral European Union (EU) by 2050. According to the so-called 'European Climate Law', the target of zero net Greenhouse Gas (GHG) emissions has to be reached by 2050 with an intermediate goal of at least 55% GHG emission reduction until 2030 compared to 1990 emission levels [1]. This puts new challenges on the energy sector and the energy intensive industries to adapt for climate neutral production [2] within a timeframe of 29 years [3]. It is widely agreed that decarbonization of industry can only be achieved by applying a mix of technologies: fuel switch, electrification, material and energy efficiency, re-use and recycling, material substitution, Carbon Capture and Storage (CCS) [4] as well as Carbon Capture and Utilization (CCU). According to a report of the International Energy Agency (IEA) decarbonization will be hardly possible without CCS and CCU. Considering GHG mitigation aspects and economic aspects the demand for Carbon Capture Utilization and Storage (CCUS) solutions in industry will grow significantly in the upcoming decades. However, the topic of CCU and CCS and the arising question on technical and economic feasibility as well as the achievable climate mitigation effect is not new as it already has been risen by [5] where three CCS technologies were compared. Since then, significant technology development has been achieved and today several CCU technology options at different technology readiness levels (TRL) are available [6]. Amongst those are horticultural production, methanol production, polymer synthesis, mineral carbonation, concrete curing, algae production as well as power-to-gas [7] or power-to-X solutions [8]. Producing synthetic methane—among other hydrocarbons—via methanation of hydrogen (H_2) and carbon dioxide (CO_2) is currently a broadly examined CCU route in scientific literature [9]. pointed out the globally rising supply for CCU technologies and that in the year 2030 there will be the potential to use 910 Mt per year for the production of CO_2-based fuels in order to use up the emitted CO_2.

Today two main routes for methanation of H_2 and CO_2 exist: biological and catalytic methanation [10,11]. The research focuses besides technological feasibility and improvements also on economic viability and environmental aspects [12–16]. Reference [16] conducted a techno-economic analysis for SNG production using catalytic methanation and found two major factors influencing the economic feasibility of SNG as an CCU option: (1) hydrogen costs; and (2) the CO_2 prices assumed as a credit. In fact, also a CO_2 price of 100 EUR/t cannot compensate the higher production costs of green hydrogen compared to H_2 from steam reforming. Comprehensive research on the techno-economics and climate impact of various CCU pathways also has been conducted by [9] showing that catalytic conversion pathways show promising economics and feasible greenhouse gas savings. Catalytic methanation requires high reactant gas purities due to the sensitivity of the metal catalyst towards contaminants, which is currently a restriction especially for large scale application as required for CCU purposes for industry. Recent experiments with real by-product gases from steel industry for example have clearly shown that without gas cleaning a constant degradation of the catalyst material occurs. A carbon filter delivered satisfying results not showing a significant catalyst degradation at lab-scale. Upscaling the lab-results for a steel mill aiming at completely substituting natural gas by integrated SNG production would still require a significant amount of catalyst [17].

In contrast to catalytic methanation the reaction in biological methanation is more robust towards impurities and intermittent reaction gas supply. However, biological methanation demands bigger reactor dimension due to lower space-time yields. This is one of the major limitations of biological methanation [18]. In summary the techno-economics and also the greenhouse gas impacts of SNG production using catalytic methanation or other CCU approaches like methanol production is well covered by recent scientific works. Ref. [19] for example showed that SNG production from biogas catalytic methanation—as a viable example for biogenic CO_2 use—leads to significant GHG-savings compared to fossil natural gas. The origin of the CO_2 makes a difference although from a "cradle-to-gate" perspective CO_2 implies—regardless the industrial source—a credit of approximately—0.8 kg $CO_2eq/kg_{CO2captured}$ which corresponds to a CO_2 uptake of—2.69 kg CO_2/kg_{SNG} [20]. The remaining question nevertheless is, if a carbon uptake or at least carbon neutrality can be justified by establishing a loop applying large scale underground methanation. Depending on assumptions concerning CAPEX and OPEX taken into the results of techno-economic analysis of CCU pathways can differ significantly—also among identical products [21].

In a recent review on the technology readiness levels of carbon capture utilization and storage concepts, the option of CCU in depleted gas reservoirs has been neglected if this option only was referred to be a CO_2 storage option [22]. However depleted gas reservoirs offer a great potential to be the biological methanation reactors of the future overcoming the limitation of scale. For geo-methanation CO_2 and H_2 are injected into a porous underground reservoir and converted to methane by microbes already present in the reservoir.

Whereas underground storage of CO_2 and also H_2 in depleted hydrocarbon reservoirs, coal seams, aquifers and salt caverns already has been discussed numerously in research [23–26] the simultaneous injection of CO_2 and H_2 for exploiting the potential of underground biological methanation still lacks recognition. Although the work of [27] focusses on H_2 underground storage the identified need for more know how on microbiological data can also be transferred to geo-methanation, which will require more field tests. Microbial activity and the conversion of H_2 to CH_4 is mostly described as potential pitfall for underground hydrogen storage [27,28] and not as a potential for using underground storage facilities for targeted geo-methanation of H_2 with additionally injected CO_2 to develop a large scale CCU possibility for future decarbonization. Reference [29] showed that carbon neutrality for hydrocarbons from high CO_2 gas condensate reservoirs can be achieved by closing the loop applying CCS. The work clearly indicates that closing carbon cycles provides carbon neutral option but it lacks of economic perspectives as well as a detailed greenhouse gas assessment and solely takes into account a CCS option ignoring potentials for CCU.

A current study reports two demonstration projects which investigate the feasibility of geo-methanation: HyChico in Argentina and Underground Sun Conversion (USC) in Austria [30]. The lack of projects investigating large scale CCU utilizing the natural presence of microbes in underground reservoirs provides the first evidence of a lack of technical and economic feasibility studies for that approach.

Existing studies strongly focus on large scale hydrogen storage in different geological formations. Reference [31] for example examined systemic aspects of large scale H_2 storage in Romania. Reference [32] studied the cost of H_2 storage in French salt caverns and concluded that this is economically feasible as the storage costs only make up a minor share of the total hydrogen costs. Another work on the economics of hydrogen underground storage has been done by Reference [33] where the storage in salt caverns has been compared with storage in buried pipes and an aboveground tank. According to a recent review several studies exist that estimate the costs of H_2 underground storage, but partly focusing on storage in salt caverns [34]. There is a lack of detailed investigations for underground storage in depleted gas reservoirs in terms of techno-economics. The existing studies strongly focus on technical aspects such as the work done by [35] or [36] who examined the behaviour of H_2 in the reservoir and defined criteria for choosing reservoirs. Reference [37] studied the environmental impacts of H_2 storage in a depleted gas reservoir.

In contrast only a few studies investigate underground methanation of H_2 and CO_2. The few available works concentrate on technical feasibility and potentials like [30] or [38]. To the knowledge of the authors there are no other recent studies on underground methanation of H_2 and CO_2 in depleted gas reservoirs available.

The research project USC investigates the in situ microbial methanation of carbon dioxide and hydrogen in depleted natural gas reservoirs and aims to develop a process chain for its industrial utilization. Results from the Underground Sun Storage research project [39] have strongly indicated that microbial consortia present in depleted biogenic gas reservoirs are capable of using hydrogen for the formation of methane. Such a technology would make it possible to generate and convert large amounts of renewable energy, both in Austria and in areas with a high potential and export this energy to densely populated areas with a lower renewable energy generation potential. In addition, this represents a huge source by potentially providing the urgently needed flexibility, which renewable energy sources currently lack. The utilization of CO_2 in that process enables the creation of a sustainable carbon cycle.

In general, there are two main use cases for the USC technology:

1. Seasonal electricity storage (use of the USC technology for large-scale long-term storage of renewable electricity)
2. CCU (use the USC technology for the production of green SNG from renewable electricity and CO_2 from the industry)—main focus of this paper.

The produced geomethane can be used in different sectors (mobility, power, heat, industry) and within these for various applications (e.g., in mobility as a fuel for ships, busses and trucks; in industry as a power source or chemical feedstock; for electricity production in gas fired power plants; or for space heating).

Whereby use case 2 is the one in focus of the analysis conducted in the current work. The novelty of the current work can be found in the following aspects:

1. Addressing underground methanation as use case for large scale CCU;
2. Assessing the techno-economics and greenhouse gas performance for this use case;
3. Assessing the technical feasibility of this approach by process simulations in demonstration-scale;
4. Assessing these aspects based on data derived from the field trial conduction in Austria within the project USC.

Additionally, the legal aspects concerning CO_2 capture and reuse are investigated to provide a holistic picture for large scale CCU. The embedding of the respective processes in the legal framework is crucial for their realisation, which is why a legal analysis of it is provided in this paper. The development of a green gas certificate system is also of great relevance. Indeed, a uniform European system for guarantees of origin must be established quickly. Therefore, the current status is also subjected to a legal analysis in this regard, and related difficulties are presented.

2. Materials and Methods

2.1. Process Simulations

For the simulations performed in this work, Aspen Plus® flowsheet simulations were developed with the goal to establish a reliable data base for the technical ability of an upscaled underground geo-methanation process. Based on these data the possibility to integrate such a process in future CCU scenarios shall be assessed. Therefore, a series of process flowsheets were mapped out representing the existing USC pilot plant in Pilsbach, Upper Austria, designed and operated by RAG Austria AG [40]. Figures 1 and 2 show a simplified version of the established flowsheets for two operating modes named BATCH and CYCLE. For both modes, all relevant process components from the pilot plant operated in Pilsbach have been included in the simulation setup.

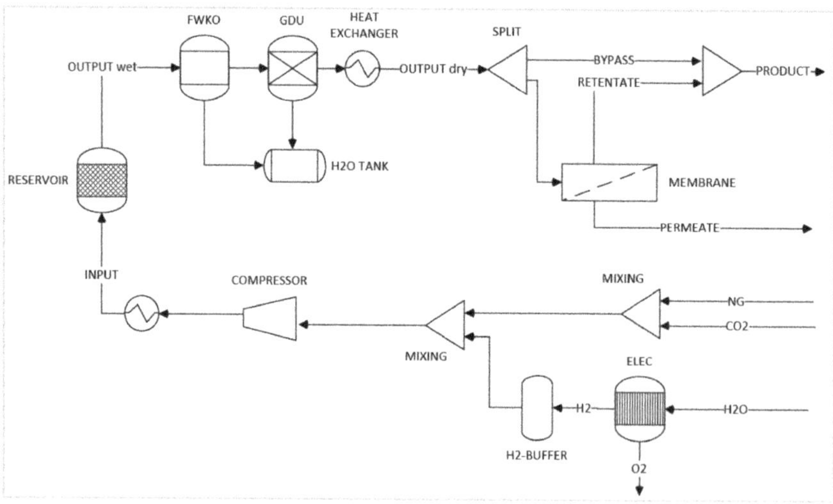

Figure 1. Simplified version of Aspen Plus® flowsheet for BATCH mode.

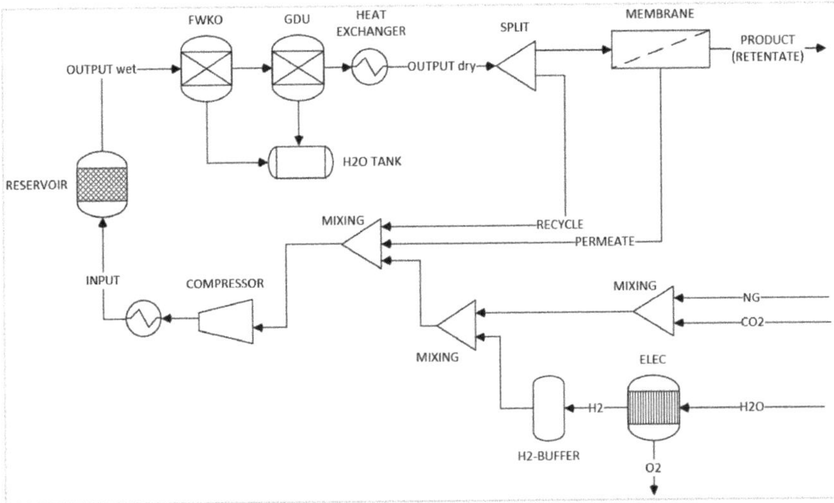

Figure 2. Simplified version of Aspen Plus® flowsheet for CYCLE mode.

The lower half of each figure represents the storing of natural gas mixed with CO_2 and H_2 produced from renewable energy sources. Starting with blending of CO_2 with Natural Gas (NG) from a nearby natural gas reservoir, H_2 produced by Electrolysis (ELEC) is added afterwards. The electrolyser is operated at 65 °C and 30 bars, with an efficiency of 70%. Once the gas is mixed, it is compressed to 45 bar (INPUT) and injected into the depleted gas reservoir acting as bio-/geo-reactor for the methanation of CO_2 and H_2 (RESERVOIR). A total operating volume (TOV) of 500,000 Nm^3 with injection/production rates up to 2000 Nm^3/h can be managed with the pilot plant. On the upper half of the figures the withdrawal of natural gas and any converted synthetic natural gas (geo-methane) from the reservoir is shown. The produced gas stream including reservoir and reaction water according to Equation (1) [41], as well as not-converted H_2, CO_2 and higher hydrocarbons (OUTPUT wet), is fed to a free water knockout (FWKO) and a gas drying unit (GDU).

$$\begin{gathered} CO_2 + 4H_2 = CH_4 + 2H_2O\ (g) \\ \Delta H_r^0 = -165\ \text{kJ/mol} \end{gathered} \tag{1}$$

For both units split fractions of 0.9 for water were implemented. The separated water is stored in a tank from where it can be further treated and possibly re-used in the overall process concept (e.g., for the PEM electrolyser). Then, the pre-treated dry gas (OUTPUT dry) is tempered and fed into a membrane-based gas separation unit with a split fraction of 0.2 for hydrogen and CO_2, and 0.8 for methane and higher hydrocarbons. While the PERMEATE contains the main share of CO_2 and H_2, the RETENTATE is rich on natural gas and geo-methane.

Depending on the operating mode, the PERMEATE is either re-injected into the reservoir after being blended with the NG, CO_2 and H_2 mixture (CYCLE mode), or it is re-injected into the Austrian gas grid with the RETENTATE (BATCH mode). For both modes, a RECYCLE resp. BYPASS stream is implemented, to limit the flow rate of dry gas fed to the installed membrane unit to 600 Nm^3/h. Furthermore, in CYCLE mode, injection and production take place at different wells (LEH002 & LESP-001A) which were drilled in the same reservoir structure but 200 m apart from each other. In BATCH mode, the geo-methane is retrieved from the very same well it was injected to (LEH002).

2.2. *Techno-Economic Assessment*

The techno-economic assessment (TEA) of the USC technology is performed for CCU use cases, two different USC plant configurations (performance of the plant) and for two different plant sizes (defined by the nominal gas volume flow into the underground pore storage). Further, the TEA is done for the timeline which starts in 2025 and end in far future (2050). To perform the TEA different methods are used.

First, the TEA is carried out in large part in accordance with the guideline "Tech-no-Economic Assessment and Life Cycle Assessment—Guidelines for CO_2 Utilization" by Zimmermann et al. (2018) [42].

Second, for calculating the future cost reduction potentials due to learning curve effects and economies of scale the Tool CoLLeCT (Component Level Learning Curve Tool), see Böhm et al. (2019) [43], an in-house development of the Energieinstitut an der JKU Linz, is used. CoLLeCT represents a tool for calculation and analysis of overall learning curves for variable assemblies by observance of the learning rates of its sub-components and their properties. Thus, it allows a detailed analysis of the cost structure as well as the com-prehension of components from comparable and well-known technologies. So, it is rendered possible to determine the cost development of complex structures on a low technology readiness level according to future production volumes. Additionally, the modular approach allows for a consideration of spill-over effects from concurrent us-ages of technologies and components.

Third, for calculating the production costs of geomethane for different use cases out of the USC plant the tool PResTiGE (Power to gas assessment tool), see Böhm et al. (2020) [44], an in-house development of the Energieinstitut an der JKU Linz, is used. PResTiGE is a toolbox for current and prospective techno-economic and environmental benchmarking of PtG systems. The tool comprises data from demo sites and benchmark systems as options for electricity storage or applications of the gaseous products H_2 or SNG at different scales, in forms that are regionally adaptable over all process steps of the PtG system and product application. The assessment results reveal the optimal PtG system configuration and implementation (i.e., with minimal cost and maximal system benefits). Sensitivities can be systematically analysed to explore the robustness of the results. The quantitative economic assessment via PResTiGE is based on the specific production costs of hydrogen or SNG, which are calculated from the total annual costs in relation to the amount of annually produced energy. The total annual costs are calculated using the so-called "annuity method" following VDI 2067.

2.2.1. Overview of the Assessed CCU Use Case

The goal of the use case CCU is to produce a maximum amount of SNG with green electricity and CO_2 from industry for the use in industry. In Figure 3 the interaction of the individual components and players are shown.

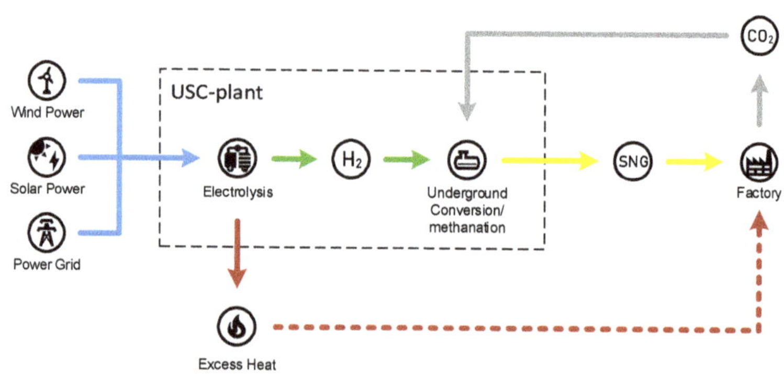

Figure 3. Overview diagram of the use case CCU.

CO_2 from the industry (from processes available the whole year) and renewable H_2 from the electrolyser are converted in the underground pore storage to geomethane. The geomethane is again used in the same industry process, where previously the CO_2 was released.

2.2.2. General Parameters and Data for the TEA

In Table 1 the most important general parameters for the TEA are listed.

Table 1. General parameters and data for the TEA.

Parameter	Sub-Parameter	Value
Interest rate		4%
Observation period		20 years
Electrolyser	investment costs [1]	Year 2025: 800–930 EUR/kW Year 2050: 200–260 EUR/kW
	plant efficiency	Year 2025: 65.0% Year 2050: 72.5%
	time hot stand-by	1 h
	power hot stand-by	2% of nominal power
	Lifetime stack	Year 2025: 6 years Year 2050: 18 years
	Lifetime BoP	30 years
	Excess heat (share of nominal power)	Year 2025: 25% Year 2050: 20%
Above Ground Methanation	Investment costs [1]	Year 2025: 410–470 EUR/kW Year 2050: 200–250 EUR/kW
	Efficiency/degree of conversion	78%
	Lifetime	20 years
	Excess heat (share of nominal power)	17%
CO_2 costs		40 EUR/t
Sale price for excess heat		Year 2025: 50 EUR/MWh Year 2050: 57 EUR/MWh
Sale price for oxygen		50 EUR/t
Electricity costs [2]		Year 2025: 45 EUR/MWh Year 2050: 75 EUR/MWh
Tariff for long-term storage service [3]		7860 EUR/MW

[1] Depending on nominal power, for detail see [44]. [2] Based on [45–52]. However, the projected electricity prices differ due to system's assumptions (among other things, to the underlying climate protection targets and their measures) there is a clear trend towards higher electricity prices. [3] Compare the tariff calculator for long-term storage service of RAG Energy Storage GmbH—https://www.rag-energy-storage.at/speicherdienstleistungen/tarifrechner.html (accessed on 19 May 2021).

2.2.3. USC Plant Configurations

Based on the real data of the test series directly at the storage facility as well as the laboratory tests, two possible configurations of a USC plant are derived, which essentially differ in terms of the stored and withdrawn gas composition (input/output), and thus by the H_2 conversion, the efficiency and the stoichiometric ratios:

- USC base (average case);
- USC advance (close to the theoretical optimum);

The gas composition (proportion of CH_4, H_2 and CO_2) of the injected and withdrawn volume flow for the two plant configurations is shown in Figure 4.

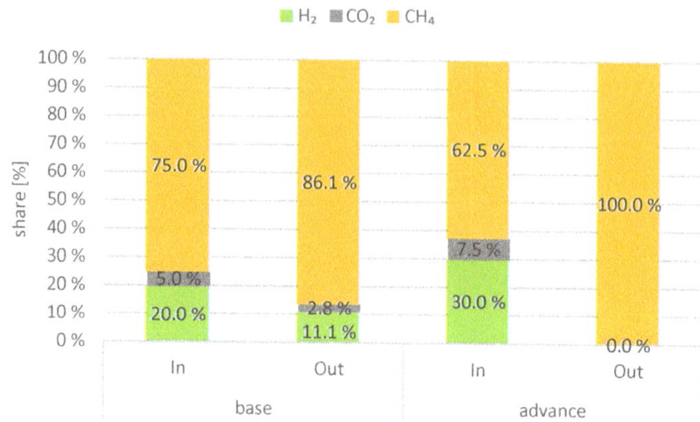

Figure 4. Gas composition (proportion of CH_4, H_2 and CO_2) during injection and withdrawal.

In the plant configuration base, 20% H_2, 5% CO_2 and 75% CH_4 are stored in the underground pore storage. After the conversion, the output volume flow consists of 11.1% H_2, 2.8% CO_2 and 86.1% CH_4. Thus, the share of CH_4 in the overall mixture is increased by 11.1 percentage points. In comparison, a significantly higher proportion of H_2 (30%) is stored in the advance plant configuration. Similar to the plant configurations base, CO_2 is added stoichiometrically (7.5%). The remaining share in the volume flow is accounted for CH_4 (62.5%). In the advance plant configuration, it is assumed, that all H_2 and CO_2 is converted to CH_4, which means that 100% CH_4 is available at withdrawal.

Based on the gas composition, further characteristic data (volume flows, mass flows, energy flows, H_2 or CO_2 demand, CH_4 production, etc.) of the USC plants are be derived.

2.2.4. USC Plant Size

The USC demo plant was designed for a gas input volume flow of 2000 Nm^3/h. In the TEA, larger plants with a volume flow of

- 25,000 Nm^3/h
- 100,000 Nm^3/h

are considered.

Depending on the USC plant size (NOTE: The term "USC plant size" refers to the nominal gas input volume flow to the underground storage) and the plant configuration (see Section 2.2.3), the required electrolysis capacity, the corresponding methane production and the converted amount of CO_2 result, see Table 2.

Table 2. Nominal electric power of the electrolyser and CH_4 production related to the plant size (volume flow gas input) and process con-figuration.

Volume Flow Gas Input (Nm^3/h)	Process Configuration	Nominal Electric Power EL (MW)	Geomethane Production (MW)	CO_2 Demand (Converted) (t_CO_2/h)
25,000	base	12.5	6.9	1.2
	advance	37.5	20.7	3.7
100,000	base	50.0	27.6	4.9
	advance	150.0	82.7	14.7

There is a strong influence of the plant size and, above all, the plant configuration on the required electrolysis capacity, the corresponding geomethane production and the converted amount of CO_2. Between the plant configuration base and advance there is a factor of three. This means that with the same plant size (volume flow of the plant) three times more geomethane can be produced in the advance than in the base process configuration. This is due to the lower conversion of H_2 in the base compared to the advance configuration (see therefore Figure 4).

2.2.5. USC Plant Investment Costs

The total investment costs of the USC pilot plant (confidential data) and their subdivision into individual cost items (e.g., costs for the electrolyser, installations above ground and underground, etc.) serve as a starting point for future cost projections.

The aim of a USC plant is the production (and storage) of SNG, which is why the Figure 5 shows the specific investment costs of USC plants in relation to the production of 1 kW_{SNG}. The costs are compared to a state of the art aboveground (catalytic) methanation (AGM) with an equivalent methanation (SNG output) and included with all necessary devices for storage in an underground pore storage (drilling, compressors, pipes, etc.).

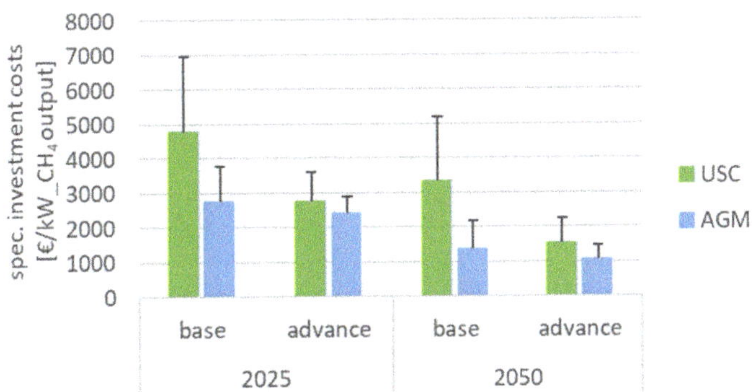

Figure 5. Specific investment costs of USC plants related to the process configuration (base and advance) and year of installation and additionally the costs for an Above Ground (catalytic) Methanation (AGM) with comparable SNG production. (A plant with a nominal volume flow of 100,000 Nm^3/h is given as a reference. The upper value of the error indicator indicates the costs for a 25,000 Nm^3/h plant).

The specific investment costs of USC plants are highly influenced by the process configuration. In base configuration the costs are approximately double as high as in the advance configuration. The reason for this can be found, among other things, in the size of the electrolyser (due to the composition of the input gas) and consequently the SNG production. Between base and advance lies a factor of three, i.e., in the advance process configuration a three times larger electrolyser is used and the SNG production is also three times higher than in the base case (see Table 2). However, there will be a reduction in investment costs in the future due to learning curve effects.

Compared to a conventional Above Ground Methanation (AGM) plants the specific investment costs of USC plants are generally higher, especially for the process configuration base. However, it should be noted that with a USC system, large quantities of gas (carrier gas required for the process; depending on the process configuration 62.5–75% of the volume flow of the plant) are additionally stored. In comparison to the AGM system, only the amount of SNG produced in the process is stored (the amount of SNG produced in USC and AGM is equal; depending on the plant configuration 20–30% of the volume flow of an USC plant).

2.3. Greenhouse Gas Assessment

For assessing the greenhouse gas performance for chosen USC scenarios the ISO 14040/44 Life Cycle Assessment (LCA) methodology has been applied. The methodology consists of four steps: (1) definition of goal and scope defining the technical system boundary as well as temporary and geographical boundaries; (2) Life Cycle Inventory (LCI) where mass and energy flows to and from the system are gathered; (3) Life Cycle Impact Assessment (LCIA) finally normalizing and attributing mass and energy flows to environmental impact categories and (4) interpretation [53,54]. The methodology of LCA is vastly discussed in the scientific literature [55–59] in the context of power-to-gas applications also [60–62]. Accordingly there will not be a detailed description of the general methodology within the current work.

The Greenhouse Gas Assessment (GHG-Assessment) is conducted using GaBi ts 10 LCA software and the corresponding Professional database as well as ecoinvent v3.6 database. Data for background processes is derived from named LCA databases. Mass and energy flows in foreground processes are based on assumptions in techno-economic evaluation according to the process design. Figure 6 shows the system boundaries for the geo-methanation process and also displays foreground and background processes. Erection of infrastructure (i.e., road to plant) and construction plant (i.e., piping, electrolyser, compressors, etc.) are not considered in the assessment considering that the specific impact broken down to the functional unit is comparably low due to long life time [63] and power-to-gas system components play a minor role for the environmental impact of produced gases [64].

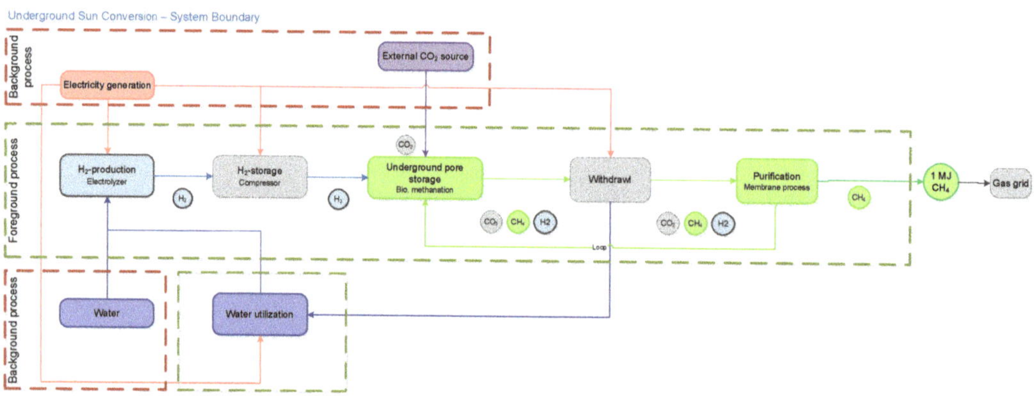

Figure 6. System boundary for the LCA of the Underground Sun Conversion process using the example of the advance scenario.

Figure 6 shows the applied "cradle-to-gate" system boundaries without considering the end use of the produced geo-methane. The function of the system is to produce synthetic methane from H_2 and CO_2. Accordingly, the functional unit is 1 MJ CH_4 produced within the geo-methanation process. Carbon dioxide as a by-product of ammonia production is assumed to be the external carbon source in order to reflect the industrial CCU use case also applied in techno-economic analysis. The geographic boundary is Austria and the temporary boundary is the timeframe from 2017 to 2021 which corresponds to the duration of the field trial.

In line with the technoeconomic analysis, LCA also examines three scenarios differing in plant configuration showing different conversion rates for CO_2 and H_2 to CH_4 in the underground storage resulting in different gas mixtures as an output. The mass balance representing the LCI for GHG-assessment is shown in Figure 7.

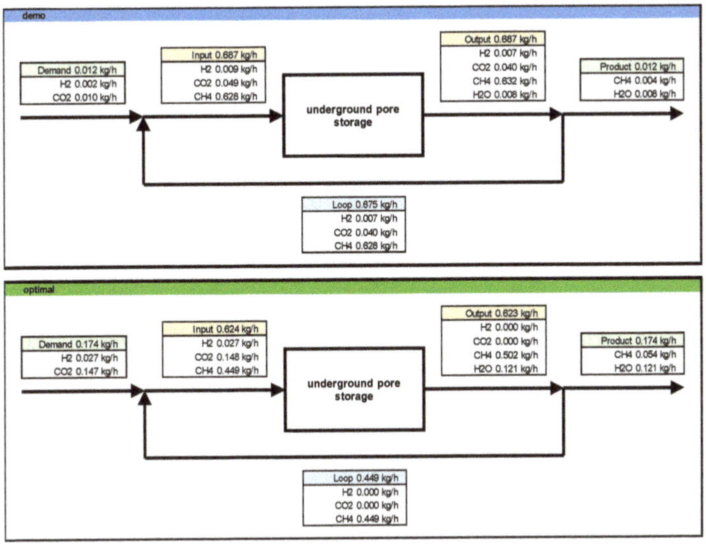

Figure 7. Gas composition (proportion of CH_4, H_2 and CO_2) during injection and withdrawal.

As the mass balance shows, a significant amount of gas is kept in the loop and injected again to the underground pore storage. The operating energy used for geo-methanation is solely made up by electricity for electrolysis, gas injection and gas withdrawal. For the scaling and the electricity consumption of the compressors, a calculation based on the real gas values of methane and hydrogen was performed [65,66]. The electricity demand for the advance case is estimated to be 728 MJ/h for electrolysis, 288 MJ/h for injection and ranges from 181 MJ/h for the advance case up to 233 MJ/h for the base case for gas withdrawal. Due to the obtained gas mixtures an allocation procedure for CH_4 and H_2 as energetically usable process outputs based on the lower heating value (LHV) was developed (see Table 3).

Table 3. Energy allocation in scenarios based on LHV of H_2 (3.0 kWh/Nm^3) and CH_4 (9.9 kWh/Nm^3) and volumetric gas composition shown in Figure 7.

Gas	Base	Advance
CH_4	96.0%	100.0%
H_2	4.0%	0.0%

Source: authors own.

All three scenarios were further analysed applying the following three different electricity sources for each of them for operating the electrolyser:

- 100% wind power
- 100% solar power
- Austrian grid mix

The other electricity consuming components (i.e., use the AT electricity mix). Table 4 shows the applied processes from LCA databases. Additionally, two different CO_2 sources have been assumed—one biogenic from biogas and the other from ammonia synthesis (see also Table 4).

For the LCIA CML 2015 method is applied, which is a problem-oriented classification of material and energy flows for impact assessment and categorizes results in so-called midpoint categories. The GHG-performance is expressed in the Global Warming Potential (GWP) measured in kg CO_2-equivalents (kg CO_2eq).

Table 4. Life cycle inventory—process and their accumulated process from the database.

Process	GaBi ts Professional/Ecoinvent v3.6 Process
Electricity	AT: Electricity from wind power ts (GaBi ts 10 Professional Database) AT: Electricity from photovoltaic ts (GaBi ts 10 Professional Database) AT: Electricity grid mix ts (GaBi ts 10 Professional Database)
Renewable electricity mix 2030	AT: Electricity from hydro power ts (GaBi ts 10 Professional Database) (56.7%) AT: Electricity from wind power ts (GaBi ts 10 Professional Database) (21.6%) AT: Electricity from photovoltaic ts (GaBi ts 10 Professional Database) (13.9%) AT: Electricity from biomass (solid) ts (GaBi ts 10 Professional Database) (5.7%) AT: Electricity from biogas ts (GaBi ts 10 Professional Database) (2.1%)
CO_2	DE: Carbon dioxide (CO_2) by-product ammonia (NH3) (economic allocation ts (GaBi ts 10 Professional Database)
Water	EU-28: Water (deionised) ts (GaBi ts 10 Professional Database)

Source: authors own based on GaBi ts 10 Professional and ecoinvent 3.6.

3. Results

3.1. Process Simulations

With the established flowsheets (Figures 1 and 2) a series of simulations was performed with varying partial conversion rates for the bio-methanation reaction taking place in the downhole reservoir according to the Sabatier equations for CO_2 methanation Equation (1) [41]. In addition, two different gas mixtures were considered for the injected gas stream, one containing 10 vol.-% H_2 and 2.5 vol.-% CO_2 (reaction stoichiometry, INPUT 1), and another one with an increased hydrogen amount of 20 vol.-% and 2.8 vol.-% of CO_2 (hydrogen surplus of 80%. INPUT 2). Finally, also the injection and production flow rates were varied between 1000 and 2000 Nm^3/h. All variations were simulated for both operating modes (BATCH and CYCLE). Table 5 shows the gas composition of the individual streams on the injection side of the simulated geo-methanation process [67]. The simulation results shown in the following figures represent the CYCLE operating mode with a flow rate of 2000 Nm^3/h.

Table 5. Gas composition of process streams on injection side of the geo-methanation process (values in vol.-% dry).

Stream	CO_2	CH_4	H_2	Other (N_2, C_mH_n)
NG	0.2	98.4	-	1.4
CO_2	100	-	-	-
H_2	-	-	100	-
INPUT Case 1	2.5	86.3	10	1.2
INPUT Case 2	2.8	76.1	20	1.1

For INPUT Case 1, Figure 8 shows the gas composition in vol.-% of the dry gas stream after the water separation units (OUTPUT dry, left) as well as the final product stream after an integrated membrane unit for gas upgrading (PRODUCT, right). A series of partial conversion rates from 0.1 to 0.9 in steps of 0.1 was considered. A value of 1.0 would represent full conversion of any CO_2 to geo-methane through the downhole geo-methanation process. For Case 1 in CYCLE mode, the methane content varies between 87.4 and 97.3 vol.-% for the dry output stream, as well as between 95.5 and 98.3 vol.-% for the final product stream. The hydrogen share decreases for the dry output stream from 9.1 to 1.1 vol.-% with increasing conversion rate. On the product side, hydrogen is already at a low level of 2.5 vol.-% with a 0.1 conversion rate due to the high efficiency of the membrane unit. For a partial conversion rate of 0.9, almost no hydrogen and CO_2 are present in the product gas anymore, leaving 98.3 vol.-% CH_4 and 1.3 vol.-% of higher hydrocarbons in the product (all values in vol.-% (dry)).

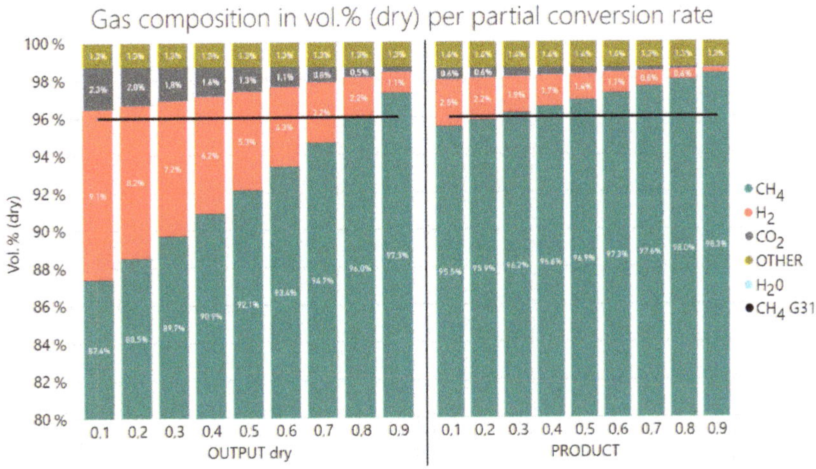

Figure 8. Gas composition of dry gas stream after water separation (**left**), and of product gas stream (**right**); all values in vol.-% (dry) for INPUT Case 1 (10% H_2 and 2.5% CO_2).

Figure 9 shows the composition of all relevant gas streams for simulation Case 1 and a selected partial conversion rate of 0.7. This value represents one specific operation mode conducted in the field. For this simulation, the gas composition from the injected gas stream with 86.3 vol.-% of methane upgraded to 94.7 vol.-% during the geo-methanation and further up to 97.6 vol.-% with the membrane unit. H_2 and CO_2 started with 10 respectively 2.5 vol.-%, and decreased to 0.8 and 0.2 vol.-% for the final product stream (all values in vol.-% (dry)).

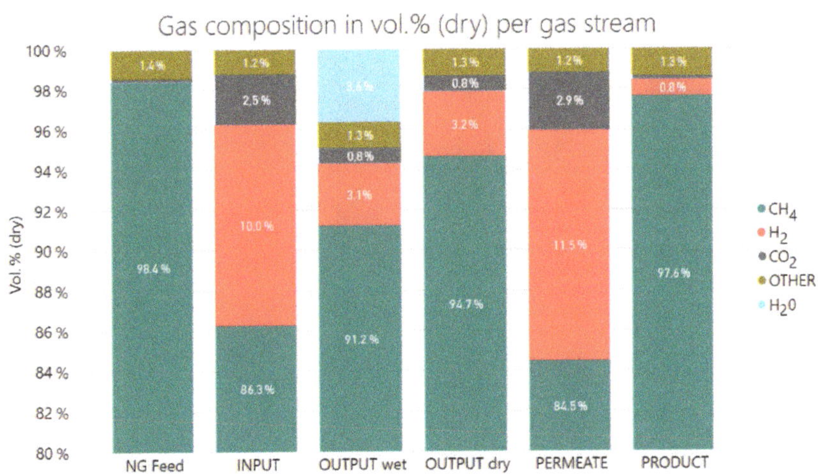

Figure 9. Gas composition per gas stream; all values in vol.-% (dry) for INPUT Case 1 (10% H_2 and 2.5% CO_2).

The second simulation case addresses an elevated amount of hydrogen to be stored in the underground gas reservoir. This INPUT Case 2 includes 20 vol.-% of H_2 and 2.8 vol.-% of CO_2 as INPUT (highly over-stoichiometric, H_2:CO_2 = 7.14:1). Figure 10 gives an overview on the gas composition of the output stream in vol.-% (dry) prior to the membrane unit (left) as well as the one after the gas upgrading took place (right). Due to the higher amount of hydrogen injected, and the resulting surplus of hydrogen for the methanation reaction, higher shares of hydrogen remain present in the produced gas streams (e.g., 12.4 and 3.4 vol.-% compared to 1.1 and 0.2 vol.-% for Case 1, all values in vol.-% (dry).

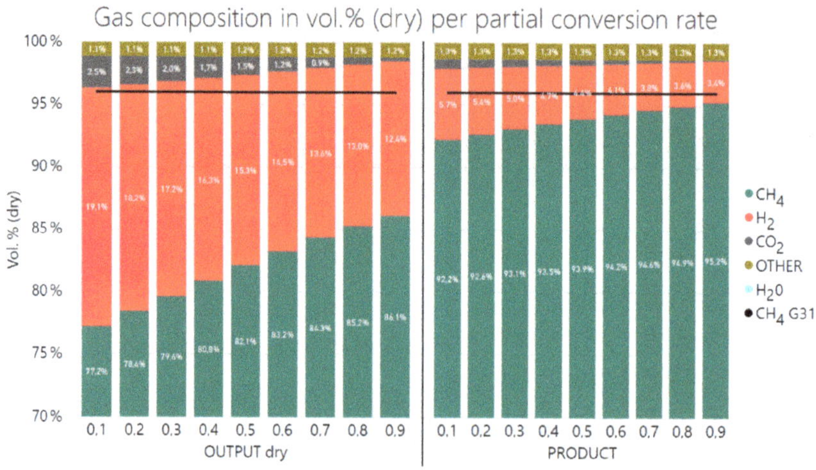

Figure 10. Gas composition of dry gas stream after water separation (**left**), and gas composition of product gas stream (**right**); all values in vol.-% (dry) for INPUT Case 2 (20% H_2 and 2.8% CO_2).

Figure 11 summarises the gas compositions for all relevant gas streams from the simulation of INPUT Case 2. Again, a partial conversion rate of 0.7 was selected. The gas composition from the injected gas stream with 76.1 vol.-% methane increased to 84.3 and 94.6 vol.-% with the geo-methanation as well as the gas upgrading unit at the surface. H_2 and CO_2 started with 20 and 2.8 vol.-%, and de-creased to 3.8 and 0.2 vol.-% for the final product stream (all values in vol.-% (dry)).

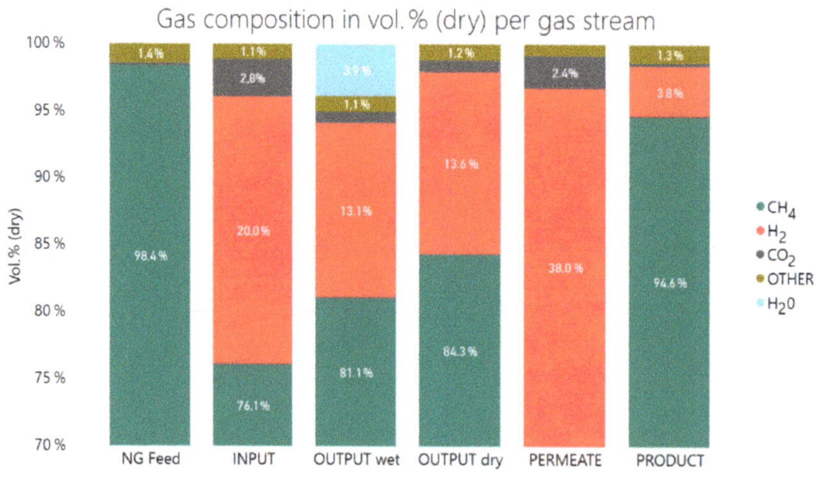

Figure 11. Gas composition per gas stream; all values in vol.-% (dry) for INPUT Case 2 (20% H_2 and 2.8% CO_2).

3.2. Techno-Economic Assessment

In this chapter the main results of the TEA, the costs for conversion of H_2 and CO_2 to geomethane and the total costs for the production of geomethane as well as the comparison to the costs of a SNG production in an Above Ground Methanation, are summarised.

3.2.1. Conversion Costs

The costs for the conversion of H_2 and CO_2 to geomethane are charged to the customer by the storage operator. The storage operator incurs costs for the construction and operation of the plant. In return, revenues are generated for the storage the carrier gas (which is from a today's perspective necessary for the USC process) and the storage of the produced geomethane. The difference is the cost for conversion to geomethane, which the storage operator charges to the customer. In Figure 12 the structure of the conversion costs is shown in detail for a 100,000 Nm^3/h USC plant for the base and advanced configuration 2025 and 2050. Additionally, for comparison, the conversion costs with an AGM (equivalent SNG output) are shown.

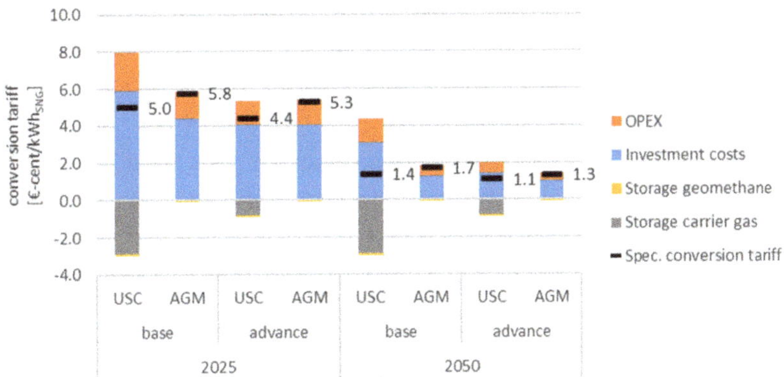

Figure 12. Cost for conversion for a 100,000 Nm^3/h USC plant and a comparable AGM.

The specific investment costs and OPEX are different for the process configurations base and advance, since they depend on the amount of SNG produced (Note: The absolute investment costs are independent of the process configuration. In base, the specific investment costs are higher than in advance, since the amount of produced gas is comparatively low).

The specific revenues for storage behave similarly. In the base configuration, a large amount of carrier gas is required and stored, and a comparatively small amount of geomethane is produced, making the specific revenues for storage very high. In comparison, in the advance configuration much more geomethane is produced and at the same time less carrier gas is stored, which means that the specific revenues for storage are significantly lower.

In Figure 12, the USC technology is compared to conventional methane production using aboveground catalytic methanation (AGM). Compared to the USC technology, the specific investment costs of an AGM are in the base configuration significantly lower and at a similar level in the advance configuration. In the AGM process, no carrier gas is required, therefore only revenues for the storage of the produced gas are generated, which are very low compared to the storage of carrier gas required for USC process.

3.2.2. Geomethane Production Costs

The total costs for geomethane production consist of the conversion costs (tariff for conversion, which is charged to the customer by the storage operator) and operational costs for producing the H_2 with an electrolyser (electricity, electricity tariffs, water), CO_2

from industry and revenues for excess heat (use in district heating or for process heat) and oxygen. In Figure 13 the structure of the production costs is shown in detail for a 100,000 Nm^3/h USC plant for the base and advanced configuration 2025 and 2050.

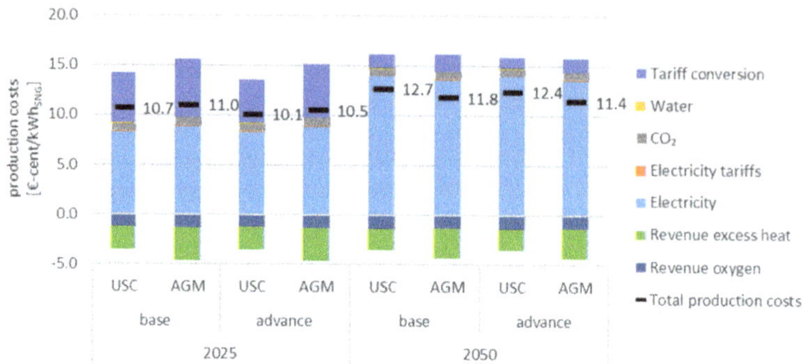

Figure 13. Geomethane production costs with a 100,000 Nm^3/h USC plant and a comparable AGM.

The largest share of the production costs is accounted for electricity, which increases in future, to power the electrolyser to produce H_2. Especially in early installations (2025), the tariff for conversion also contributes a significant part to the total cost. Not to be neglected should be the revenues generated by the sale of excess heat and oxygen from the electrolysis process.

In Figure 14 the development of the specific production costs of geomethane for two different plant sizes (injection and withdrawal volume flows of 25,000 and 100,000 Nm^3/h) up to the year 2050 are shown, whereby a distinction is also made between the plant configurations base and advance. For comparison, the production costs of SNG produced with a conventional Above Ground Methanation with equivalent SNG capacity are shown.

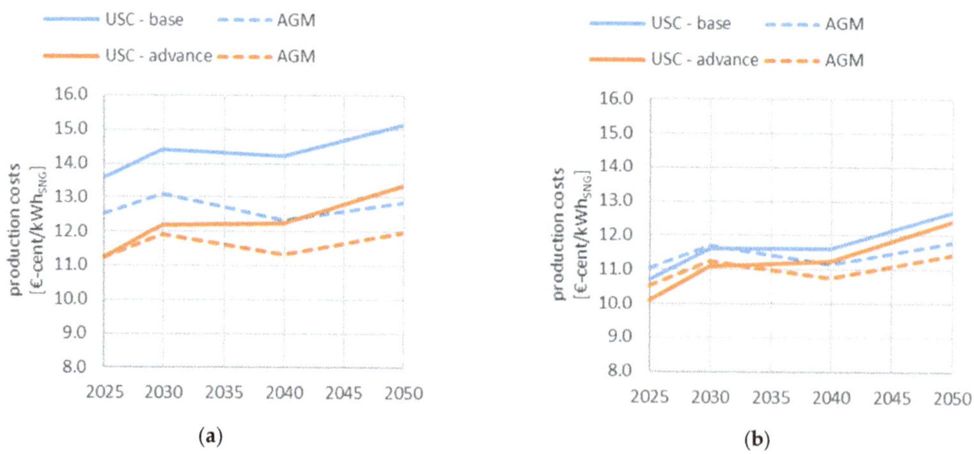

Figure 14. Geomethane production costs. (**a**) plant size 25,000 Nm^3/h (**b**) plant size 100,000 Nm^3/h.

Both, plant configuration and size have a significant impact on production costs, i.e., the larger the plant or the better the plant configuration (or plant performance) the lower the costs. In general, the production costs of AGM are lower than those of USC. In an advance plant configuration, the production costs of USC and AGM are in a similar range. In the base configuration, the costs of USC are higher than with AGM.

The geomethane production costs are in the range of 10.1–13.6 cent/kWh in 2025 and 12.4–15.2 cent/kWh in 2050. Especially for comparatively large plant with 100,000 Nm^3/h, both plant configurations show similar costs. The production costs will remain at a similar level in the future as in 2025 due to the high full load hours of the plant and the rising electricity costs in the future. The high full load hours mean that the share of conversion costs in the total production costs is rather low compared to the share of electricity costs. Therefore, the electricity costs have a higher influence on the total costs. This means that if the share of electricity costs increases more than the share of conversion costs (mainly investment costs) of the plant decreases due to learning curve effects, the production costs will remain the same (or increase slightly) in the future.

3.3. Greenhouse Gas Assessment

The results for the GWP in Figure 15 clearly show that the electricity needed for compressors are a major contributor to the GWP in case of base case, where the electricity is for compressing the gas is responsible for 56% if AT electricity mix is used for electrolysis up to 77% if wind power is used for hydrogen production. As the GWP impact of electricity used for compressing operations decreases, the significance of electricity used for hydrogen production increases. This is majorly due to increased conversion efficiencies in the underground pore storage and therefore less gas is compressed in the loop (see Figure 7). As expected, renewable power sources are clearly favorable in terms of GWP. The impact of water use and CO_2 source are neglectable in terms of GWP of synthetic methane produced applying USC.

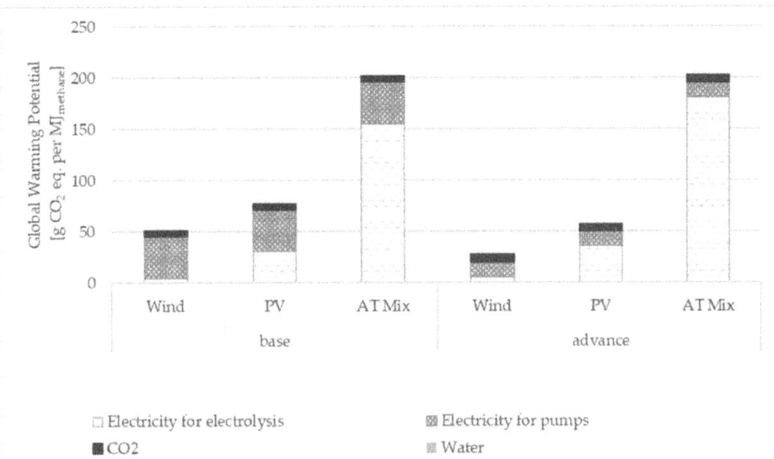

Figure 15. Global Warming Potential (GWP) for USC Scenarios (source: authors own based on GaBi ts 10 and ecoinvent v3.6).

Figure 16 shows the comparison of USC Scenarios with relevant benchmark processes regarding their GWP. It is shown that in the advance case using wind power the obtained GWP of 1 MJ synthetic methane has a comparable GWP as upgraded biogas and conventional (above ground) methanation applying CO_2 from biogas upgrading and wind power (PtG with biogas CO_2 and wind el.). All scenarios of the base and advance case show a significant GHG-saving ranging from approximately 26% up to 90% for the wind power operated advance case.

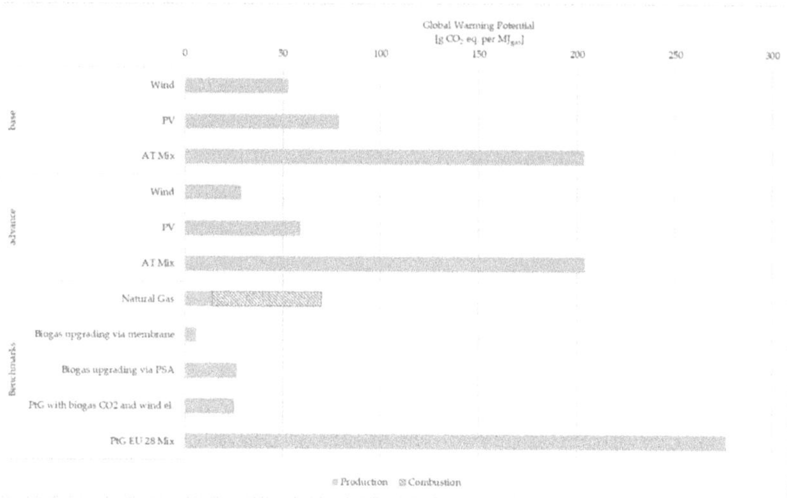

Figure 16. Global Warming Potential (GWP) for USC Scenarios compared to relevant benchmark processes (source: authors own based on GaBi ts 10 and ecoinvent v3.6 and [68–70]).

All benchmark processes refer to a "cradle-to-gate" system boundary, except natural gas for which the GHG emission factor for combustion is considered. This is due to the fact that the "cradle-to-gate" boundary leads to the wrong conclusion as for fossil energy carriers the GHG emissions for their production a comparably low and the majority of GHG emissions occur during the use phase. So, it is common to use the "cradle-to-use" system boundary for fossil fuel comparators [71].

4. Discussion

4.1. Interpretation of Process Simulations and Comparison with Field Experiments

Considering the dry output stream of Case 1, the minimum required value for methane of 96 vol.-% according to ÖVGW G31 [72] can only be met with a high CO_2 partial conversion rate of 0.8 and above. The same applies for the other gases in the product stream as the corresponding values for H_2 and CO_2 stayed below their maximum values. With the updated norm G B210 released in 2021 [73], the new upper limit for H_2 of 10 vol.-% can be achieved with every conversion rate simulated. The same applies for CO_2 with a maximum value of 2.5 vol.-%. Nevertheless, the specification window including the Wobbe index, the higher heating value as well as the relative density of the gas mixture cannot be met with a conversion rate below 0.8. The integration of a membrane unit would result in the successful upgrading of gas mixtures with conversion rates above 0.2. Such gas streams would meet the specifications of the updated norm G B210 for injection into the Austrian gas grid, respectively the EN 16726 norm for gas infrastructure and gas quality [74].

For Case 2 with an elevated share of hydrogen (20 vol.-% instead of 10 vol.-%), the implementation of one membrane unit for gas upgrading results in a final product stream which does not meet the required specifications of ÖVGW G B210, independent of the downhole conversion rate. Although the thresholds of the individual gas components can be met, all product gas compositions are outside of the specification window. Additional gas treatment or a different gas upgrading system is required, to meet the given specifications of the Austrian gas grid (e.g., catalytic methanation unit on surface [75]).

The simulation results for both cases and operating modes are well aligned with the actual measurements obtained in the field. Conversion throughout the reservoir is heterogenous as the microbes are also not dis-tributed homogenously throughout the reservoir. Additionally, the conversion rates depend on the time the gas remains in the reservoir. Therefore, different conversion rates ranging from 0.1 up to 0.9 were observed for different regions in the gas reservoir.

4.2. Techno-Economic Assessment

As discussed in Böhm et al. (2021) [76], the production of H_2 with an electrolyser and (district) heating systems have several synergies, that should be exploited in terms of achieving a high primary energy efficiency and further to increase the macro- and business economic efficiency.

In Figure 17 the economic effect in terms of geomethane production costs with and without the revenue for excess heat for a 100,000 Nm^3/h USC plant (advance configuration) and a comparable AGM are shown. If the excess heat from the electrolyser is not used, for example in district heating systems, the production costs will rise about 20%. In case of AGM, the SNG production costs will rise about 30%, since also excess heat form the catalytic methanation is available.

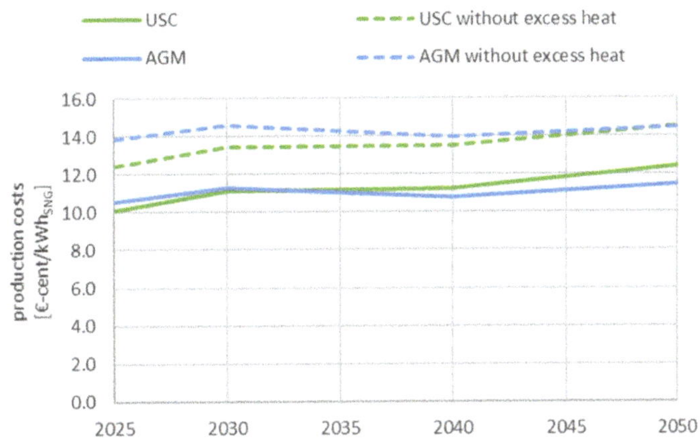

Figure 17. Geomethane production costs with and without revenue for excess heat for a 100,000 Nm^3/h USC plant (advance configuration) and a comparable AGM.

In general, the calculated geomethane production costs are in a similar range compared the other calculations done for the costs for producing of conventional SNG, e.g., [16] or [10]. However, a direct comparison of the geomethane production costs with conventional SNG production costs is not reasonable, because no long-term storage, like it is done in the USC technology, is gas taken into account. Additionally, as shown in Sections 3.2.1 and 3.2.2, electricity and investment costs, and in particular the future development of these, have a significant influence on SNG production costs. Further, if the value for storing gas increases due to the growing demand for seasonal storage, the gas storage tariff will also rise. This will result in higher revenues for storing the carrier gas, which in turn will reduce the cost of converting H_2 and CO_2 in the USC process. However, these developments depend very strongly on the future energy and climate policy framework conditions.

Further questions have arisen in the course of this project regarding the relevance of the USC technology from an overall energy system perspective such as the need for seasonal electricity storage and the pricing of this as a system service, the use of CO_2 from industrial processes in terms of a carbon cycle economy. These and others will be investigated in the follow-up projects Underground Sun Storage 2030 [77], Carbon Cycle Economy Demonstration [78] and USC FlexStore [79].

4.3. Greenhouse Gas Assessment

From a GHG-perspective two major preconditions have to be fulfilled to achieve a favourable GHG-performance and to maximize GHG-savings compared to fossil counterparts and EU-28 mix based PtG products: (1) renewable electricity sources have to be used for hydrogen production (i.e., wind and solar power); and (2) conversion efficiencies of CO_2 and H_2 in the geo-methanation process have to be increased significantly, and simultaneously the amount of gas kept in the loop has to be decreased in order to save electricity for compressing.

As the electricity demand for compressing has been identified main driver for the GHG-emissions related to synthetic methane produced in geo-methanation process the impact of the share of backflow (gas kept in the loop) on the total system output has been issue of further investigation. Due to a constant compressor electricity demand in all three cases due to the constant output rate of 1000 Nm^3 per h, while due to increasing backflow the overall product gas output is decreasing. As shown in Figure 18, the GWP from the compressors would be neglectable, due to the exponential growth of GWP with increasing backflow rates, if there would be no backflow.

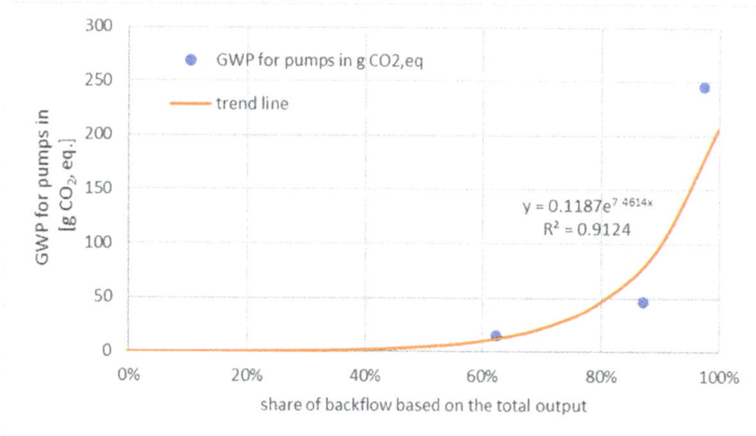

Figure 18. Relationship between backflow and GWP from compressor usage (Source: Energieinstitut an der JKU).

For a future application the aim is to reach low backflow rates, therefore the base and advance case are much more representative for the technology rollout than the demo plant.

A limitation of GHG-assessment results from the "cradle-to-gate" boundary for synthetic methane and the use of industrial CO_2 for geo-methanation. Only a temporary CO_2 storage is achieved and in case of feeding in the synthetic methane to the gas grid a combustion has to be assumed as use case. Accordingly, the carbon is again released to the atmosphere and does not account to be carbon neutral due to its fossil origin. Therefore, the aim of the USC geo-methanation is to close the carbon cycle so that the amount of CO_2 injected to the underground porous storage equals the amount of CO_2 released by burning the geo-methane. This concept fit the approach of converting "spent carbon" emissions into "working carbon" emission which leads to reduced carbon emissions in the overall economy as the CO_2 is recycled and kept in the loop [80].

4.4. *Legal Aspects*

4.4.1. Carbon Capture and Utilization

CCU enables CO_2 not to be emitted directly, but to shift the emission. Depending on the specific use of the carbon dioxide, a shorter, longer or possibly permanent avoidance of carbon dioxide emissions is made possible. There is no specific comprehensive legal basis for CCU at European level. However, there are provisions in particular in the ETS Directive [81] and the Monitoring and Reporting Regulation (MRR) [82] which are of great importance for CCU applications.

The European emissions trading system (EU ETS) aims to reduce greenhouse gas emissions and applies to emissions from activities listed in Annex I of the ETS Directive and greenhouse gases listed in Annex II of the Directive which include CO_2. The source of CO_2 plays an important role in the legal assessment, since depending on whether it is fossil or biogenic, certain rules are applied or not. According to Annex I no. 1 ETS Directive installations exclusively using biomass are not covered by this Directive and Annex IV of the ETS Directive and Article 38(2) MRR state that the emission factor of biomass shall be zero. Therefore, no allowances have to be surrendered for emissions from biomass.

Based on the ETS Directive and in particular Article 14(1), the Commission implementing Regulation on the monitoring and reporting of greenhouse gas emissions pursuant to the ETS Directive was adopted (MRR). Together with the Commission implementing Regulation on the verification of data and on the accreditation of verifiers pursuant to the ETS Directive [83] (AVR) they regulate the ETS compliance cycle. Monitoring, Reporting and Verification (MRV) provide traceability and transparency, creating trust in emissions trading and thus ensuring enforcement of the EU ETS [84].

The EU ETS aims at reducing greenhouse gas emissions and it has already been stated that CCU provides the possibility to retain CO_2 emissions (depending on the actual deployment pathway for a shorter or longer retention). Thus, whether and if so, to what extent CCU applications are taken into account in the EU ETS, is to be assessed.

As described above Article 2(1) ETS Directive stipulates that the Directive shall apply to emissions from the activities listed in Annex I of the ETS Directive and greenhouse gases listed in Annex II ETS Directive. Article 3(b) ETS Directive defines emissions as 'the release of greenhouse gases into the atmosphere from sources in an installation or the release from an aircraft performing an aviation activity listed in Annex I of the gases specified in respect of that activity' [85]. In order to speak of emissions, the release of greenhouse gases into the atmosphere is therefore necessary. The question arises whether allowances have to be surrendered for re-used CO_2 and thus not (immediately) emitted CO_2.

The ETS regime sets incentives for long-term geological storage (in a under the CCS Directive [86] permitted storage site) by providing exemptions of having to surrender of ETS allowances (cf. Article 12(3a) ETS Directive). With regard to the question of how CCU is handled under the EU ETS regime, the following not only looks at the current status of the legislation, but also presents the previously existing legal situation and the development towards the current legal situation. For CCU, there is no provision in the ETS Directive comparable to that in Article 12(3a) ETS Directive for CCS. However, the MRR also contains provisions on transferred CO_2. In an older version of the MRR (EU 601/2012 [87]—no longer in force) Article 49(1) contained the allowance deduction possibility in the case of CCS as defined in the CCS Directive. Subsequently, it was stated that 'for any other transfer of CO_2 out of the installation, no subtraction of CO_2 from the installation's emissions shall be allowed.' Recital 13 of the MRR 601/2012 stated that transfers of inherent or pure CO_2 should, in order to close potential loopholes, only be allowed subject to very specific conditions. It further stated that concerning the transfer of pure CO_2 the condition was that it should only occur for the purposes of storage in a geological site pursuant to the Union's greenhouse gas emission allowance trading scheme. Interestingly, it was also noted in the recital that those conditions should not, nevertheless, exclude the possibility of future innovations. Thus, any deductibility for transferred CO_2 beyond the exception referred to above was explicitly excluded. With regard to the production of precipitated calcium carbonate (PCC), MRR 601/2012 contained a specific provision, which enshrined that where CO_2 is used in the plant or transferred to another plant for the production of PCC, that amount of CO_2 shall be considered emitted by the installation producing the CO_2 (cf. Annex IV Section 10. B. MRR 601/2012).

Article 49(1) and Annex IV Section 10 of the MRR 601/2012, more precisely, their validity, was the subject of a preliminary ruling requested by the Administrative Court of Berlin (Verwaltungsgericht, Berlin, Germany). The so-called Schaefer Kalk case (Schaefer Kalk GmbH & Co. KG v Federal Republic of Germany) [88] concerned the refusal to allow

Schaefer Kalk GmbH & Co. KG to subtract from the emissions subject to the monitoring obligation the carbon dioxide produced in an installation for the calcination of lime transferred to a precipitated calcium carbonate installation. Schaefer Kalk was of the opinion that CO_2 that is transferred to a plant not subject to the EU ETS, where it is used for the production of PCC may be deducted from the emissions reported under its monitoring plan, as this transferred CO_2 is chemically bound in the PCC and, due to the lack of release into the atmosphere, does not correspond to emissions within the meaning of the ETS Directive. This dispute ended up before the Berlin Administrative Court, which doubted the validity of the aforementioned provisions and therefore stayed the proceedings and referred the questions for a preliminary ruling to the European Court of Justice. First of all, the Court stated that it follows from the very wording of Article 3(b) of the ETS Directive 'that, for there to be an emission within the meaning of that provision, a greenhouse gas must be released into the atmosphere'. Furthermore it found that, the MRR is based on the ETS Directive and has to be within its boundaries. The provision in Article 12(3a) of the ETS Directive does not mean that the EU legislator assumed that operators are not obliged to surrender only in the case of permanent geological storage, according to the Court. It stated that it has to be verified whether there is a release of CO_2 into the atmosphere. Pursuant to the Court, it is undisputed that the CO_2 used for the production of PCC is chemically bound in this stable product. The Court held that Article 49(1), second sentence, and Annex IV, Section 10, Subsection B, of MRR 601/2012 were excessive, and therefore ruled in its judgment of 19 January 2017 that they were 'invalid in so far as they systematically include the carbon dioxide (CO_2) transferred to another installation for the production of precipitated calcium carbonate in the emissions of the lime combustion installation, regardless of whether or not that CO_2 is released into the atmosphere.' [88].

The MRR has been adapted accordingly. The phrase 'for any other transfer of CO_2 out of the installation, no subtraction of CO_2 from the installation's emissions shall be allowed' in Article 49(1) MRR has been removed. Instead, in addition to the already existing deductibility of emissions relating to CCS, the provision that the operator shall subtract from the emissions of the installation any amount of CO_2 originating from fossil carbon in activities covered by Annex I to the ETS Directive that is not emitted from the installation, but transferred out of the installation and used to produce PCC, in which the used CO_2 is chemically bound, has been included in Article 49(1)(b) MRR. Furthermore recital 17 of the revised MRR now states that CO_2 that is transferred for the production of PCC and chemically bound in it should not be considered as released into the atmosphere. However, beyond this specific case, CCU is not rewarded in a comparable way in the ETS regime. In March 2018 the ETS Directive has been revised [89]. According to recital 14 in the ETS Directive, the main long-term incentive of the ETS Directive for, inter alia, breakthrough innovations in low-carbon technologies and processes, including environmentally safe carbon capture and utilization, is its carbon price signal and the fact that allowances will not need to be surrendered for CO_2 emissions that are avoided or permanently stored.

The fact that the phrase 'for any other transfer of CO_2 out of the installation, no subtraction of CO_2 from the installation's emissions shall be allowed' has been removed entirely and that it is also stated in the recital of the ETS Directive that no allowances are to be surrendered for avoided CO_2 emissions as well as the considerations of the court give the impression that not only the one explicitly mentioned case is to benefit from the exemption, but also others that would fulfil the relevant criteria. However, even though CCU can certainly provide a contribution to achieving climate and energy policy goals, a far-reaching exemption from the obligation to surrender certificates is fraught with many (legal and also technical) problems and uncertainties and cannot be implemented so easily without further ado. The question of when CO_2 is permanently avoided or 'permanently bound' and the associated problem of 'internal carbon leakage', which has to be assessed specifically for each case, is only mentioned here as an example [90]. Therefore, and since there is no further explicit mention in the MRR or ETS Directive it has to be assumed that surrendering ETS allowances when using captured CO_2 for other purposes (than the one explicitly addressed) is still necessary. For a proper integration of CCU into the ETS regime clear CO_2 accounting rules and a life cycle analysis to establish the final CO_2 emission reduction effect are required [91].

The Commission announced in the Circular Economy Action Plan that they 'will explore the development of a regulatory framework for certification of carbon removals based on robust and transparent carbon accounting to monitor and verify the authenticity of carbon removals' in order to 'incentivise the uptake of carbon removal and increased circularity of carbon' [92]. Already anchored in the current legal framework is the Innovation Fund, which aims at bringing low-carbon technologies to the market. The 'Fit for 55' package contains proposals for adapting the EU ETS, which also include provisions regarding CCU. It now remains to be seen how the legal framework will develop and which provisions will actually be adopted and in what form.

4.4.2. Certificates for Green Gases

With the amendment of the Renewable Energy Directive (EU) 2018/2001 (RED II), the scope of guarantees of origin (GO) is extended to renewable gases (including hydrogen), creating new possible verification paths. For sector coupling, the design of the guarantees of origins transitions when converting from one form of energy to another (conversion) is of particular relevance. With the RED II, a fundamental change is now emerging. Article 19 of RED II extends the scope of guarantees of origin for gases to biomethane, synthetic gases and hydrogen (cf. recital 59 RED II).

A uniform European system for guarantees of origin must be established. It must be possible to serve demand markets quickly and effectively and without great expense. Producers and suppliers of climate-neutral gases must be able to use the option of indirectly passing on proof of origin and sustainability information to end consumers and thus generate additional value. To this end, a standardised European guarantee of origin system for these gases should be introduced quickly.

Pursuant to Article 15(1) RED I, guarantees of origin for electricity, heating and cooling produced from renewable energy sources were introduced. Now Article 19(1) RED II extends the system of guarantees of origin to other forms of energy from renewable energy sources.

According to Article 19(1) RED II, guarantees of origin are (electronic) documents that serve as proof of the amount or share of renewable energy in the energy mix of an energy supplier vis-à-vis the end customer. According to Article 19(7) RED II, guarantees of origin contain, among other things, at least information on the energy source, the form of energy, namely electricity, gas, heating or cooling. They are also tradable throughout Europe and can be transferred and invalidated independently of the physical delivery of the energy quantity concerned. Article 19 RED II extends this system of guarantees of origin to renewable gases. According to Article 19(7) lit. b ii RED II, guarantees of origin for renewable gases, including hydrogen, shall now also be explicitly possible. This means that, according to recital 59 RED II, the aim is to achieve uniform verification of the origin of gas from renewable energy sources and more intensive cross-border trade in such gas [93].

While guarantees of origin have so far been established primarily in the electricity sector for the purpose of providing transparent and reliable information to consumers in order to exclude double marketing of ecological properties, comprehensive GOs introduced across sectors can potentially pursue more diverse goals. The information provided by a GO can be the basis for statistical purposes, for accelerating the market integration of renewable energies and for steering or incentivising policy measures (for example, meeting quotas). More precisely defined interfaces of GOs to European emissions trading are also necessary. The complexity of a future cross-sectoral verification system poses new challenges for market and regulatory actors. On the other hand, there are far-reaching opportunities in the form of new differentiation and marketing possibilities for suppliers, new sources of income for renewable energy producers and additional verification options for energy consumption in industry, mobility and the building sector. GOs can thus play an important role as an instrument in the operational implementation of the energy transition.

A uniform European system for proof of origin and GHG intensity is the most important prerequisite for cross-border trade in renewable and decarbonised gases.

Another big question is how an 'electricity guarantee of origin' can be converted into a 'gas guarantee of origin'. This means whether and in what form energy conversions are accompanied by corresponding conversions of guarantees of origin which has far-reaching implications for the issuing and use of GOs [94].

For conversion processes, there are no specific regulations in RED II on how to deal with these conversions. More specific regulations are to be made in this regard in the revision process of EN 16325 as to whether and under what conditions a transfer of the green attribute can take place during conversion from one form of energy to another [95].

In any case, it should be avoided that the property 'renewable energy' is marketed twice. Double marketing defines the multiple sale of the renewable energy property to different consumers. This could be done by devaluing GOs for energy inputs (e.g., electricity) and issuing 'conversion' GOs for energy outputs (e.g., gas). The auxiliary energy input for conversion processes must be taken into account. 'Conversions GOs' may be particularly necessary for Power-to-X applications such as the conversion of electricity (grid supply) into hydrogen, liquid fuels and heating or cooling [94].

Other conversion options exist in the conversion of gases and liquid fuels to electricity or in the combustion of gases and liquid fuels for heat and cold generation. Furthermore, GO conversions could occur during the methanation of hydrogen, provided that separate GO systems for hydrogen and other gases are permanently established. In contrast, the initial issuance of GOs in direct conversion, such as wind turbines and electrolysers for the production of hydrogen, only takes place after the conversion process, which means that GO conversion is not necessary [94].

A more difficult question is whether conversion of GO is sufficient to 'inherit' the renewable property when energy is purchased from the grid to produce another form of energy, such as electricity purchased from the public grid for use in an electrolyser. In the transport sector, the RED II goals provide for additional requirements as a prerequisite for creditability of the renewable energy property that electricity purchased from the grid must fulfil. The documentation of these requirements currently goes beyond the scope of GOs, since according to Article 19 of RED II they initially only serve to inform consumers. A function with regard to the fulfilment of renewable energy expansion targets is not currently envisaged. In principle, however, it is also possible to expand the purpose of GOs and to impose different requirements on GOs for different verification purposes. For example, one could define which properties a GO must have in order to enable the inheritance of renewable properties through GO conversion when energy is purchased from the grid.

Another problem or challenge is grid and storage losses and how to handle them. So far, losses that occur during intermediate storage and transport of energy are not taken into account when issuing and invalidating GOs. Losses in this case are implicitly covered by untracked energy volumes. As the share of renewable energy generation covered by GO in the energy system increases, a uniform regime for dealing with losses also becomes more important [96].

Expanding the GO system brings issues such as the design of GOs for gases with regard to the qualitative information content as well as corresponding labelling regulations and residual mix calculations, as they are currently common in the electricity sector. Furthermore, it is unclear how the mixing of gases (especially the admixture of hydrogen to biomethane in the gas grid) or the grid and storage losses as well as the interfaces between GOs and existing verification systems, such as sustainability certificates for biomass or mass balancing systems for biomethane, are to be handled. It also needs to be clarified whether the conversion of GO is sufficient to 'inherit' the renewable property when energy is drawn from the grid to produce another form of energy, such as electricity drawn from the public grid for use in an electrolyser.

Although the expansion of the scope of GO systems creates additional technical and administrative effort in the short to medium term, new potential uses also arise in the area of environmental accounting for the various actors in the electricity, heat and transport sectors.

5. Conclusions and Outlook

The comparison between the process simulations performed in Aspen Plus® and the experiments in the field revealed high variations in the partial conversion rate for the geo-methanation taking place in the underground gas reservoir. Independent of the achieved conversion rate, a gas upgrading system such as a reverse-selective membrane unit is required for a basic scenario to achieve the specifications of the Austrian gas grid for injection (ÖVGW G B210). For cases with an increased amount of hydrogen to be stored and converted seasonally, the gas upgrading system needs to be extended, as with a single membrane unit the specifications cannot be met (e.g., by implementing a surface catalytic post-methanation).

The techno-economic assessment has shown that the production costs of geo-methane with a USC plant can be, under certain conditions (i.e., advance process configuration or comparatively large plant), similar or lower than the costs of SNG produced with a conventional aboveground methanation plant.

However, it must be taken into account that in order to produce geomethane with a USC plant, according to current knowledge, large quantities of carrier gas are required by the process and must be stored simultaneously with the hydrogen and CO_2 used for underground methanation. Therefore, the USC technology is particularly suitable when large quantities of gas have to be stored already. This commonly occurs for reasons of system relevance or supply security. As an additional benefit of this storage process, geomethane can be produced from renewable hydrogen and CO_2, thus contributing to the achievement of climate targets.

The greenhouse gas assessment also shows that geomethane produced in a USC plant leads to greenhouse gas savings compared to fossil natural gas and synthetic natural gas produced by using H_2 from electrolysis driven by an EU28 electricity grid mix. Applying renewable electricity and aiming at a low backflow rate in order to reduce needed compressing capacity are prerequisites to achieve a favourable GHG-performance.

While the ETS regime sets incentives for CCS by providing exemptions of having to surrender of ETS allowances, CCU is not equally incentivised. Except for the possibility of deduction from the plant's emissions for transferred CO_2 used to produce precipitated calcium carbonate, CCU is not rewarded in a comparable way in the ETS regime. It must be noted, however, that a far-reaching exemption from the obligation to surrender certificates would be fraught with many legal and technical problems and uncertainties. In any case, clear CO_2 accounting rules and a life cycle analysis are required for a proper integration of CCU into the ETS regime.

In the field of green gas certification, there is no uniform European classification (terminology) for climate-neutral gases throughout the EU. The introduction of a uniform European system for guarantees of origin for gases is necessary. Nevertheless. there are, of course, also some challenges in expanding the GO system; for example, if it is extended to other forms of energy. However, additional issues exist concerning the design of GOs for gases with regard to the quality or furthermore, how the mixing of gases (admixture of hydrogen to biomethane) is to be handled when feeding-in to the gas grid. It also needs to be clarified whether the conversion of GO is sufficient to 'inherit' the renewable property when energy is drawn from the grid to produce another form of energy.

Uniform green gas certification within Europe is necessary, but certainly brings some challenges.

Author Contributions: Conceptualization, A.Z.; methodology, A.Z., K.F.-F., P.W.-Z., A.V. and M.-T.H.; validation, A.Z., K.F.-F., P.W.-Z., A.V. and M.-T.H., M.L., S.B. and M.P.; formal analysis, A.Z., K.F.-F., P.W.-Z., A.V. and M.-T.H.; investigation, A.Z., K.F.-F., P.W.-Z., A.V. and M.-T.H.; writing—original draft preparation, A.Z., K.F.-F., P.W.-Z., A.V., M.-T.H.; writing—review and editing, A.Z., K.F.-F., P.W.-Z., A.V., M.-T.H., M.L., S.B. and M.P.; supervision, A.Z. All authors have read and agreed to the published version of the manuscript.

Funding: This research was funded by the Austrian Research Promotion Agency (FFG), grant number 855231.

Institutional Review Board Statement: Not applicable.

Informed Consent Statement: Not applicable.

Data Availability Statement: Not applicable.

Acknowledgments: The work performed in this article was conducted as part of the research project "Underground Sun Conversion (USC)", funded by the FFG and including the following partners: RAG Austria AG, Chair of Process Technology and Industrial Environmental Protection as well as Chair of General and Analytical Chemistry at Montanuniversität Leoben, Department of Agrobiotechnology at IFA-Tulln, acib the Austrian Centre of Industrial Biotechnology, Energieinstitut at Johannes Kepler Universität Linz, and AXIOM Angewandte Prozesstechnik Ges.m.b.H. The authors would also like to acknowledge the work of Hanna Weiss during her Bachelor thesis and Daniel C. Rosenfeld for his work in the field of LCA in the project during his employment at the Energieinstitut at Johannes Kepler Universität Linz.

Conflicts of Interest: The authors declare no conflict of interest. This paper reflects only the author's view, and the founding sponsors had no role in the design of the study, in the collection, analyses, or interpretation of data, in the writing of the manuscript, and in the decision to publish the results.

References and Note

1. Regulation (EU) 2021/1119 of THE European Parliament and of the Council of 30 June 2021 Establishing the Framework for Achieving Climate Neutrality and Amending Regulations (EC) No 401/2009 and (EU) 2018/1999 ('European Climate Law') OJ 2021 L 243/1. Available online: https://eur-lex.europa.eu/legal-content/EN/TXT/?uri=celex%3A32021R1119 (accessed on 8 November 2021).
2. Gabrielli, P.; Gazzani, M.; Mazzotti, M. The Role of Carbon Capture and Utilization, Carbon Capture and Storage, and Biomass to Enable a Net-Zero-CO_2 Emissions Chemical Industry. *Ind. Eng. Chem. Res.* **2020**, *59*, 7033–7045. [CrossRef]
3. Papadis, E.; Tsatsaronis, G. Challenges in the decarbonization of the energy sector. *Energy* **2020**, *205*, 118025. [CrossRef]
4. Rissman, J.; Bataille, C.; Masanet, E.; Aden, N.; Morrow, W.R.; Zhou, N.; Elliott, N.; Dell, R.; Heeren, N.; Huckestein, B.; et al. Technologies and policies to decarbonize global industry: Review and assessment of mitigation drivers through 2070. *Appl. Energy* **2020**, *266*, 114848. [CrossRef]
5. Markewitz, P.; Kuckshinrichs, W.; Leitner, W.; Linssen, J.; Zapp, P.; Bongartz, R.; Schreiber, A.; Müller, T.E. Worldwide innovations in the development of carbon capture technologies and the utilization of CO_2. *Energy Environ. Sci.* **2012**, *5*, 7281–7305. [CrossRef]
6. IEA. *Energy Technology Perspectives 2020-Special Report on Carbon Capture Utilisation and Storage*; IEA: Paris, France, 2020.
7. Patricio, J.; Angelis-Dimakis, A.; Castillo-Castillo, A.; Kalmykova, Y.; Rosado, L. Method to identify opportunities for CCU at regional level—Matching sources and receivers. *J. CO_2 Util.* **2017**, *22*, 330–345. [CrossRef]
8. Farfan, J.; Fasihi, M.; Breyer, C. Trends in the global cement industry and opportunities for long-term sustainable CCU potential for Power-to-X. *J. Clean. Prod.* **2019**, *217*, 821–835. [CrossRef]
9. Do, T.N.; You, C.; Kim, J. A CO_2 utilization framework for liquid fuels and chemical production: Techno-economic and environmental analysis. *Energy Environ. Sci.* **2022**, *15*, 169. [CrossRef]
10. Götz, M.; Lefebvre, J.; Mörs, F.; McDaniel Koch, A.; Graf, F.; Bajohr, S.; Reimert, R.; Kolb, T. Renewable Power-to-Gas: A technological and economic review. *Renew. Energy* **2016**, *85*, 1371–1390. [CrossRef]
11. Ghaib, K.; Ben-Fares, F.-Z. Power-to-Methane: A state-of-the-art review. *Renew. Sustain. Energy Rev.* **2018**, *81*, 433–446. [CrossRef]
12. Romeo, L.M.; Bailera, M. Design configurations to achieve an effective CO_2 use and mitigation through power to gas. *J. CO_2 Util.* **2020**, *39*, 101174. [CrossRef]
13. Schildhauer, T.J.; Calbry-Muzyka, A.; Witte, J.; Biollaz, S.; Jansohn, P. Producing Renewable Methane–Demonstration of CCU from Biomass. In Proceedings of the 14th Greenhouse Gas Control Technologies Conference, Melbourne, Australia, 21–26 October 2018. [CrossRef]
14. Chwoła, T.; Spietz, T.; Więcław-Solny, L.; Tatarczuk, A.; Krótki, A.; Dobras, S.; Wilk, A.; Tchórz, J.; Stec, M.; Zdeb, J. Pilot plant initial results for the methanation process using CO_2 from amine scrubbing at the Łaziska power plant in Poland. *Fuel* **2020**, *263*, 116804. [CrossRef]

15. Meylan, F.D.; Piguet, F.-P.; Erkman, S. Power-to-gas through CO_2 methanation: Assessment of the carbon balance regarding EU directives. *J. Energy Storage* **2017**, *11*, 16–24. [CrossRef]
16. Szima, S.; Cormos, C.-C. CO_2 Utilization Technologies: A Techno-Economic Analysis for Synthetic Natural Gas Production. *Energies* **2021**, *14*, 1258. [CrossRef]
17. Wolf-Zoellner, P.; Medved, A.R.; Lehner, M.; Kieberger, N.; Rechberger, K. In Situ Catalytic Methanation of Real Steelworks Gases. *Energies* **2021**, *14*, 8131. [CrossRef]
18. Thema, M.; Weidlich, T.; Hörl, M.; Bellack, A.; Mörs, F.; Hackl, F.; Kohlmayer, M.; Gleich, J.; Stabenau, C.; Trabold, T.; et al. Biological CO2-Methanation: An Approach to Standardization. *Energies* **2019**, *12*, 1670. [CrossRef]
19. Guilera, J.; Filipe, M.; Montesó, A.; Mallol, I.; Andreu, T. Carbon footprint of synthetic natural gas through biogas catalytic methanation. *J. Clean. Prod.* **2021**, *287*, 125020. [CrossRef]
20. Bargiacchi, E.; Thonemann, N.; Geldermann, J.; Antonelli, M.; Desideri, U. Life Cycle Assessment of Synthetic Natural Gas Production from Different CO_2 Sources: A Cradle-to-Gate Study. *Energies* **2020**, *13*, 4579. [CrossRef]
21. Centi, G.; Perathoner, S.; Salladini, A.; Iaquaniello, G. Economics of CO_2 Utilization: A Critical Analysis. *Front. Energy Res.* **2020**, *8*, 241. [CrossRef]
22. Regufe, M.; Pereira, A.; Ferreira, A.; Ribeiro, A.; Rodrigues, A. Current Developments of Carbon Capture Storage and/or Utilization–Looking for Net-Zero Emissions Defined in the Paris Agreement. *Energies* **2021**, *14*, 2406. [CrossRef]
23. Lankof, L.; Tarkowski, R. Assessment of the potential for underground hydrogen storage in bedded salt formation. *Int. J. Hydrogen Energy* **2020**, *45*, 19479–19492. [CrossRef]
24. Lemieux, A.; Shkarupin, A.; Sharp, K. Geologic feasibility of underground hydrogen storage in Canada. *Int. J. Hydrogen Energy* **2020**, *45*, 32243–32259. [CrossRef]
25. Tarkowski, R. Underground hydrogen storage: Characteristics and prospects. *Renew. Sustain. Energy Rev.* **2019**, *105*, 86–94. [CrossRef]
26. Gabrielli, P.; Poluzzi, A.; Kramer, G.J.; Spiers, C.; Mazzotti, M.; Gazzani, M. Seasonal energy storage for zero-emissions multi-energy systems via underground hydrogen storage. *Renew. Sustain. Energy Rev.* **2020**, *121*, 109629. [CrossRef]
27. Dopffel, N.; Jansen, S.; Gerritse, J. Microbial side effects of underground hydrogen storage–Knowledge gaps, risks and opportunities for successful implementation. *Int. J. Hydrogen Energy* **2021**, *46*, 8594–8606. [CrossRef]
28. TNO. *Large-Scale Energy Storage in Salt Caverns and Depleted Fields (LSES)–Project Findings*, 2020th ed.; TNO: The Hague, The Netherlands, 2020.
29. Jukić, L.; Vulin, D.; Kružić, V.; Arnaut, M. Carbon-Negative Scenarios in High CO_2 Gas Condensate Reservoirs. *Energies* **2021**, *14*, 5898. [CrossRef]
30. Strobel, G.; Hagemann, B.; Huppertz, T.M.; Ganzer, L. Underground bio-methanation: Concept and potential. *Renew. Sustain. Energy Rev.* **2020**, *123*, 109747. [CrossRef]
31. Iordache, I.; Schitea, D.; Gheorghe, A.V.; Iordache, M. Hydrogen underground storage in Romania, potential directions of development, stakeholders and general aspects. *Int. J. Hydrogen Energy* **2014**, *39*, 11071–11081. [CrossRef]
32. Le Duigou, A.; Bader, A.-G.; Lanoix, J.-C.; Nadau, L. Relevance and costs of large scale underground hydrogen storage in France. *Int. J. Hydrogen Energy* **2017**, *42*, 22987–23003. [CrossRef]
33. Bünger, U.; Michalski, J.; Crotogino, F.; Kruck, O. Large-scale underground storage of hydrogen for the grid integration of renewable energy and other applications. *Compend. Hydrogen Energy* **2016**, *4*, 133–163. [CrossRef]
34. Zivar, D.; Kumar, S.; Foroozesh, J. Underground hydrogen storage: A comprehensive review. *Int. J. Hydrogen Energy* **2020**, *46*, 23436–23462. [CrossRef]
35. Amid, A.; Mignard, D.; Wilkinson, M. Seasonal storage of hydrogen in a depleted natural gas reservoir. *Int. J. Hydrogen Energy* **2016**, *41*, 5549–5558. [CrossRef]
36. Matos, C.R.; Carneiro, J.F.; Silva, P.P. Overview of Large-Scale Underground Energy Storage Technologies for Integration of Renewable Energies and Criteria for Reservoir Identification. *J. Energy Storage* **2019**, *21*, 241–258. [CrossRef]
37. Tschiggerl, K.; Sledz, C.; Topic, M. Considering environmental impacts of energy storage technologies: A life cycle assessment of power-to-gas business models. *Energy* **2018**, *160*, 1091–1100. [CrossRef]
38. Perez, A.; Pérez, E.; Dupraz, S.; Bolcich, J. Patagonia Wind-Hydrogen Project: Underground Storage and Methanation. In Proceedings of the 21st World Hydrogen Energy Conference 2016, Zaragoza, Spain, 13–16 June 2016.
39. RAG Austria AG. Underground.Sun.Storage. Available online: https://www.underground-sun-storage.at/ (accessed on 15 November 2021).
40. RAG Austria AG. Underground Sun Conversion. Available online: https://www.underground-sun-conversion.at (accessed on 9 November 2021).
41. Sabatier, P.; Senderens, J.B.; Acad, C.R. New methane synthesis. *Compt. Rend. Acad. Sci.* **1902**, *134*, 514–516.
42. Zimmermann, A.; Berlin, T.U.; Wunderlich, J.; Buchner, G.; Müller, L.; Armstrong, K.; Michailos, S.; Marxen, A.; Naims, H.; Mason, F.; et al. *Techno-Economic Assessment & Life-Cycle Assessment Guidelines for CO_2 Utilization*; CO2Chem Media and Publishing Ltd.: Sheffield, UK, 2018. [CrossRef]

43. Böhm, H.; Goers, S.; Zauner, A. Estimating future costs of power-to-gas–A component-based approach for technological learning. *Int. J. Hydrogen Energy* **2019**, *44*, 30789–30805. [CrossRef]
44. Böhm, H.; Zauner, A.; Rosenfeld, D.C.; Tichler, R. Projecting cost development for future large-scale power-to-gas implementations by scaling effects. *Appl. Energy* **2020**, *264*, 114780. [CrossRef]
45. Agora Energiewende; Aurora Energy Research. In 65 Prozent Erneuerbare bis 2030 und ein Schrittweiser Kohleausstieg: Auswirkungen der Vorgaben des Koalitionsvertrags auf Strompreise, CO_2-Emissionen und Stromhandel. 2018. Available online: https://static.agora-energiewende.de/fileadmin/Projekte/2018/65_EE_und_Kohleausstieg/142_Stromsektor-2030_65-Prozent-EE-und-schrittweiser-Kohleausstieg_WEB.pdf (accessed on 17 November 2021).
46. Greimel, F.; Neubarth, J.; Fuhrmann, M.; Führer, S.; Habersack, H.; Haslauer, M.; Hauer, C.; Holzapfel, P.; Auer, S.; Pfleger, M.; et al. *SuREmMA: Sustainable River Management-Energiewirtschaftliche und Umweltrelevante Bewertung Möglicher Schwalldämpfender Massnahmen*; Federal Ministry for Agriculture and Forestry, Environment and Water Management: Vienna, Austria, 2017.
47. Energy Brainpool GmbH & Co. KG. Update: Trends der Strompreisentwicklung-EU Energy Outlook 2050. Available online: https://blog.energybrainpool.com/update-trends-der-strompreisentwicklung-eu-energy-outlook-2050/ (accessed on 29 November 2018).
48. Deloitte. Strommarktstudie 2030-Ein Neuer Ausblick für Die Energiewirtschaft; Deloitte. 2018. Available online: https://www2.deloitte.com/content/dam/Deloitte/de/Documents/energy-resources/Deloitte-Strommarktstudie-2030.pdf (accessed on 19 November 2021).
49. Haas, R.; Resch, G.; Burgholzer, B.; Totschnig, G.; Lettner, G.; Auer, H.; Geipel, J. *Stromzukunft Österreich 2030-Analyse der Erfordernisse und Konsequenzen Eines Ambitionierten Ausbaus Erneuerbarer Energien*; IG Windkraft, Kompost & Biogas Verband Österreich, IG-Holzkraft: Wien, Austria, 2017.
50. Öko-Institut e.V. *Strompreis- und Stromkosteneffekte Eines Geordneten Ausstiegs aus der Kohleverstromung*; Öko-Institut e.V.: Berlin, Germany, 2019.
51. FH Erfurt. *Strommarkt 2050-Analyse Möglicher Szenarien der Entwicklung des Deutschen und Mitteleuropäischen Strommarktes bis zum Jahr 2050*; FH Erfurt: Erfurt, Germany, 2018.
52. Frontier Economics. Economic Trends Resarch, Georg Consulting, Visionometrics. Folgenabschätzung des CO2-Sektorziels Für Die Energiewirtschaft im Klimaschutzplan 2050: Studie im Auftrag der RWE AG. 2018. Available online: https://www.frontier-economics.com/media/2263/frontier-et-al-folgenabschatzung-ksp2050-endbericht-2.pdf (accessed on 19 November 2021).
53. *ISO 14044*; Umweltmanagement-Ökobilanz. Anforderungen und Anleitungen. Deutsches Institut für Normung, 2006. Available online: https://www.beuth.de/de/norm/din-en-iso-14044/279938986 (accessed on 19 November 2021).
54. Klöpffer, W.; Grahl, B. *Ökobilanz (LCA): Ein Leitfaden für Ausbildung und Beruf*; WILEY-VCH: Weinheim, Germany, 2009.
55. Klöpffer, W. Life Cycle Assessment. From the Beginning to the Current State. *Environ. Sci. Pollut. Res.* **1997**, *4*, 223–228. [CrossRef]
56. Guinée, J.; De Haes, H.U.; Huppes, G. Quantitative life cycle assessment of products: 1:Goal definition and inventory. *J. Clean. Prod.* **1993**, *1*, 3–13. [CrossRef]
57. Rebitzer, G.; Ekvall, T.; Frischknecht, R.; Hunkeler, D.; Norris, G.; Rydberg, T.; Schmidtg, W.-P.; Suh, S.; Weidemai, B.P.; Pennington, D.W. Life cycle assessment. Part1: Framework, goal, scope, definition, inventory analysis, and applications. *Environ. Int.* **2004**, *30*, 701–720. [CrossRef]
58. Reap, J.; Roman, F.; Duncan, S.; Bras, B. A survey of unresolved problems in life cycle assessment. *Int. J. Life Cycle Assess.* **2008**, *4*, 374. [CrossRef]
59. Finnveden, G.; Hauschild, M.Z.; Ekvall, T.; Guinée, J.; Heijungs, R.; Hellweg, S.; Koehler, A.; Pennington, D.; Suh, S. Recent developments in life cycle assessment. *J. Environ. Manag.* **2009**, *91*, 1–21. [CrossRef]
60. Lotrič, A.; Sekavčnik, M.; Kuštrin, I.; Mori, M. Life-cycle assessment of hydrogen technologies with the focus on EU critical raw materials and end-of-life strategies. *Int. J. Hydrogen Energy* **2021**, *46*, 10143–10160. [CrossRef]
61. Schreiber, A.; Peschel, A.; Hentschel, B.; Zapp, P. Life Cycle Assessment of Power-to-Syngas: Comparing High Temperature Co-Electrolysis and Steam Methane Reforming. *Front. Energy Res.* **2020**, *8*, 287. [CrossRef]
62. Zhang, X.; Bauer, C.; Mutel, C.L.; Volkart, K. Life Cycle Assessment of Power-to-Gas: Approaches, system variations and their environmental implications. *Appl. Energy* **2017**, *190*, 326–338. [CrossRef]
63. Ivanov, O.L.; Honfi, D.; Santandrea, F.; Stripple, H. Consideration of uncertainties in LCA for infrastructure using probabilistic methods. *Struct. Infrastruct. Eng.* **2019**, *15*, 711–724. [CrossRef]
64. Bareiß, K.; de la Rúa, C.; Möckl, M.; Hamacher, T. Life cycle assessment of hydrogen from proton exchange membrane water electrolysis in future energy systems. *Appl. Energy* **2019**, *237*, 862–872. [CrossRef]
65. Unternehmensberatung Babel. Realgasfaktoren für Wasserstoff. Available online: http://www.unternehmensberatung-babel.de/industriegase-lexikon/industriegase-lexikon-n-bis-z/realgasfaktor/realgasfaktoren-wasserstoff.html (accessed on 29 January 2021).
66. Engineering Units. Determine Compressability Factor, Z Factor. Available online: http://engineeringunits.com/compressibility-factor-z-factor/ (accessed on 29 January 2021).
67. Bauer, S. *Underground Sun Conversion Final Report*; Rag Austria AG: Vienna, Austria, 2021.
68. Zhang, X.; Witte, J.; Schildhauer, T.; Bauer, C. Life cycle assessment of power-to-gas with biogas as the carbon source. *Sustain. Energy Fuels* **2020**, *4*, 1427–1436. [CrossRef]

69. Juhrich, K. *CO_2-Emissionsfaktoren für Fossile Brennstoffe*; Umweltbundesamt: Dessau-Roßlau, Germany, 2016.
70. Reiter, G.; Lindorfer, J. Global warming potential of hydrogen and methane production from renewable electricity via power-to-gas technology. *Int. J. Life Cycle Assess.* **2015**, *20*, 477–489. [CrossRef]
71. Zhou, Y.; Swidler, D.; Searle, S.; Baldino, C. *Life-Cycle Greenhouse Gas Emissions of Biomethane and Hydrogen Pathways in the European Union*; White Paper; International Council on Clean Transportation: Berlin, Germany, 2021.
72. Österreichische Vereinigung für das Gas-und Wasserfach. *Erdgas in Österreich—Gasbeschaffenheit Richtlinie G31*; Österreichische Vereinigung für das Gas-und Wasserfach: Vienna, Austria, 2001.
73. Österreichische Vereinigung für das Gas-und Wasserfach. *Erdgas in Österreich—Gasbeschaffenheit Richtlinie G B210*; Österreichische Vereinigung für das Gas-und Wasserfach: Vienna, Austria, 2021.
74. *European Standard EN 16726:2015+A1:2018 (E)*; Gas Infrastructure—Quality of Gas—Group H. Europoean Committee for Standardization: Brussels, Belgium, 2018.
75. Biegger, P.; Kirchbacher, F.; Medved, A.R.; Miltner, M.; Lehner, M.; Harasek, M. Development of Honeycomb Methanation Catalyst and Its Application in Power to Gas Systems. *Energies* **2018**, *11*, 1679. [CrossRef]
76. Böhm, H.; Moser, S.; Puschnigg, S.; Zauner, A. Power-to-hydrogen & district heating: Technology-based and infrastructure-oriented analysis of (future) sector coupling potentials. *Int. J. Hydrogen Energy* **2021**, *46*, 31938–31951. [CrossRef]
77. RAG Austria AG. Underground Sun Storage 2030. Available online: https://www.uss-2030.at/ (accessed on 16 November 2021).
78. RAG Austria AG. Carbon Cycle Economy Demonstration. Available online: https://www.rag-austria.at/forschung-innovation/carbon-cycle-economy-demonstration.html (accessed on 16 November 2021).
79. RAG Austria AG. USC FlexStore. Available online: https://www.eranet-smartenergysystems.eu/global/images/cms/Content/Fact%20Sheets/ERANetSES_FactSheet_JC19_USC-FlexStore.pdf (accessed on 15 November 2021).
80. Rodin, V.; Lindorfer, J.; Böhm, H.; Vieira, L. Assessing the potential of carbon dioxide valorisation in Europe with focus on biogenic CO_2. *J. CO_2 Util.* **2020**, *41*, 101219. [CrossRef]
81. Directive 2003/87/EC of the European Parliament and of the Council of 13 October 2003 Establishing a System for Greenhouse Gas Emission Allowance Trading within the Union and Amending Council Directive 96/61/EC, OJ 2003 L 275/32. Commission Delegated Regulation (EU) 2021/1416, OJ 2021 L 305/1. Available online: https://eur-lex.europa.eu/legal-content/EN/ALL/?uri=CELEX%3A32003L0087 (accessed on 15 November 2021).
82. Commission Implementing Regulation (EU) 2018/2066 of 19 December 2018 on the Monitoring and Reporting of Greenhouse Gas Emissions Pursuant to Directive 2003/87/EC of the European Parliament and of the Council and Amending Commission Regulation (EU) No 601/2012, OJ 2018 L 334/1. Commission Implementing Regulation (EU) 2020/2085, OJ 2020 L 423/37. Available online: https://eur-lex.europa.eu/legal-content/EN/TXT/?uri=uriserv:OJ.L_.2018.334.01.0001.01.ENG (accessed on 15 November 2021).
83. Commission Implementing Regulation (EU) 2018/2067 of 19 December 2018 on the Verification of Data and on the Accreditation of Verifiers Pursuant to Directive 2003/87/EC of the European Parliament and of the Council, OJ 2018 L 334/94. Available online: https://eur-lex.europa.eu/legal-content/EN/TXT/?uri=CELEX%3A32018R2067 (accessed on 15 November 2021).
84. European Commission. *EU ETS Handbook*; European Commission: Brussels, Belgium, 2015.
85. An Installation Is According to Article 3(e) ETS Directive 'A Stationary Technical Unit Where one or more Activities Listed in Annex I are Carried Out and any other Directly Associated Activities Which Have a Technical Connection with the Activities Carried Out on That Site and Which Could Have an Effect on Emissions and Pollution'.
86. Directive 2009/31/EC of the European Parliament and of the Council of 23 April 2009 on the Geological Storage of Carbon Dioxide and Amending Council Directive 85/337/EEC, European Parliament and Council Directives 2000/60/EC, 2001/80/EC, 2004/35/EC, 2006/12/EC, 2008/1/EC and Regulation (EC) No 1013/2006 OJ 2009 L 140/114. Regulation (EU) 2018/1999 of the European Parliament and of the Council, OJ 2018 L 328/1. Available online: https://eur-lex.europa.eu/legal-content/EN/TXT/?uri=celex%3A32009L0031 (accessed on 15 November 2021).
87. Commission Regulation (EU) No 601/2012 of 21 June 2012 on the Monitoring and Reporting of Greenhouse Gas Emissions Pursuant to Directive 2003/87/EC of the European Parliament and of the Council, OJ 2012 L 181/30. Available online: https://eur-lex.europa.eu/legal-content/EN/ALL/?uri=CELEX%3A32012R0601 (accessed on 15 November 2021).
88. *ECJ 19.01.2017 C 460/15; ECLI:EU:C:2017:29*; Judgment of the Court (First Chamber) of 19 January 2017. Schaefer Kalk GmbH & Co. KG: Verbandsgemeinde Diez, Germany, 2017.
89. Directive (EU) 2018/410 of the European Parliament and of the Council of 14.03.2018 Amending Directive 2003/87/EC to Enhance Cost-Effective Emission Reductions and Low-Carbon Investments, and Decision (EU) 2015/1814, OJ 2018 L 76/3. Available online: https://eur-lex.europa.eu/legal-content/DE/TXT/?uri=CELEX%3A32018L0410 (accessed on 15 November 2021).
90. Porteron, S.; de Bruijne, E.; Le Den, X.; Zotz, F.; Olfe-Kräutlein, B.; Marxen, A. *Identification and Analysis of Promising Carbon Capture and Utilisation Technologies, Including Their Regulatory Aspects*; European Commission: Brussels, Belgium, 2019.
91. IOGP. *The Potential for CCS and CCU in Europe*; IOGP: Madrid, Spain, 2019.
92. European Commission. *Circular Economy Action Plan*; European Commission: Brussels, Belgium, 2020.
93. Castro, J.; Landspersky, M.; Nürnberger, K.; Schmidt, P.; Weindorf, W.; Kalis, M.; Yilmaz, Y. *Ad hoc Beratung bei der Umsetzung der Monitoring Verordnung für Die 4. Phase des EU Emissionshandels*; Umweltbundesamt: Dessau-Roßlau, Deutschland, 2020.

94. Purkus, A.; Sakhel, A.; Werner, R.; Maaß, C. *Herkunftsnachweise für Erneuerbare Energien Jenseits des Stromsetors, Chancen und Herausforderungen*; HIR Hamburg Institut Research: Hamburg, Germany, 2020.
95. Verwimp, K.; Moody, P.; Van Stein Callenfels, R.; Kovacs, A.; Vanhoudt, W.; Barth, F.; Pedraza, S.; Lehtovaara, M.; Klimscheffskij, M.; White, A. Identification of the Main Challenges Which Currently Exist in the Management of Guarantee of Origin System (Task 1.3). In *Technical Support for RES Policy Development and Implementation*; European Commission: Brussels, Belgium, 2020.
96. Cornélis/Lenzen. Guarantees of Origin and Disclosure. Core Theme 3 Report; (Horizon 2020 Project, Grant Agreement N°101035887); Concerted Action-Renewable Energy Sources Directive. 2020. Available online: https://www.ca-res.eu/fileadmin/cares/PublicArea/CA-RES3FinalPublication/CARES3_Final_CT3_Summary.pdf (accessed on 17 November 2021).

energies MDPI

Article

An Overview of Promising Alternative Fuels for Road, Rail, Air, and Inland Waterway Transport in Germany

Janos Lucian Breuer [1,2,*], Juri Scholten [3], Jan Christian Koj [4], Felix Schorn [1,2], Marc Fiebrandt [3], Remzi Can Samsun [1], Rolf Albus [3], Klaus Görner [3], Detlef Stolten [2,5] and Ralf Peters [1]

1 Institute of Energy and Climate Research—Electrochemical Process Engineering (IEK-14), Forschungszentrum Jülich GmbH, 52428 Jülich, Germany; f.schorn@fz-juelich.de (F.S.); r.c.samsun@fz-juelich.de (R.C.S.); ra.peters@fz-juelich.de (R.P.)
2 Chair for Fuel Cells, RWTH Aachen University, 52072 Aachen, Germany; d.stolten@fz-juelich.de
3 Gas- und Wärme-Institut Essen e.V., 45356 Essen, Germany; juri.scholten@gwi-essen.de (J.S.); marc.fiebrandt@gwi-essen.de (M.F.); rolf.albus@gwi-essen.de (R.A.); klaus.goerner@uni-due.de (K.G.)
4 Institute of Energy and Climate Research—Systems Analysis and Technology Evaluation (IEK-STE), Forschungszentrum Jülich GmbH, 52428 Jülich, Germany; j.koj@fz-juelich.de
5 Institute of Energy and Climate Research—Techno-Economic System Analysis (IEK-3), Forschungszentrum Jülich GmbH, 52428 Jülich, Germany
* Correspondence: ja.breuer@fz-juelich.de

Abstract: To solve the challenge of decarbonizing the transport sector, a broad variety of alternative fuels based on different concepts, including Power-to-Gas and Power-to-Liquid, and propulsion systems, have been developed. The current research landscape is investigating either a selection of fuel options or a selection of criteria, a comprehensive overview is missing so far. This study aims to close this gap by providing a holistic analysis of existing fuel and drivetrain options, spanning production to utilization. For this purpose, a case study for Germany is performed considering different vehicle classes in road, rail, inland waterway, and air transport. The evaluated criteria on the production side include technical maturity, costs, as well as environmental impacts, whereas, on the utilization side, possible blending with existing fossil fuels and the satisfaction of the required mission ranges are evaluated. Overall, the fuels and propulsion systems, Methanol-to-Gasoline, Fischer–Tropsch diesel and kerosene, hydrogen, battery-electric propulsion, HVO, DME, and natural gas are identified as promising future options. All of these promising fuels could reach near-zero greenhouse gas emissions bounded to some mandatory preconditions. However, the current research landscape is characterized by high insecurity with regard to fuel costs, depending on the predicted range and length of value chains.

Keywords: Power-to-Gas; Power-to-Liquid; hydrogen; transport; future mobility concepts; LCA; environmental impacts; synthetic fuels; synthetic natural gas; technology readiness level

Citation: Breuer, J.L.; Scholten, J.; Koj, J.C.; Schorn, F.; Fiebrandt, M.; Samsun, R.C.; Albus, R.; Görner, K.; Stolten, D.; Peters, R. An Overview of Promising Alternative Fuels for Road, Rail, Air, and Inland Waterway Transport in Germany. *Energies* **2022**, *15*, 1443. https://doi.org/10.3390/en15041443

Academic Editor: Wei-Hsin Chen

Received: 22 December 2021
Accepted: 8 February 2022
Published: 16 February 2022

1. Introduction

Anthropogenic climate change requires a comprehensive structural change in the energy sector to be enacted [1]. With the Paris climate conference in 2015, a limit for global warming was set to a maximum of 2 K and 1.5 K compared to the pre-industrial period [2]. The annual increase in greenhouse gas emissions resulting from rising energy consumption and a growing world population also requires rapid and targeted actions. In Germany, for instance, greenhouse gases in the energy and industrial sectors were reduced by 45% and 34%, respectively, compared to 1990 [3]. In contrast, greenhouse gas emissions of the transport sector stay at the same level as 1990 despite increasing levels of engine efficiencies [3]. In order to reach the goals in the transport sector, several alternative fuels have been researched and developed in recent years, all with different properties and technical maturity. Current literature either assesses single fuels or only single criteria of these fuels like technical maturity of fuel production [4,5], net production cost [6], import [7], well-to-wheel efficiencies [8], or environmental impacts [9,10]. The literature lacks a holistic analysis considering all criteria, spanning from production to utilization, and all fuel and drivetrain options. This review aims to connect all information of different fuel and drivetrain options, elaborate on the advantages and disadvantages and identify the most promising fuels for specific applications in the transport sector. For this purpose, it classifies different fuels and drivetrain options for road, rail, air, and inland waterway transport in terms of technical maturity, costs, and environmental impacts. For the three mentioned fields, different criteria will be defined and subsequently analyzed and discussed based on existing literature. The chosen criteria will be explained in Section 2 in detail.

Figure 1 provides an overview of fuel pathways. The pathways are subdivided into primary fuels, conversion, fuels, and drive systems. This work focuses on alternative fuels from RE sources, seen on the top right-hand side of Figure 1. These fuels are subdivided into synthetic fuels obtained using renewable electricity, also known as the Power-to-Fuel, Power-to-Gas or Power-to-Liquid (PtL) concepts, and biomass-based fuels. The latter is subdivided into conventional biofuels such as fatty acid methyl esters (FAMEs) or hydrotreated vegetable oil (HVO) derived from feed crops or advanced biofuels from lignocellulose [11]. Electricity-based fuels are further subdivided into methanol and higher alcohols, ethers such as oxymethylene ether (OME), and hydrocarbons like synthetic gasoline and diesel [4]. Starting from biomass and renewable electricity as resources, different pathways may lead to the same product fuel. One example is the production of methane from biomass via fermentation and subsequently biogas upgrading, as described in [12], or synthetic electricity-based production via the PtG pathway [13]. The produced fuels are then used in different propulsion systems which are subdivided into electric and internal combustion powertrains. Internal combustion drivetrains are operated with liquid or gaseous fuels, whereas electric engines utilize electricity, which is either stored in batteries or converted from an energy carrier such as hydrogen by means of a fuel cell (see Figure 1). A combination of electric and internal combustion powertrains, as in hybrid-electric vehicles, is also common.

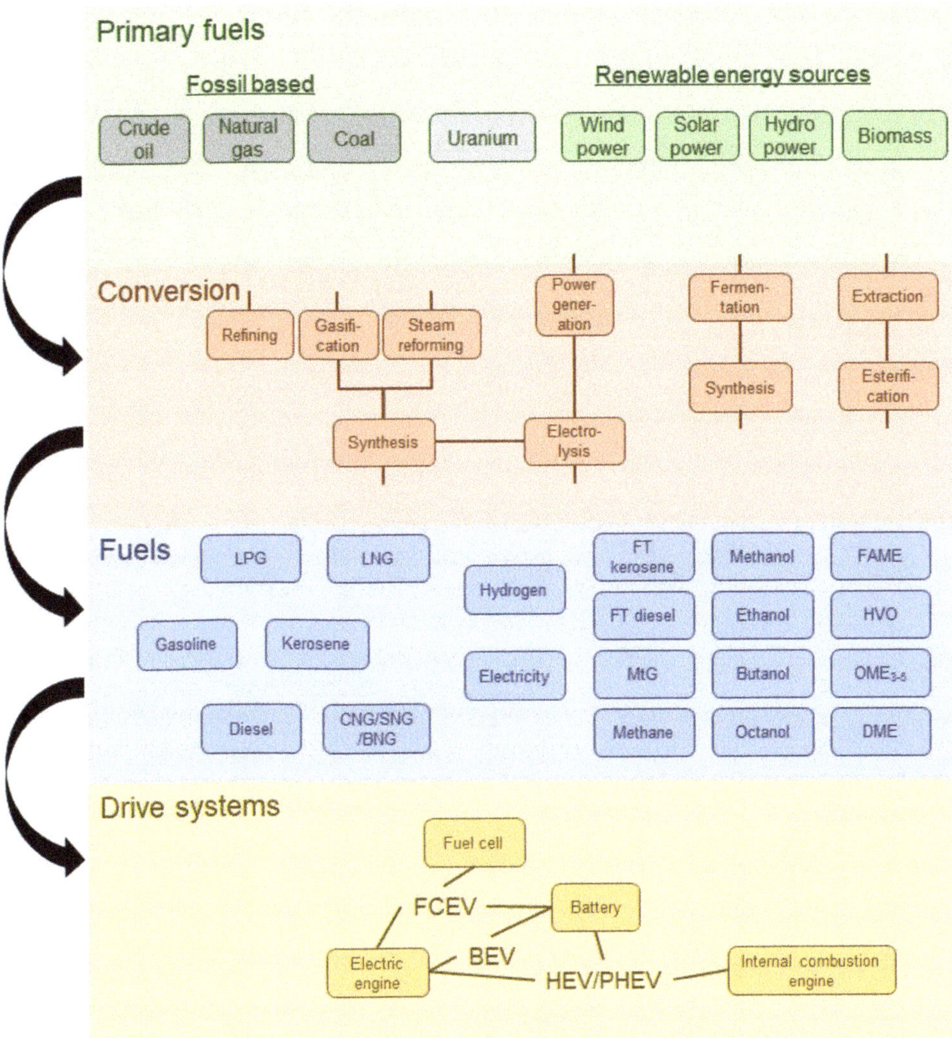

Figure 1. Overview of alternative fuel pathways. Source: Own elaboration based on Bruchof [14]. LPG: Liquified petroleum gas, LNG: Liquefied natural gas, CNG: Compressed natural gas, SNG: Synthetic natural gas, BNG: Bio natural gas, FT: Fischer–Tropsch, FAME: Fatty acid methyl ester, HVO: Hydrotreated vegetable oils, OME: Oxymethylene ether, DME: Dimethyl ether, FCEV: Fuel cell–electric vehicle, BEV: Battery–electric vehicle, HEV: Hybrid-electric vehicle, PHEV: Plug-in hybrid-electric vehicle.

2. Materials and Methods

In this work, alternative fuels are investigated on the production side, as well as on the application site. Additionally, the environmental impacts of promising fuels are discussed. The production side is assessed with respect to the technical maturity of fuel production and fuel production costs, considering domestic production as well as imports. Technical maturity is assessed via technology readiness level, which is described in detail in the next section. The application side is analyzed in terms of potential mission ranges using the fuels in different vehicles and the reduction in their possibilities. For this purpose, the potential mission ranges of different vehicles are analyzed for Germany as a case study. The achievable mission range of the fuels used in propulsion systems depends on tank-to-wheel (TtW) efficiency, the consumption, the amount of saved fuel compared to a conventional drivetrain, and the fuel's heating value. The criteria TtW efficiency and heating value are used in this work to assess the possible mission range of different fuel applications. TtW efficiencies are employed instead of maximum efficiencies for assessing the fuels, as the efficiency–load curves of the different drivetrains vary quite significantly and the load demand of different vehicle classes may also vary.

2.1. Technology Readiness Level

The technology readiness level (TLR) concept was primarily developed by the National Aeronautics and Space Administration (NASA) [15]. In the meantime, the United States Department of Defense [16], the European Space Agency [17], and the European Commission [18] adapted the method. The European Space Agency employs ISO standard 16290 Space systems—Definition of the Technology Readiness Levels (TRLs) and their criteria assessment [19]. The European Commission has also developed guidance for the application of TRL to RE technologies [20]. Table 1 lists the TRL definitions according to Rose et al. [20]. It is also noteworthy that the Joint Research Centre (JRC) expanded the TRL assessment to a commercial readiness level (CRL) [5], which was developed by the Australian Renewable Energy Agency [21] and takes into account the fact that high TRLs among particular technologies do not automatically result in market adoption, e.g., due to excessively high capital costs or regulatory burdens [5].

Table 1. TRL definitions.

TRL	Definition
1	Identification of new concept, applications, and barriers
2	Definition of application, consideration of interfaces, and commercial offer
3	Proof of concept prototype ready: concept is laboratory tested
4	Integrated small-scale prototype with auxiliary systems laboratory validated
5	Large-scale prototype completed with auxiliaries, refined commercial assessment
6	Technology pilot demonstrated in relevant environment, manufacturing strategy defined
7	Pilot demonstrated in operational environment, manufacturing approach demonstrated
8	Technology in its final form, low-rate production
9	System fully operational and ready for commercialization

Notes. Source: Rose et al. [20].

2.2. Identification of Required Mission Ranges for Different Means of Transport

The required mission range for road transport in Germany was analyzed using a dataset from the Federal Ministry of Transport and Digital Infrastructure (BMVI) [22]. The BMVI [22] determined, amongst other variables, the daily driven distances of the following vehicle classes: passenger cars, light duty vehicles, motorcycles, coaches, trucks, articulated trucks, other tractors, other motor vehicles, and agricultural tractors, via a survey in 2010. Even though the dataset is from 2010, it is the only currently available source containing the daily driven distance for all road transport vehicle classes in the necessary accuracy. Comparable is only the study series "Mobilität in Deutschland", which was performed for the years 2002, 2008, and 2017 [23]. However, it only contains information about the vehicle class passenger car. Yet it shows that the following used daily driven distance is, at least for passenger cars, mostly constant from 2002 to 2017. The daily driven distance for passenger cars per person was 37 km in 2002 and 39 km in 2017 with a constant occupancy of 1.5 persons per car [23]. For other vehicle classes, other statistics from the BMVI [24] show, that driven distance will change slightly. For example, the average transport distance from trucks was 89 km in 2002 and 93 km in 2017. The explained literature justifies the use of the 2010 survey from the BMVI.

Figure 2 shows the mission range distribution of different means of transport based on the BMVI survey from 2010 [22]. To create the different distribution curves, the daily driven distance samples for each vehicle class from the dataset were assigned to 10 km classes in the range of 0 to 2500 km, normalized with the absolute number of samples, and summed up for each 10 km class, starting at 0 km. Daily mileage values of 0 km were ignored during this analysis. As an example, a Tesla Model S with a WLTP mission range of 610 km [25] would be able to cover 99.4% of all daily distances from passenger cars (see Figure 2). Figure 2 illustrates the high required mission range for articulated trucks, coaches, and other trucks. Public urban buses are not covered in the analyzed dataset, but probably require low daily ranges, as they mostly operate in urban areas. Trucks are operated as either rigid trucks without an attached trailer or trailer trucks with an attached trailer. Analyzing the dataset of Breuer et al. [26], which was published in the article by Breuer et al. [27] indicates that trailer trucks mostly operate on highways, similar to articulated trucks, whereas rigid trucks operate in urban areas. Considering this finding while investigating the results presented in Figure 2, leads to the conclusion that the curve of rigid trucks is most likely on the left-hand side of the corresponding one from all trucks, whereas the curve from trailer trucks is on the right-hand side. As a result, trailer trucks probably require drive systems with higher mission ranges and rigid trucks systems with lower ones. As can be seen in Figure 2, the road transport classes of light duty vehicles, passenger cars, motorcycles, other tractors, and agricultural tractors all have low requirements in the case of achievable mission ranges.

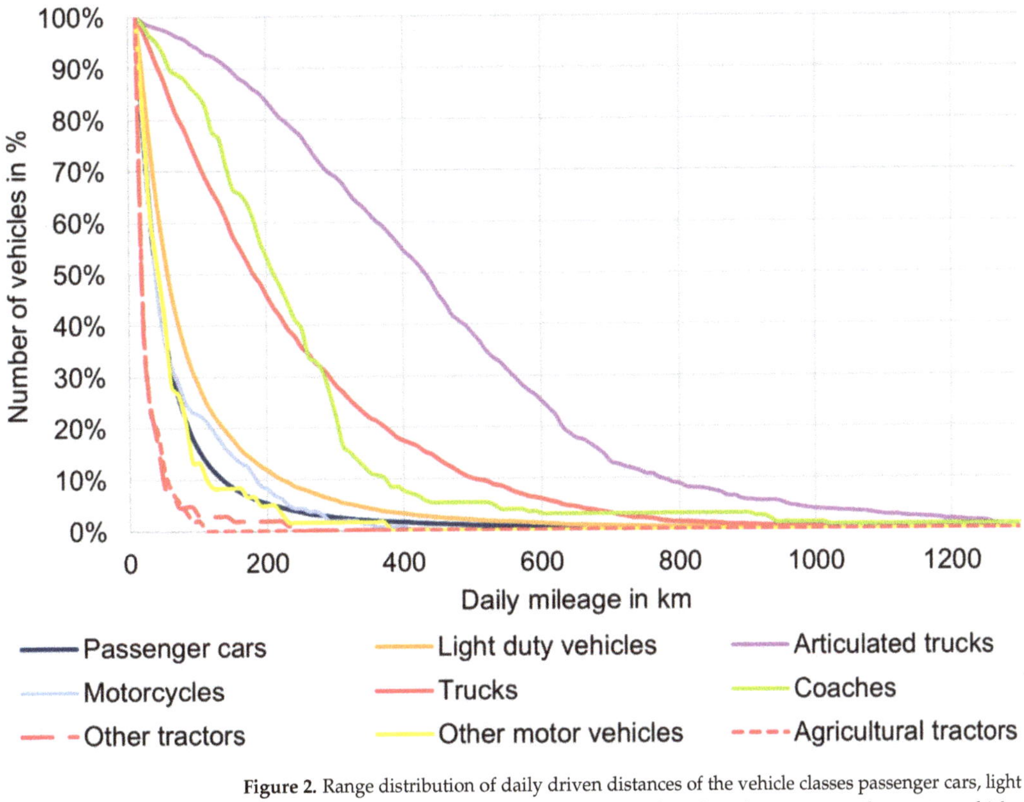

Figure 2. Range distribution of daily driven distances of the vehicle classes passenger cars, light duty vehicles, motorcycles, coaches, trucks, articulated trucks, other tractors, other motor vehicles, and agricultural tractors. Source: Own analysis based on [22].

McKinsey & Company [28] analyzed global air transport in terms of CO_2 emissions considering the different aircraft classes of commuter, regional, short-range, medium-, and long-range aircraft, as well as different mission ranges. Deviating required mission ranges from the results from McKinsey & Company [28] leads to maximum mission ranges of up to 500 km for commuter, 2000 km for regional, 4500 km for short-range, and over 10,000 km for medium- and long-range aircraft. Commuter and regional aircraft have a lower required mission range, short-range aircraft with 4500 km having a higher mission range, and medium- and long-range aircraft with >10,000 km having the highest required mission ranges. The global fleet consists of 4% commuter aircraft, 13% regional aircraft, 53% short-range aircraft, and 30% medium- and long-range aircraft. Aircraft require fuel with high gravimetric and volumetric densities, as well as high powertrain efficiencies.

Rail transport in Germany is mostly electrified in terms of transport performance. As of 2019, 53% of all rail sections were equipped with overhead catenary lines [29]. In the case of transport performance, in 2019, long-distance rail passenger transport totaled 99%, regional rail transport 79%, and rail freight transport 87% electric operation [29]. This leads to the conclusion that non-electric operation with diesel is mostly used in sections with low transport performances. Alternative fuels might be a more economical solution compared to overhead catenary lines. A high gravimetric energy density of a possible fuel seems to be more important for rail transport than a high volumetric energy density, as the maximum load in each rail section is limited. Rail sections in Germany are divided into different track classes defined by DIN EN 15528, which limits the maximum weight of operating trains to between 6.4 t/m and 8 t/m [30].

Inland waterway transport is subdivided into freight and passenger transport. Freight transport is separated into the ship classes of cargo barges, liquid cargo barges, pushed barges, and pushed tankers, with the latter two being operated by pusher boats. Passenger transport primarily takes place using day trip and cabin vessels. In addition to these, small watercraft like sporting boats are used on inland waterways.

Figure 3 shows the share of transport performance (tkm) and volume of transport (t), as well as the average trip distance of different distance classes for inland waterway freight transport in Germany in 2016, based on data from the Federal Statistical Office (Destatis). This covers the ship classes of cargo barges, liquid cargo barges, pushed barges, and pushed tankers. As can be seen in Figure 3, 42% of freight transport ships fall into the distance class of >500 km per trip, with an average trip distance of 627 km. In the case of volume of transport (t), the value is 17% lower, because this value does not cover the traveled distance. In the case of transport performance, only 11% would be covered with a mission range of 100 km, whereas 89% require a mission range of >100 km. The consumed fuel is proportional to transport performance or the distance traveled, and less to the weight of transported goods. Unfortunately, the distance class of >500 km is not defined in greater detail. Destatis also provides transport performance and transported good data for the different ship classes. Based on Destatis data, the average trip distances for cargo barges, liquid cargo barges, and pushers were calculated to be 296 km, 225 km, and 158 km with transport performances of 34 bn tkm, 11 bn tkm, and 7 bn tkm. These average trip distances are not classified for distance classes as in Figure 3, and therefore may appear smaller. Pusher boats seem to operate over shorter distances, liquid cargo barges with 225 km over medium trip distances, and cargo barges over longer ones. However, the transport performance of cargo barges is about five times larger than that of pusher boats and three times greater than the transport performance of liquid cargo barges.

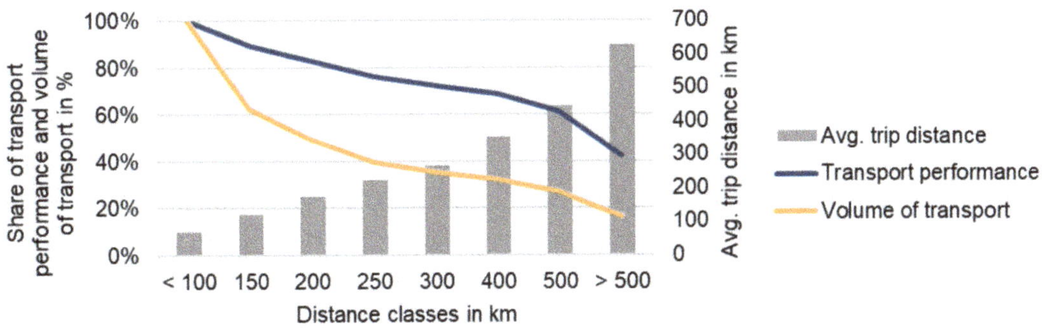

Figure 3. Share of transport performance (tkm) and volume of transport (t), as well as the average trip distance of different distance classes for inland waterway freight transport in Germany in 2016. Source: Own elaboration based on [31,32].

Similar to rail transport, gravimetric energy density seems to be more important for inland waterway transport, as increasing the weight would also increase the energy consumption. Based on Figure 3, inland waterway freight transport is classified as a sector with a higher required mission range. Compressed natural gas (CNG) is more attractive for pusher boats due to their smaller sizes and lower daily distances [33]. In the case of inland waterway passenger transport, energy density and mission range for small craft like sport boats and ferries appear to be less important, whereas for cabin vessels, mission range could be a more important criterion.

In this section, the methodology of this work was explained. First, the technology readiness level was explained as the basis for a technological assessment. Furthermore, mission ranges of the different vehicles classes in road, air, rail, and inland waterway transport were analyzed and identified in the framework of the methodology. The analysis showed, that requirements on the drive systems and subsequently the fuel vary strongly even inside of each of the four sectors.

3. Potential, Technical Maturity, and Costs of Alternative Fuel Production

As outlined in the introduction, renewable alternative fuels are divided into biomass- and electricity-based ones. Both fuel pathways will be discussed in this section. First, the capacity of biomass-based fuels to cover energy demand in the transport sector in Germany is discussed. Second, the technical maturities of the different fuel pathways will be investigated.

3.1. Potential of Biomass-Based Alternative Fuels

Studies such as those by Robinius et al. [34] show that in order to reduce CO_2 emissions in Germany by 80% or 95% by 2050 against 1990 levels, CO_2 emissions from the transport sector must be reduced by 76% or 100% from 2020 to 2050. In the case of the 95% target, a complete abandonment of fossil fuels in the transport sector is essential. The Germany-wide potential of biomass is not sufficient to fully cover this demand. Electricity-based fuels can be produced locally or imported to close the gap [34]. Indeed, the global potential of wind and solar energy is already more than sufficient to meet global energy demand [35]. According to this concept, hydrogen can be produced in advantageous regions with high wind and/or solar energy potential. Studies by the Hydrogen Council, for instance, forecast a price of EUR 1.4–2.3/kg_{H2} in 2030 [36,37], with further current and future expected prices being discussed in Section 3.3.2 Review of Total Costs below. The following discussion of biomass potential reveals that electricity-based fuels are vital, in addition to biomass-based ones. Therefore, a selection of electricity-based fuels within their category, regardless of the achievable price of biomass-based fuels, is necessary.

Biofuels are subdivided into conventional and advanced types. Conventional biofuels contain the first-generation biofuels ethanol and biodiesel from eatable crops, whereas advanced biofuels constitute the second-, third-, and fourth-generation forms. These fuels are obtained from nonfood, sustainably-grown feedstocks, and agricultural wastes. The second-generation biofuels also encompass fuels from cellulosic biomass, whereas the third- and fourth-generation ones contain fuels from both natural and genetically-engineered algae biomass [11]. According to Ziolkowska [11], the latter are based on algae biomass at demonstration level (TRL 4–5), whereas the first generation is well-established in the market (TRL 9) and the second (TRL 9) is gaining increasing market shares.

The viewpoint on biofuels in EU countries changed with the introduction of RED II in 2018. Since then, biofuels, which are not based on food or feed crops, are promoted with increased credibility for the national targets of RE usage in transport. Furthermore, targets to achieve specific shares of advanced biofuels and renewable electricity ones are binding, whereas the targets for food-based fuels are optional [38]. RED II also limited the share of food and feed crops in EU countries as a function of total energy consumption in the transport sector to 7% of energy consumption from road and rail transport in 2020. The possibility of member states voluntarily reducing this threshold is ambiguous. Furthermore, the threshold must be reduced to 0% by 2030, with the exception of feedstocks with certified low risks for Indirect Land Use Change (ILUC) [39]. According to the EU legislation, conventionally-produced biofuels are not regarded as an option for mass application in the transport sector.

The production pathway of a fuel is essential for the assessment of its ILUC risk. Both of the most common biofuels, namely FAME and HVO, can have a high or low ILUC risk, depending on their feedstock. Feedstocks that are produced by means of BtL processes from residue or waste oil are noncritical with respect to their ILUC risk. However, FAME or HVO obtained from vegetable oil are ILUC-critical [40]. The European Commission published a report [41] assessing global increases in ILUC areas devastating plants while maintaining large quantities of carbon stock and biodiversity. The report indicates that harvesting areas of biofuel feedstock plants increased globally by 2.3% for maize, 1.2% for sugar beets, 4.0% for palm oil and 3.0% for soya beans between 2008 and 2016. These growths rates are not only related to increased biofuel production but other factors as well. Further valuations can be noted in the report [41]. The European Commission [42] published criteria to identify resources with high and low ILUC risks. These criteria identify palm oil as a resource with a high ILUC risk. The status quo of the JRC Biofuel Program in 2014 and other earlier published literature regarding feedstocks for biofuels like ethanol or FAME did not take into account ILUC or other issues such as conflict affecting food production as a limiting factor for fuel and decarbonization strategies [40].

As an interim conclusion, it can be stated that biofuels possess an ambivalent positioning in the field of alternative fuels. Furthermore, resources for biofuels vary significantly across different locations and are limited to a greater extent than electricity-based fuels. Thus, the following section discusses the biomass potential for the case of Germany.

In 2017, 1668 PJ of diesel, 791 PJ of gasoline, and 428 PJ of jet fuel were consumed in Germany, according to the Mineralöl Wirtschafts Verband e.V. (MWV) [43]. Additionally, 81 PJ of biodiesel, 0.04 PJ of vegetable oils, 31 PJ of ethanol, 2 PJ of biomethane, 43 PJ of electricity, and 6 PJ of natural gas were consumed within the country's transport sector in 2018/2019 [24,44]. Based on data from the Federal Office for Agriculture and Food (BLE) [45], Fehrenbach [46] concluded that the 81 PJ of biodiesel consisted of 27.5 PJ produced in Germany, with 19 PJ imported as palm oil biodiesel, 31.5 PJ coming from waste oils, and 1.6 PJ from grain straw and industrial waste. Peters et al. [12] reviewed the potential of biofuels. Based on data from Billig et al. [47], Peters et al. [12] state that 270 PJ of methane could be directly produced by biogas plants in Germany in 2050. If the biogas plant-generated CO_2 is combined with renewable hydrogen, an additional value of 205 PJ can be achieved [47]. The maximum amount of methane produced from this concept in 2050 is estimated to be 750 PJ, also drawing on CO_2 from the cement industry and combining it with renewable hydrogen, alongside the mentioned sources [47]. In their calculated scenario, Billig et al. [47] assume a 100% waste stream feedstock for biogas plants in 2050 based on the 15.9–20.5 Mt (dry matter) of unused biomass in Germany identified by Brosowski et al. [48]. According to Brosowski et al. [49], the biomass potential of agricultural byproducts in Germany is 17.6 Mt (dry matter), comprising 52% manure and 48% grain straw. Municipal waste biomass potential is about 0.3 Mt [49]. Technical biomass potential is 141 PJ for grain straw, 70 PJ for manure, and ~4 PJ for municipal waste [49]. Between the 270 PJ calculated by Peters et al. [12] based on Billig et al. [47] and the 218 PJ of Brosowski et al. [49] is a gap of 52 PJ that may result in different conversion efficiencies. Fehrenbach [46] calculated a potential unused biomass quantity of 250 PJ based on data from Fehrenbach et al. [50] and combined it with an efficiency of 50%, leading to a biofuel potential of 130 PJ. Comparing the 130–270 PJ of the three mentioned sources [12,46,49] with the transport sector's fossil energy demand of 2887 PJ [43] (excluding renewables, electricity, and natural gas) leads to a biomass potential of 4.5–9.4%. This value will rise to 26% (750 PJ) by upgrading CO_2 from biogas plants and the cement industry, as described in Billig et al. [47]. The RED II limits will reduce the palm oil biodiesel share to 0% in 2030 [46]. Fehrenbach [46] mentioned that the RED II limits will not resolve the ILUC risk issues with biofuels. Other resources that are not classified as having a high ILUC risk according to the European Commission [42] could replace palm oil [46]. The phase-out of palm oil could prompt a boost in rapeseed production, which will then produce further ILUC effects [46]. As noted earlier, the palm oil-based biodiesel share of total biodiesel is about 23%. The potential of first-generation, conventional biomass-based fuels' potential to satisfy transport sector energy demand is low [51,52]. Furthermore, their sustainability and macroeconomic benefit due to conflicts regarding land use for growing food, as discussed above, is questionable [51]. Assuming that conventional biofuels will drop to 0% after 2030 and advanced biofuels such as biogas methane will replace them, the calculated biomass potentials shares on total transport sector energy demand will be reduced to 0.6–5.5% (18–158 PJ) and a maximum of 22.1% (638 PJ). This share is increasable by importing biomass from other countries. In terms of biofuel imports, the current literature reveals uncertainty and concerns regarding large quantities of sustainably-produced sources [51,53,54]. With respect to land use efficiency, biofuels are, with a difference by a factor of up to 1000, significantly lower than e-fuels [53].

In its 2011 Technology Roadmap, the International Energy Agency (IEA) forecasts a marked increase in the importance of biofuels [55]. It is stated that a total global energy demand of the transport sector, including road, aviation, and shipping, of 116 EJ in 2050, could be satisfied using 32 EJ biofuels (27.5%), equivalent to around 100 Megahectares (Mha) of land used for feedstock. Comparing this area with the EU's use of cropland, totaling 97 Mha in 2015 [56], underscores the pressing need for acreage for the high penetration of biofuels, in accordance with the IEA Roadmap for Biofuels.

3.2. TRL of Fuel Production Pathways

In this section, the technical maturity of fuel production pathways is discussed using TRL as a performance indicator. The TRL assessment of alternative fuel production pathways is illustrated in Figure 4. The striped areas represent ranges that are either caused by different process pathways or different TRL assessments. The literature sources are listed in Table A1 in Appendix A. The TRL evaluation of the fuel pathways in Figure 4 is divided into synthetic production from CO_2 and renewable electricity, conventional biofuel production, and advanced biofuel production. As noted in the previous section, conventional biofuels include fuels from edible crops, whereas advanced biofuels encompass those from nonfood, sustainably-grown feedstocks, and agricultural wastes.

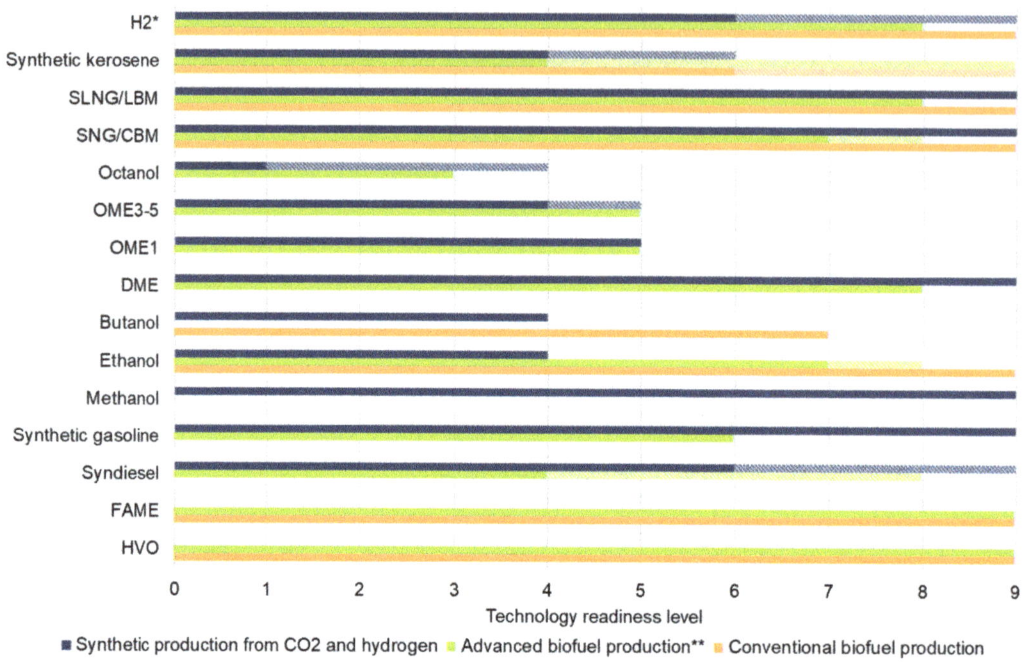

Figure 4. TRL of alternative fuels from synthetic production from hydrogen and renewable electricity, conventional biofuel production, and advanced biofuel production. Source: Own elaboration based on literature. The literature sources are listed in Table A1. * Production from renewable electricity. ** If manure, sludge or waste is used for conventional biofuel production processes, the product will be classified as advanced biofuel. These pathways are not covered by advanced biofuel production in this diagram. SLNG: Synthetic liquefied natural gas, LBM: Liquefied biomethane, SNG: Synthetic natural gas, CBM: Compressed biomethane, OME: Oxymethylene ether, DME: Dimethyl ether, FAME: Fatty acid methyl ester, HVO: Hydrotreated vegetable oils.

Hydrogen can be produced from upgraded biogas from municipal organic waste, wet manure, sewage sludge, maize, or double cropping with a TRL of 9 [5]. Production from municipal organic waste, wet manure, and sewage sludge would via definition count as advanced biofuel production, but because the process is equal to the one for maize, it is illustrated as conventional biofuel production in Figure 4. Alternatively, hydrogen is produced from farmed wood via gasification with a TRL of 8 as advanced biofuel from renewable electricity with TRL 9 [5]. However, the high TRL assessment of the hydrogen production from renewable electricity originates in using an alkaline electrolyzer. The production of hydrogen via the polymer electrolyte membrane electrolysis also has a high TRL of 9, while the production via the solid oxide electrolysis cell has a TRL of 6–7 [57].

Synthetic kerosene includes all fuels that are certified by legal ruling ASTM D7566-20b [58]. In the reviews of Wormslev and Broberg [59], the International Civil Aviation Organization (ICAO) [60], and Sustainable Aviation [61], emerging sustainable aviation fuel production plants are presented. Literature references and technical assessments of all three were, amongst other sources, used to assess the technical maturity of sustainable alternative aviation fuels in this study. Synthetic Paraffinic Kerosene (SPK) obtained via the Fischer–Tropsch (FT) process, was evaluated by Schemme et al. [4] as having a TRL of 6. An alternative pathway utilizes synthetic production, with methanol as an intermediate product. The production of methanol from CO_2 and H_2 was assessed by Schemme et al. [4] to have a TRL of 9. Schmidt et al. [62] determined that methanol to olefin conversion and subsequent distillate synthesis could be demonstrated in the 1980s, drawing on Tabak et al. [63] and Tabak and Yurchak [64]. Both process steps are essential to the production of synthetic kerosene from methanol. Ruokonen et al. [65] analyzed methanol-based pathways to transport fuels and rated them as having a TRL of 8. In turn, K. Zech et al. [66] evaluated the TRL of methanol-based kerosene as being 7–9. However, they note that the entire process was never tested but rather single sections of it were tested on demonstration level. Therefore, the Methanol-to-Kerosene process is determined to have a TRL of 4. The TRL of biomass-based FT-SPK achieved via gasification and the FT process, is rated as having a TRL of 9, because Fulcrum [67] and Red Rock [68] have plants under construction that will have outputs of 30 kt/year and 45 kt/year, respectively. Another biomass-based fuel is Hydroprocessed Esters and Fatty Acids Synthetic Paraffinic Kerosene (HEFA-SPK) from bio-oils, animal fat, and recycled oils. The production of HEFA-SPK is determined to have a TRL of 9, as it is already used as a commercial process by World Energy Paramount (former AltAir Paramounts LLC) [60] and Neste Oyj [69]. The sustainable aviation fuel Hydroprocessed Fermented Sugars to Synthetic Isoparaffins (HFS-SIP) is produced via the microbial conversion of sugars to hydrocarbon and is also produced in a commercial process by Amyris in Brazil [60]. Therefore, the production of HFS-SIP is rated as having a TRL of 9. Alcohol-to-Jet Synthetic Paraffinic Kerosene (ATJ-SPK) from agricultural waste products (i.e., stover, grasses, forestry slash, and crop straws) can either be produced with isobutanol or ethanol as the intermediate product. LanzaTech [70] operates a pilot plant for ATJ-SPK production, while Ekobenz operates a commercial plant with a production of 22.5 kt/year [71]. In turn, Lanzatech and Swedish Biofuels AB have planned plants at the commercial scale for the coming years [72,73]. Catalytic hydrothermolysis synthetic jet fuel (CHJ) from triglyceride-based feedstocks (plant oils, waste oils, algal oils, soybean oil, jatropha oil, camelina oil, carinata oil, and tung oil) is evaluated to have a TRL of 6–7, because ARA and euglena operate a plant on the demonstration scale [74]. The sustainable aviation fuel High Hydrogen Content Synthetic Paraffinic Kerosene (HHC-SPK), produced from biologically derived hydrocarbons such as algae, is assessed as having TRL of 4, because Ishikawajima-Harima Heavy Industries Co., Ltd. developed and certified a process that is likely on the laboratory scale [75].

As can be seen in Figure 4, HVO and FAME have both been analyzed to be TRL 9 by Müller-Langer et al. [76] and Prussi et al. [5]. Advanced or conventional classification depends on the feedstock. In the case of waste oil or algae as feedstock, the product fuel constitutes an advanced biofuel. The key process of most synthetic diesel production processes is the already-mentioned FT process. Advanced biofuel synthetic diesel is produced from lignocellulose via pyrolysis (TRL 6 [5,76]), gasification (TRL 8 [5]), or hydrothermal liquefaction and upgrading (TRL 4 [5]). Synthetic production based on CO_2 and H_2 can either be performed with the reverse-water-gas-shift reaction in between, limiting the TRL to 6 [4], or with the intermediate product methanol, assessed to have a TRL of 9 [5]. Synthetic gasoline can be either produced via the Methanol-to-Gasoline processes, rated to have a TRL of 9 by Schemme et al. [4], or as advanced biofuel from lignocellulose, assessed as having a TRL of 8 by Prussi et al. [5]. Ethanol is produced as a conventional biofuel from sugar and evaluated with a TRL of 9 by Müller-Langer et al. [76] and Prussi et al. [5]. The production as advanced biofuel based on lignocellulose is determined to have a TRL of 7 by Müller-Langer et al. [76] and of 8 by Prussi et al. [5]. Alternatively, ethanol is produced utilizing CO_2 and H_2. This pathway has been assessed by Schemme et al. [4] to have a TRL of 4. The higher alcohol butanol is produced as a conventional biofuel via fermentation and with a TRL of 7 according to Prussi et al. [5]. Alternatively, butanol is produced from CO_2 and H_2 in the framework of the Power-to-Fuel concept with a TRL of 4 [4]. 2-octanol has also been produced from lignocellulose on the laboratory scale, according to Leitner et al. [77]. Therefore, it is evaluated as having a TRL of 3. In the framework of the PtL concept, octanol is either produced as 1-octanol with a TRL of 1 or as iso-octanol with a TRL of 4 [4,78]. For the alternative fuels in the ethers category, the production of dimethyl ether (DME) exhibits the highest technical maturity. The production as advanced biofuel from lignocellulose is rated as having a TRL of 8 by Prussi et al. [5], whereas production from CO_2 and H_2 is determined to have a TRL of 9 by Prussi et al. [5] and Schemme et al. [4]. For longer ether-chains, the technical maturity is lower. Synthetic $OME_{3\text{-}5}$ from CO_2 and H_2 can be produced through different pathways with varying technical maturity levels in the range of a TRL of 4–5 [4,79], whereas OME_1 production is determined to have a TRL of 5 [4]. The production of OME_1 and $OME_{3\text{–}5}$ from lignocellulose as advanced biofuel is assessed as TRL 5 by Prussi et al. [5]. For $OME_{3\text{–}5}$ synthesis, Prussi et al. [5] assessed the more mature pathway via OME_1 and trioxane. The production of methane and liquefied methane has, in general, a high degree of technical maturity. Using the conventional biofuel production route, compressed biomethane (CBM) and liquefied biomethane (LBM) are produced at TRL 9 [5,76]. If a nonfood feedstock such as manure were to be used, the produced fuel would be assessed as an advanced biofuel, as explained previously. The potential of this resources is low, and therefore no more specific subdivision is presented in Figure 4. Synthetic natural gas (SNG) and synthetic liquefied natural gas (SLNG) as advanced biofuel from lignocelluloses such as straw are produced via gasification with TRL 7–8 [5,76]. The electricity-based production of SLNG and SNG is rated as TRL 9 by Prussi et al. [5]. The German car manufacturer Audi AG also operates an SNG plant with a possible product stream of 200 kg/h [80].

3.3. *Costs*

According to Germany's National Hydrogen Strategy [81], the German government forecasts a demand of 90–110 TWh of green hydrogen in 2030 to meet decarbonization targets. The hydrogen is envisioned to be used in industry, as well as specific applications in the transport sectors. Of the total demand for green hydrogen, 14 TWh are planned to be synthesized in Germany, corresponding to 20 TWh of renewable electricity, mainly from on- and offshore wind turbines. Accordingly, around 85% of green hydrogen demand must be imported. [81] In the long term, the imported quantities of green energy carriers are forecast to rise even higher. The demand of imported green energy carriers in 2050, to be used in all sectors, is estimated to be 150–900 TWh [82]. However, Merten et al. [83] state that there is no unanimous opinion regarding import strategies. The authors list arguments against large amounts of energy imports, such as high RE potential in Germany, overestimation of RE potential in typical export regions, political instability in export regions, delay of energy transition in export countries, as well as the necessity of efficiently using excess German RE energy. In contrast, Pfennig et al. [6] state that a solely national supply of electricity-based fuels from Germany is not reasonable. This is due to significantly higher onshore electricity generation costs by around a factor of two in Germany, in comparison to the North and Baltic Sea areas and the Middle East and North Africa (MENA) region. This assumption is shared by most studies that optimize future energy systems, taking into account energy carrier imports from regions with high RE full load hours (FLH), such as Robinius et al. [34], dena [84]. Therefore, it is likely that a mix of different green energy carriers will be imported, as is common practice today. [85] Due to the fact that fuel production will experience fundamental changes independently of the fuel choice and mix, an overview of the currently described costs of energy carriers that can be utilized in the transport sector is provided. For the import of energy carriers, additional costs accrued through transportation arise, which vary between different energy carriers. Before reviewing production and end user costs, the following section presents an overview of transportation costs.

3.3.1. Import Costs of Energy Carriers

Perner et al. [85] identify electricity generation costs as being the main cost factor of electricity-based fuels. Consequently, comparative system analyses like that of Robinius et al. [34], Fasihi and Breyer [86] determine regions with high RE full load hours to ensure low energy prices. Worldwide detailed RE generations costs were determined by Fasihi and Breyer [86] and used for scenario-based import routes by Hank et al. [87]. The

H2AtlasAfrica [88], funded by the Federal German Ministry of Education and Research, however, is focusing on hydrogen production from RE sources in West Africa.

Most other studies assume location-specific RE FLH and do not consider international locations for production [85,89,90]. Transport costs sum up to a relatively small share of cross-border prices for fuels, which allow transportation in existing infrastructure. However, transport costs for different fuel options substantially differ. In particular, if current infrastructure-compatible PtL fuels are compared for the transport of H_2, new and costly infrastructure must be installed. In various studies, high H_2 transport costs have been found to be economically decisive in favor of H_2 derivatives such as liquid organic hydrogen carrier (LOHCs), SNGs, ammonia, or other PtL fuels that offer infrastructure compatibility [7,91]. The dilemma between costs for the scale-up of transport infrastructure for H_2 and the less efficient process chains of PtL fuels also leads to the question of domestic H_2 production, which could reduce transportation cost efforts and conversion losses before and after transportation [83].

However, if H_2 import is considered, it can be transported in liquid form (LH_2), bound to an LOHC, or as a derivative, such as ammonia, methane, or methanol [7]. H_2 requires between 25% and 35% of its initial energy to be liquefied. Liquefied natural gas (LNG) consumes around 10% of its initial energy for liquefaction. The conversion of H_2 into the derivative ammonia consumes 7–18% of its initial energy [91]. The conversion back into H_2 in high purity consumes about the same amount of energy again. However, the liquefaction temperature of ammonia is −33 °C and the H_2 liquefaction temperature is −253 °C, resulting in lower cooling efforts for ammonia. In the two required steps of exothermal hydrogenation and endothermal dehydrogenation, between an LOHC and H_2, the processes combined consume between 35% and 40% of H_2 initial energy. Another disadvantage is the necessity of transporting the dehydrogenated LOHC back to a suitable hydrogenation station. Similar issues arise for vessels that transport LH_2, which must return empty. Furthermore, storing H_2 at export and destination terminals is costly. IEA's The Future of Hydrogen report [91] provides a detailed overview of the advantages and disadvantages of transporting H_2 in the liquid phase, converted into ammonia or within an LOHC. Schindler [89] determines the costs of different H_2 transportation possibilities from Morocco to Germany, which amounts to a shipping distance of around 3400 km. The costs of liquefaction are assumed to be 2.6 EURct/kWh$_{GH2,LHV}$. Gaseous hydrogen transportation costs are estimated to be 1 EURct/MWh$_{GH2,LHV}$/km for pipeline transport and 9.5 EURct/MWh$_{GH2,LHV}$/km for transportation by truck. In contrast, costs for the transportation of liquefied hydrogen are 0.5 EURct/MWh$_{LH2,LHV}$/km for ship transport and 2.1 EURct/MWh$_{LH2,LHV}$/km by truck [89]. Kreidelmeyer et al. [92] directly compare the transportation costs for H_2, SNG and PtL energy carries for an equal distance of 4000 km for 2020. The total transportation costs split up the energy demand costs for conditioning (e.g., compression or liquefaction), transport energy, and CAPEX, as well as the OPEX of the transport infrastructure. The calculated transportation cost values refer to the MENA transportation route in Germany and are rated as optimistic, with 2.3 EURct/kWh$_{SNG,LHV}$ for SNG through a 100-bar pipeline and 3.6 EURct/kWh$_{H2,LHV}$ for H_2, also via a 100 bar pipeline. The estimated value for unpressurized pipeline-transported PtL energy carriers is 1.5 EURct/kWh$_{PtL,LHV}$. Corresponding pessimistic values are not presented [92]. These values lead to specific transportation costs of 0.575 EURct/MWh$_{SNG,LHV}$/km, 0.9 EURct/MWh$_{H2,LHV}$/km, and 0.475 EURct/MWh$_{PtL,LHV}$/km for SNG, H_2, and PtL carriers, respectively. Schorn et al. [7] compare the import costs of the energy carriers H_2 and methanol for four different favorable global trade route locations for renewable electricity production, drawing on baseload hydrogen prices from the Hydrogen Council [36]. For Germany, an import scenario from Saudi Arabia via an LH_2 vessel is determined. The results show that the additional costs for converting hydrogen into methanol are outbalanced by the significantly lower shipping cost of methanol, in contrast to liquid hydrogen transport. For the reference year 2030, import costs for both energy carriers were determined to be 6.5–10.8 EURct/kWh$_{LHV}$, including production and transport.

Perner et al. [85] calculated liquid shipping transport costs for LNG for different export regions, which include the North and Baltic seas, North Africa and the Middle East, and Iceland, assuming that the liquefaction costs of 0.69 $EURct/kWh_{LNG,LHV}$ in 2020 will reduce to 0.61 $EURct/kWh_{LNG,LHV}$ in 2050. The authors classified these values as optimistic. Regasification costs are estimated to be 0.15 $EURct/kWh_{LNG,LHV}$ and remain constant between 2020 and 2050. Including the transport costs from Iceland (a 2300 km maritime distance to Hamburg), the determined transport cost range is 0.91 $EURct/kWh_{LNG,LHV}$ in 2020 and 0.84 $EURct/kWh_{LNG,LHV}$ in 2050. Equal considerations for the export region North Africa (3600 km maritime distance to Hamburg) result in transport costs of 0.96 $EURct/kWh_{LNG,LHV}$ in 2020 and 0.88 $EURct/kWh_{LNG,LHV}$ in 2050. For the Middle East region (11,000 km maritime distance to Hamburg), the transport cost values are 1.19 $EUR/kWh_{LNG,LHV}$ in 2020 and 1.12 $EUR/kWh_{LNG,LHV}$ in 2050. Pfennig et al. [6] differentiate between the costs for PtL production with different CO_2 sources that commonly appear as by-products and CO_2 that is captured from air (direct air capture—DAC) and LH_2 onshore wind and PV hybrid systems. Shipping transportation costs are estimated to be 0.13 $EURct/kWh_{PtL,LHV}$ for PtL for import from Morocco (Region of Tarfaya) to Germany. In the study, different production locations such as Egypt, Somalia, Brazil, and Morocco were compared, with the latter turning out to be the most cost-competitive. Energy costs account for a share of 67% of the total costs in the calculation by Pfennig et al. [6]. For LH_2, Pfennig et al. [6] calculate transportation costs of 0.27 $EURct/kWH_{LH2,LHV}$ for an equal shipping transport route. A specific transportation cost value for PtL is determined for an equal route with 0.13 $EURct/kWH_{PtL,LHV}$ for 2050. Thus, the transport costs of LH_2 exceed PtL by a factor of about two.

Merten et al. [83] determine the distance-specific transportation costs of H_2 vessels and pipeline transport. The advantages of pipeline transport include high capacity, high efficiency, and low OPEX. Its disadvantages include high CAPEX and the high transport quantities that are necessary to refinance investment, which can be economically disadvantageous for technology scale-up, starting with low H_2 production. Vessel transport offers advantages in the ramp-up phase of H_2 production through easier scalability and the meeting requirements of decentralized H_2 production. Furthermore, longer transport distances can be covered with lower increases in transport costs [83]. For shorter transportation distances of around 1000 km pipeline, transport is estimated with a cost interval of 1.0–3.4 $EURct/kWh_{H2,LHV}$ and vessel transport with 3.0–8.3 $EURct/kWh_{LH2,LHV}$, with a cost advantage for pipeline transport. According to the determined correlation, vessel transport would be advantageous for distances greater than 5000 km. An IEA study [91] calculates differing break-even transport distances, which are beneficial for vessel transport for distances higher than 1800 km. For lower transport distances, a gas pipeline would be the most cost-efficient. The assumed specific transport cost curves are illustrated in Figure 5. Hank et al. [87] also determine the distance-dependent cost functions, but for up to 18,000 km.

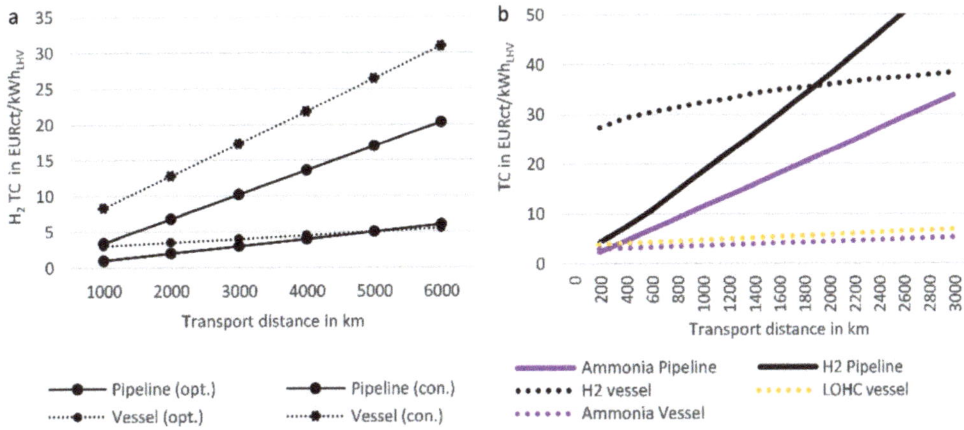

Figure 5. Comparative specific transport cost intervals for (**a**) H_2 and (**b**) H_2, ammonia, and LOHC via pipeline and vessel. Source: Own elaboration based on [83,91,93].

The estimated cost assumptions strongly depend on the transported quantities of H_2, the infrastructure utilization rate, and site-specific features. Robinius [94] already presented a three-dimensional correlation of H_2 costs at the destination depending on mass flow and transport distance, and cheapest transport type for the distribution purpose, with the longest transport distance of 500 km, which is therefore inadequate for overseas import considerations. In this context, it should be mentioned that the commercial readiness level of LH_2 via vessels is low. This is underlined by the high spread of cost range for transport costs, which other studies also conclude [83]. Kawasaki develops a large-scale vessel based on cargo containment systems (CCS) with a storage capacity of 40,000 m^3. The system's forerunner was the SUISO FRONTIER, which has a storage capacity of 1250 m^3. Kawasaki announced the development of a 160,000 m^3-capacity vessel using four CCS [95]. However, the current TRL cannot be evaluated. A TRL of H_2 transport via pipelines is rated higher, with 5000 km of pipelines operated worldwide, with around 400 km in Germany, 2600 km in the U.S., and 600 km in Belgium. The lifetime of H_2 pipelines varies between 40 and 80 years and their installation requires high investment costs. The rededication costs of natural gas pipelines to H_2 ones are below those of the new installation, and bypass the problems of acquiring way rights. Ammonia pipelines carry lower CAPEX and also have high TRL levels, and are therefore a promising transport alternative. For instance, a 4830 km pipeline is operated in the U.S. and a 2400 km one between Russia and Ukraine [91]. For all studies, which analyze the Regions of North Africa or Morocco, which are the most researched areas for synthetic energy carrier production, transportation cost values were extracted and are summarized in Figure 6. These regions are also advantageous because transportation is feasible via pipeline and vessel.

As seen in Figure 6, the transportation costs strongly vary for both transport modes; pipeline and vessel transport. The cost interval for LH_2 vessel transport reaches from 1.05 EURct/kWh$_{LH2,LHV}$ to 20.9 EURct/kWh$_{LH2,LHV}$ for equal transport distances, which is a difference of around factor 20. The two cost values of transport via LOHC carrier are relatively close to each other with 6.6 EURct/kWh$_{LHV}$ and 8.9 EURct/kWh$_{LHV}$. SNG transportation cost show lower scatter, with 0.16 EURct/kWh$_{LSNG,LHV}$ being the lowest and 2.1 EURct/kWh$_{SNG,LHV}$ being the highest value, considering only transportation cost values. The highest cost value of 2.1 EURct/kWh$_{LSNG,LHV}$ includes cost for liquefaction, while the lowest cost value does not. The transportation cost of PtL fuels reveals the lowest total cost and lowest spread of costs, with a range of 0.02 EURct/kWh$_{LHV}$ to 0.13 EURct/kWh$_{LHV}$ for vessel transport, which corresponds to a difference of factor 6.5. The extracted transportation cost values show a clear advantage for especially PtL fuels as well as SNG. The discussed cost values represent transport cost values for one specific region and cannot be transferred to other export regions' transportation costs. How transportation costs affect the total cost is discussed in the following section.

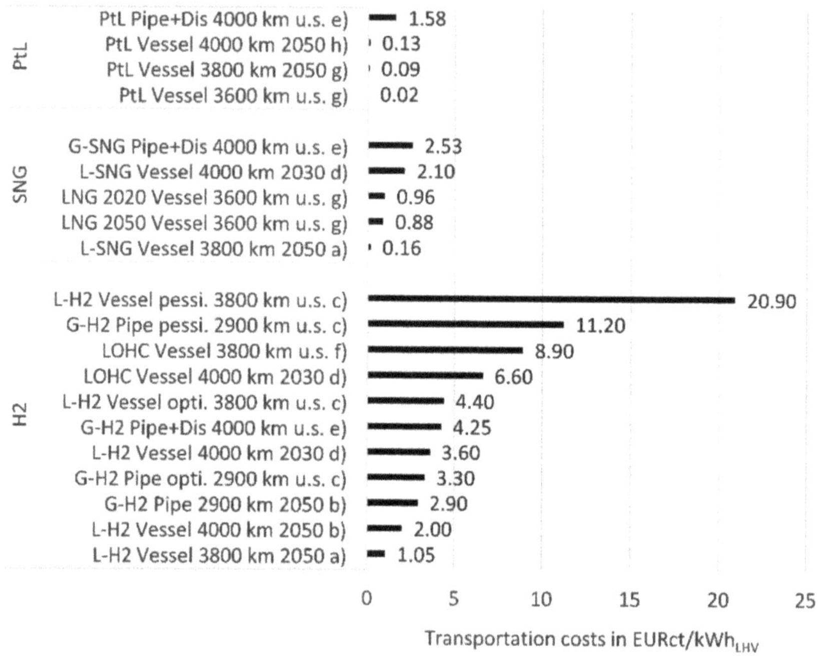

Figure 6. Comparative specific transport cost for H_2, SNG, and PtL fuels for the region of North Africa/Morocco under mention of used transportation technology, assumed distance, and referred year of cost value. All cost values refer to lower heating values. Source: Own elaboration based on (a) [96]; (b) [89]; (c) [83]; (d) [87]; (e) [92]; (f) [55]; (g) [85]; (h) [6]. PtL: Power-to-Liquid, SNG: Synthetic natural gas, LOHC: Liquid organic hydrogen carrier.

3.3.2. Review of Total Costs

This section starts with the current prices of different energy carriers and furthermore aims to provide insight into future energy carrier prices. First, the different production costs of electricity-based fuels are discussed and compared with biofuel prices. For that, only production without transport and taxes are considered. Additionally, the boundary conditions considered in the reviewed articles are similar, so that a comparison is possible. Second, the review is extended to include transport as well as taxes. Only synthetic fuels are considered in that part and are classified into H_2, SNG, and PtL fuels, and domestic production, as well as imports, are considered. The reviewed literature studies are summarized in Table A2 in Appendix A.

Table 2 presents an overview of the specific domestic production costs of alternative fuels, including biomass-based fuels, synthetic fuels from PtL processes, and hydrogen, as well as synthetic methane from CO_2 and renewable hydrogen. The values are shown in EURct/kWh and EUR/L_{DE}. The latter is the specific price for the energy, which represents one liter of diesel. The lower heating value of diesel is 9.96 kWh/L [97]. Schemme et al. [4] investigated the costs of synthetic fuels produced in the framework of the PtL concept. Amongst others, they assume a price of 4.6 EUR/kg_{H2} based on Robinius et al. [98], which is equal to 1.38 EUR/L_{DE}. Furthermore, Peters et al. [13] calculated the price for synthetic methane, but assumed CO_2 costs of 35 EUR/tCO_2 instead of 70 EUR/tCO_2 such as Schemme et al. [4]. The recent market prices of the biofuels FAME, HVO, bioethanol, and biomethane are

88 EURct/L_{DE}, 73 EURct/L_{DE}, 112 EURct/L_{DE}, and 69–72 EURct/l_{DE} [99,100]. The hydrogen price calculated by Robinius et al. [98] is based on an electricity price of 6 EURct/kWh, and is equal to 69 EURct//l_{DE}. The German consumer electricity price in 2018 was 30.19 EURct/kWh [101], which corresponds to 3.01 EUR//l_{DE}. As is shown in Table 2, the production of hydrogen, methanol, DME, and MtG is most beneficial from an economic viewpoint amongst synthetic fuels. The medium price segment of synthetic fuels includes ethanol, and SNG, as well as FT diesel and gasoline. Of biomass-based fuels, HVO and biomethane have the lowest prices, whereas the generally lower price of biofuels, compared to synthetic ones, must be considered. Detz et al. [102] state that learning curves could arise for the production of synthetic fuels, making them more competitive.

Table 2. Current costs of alternative fuel production.

Source	Fuel	Cost in EUR/L_{DE}	Cost in EURct/kWh	Reference
Renewable electricity	Hydrogen	1.38	13.85	[98]
	Methanol	1.89 [a]	18.97 [a]	[4]
	Ethanol	2.22 [a]	22.29 [a]	[4]
	Butanol	2.53–2.60 [a]	25.40–26.10 [a]	[4]
	Octanol	2.85 [a]	28.61 [a]	[4]
	Dimethyl ether	1.85 [a]	18.57 [a]	[4]
	Oxymethylene ether$_1$	2.64 [a]	26.40 [a]	[4]
	Oxymethylene ether$_{3–5}$	3.46–3.96 [a]	34.74–39.76 [a]	[4]
	Methanol-to-Gasoline	1.88 [a]	18.87 [a]	[4]
	Fischer–Tropsch-Diesel/Gasolines/Kerosene	2.30 [a]	23.09 [a]	[4]
	SNG	2.25 [b]	22.59 [b]	[13]
Biomass	FAME	0.88	8.85	[99]
	HVO	0.73	7.34	[99]
	Bioethanol	1.12	11.28	[99]
	Biomethane	0.69–0.72	6.90–7.20	[100]
	Electricity	0.60 [c]–3.01 [d]	60 [c]–30.19 [d]	[98,101]

Notes. [a] 70 EUR/t_{CO2}; [b] 35 EUR/t_{CO2}; [c] Electricity price renewable 6 EURct/kWh [98]; [d] Electricity price 2018 Germany 30.19 EURct/kWh [101]. Source: Own elaboration based on [4,13,98,100,101,103]. SNG: Synthetic natural gas, FAME: Fatty acid methyl ester, HVO: Hydrotreated vegetable oils.

The potential of biofuels, which was outlined above in a separate section, should be drawn upon as much as possible, because they possess, in general, lower production costs. This principle applies as long as the ILUC conditions can be met. Additionally beneficial is the production of HVO and biomethane, as both production pathways are technically mature (see Figure 4) and carry the lowest production costs (see Table 2). Amongst electricity-based fuels, the fuels hydrogen, DME, MtG, methanol, and FT diesel should be preferred on the basis of their lower production costs. The production of ethanol and synthetic jet fuel is limited by the low TRL of their production pathways (see Figure 4). The biomass-based production of jet fuel has a higher TRL, and therefore constitutes a promising alternative. Other alternatives to synthetic jet fuel will be discussed in the following sections.

After comparing current alternative fuel prices, the future price ranges cited in the current literature are reviewed in the following, with a focus on synthetic fuels from CO_2 and renewable hydrogen. Aside from transportation costs, end user price is decisive for market penetration, and for this reason, this section presents the results of a review of the literature regarding the costs that arise across the entire value chain. Tax and levies (T/L) strongly depend on political objectives, which is why they are subjected to independent consideration. In order to be able to compare different studies, three dominant cost levels are drawn on, if available in the literature. These are fuel production costs (PCs), transport costs (TCs = cross-border price), and third level (including T/L and end user price). In order to provide an overview of current and future expected costs, ranges with the lowest and highest specific cost values for 2020, 2030, and 2050 are extracted, which aggregate the assumptions concerning the full load hours of RES, with the choice of RE and electrolysis, CAPEX, OPEX, depreciation rate, process efficiencies, price of CO_2, and water sources, weighted average cost of capital (WACC), as well as the transportation and distribution costs.

Future Cost Ranges of Hydrogen from Renewable Sources

Figure 7 displays the first results of the conducted literature review, with H_2 cost ranges given for the three defined price levels and reference years. With respect to gaseous H_2 and LH_2 PC, the differences in Germany (H_2 domestic) and favorable locations (H_2 import) such as Morocco, Iceland, or the MENA region appear in all analyzed years. Robinius et al. [34] identify neighboring EU states such as Norway, the UK, and Ireland with high RE potential and optimized country-specific import prices. Most other studies have focused on North Africa and the MENA region for energy carrier importation. In more recent studies, export regions have more widely varied [34]. The already mentioned H2AtlasAfrica [88] is calculating H_2 PC in West Africa considering onshore, offshore, and photovoltaic RE sources. Jensterle et al. [104] identified countries with the highest medium- and long-term suitability for exporting large quantities of H_2 with consideration of the already named conditions, as well as soft factors like governmental interest, availability of skilled labor, different aspects of security, and the acceptance of the local population. The countries with the highest suitability for a 2030 perspective are Iceland, Canada, Morocco, Norway, Tunisia, and Turkey. With regard to 2050, the countries of Egypt, Algeria, Argentina, Australia, Canada, Kazakhstan, Russia, and Saudi Arabia were identified. If transportation is neglected, the advantages of higher full load hours in equatorial and coastal regions come into play, where high wind and PV FLH are ensured and appear to overweigh the higher capital and transport costs. In 2030, the high TC for hydrogen (liquefaction effort or gaseous pipeline transport) results in end user cost ranges for imported hydrogen of 15.5–22.8 EURct/$kWh_{H2,LHV}$, which are optimistic and pessimistic cost values for production in the MENA region using wind, as well as PV hybrid systems and subsequently pipeline transport, as calculated by Kreidelmeyer et al. [92]. The corresponding domestic production cost ranges are optimistic cost values of 16.5 EURct/$kWh_{H2,LHV}$ for the case of onshore wind in Germany. The upper bound of 24.5 EURct/$kWh_{H2,LHV}$ corresponds to pessimistic assumptions for offshore wind in the North or Baltic seas [92]. For 2050, the lowest H_2 PCs were determined by Eichhammer et al. [90] with an H_2 cost of 3.6 EURct/$kWh_{H2,LHV}$, whereas electrolyzers are supplied with RE in Morocco and grid connections are not taken into account. The highest regarded PC in favorable regions was calculated by Pfennig et al. [6] with 10.9 EURct/$kWh_{LH2,LHV}$ for Morocco, including liquefaction and wind and PV hybrid supply systems.

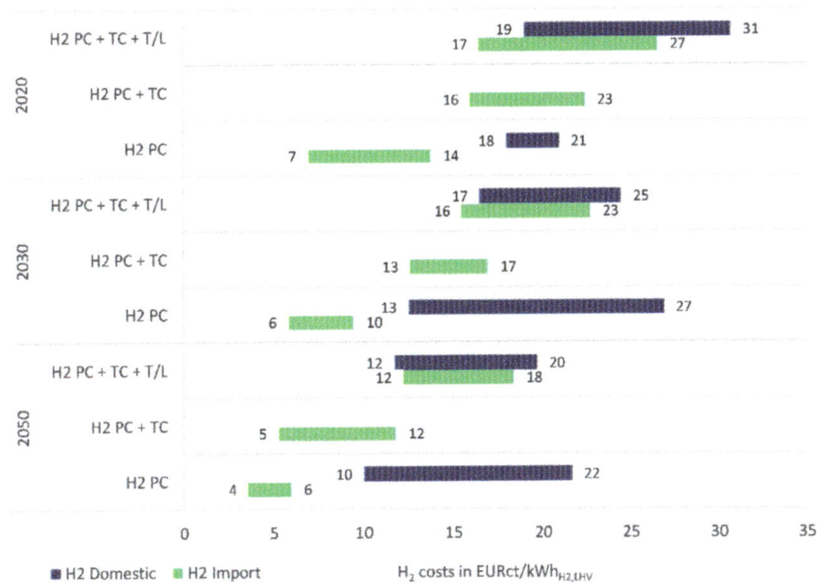

Figure 7. Result of the conducted literature review showing the cost ranges of H_2 and LH_2 production in Germany and abroad, the transport costs and taxes/levies. Source: Own elaboration based on [4,6,34,85,87–90,92,105–109]. All extracted cost values are provided in Figure A1 in Appendix A. H_2 PC: Hydrogen production costs (domestic or import), H_2 PC + TC: Hydrogen import costs, including transport, H_2 PC + TC +T/L: End user prices of hydrogen (domestic and import), including tax and levies.

The lower bounded value of 9.5 EURct/kWh$_{H2,LHV}$ of the PC range for domestic production was determined by Perner et al. [110] for offshore wind in the North and Baltic seas. The highest PC result was for LH_2 and calculated by Pfennig et al. [6] for the case of southern Germany, with 21.7 EURct/kWh$_{LH2,LHV}$. For import (PC + TC), the lowest cross-border price was calculated by Gerhardt et al. [106] for Morocco with 5.3 EURct/kWh$_{H2,LHV}$ and gaseous pipeline transport to Germany (WACC = 6%). A 10% WACC would increase the cost to 6.5 EURct/kWh$_{H2,LHV}$ and liquefied vessel transport to 7.7 EURct/kWh$_{LH2,LHV}$. The highest cross-border price was determined for liquefied H_2 vessel transport from the UK, at 11.8 EURct/kWh$_{LH2,LHV}$, by Robinius et al. [34] for the case of onshore wind in coastal regions. For the end user price of H_2, Kreidelmeyer et al. [92] determined a cost range of 12.2 to 18.4 EURct/kWh$_{H2,LHV}$, with the WACC varying between 6% and 12%, and the FLH of the electrolysis estimated to be 5000 h. Furthermore, the reference cases for domestic production are provided, which results in a low end user price of 11.7 EURct/kWh$_{H2,LHV}$ for the case of electrolysis supplied by the German electricity grid. The highest H_2 end user price of 19.7 EURct/kWh$_{H2,LHV}$ was determined for an off-grid system combining offshore wind and electrolyzers close to the German coast and includes a gaseous pipeline transport segment of 500 km and 3000 FLH for electrolysis [92]. In general, the ranges between the highest and lowest costs are smaller for import cases due to limitations across the favorable locations.

This implies lower cost ranges, which constitutes an uncertain conclusion. Most studies focus on techno-economic analyses with a minor focus on extenuating political circumstances. Recent studies such as that by Merten et al. [83] as well as Jensterle et al. [104] take these circumstances into account. However, it is difficult to include these findings in techno-economically-optimized models, which is why researchers advise building connections between the models and political aspects at an early stage, e.g., to synchronize local decarbonization strategies and intentions in the context of export. The range spread of domestic production compared to the ranges of import prices is slightly higher due to differing RE FLH in the north and south Germany, as well as the North and Baltic seas, which leads to differing electricity costs and necessities of conversion.

The reference PC of fossil H_2 based on natural gas, taking into account increasing rates of CO_2 taxation, are 4.5 EURct/kWh$_{H2,LHV}$ for 2020, 5.5 EURct/kWh$_{H2,LHV}$ for 2030, and 7.5 EURct/kWh$_{H2,LHV}$ for 2050 [92]. For optimistic assumptions, the estimated PC of renewable H_2 for import can be rated as competitive in comparison to its fossil alternatives in 2050. For domestic production, on the other hand, a stronger reallocation of CO_2 source costs could ensure competitive price ranges for renewable H_2 compared to fossil H_2. The availability of data and results of H_2 costs for the considered cost levels in the current literature was assessed to be sufficient for this work for a comparison of domestic and import costs. The determined current and future cost structures indicate that domestic H_2 production should be regarded as a possible and competitive pathway, and therefore investigated with equal effort. Aside from end user prices, studies from the Wuppertal Institute [83], Jensterle et al. [104], and Michalski et al. [105] point out the beneficial macroeconomic effects of domestic production. Jensterle et al. [104] also indicate that green hydrogen export regions or countries must match an ideal of low import costs, high RE energy potential, and country-specific political and economic conditions. Furthermore, Terlouw et al. [111] identified favorable H_2 production regions in Europe, which are located in the North and Baltic seas, the south of Spain, the south of France, and the south of Italy, as well as neighboring states. Moreover, it is stated that for production in the EU, a high share of excess energy used is a decisive factor in economic competitiveness.

Future Cost Ranges of SNG from Renewable Sources

The second product investigated in the literature study of future RE carriers is SNG. The results were determined in a similar way to those of H_2 and are shown in Figure 8. With respect to the cost development of end user prices, including taxes and levies (SNG PC + TC + T/L) for the import pathway, cost ranges drop from 23.2 to 42.1 EURct/kWh$_{SNG,LHV}$ in 2020 to 20.3 to 36.1 EURct/kWh$_{SNG,LHV}$ in 2030, and 16.8 to 28.8 EURct/kWh$_{SNG,LHV}$ in 2050. The ranges between the lowest and highest price of imported SNG are 12.0, 15.8, and 12.0 EURct/kWh$_{SNG,LHV}$ for 2020, 2030, and 2050, respectively, which is almost a doubling of the cost ranges compared to H_2 for the same reference years. The larger spreads reveal higher cost insecurities due to longer production pathways, as well as strongly deviating assumptions for costs and the availability of CO_2 sources. For 2050, the lower bound of the SNG PC + TC cost range is 8.8 EURct/kWh$_{SNG,LHV}$ for domestic (offshore wind in the Baltic sea) and 8.0 EURct/kWh$_{SNG,LHV}$ for import cases (the RE source is hybrid wind onshore and PV plus import via pipeline from the optimistic MENA scenario) [89,112]. The upper bound for domestic production and distribution in 2050 is 26.1 EURct/kWh$_{SNG,LHV}$. To achieve this cost value, electrolysis is enabled by the German power grid. The upper bound of import costs is 14.2 EURct/kWh$_{SNG,LHV}$ [108], which indicates that, especially for energy carriers with low transportation costs, production in regions with high RE full load hours can have a positive impact.

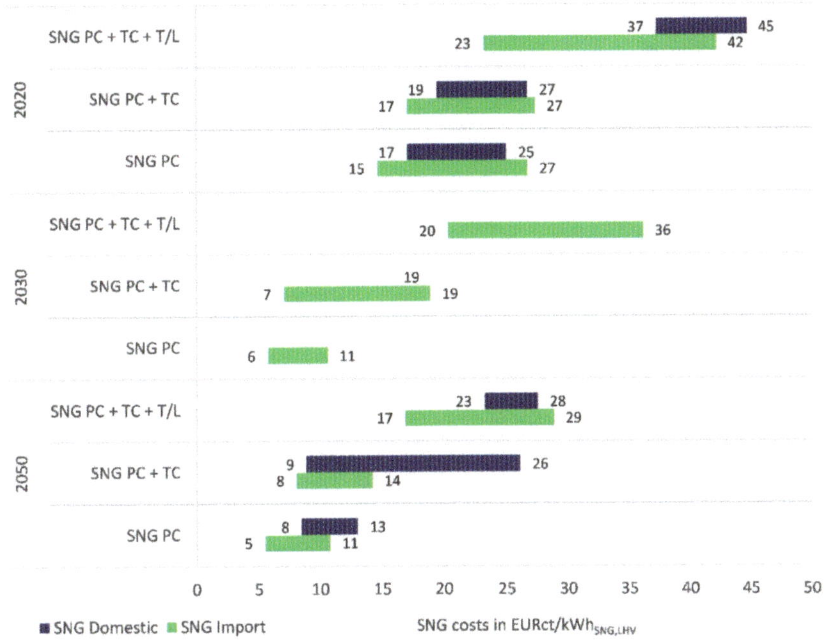

Figure 8. Cost ranges for domestic SNG production and abroad, transport costs, and taxes/levies. Source: Own elaboration based on [85,89,90,92,107–109,113]. All of the extracted cost values are provided in Appendix A in Figure A2. SNG PC: Synthetic natural gas production cost (domestic or import), SNG PC + TC: Synthetic natural gas import cost including transport, SNG PC + TC +T/L: End user prices of synthetic natural gas (domestic and import), including taxes and levies.

Future Cost Ranges of PtL-Fuels from Renewable Sources

The following, final section of the literature study analyses the production cost of PtL products. Eight studies investigate unspecified PtL fuel production, mostly located in North Africa and the MENA region. The results of these studies exhibit similar trends to those seen for SNG, as illustrated in Figure 9. Compared to SNG, PtL shows lower process efficiency in fuel production but also lower transport costs for the products, as they are liquid and chemically similar to currently utilized energy carriers. End user price ranges decrease from 28.2 to 50.9 EURct/kWh$_{LHV}$ in 2020, to 24.6 to 43.4 EURct/kWh$_{LHV}$ in 2030, and 20.1 to 34.4 EURct/kWh$_{LHV}$ in 2050 (see Figure 9). The ranges of PC + TC decrease from 20.0 EURct/kWh$_{LHV}$ to 28.0 EURct/kWh$_{LHV}$ in 2020, to 10.0–20.0 EURct/kWh$_{LHV}$ in 2030, and 8.0–19.1 EURct/kWh$_{LHV}$ in 2050. The PC range of PtL fuel costs in 2050 also decreases to 5.1–13.9 EURct/kWh$_{LHV}$, which is a competitive cost value to energy carriers with significantly higher synthesis efficiencies like H_2. For domestic production, three cost ranges were extracted that show a decreasing cost range for PC + TC from 22.1 to 26.2 EURct/kWh$_{LHV}$ in 2030 to 14.0 to 29.0 EURct/kWh$_{LHV}$ in 2050. Both cost ranges are significantly higher for domestic production compared to imported fuels due to a lower FLH of RES and more conversion losses in the supply chain. The PC cost range for 2050 of 16.1 EURct/kWh$_{LHV}$ to 19.2 EURct/kWh$_{LHV}$ indicates uncompetitive domestic production costs for PtL.

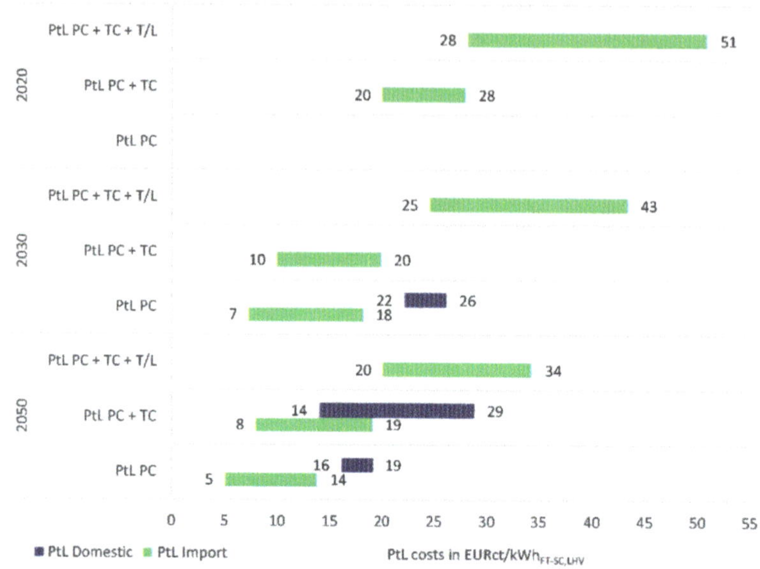

Figure 9. Cost ranges of unspecified PtL fuel production domestic and abroad, transport costs, and taxes/levies. Source: Own elaboration based on [6,85,89,92,105,107,108,112]. All extracted cost values are provided in Appendix A in Figure A3. PtL PC: Power-to-Liquid production cost (domestic or import); PtL PC + TC: Power-to-Liquid import cost including transport; and PtL PC + TC + T/L: End user prices of Power-to-Liquid product (domestic and import), including taxes and levies.

Finally, in order to analyze the observed cost ranges of H_2, SNG, and unspecified PtL fuels, Figure 10 compares the import cost ranges of the different energy carriers. With respect to the PC for all three energy carrier types, strong cost reductions are expected. According to the analyzed studies, H_2 is the cheapest energy carrier, with a cost range of 3.5–10.9 EURct/$kWh_{H2,LHV}$, whereas SNG has a cost range of 5.3–11.8 EURct/$kWh_{SNG,LHV}$, with the unspecified PtL fuels exhibiting a range of 5.1–13.8 EURct/kWh_{LHV}. For PC + TC, these ranges increase to 5.3–11.8 EURct/$kWh_{H2,LHV}$ for H2, 8.0–14.2 EURct/$kWh_{SNG,LHV}$ for SNG, and 8.0–19.2 EURct/kWh_{LHV} for unspecified PtL. The lower bounds of PC + TC converge with 5.1, 8.0, and 8.0 ct/kWh_{LHV}, respectively. For PC + TC + T/L, the ranges further increase to 12.2–18.4 EURct/$kWh_{H2,LHV}$ for H_2, to 16.8–28.8 EURct/$kWh_{SNG,LHV}$ for SNG, and to 20.1–4.4 EURct/kWh_{LHV} for PtL fuels. In 2050, the PC range of SNG and PtL fuels is characterized by a larger spread in comparison to H_2, due to longer process chains with CO_2 input, which strongly differs for different CO_2 sources. Multiple varying assumptions produce the spread of cost ranges. The CAPEX of electrolysis differ and have a strong influence on energy costs. Politically-insecure export regions also lead to higher capital costs and increasing interest rates, from 4% to 12% WACC, due to higher investment risks. This disadvantages SNG and PtL fuel production because of higher total investment costs compared to H_2. The investment costs for DAC technology, if used, further increase the total investment. From an energetic perspective, the increased demand for RE of around 700 kWh_{el}/tCO_2 used for DAC is moderate [92]. However, the CAPEX for DAC technology is high, at 1.416 EUR/tCO_2/a in 2020, and decreasing to 1.033 EUR/tCO_2/a in 2050 [92], resulting in high capital costs for realizing carbon-neutral e-fuels. On the other hand, transportation expenditures, as well as the uncertainty of transport costs for H_2 via pipeline or vessel as a result of the technical developments still required, constitute a weakness of the H_2 value chain.

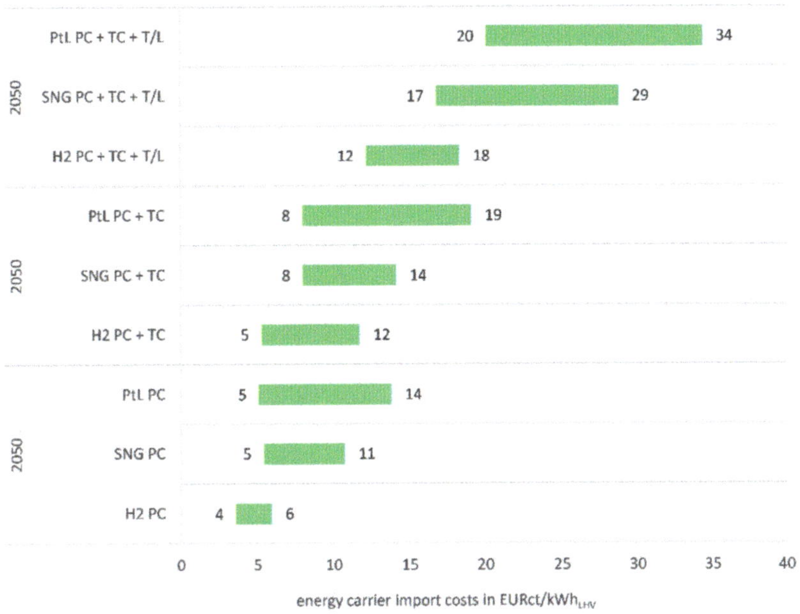

Figure 10. Cost intervals of H_2, SNG, and PtL for the import case. Source: Own elaboration based on literature. Literature sources are listed in Figures 7–9. PtL: Power-to-Liquid, SNG: Synthetic natural gas, PC: Production cost, TC: Transport cost, T/L: Taxes and levies.

3.4. Interim Conclusion

This section showed, that, even if biomass-based fuels are beneficial from an economic view, caused by the limited potential of these, other alternatives are needed additionally. Additionally, the ILUC risk of biomass-based fuels should be considered. The alternative fuels H_2, synthetic or biomass-based CNG/LNG, methanol, DME, MtG gasoline, FT diesel, and kerosene, as well as HVO have been identified in this sections as the best options analyzing the technological readiness of the production pathway as well as the production costs. FT products and synthetic natural gas have the disadvantage of, in comparison, higher production costs. Biomass-based ethanol is also beneficial from this section's analyzed criteria.

Furthermore, the cost analysis showed the insecurities around the regarded cost level production costs, cross-border prices, and end user prices in the current research landscape. Cost insecurity increases with the predicted range and length of the value chains. At present, cost comparisons indicate that lower production costs of H_2 are nearly compensated by higher transport costs in comparison to other fuels that offer existing infrastructural compatibility. Furthermore, the domestic production of H_2 is considered cost-competitive to LH_2 imports.

4. Assessment of Alternative Fuel Utilization

After investigating the production and costs of alternative fuels, the following sections focus on their utilization on the vehicle side. First, existing regulations are reviewed with consideration to maximum official and experimental blending rates. Second, the achievable range of the fuel–drivetrain combination will be reviewed considering lower heating values and TtW efficiencies.

4.1. Drop-In Possibility of Alternative Fuels

Conventional diesel fuels in Europe are regulated by DIN EN 590 [114], whereas conventional gasoline fuels are certified with DIN EN 228 [115]. Sustainable aviation fuels are certified by legal rule ASTM D7566-20b [58]. The Drop-In possibility in the existing vehicle fleet in this study was assessed through blending percentages by legal rules or experiments. Fuel blends that fulfil special legal rules but require the approval of the manufacturer or are not classified in the literature for most vehicles in the fleet are also listed as experimental. The results are shown in Figure 11. The intervals are striped, similar to those for the TRL assessment. Flexfuel and Dual-Fuel concepts are not included in this chapter, as a new vehicle or retrofit of an existing vehicle is necessary.

As previously noted, the blending of sustainable aviation fuels with conventional kerosene is regulated by legal rule ASTM D7566-20b [58]. Sustainable fuels and their common abbreviations are listed in Table 3. Methanol-to-Kerosene processes are limited by the low share of aromatics. However, 100% ATJ-SPK was used in an engine by Schripp et al. [116] and they did not observe major problems during the operation. Meanwhile, fully synthetic Coal-to-Liquid kerosene manufactured by Sasol in South Africa has been certified by the UK Ministry of Defence, Defence Standard 91-91, and has also been recognized by the ASTM D1655 [117,118]. Coal-to-Liquid kerosene should be equal to FT-SPK because both fuels are produced via the FT process. Subsequently, neat synthetic kerosene is fully used in British military aircraft since 2008 [119]. According to Bauen et al. [120], synthetic FT kerosene can, in principle, be used in any blend ratio for jet fuel. Schmidt et al. [62] concluded in their review that 100% Methanol-to-Kerosene could theoretically replace Jet-A1 by 100%, but the process was not commercialized and the product fuel was never subject to the approval procedure of the American Society for Testing and Materials (ASTM).

Table 3. Sustainable aviation fuels and their abbreviations.

Abbreviation	Sustainable Aviation Fuel
FT-SPK	Fischer–Tropsch hydroprocessed synthesized paraffinic kerosene
HEFA-SPK	Synthesized paraffinic kerosene from hydroprocessed esters and fatty acids
HFS-SIP	Synthesized iso-paraffins from hydroprocessed fermented sugars
FT-SPKA/A	Synthesized kerosene with aromatics derived by the alkylation of light aromatics from non-petroleum sources
ATJ-SPK	Alcohol-to-Jet synthetic paraffinic kerosene
CH-SK, or CHJ	Catalytic hydrothermolysis synthesized kerosene
HHC-SPK or HC-HEFA-SPK	Hydroprocessed hydrocarbons, esters, and fatty acid synthetic paraffinic kerosene

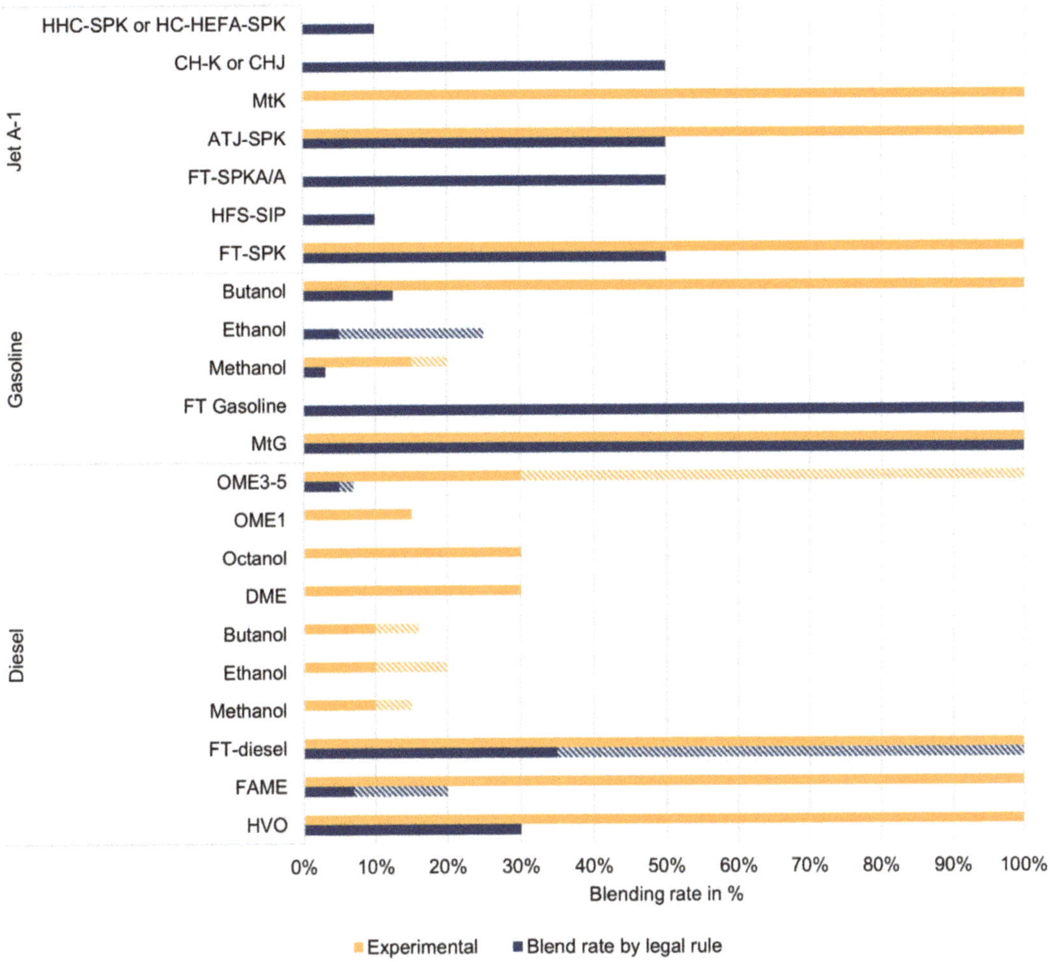

Figure 11. Blending percentages of alternative fuels in diesel, gasoline, and jet A-1. Source: Own elaboration based on [4,58,62,109,114–136]. Sustainable aviation fuel abbreviations: see Table 3. MtK: Methanol-to-Kerosene, FT: Fischer–Tropsch, MtG: Methanol-to-Gasoline, OME: Oxymethylene ether, DME: Dimethyl ether, FAME: Fatty acid methyl ester, HVO: Hydrotreated vegetable oils.

The maximum blending shares are 50% for FT SPK, 50% for HEFA SPK, 10% for SIP, 50% for SPK/A, and 50% for ATJ-SPK [58]. Blending rates of synthetic kerosene obtained via the FT and MtK process are limited through its aromatic compounds. Besides aromatic compounds, the limitation of the oxygen content is another challenge of synthetic kerosene [137].

For gasoline fuel, the highest blends are possible with FT gasoline and synthetic gasoline through the MtG process (see Figure 11). Schemme et al. [4] state that 100% MtG gasoline fulfills EN 228, whereas Bauen et al. [120] state that MtG gasoline can, in principle, be used as a substitute, in any blend ratio, for gasoline. Blending conventional fuel with high shares (100% assumed in Figure 11) of FT-gasoline still fulfills EN 228 according to Kramer et al. [109]. In the case of alcohol-based alternative fuels as blends for conventional gasoline, methanol-based blends of up to 3% or ethanol blends of up to 5% and 10% are covered by DIN EN 228 [115]. Higher blends of up to 15% ethanol or 12.5% butanol are also possible in the United States under the legal rules ASTM D4806-13a [121] and ASTM D7862-13 [122]. The INWL [123] reports that gasoline–methanol blends of up to 15% are usable in conventional engines and up to 20% in modern engines. Wei et al. [124] have researched the impacts of pure butanol in an SI engine in a test rig.

For CI engines, the highest blending permitted by legal rule is possible for FT diesel. According to Kramer et al. [109], blending of up to 35% fulfills EN 590, whereas United States legal rule ASTM D975-20c [125] covers 100% FT diesel utilization. FT diesel is certified in Europe by legal rule DIN EN 15940 [126]. Bauen et al. [120] state that in theory, up to 100% FT diesel could be used as a substitute for conventional diesel. The biomass-based alternative diesel fuel FAME can be blended into diesel at 7% in Germany under DIN EN 590 [114]. FAME is classified in the United States by ASTM D7467-20a [138] and can be blended with conventional diesel in concentrations of 6–20%. In Europe, the blending of conventional diesel with FAME is possible with shares of 20, 30, and 100% under the legal rules DIN EN 16734 [139], DIN EN 16709 [127], and DIN EN 14214 [128], if the blends are approved by the vehicle manufacturer. HVO blending of up to 30% fulfills EN 590 according to Bohl et al. [129]. Meanwhile, Kuronen et al. [130] report usage of 100% HVO in city buses. Up to 10% methanol was also used as a diesel blend in a CI engine by Damyanov [133] and up to 15% by Sayin et al. [131]. Damyanov [133] reports the utilization of diesel–ethanol blends of 10% and 20%. Furthermore, Damyanov [133] describes usage of diesel–butanol blends of 10% and 20%, whereas Sayin [132] reports a 10% share of butanol in a diesel–butanol blend. Bauen et al. [120] state that in the United States, standard waterborne engines with diesel–butanol blends of up to 16% (iB16) have been tested. In the case of octanol, Rajesh Kumar et al. [134] report the use of a 30% octanol–diesel blend. The alternative fuel OME_1 was tested in a vehicle with engine modifications on the road as a 15% OME_1–diesel blend by Continental [135]. OME_{3-5} as a diesel blend fulfills EN 590 with a maximum OME_{3-5} share of 5–7% [136]. However, Beidl et al. [136] report OME_{3-5}-diesel blends with OME_{3-5} shares from 30–100%.

Overall, blending rates of up to 100% are possible for synthetic aviation fuels, although these are limited to 50% by legal regulations. Based on the literature reviewed in these sections, blending of up to 100% synthetic gasoline from the MtG or FT processes in the existing vehicle fleet is possible (see Figure 11). High blending percentages in conventional diesel fuels can also be achieved with FT diesel, HVO, and FAME.

4.2. Heating Value of Alternative Fuels

The reachable mission range of a vehicle primarily depends on TtW or tank-to-propeller efficiency, as well as the gravimetric and volumetric energy density of the fuel and engine load behavior of the respective vehicle. Efficiency as a function of the engine load varies widely for the different drive systems. In this section, the heating value of the different alternative fuels will be discussed. The TtW efficiency of different vehicles in road, air, rail, and inland waterway transport will be discussed in dedicated sections for each mode of transport.

Figure 12 shows the gravimetric and volumetric energy densities of different conventional and alternative fuels. Aside from the energy density, the TtW efficiency is equally important for the possible mission range of vehicle applications. As the TtW efficiency strongly depends on the application and engine loads, it will be discussed separately for each vehicle class in the upcoming sections. Figure 12 depicts the superiority of liquid fuels with respect to volumetric energy density, whereas hydrogen has by far the highest gravimetric energy density, with 120 MJ/kg. The highest volumetric densities feature the heavy fuel oil, Jet A-1, and diesel with 39–43 GJ/m^3, 33–36 GJ/m^3, and 36 GJ/m^3. Biodiesel, synthetic FT diesel, synthetic MtG gasoline, liquified petroleum gas (LPG), and the higher alcohols octanol and butanol are all in the range of 25–34 GJ/m^3, whereas the volumetric energy density of the remaining liquid fuels LNG, OME_x, DME, gasoline E85, and the lower alcohols ethanol and methanol are within the range of 16–23 GJ/m^3. Methanol has the lowest volumetric energy density of the liquid fuels at 16 GJ/m^3 and the second lowest gravimetric energy density. Only OME_{3-5} has a lower gravimetric energy density in the field of liquid fuels, with about 19 MJ/kg. Liquid hydrogen, which is a gaseous fuel but stored in liquid form, has a volumetric energy density of 8 GJ/m^3. For gaseous fuels, the gravimetric energy density strongly depends on the pressure. CNG has a volumetric energy density in the range of 36 MJ/m^3–9 GJ/m^3, whereas hydrogen has a volumetric energy density of 10 MJ/m^3–5 GJ/m^3. However, batteries have the lowest gravimetric and volumetric energy densities, with 0.94 MJ/kg and 3 GJ/m^3, respectively. Even the batteries at the research level of development have, with 1.71 MJ/kg and 5 GJ/m^3, low values. In particular, the high gravimetric energy density suitable for applications like air transport is strongly limiting, which will be discussed in detail in the air transport section.

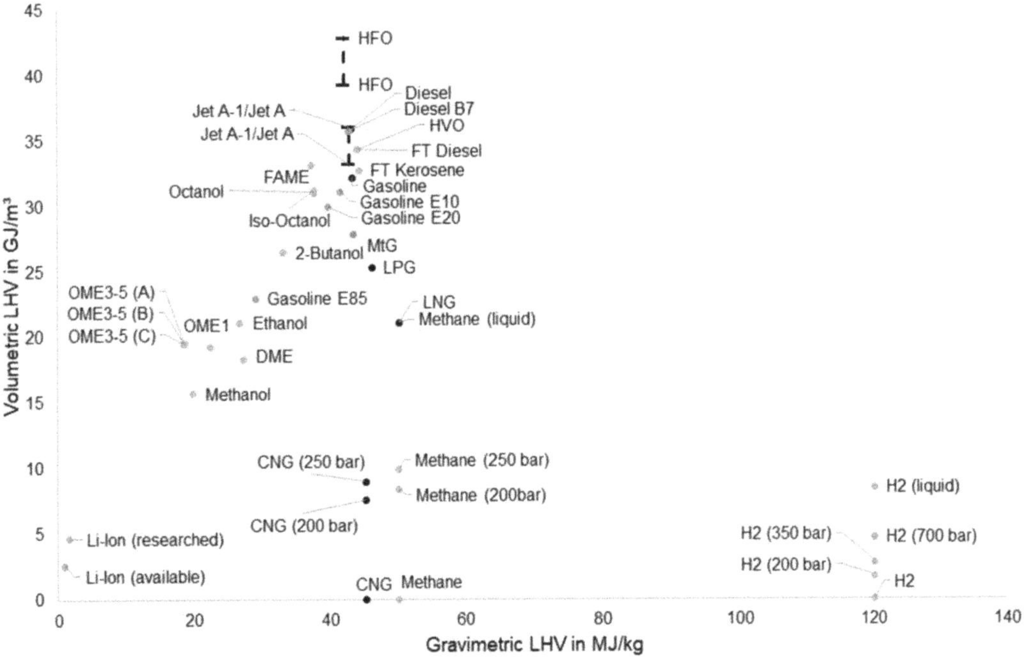

Figure 12. Lower heating value of alternative fuels. Source: Own elaboration based on [97,140–150]. HFO: Heavy fuel oil, FAME: Fatty acid methyl ester, HVO: Hydrotreated vegetable oils, FT: Fischer–Tropsch, MtG: Methanol-to-Gasoline, LPG: Liquefied petroleum gas, LNG: Liquefied natural gas, OME: Oxymethylene ether, DME: Dimethyl ether, CNG: Compressed natural gas, LHV: Lower heating value.

4.3. Drivetrain Efficiencies

In this section, the maximum TtW efficiencies and, if possible, the TtW energy consumption of different applications, will be discussed. Table 4 provides a general overview of the TtW efficiencies of different powertrains. In general, BEVs have the highest efficiency, at 81–95% [109,151], followed by fuel cell–electric vehicles with 49–62%. Aircraft turbines have maximum efficiencies of about 50% [152], whereas ship engines can achieve around 44–56% [153]. Low efficiencies can be found in small two-stroke engines with about 24% [153]. Notable is also the increasing consumption with lower temperatures. The consumption of BEVs increases in winter by about 50%, whereas that of conventional engines only increases by about 10% [154]. However, the stated efficiencies are maximum efficiencies. Efficiencies during utilization of the powertrains will strongly depend on the field of application. The different propulsion systems have strongly differing part load behaviors. The fuel cell system has its maximum efficiency with 10–20% load, which slightly decreases after that point until 100% load [155]. Diesel engines have low efficiencies that increase with increasing loads and a maximum efficiency at approximately 100% load [97]. The efficiency–load curve of electric engines is similar to that of fuel cell systems, with the difference being that the efficiency of electric engines will not decrease with increasing load but remain constant [156].

Table 4. Efficiencies of different drivetrains.

Drivetrain	Fuel	Field of Application	Max. Efficiency
Otto engine	Gasoline	Motorcycle	31% [153]
		Passenger cars, commercial vehicles	36% [109,153]
	Gasoline 2 stroke	Small engines	24% [153]
	LPG	Commercial vehicles	36.5% [109]
	CNG	Commercial vehicles	37% [109]
Diesel engine	Diesel	Passenger cars	43% [153]
	Diesel	Commercial vehicles, trucks	42–45% [109,153]
	LNG	Commercial vehicles	42% [109]
	FT diesel, HVO, OME_{3-5}, DME	Commercial vehicles	42% [109]
Large diesel engine with high RPM	Diesel	Shipping	44% [153]
Diesel engines with medium RPM	Diesel	Shipping	45% [153]
Cross-head engines	HFO 2 stroke	Shipping	54% [153]
Turbines	Jet fuel	Air transport	50% [152]
Battery–electric	Electric energy	-	81–95% [109,151]
Fuel cell–electric	H_2	-	49–62% [a]

Notes. [a] Polymer electrolyte fuel cell with 60–65% efficiency [155,157,158]; electric system with 81–95% efficiency [109,151]. Source: Own elaboration based on [109,151,153,155,157,158]. LPG: Liquefied petroleum gas, CNG: Compressed natural gas, LNG: Liquefied natural gas, FT: Fischer–Tropsch, HVO: Hydrotreated vegetable oils, OME: Oxymethylene ether, DME: Dimethyl ether, HFO: Heavy fuel oil, RPM: Rounds per minute.

The consumption values of the different vehicle classes, therefore, depend, on the one hand, on the operating behavior of the vehicle class and, on the other, on the load behavior of the vehicles. The following sections discuss the efficiencies of the different technology options when used in road, inland waterway, rail, and air transport.

4.3.1. Road Transport

Figure 13 shows the energy consumption of different drivetrain–fuel combinations for passenger cars (PC), light duty vehicles (LDV), heavy-duty vehicles (HDV), buses and coaches based on Helgeson and Peter [32] and HBEFA 4.1 [159]. These consumption values are averages and include the average operating behavior of the different vehicle classes. The consumption values are always referenced to the diesel value of the respective class, with motorcycle values being the exception. These are referenced to the motorcycle gasoline value. Gasoline-powered light duty vehicles are excluded in Figure 13. As can be seen in Figure 13, the battery–electric drivetrain features the highest consumption reduction, at 50–65% compared to conventional diesel engines for coaches, buses, PC, LDV, and HDV, whereas electric motorcycles (MC) feature an 84% consumption reduction compared to conventional gasoline-powered MC. Fuel cell–electric vehicles offer a consumption reduction of about 40% for PC, LDV, and HDV, and about 22% for coaches and buses. This difference could also arise from the different literature sources. For coaches and buses, the reduction was calculated on the basis of EURO VI diesel vehicles in HBEFA 4.1 The utilization of CNG or LNG in a gas engine leads to a higher general energy consumption [160]. For LDV the energy consumption increases to about 42%, whereas for PC, coaches, buses, and HDV, the consumption increases by around 13–24%. The utilization of dual-fuel, high pressure direct injection diesel engines, using CNG or LNG and a small amount of diesel, leads to a 3% increase in energy consumption. Hybrid technologies have the potential to reduce energy consumption, but this reduction strongly depends on the electric mode driven range and also on the vehicle's consumption profile. Drawing on values calculated for 2020 by Helgeson and Peter [32], plug-in hybrid drivetrains could reduce energy consumption by 20–43% for diesel-hybrid vehicles and by 31–50% for Otto-hybrid ones. For gas-engine hybrid vehicles, the reduction potential is between 28% and 72%. The operation of fuel cell electric vehicles as plug-in hybrid vehicles can reduce the energy consumption by about 15%. For BEVs, the energy consumption is primarily influenced by the maximum mission range and battery size, respectively. Weiss et al. [161] state that the energy consumption of BEVs is primarily influenced by vehicle mass. According to Weiss et al. [161], a 10 kWh increased battery capacity leads to an increase in PC mission range of 40–50 km, whereas energy consumption increases by 0.7–1 kWh/100 km. Helgeson and Peter [32] forecast the development of different drive systems for PC, LDV, and HDV. They forecast a strongly decreasing energy consumption for CNG, gasoline, and diesel, as well as CNG-hybrid-powered PC. For LDV, the energy consumption of diesel, diesel-hybrid, and fuel cell drivetrains decreases, whereas for HDV the energy consumption of the diesel engine and the CNG/LNG-powered dual-fuel engine is decreasing.

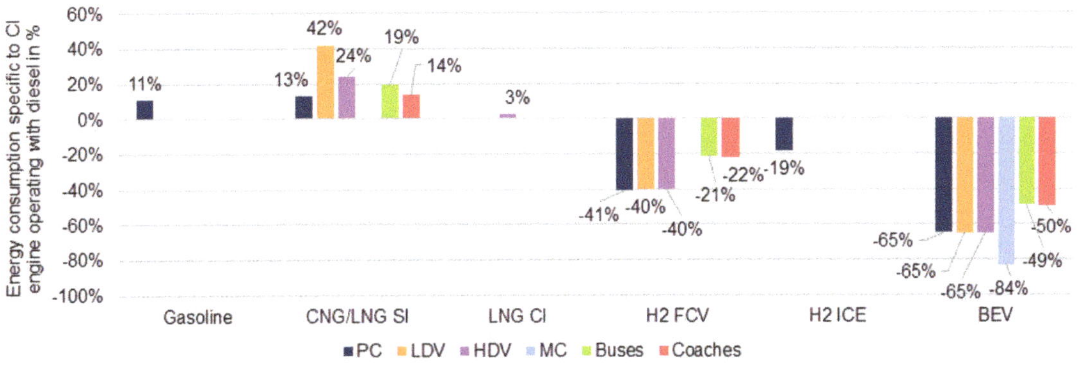

Figure 13. Energy consumption of different powertrains for passenger cars (PC), light duty vehicles (LDV), heavy-duty vehicles (HDV), motorcycles (MC), buses, and coaches. Source: Own elaboration based on Helgeson and Peter [32] and HBEFA 4.1 [159]. The consumption values are relative to the diesel consumption of the respective vehicle class. The motorcycle values are not diesel- but gasoline-specific. CNG: Compressed natural gas, LNG: Liquefied natural gas, SI: Spark ignition, CI: Compression ignition, FCV: Fuel cell vehicle, ICE: Internal combustion engine, BEV: Battery–electric vehicle, PC: Passenger car, LDV: Light duty vehicle, HDV: Heavy duty vehicle, MC: Motorcycle.

4.3.2. Inland Waterway Transport

In inland waterway transport, mostly diesel engines with higher RPMs are used. These engines have a maximum efficiency of 44% [153], whereas electric drivetrains have maximum efficiencies of 85% [109]. These values were used to estimate energy consumption. For LNG, diesel-hybrid and diesel with selective catalytic reaction (SCR) and diesel particulate filter (DPF) after treatment were also no class-specific values used. Otten et al. [162] state that diesel ships with SCR/DPF consume 1% more energy in comparison to conventional ships classified by the old emission regulation of the Commission for the Navigation of the Rhine (CCNR2), whereas diesel-hybrid ships consume 5% less. Efficiencies for inland waterway fuel cell ships were obtained from Zerta et al. [155].

The resulting energy consumption reductions are presented in Figure 14. The fuel cell drivetrain exhibits the largest energy consumption reduction for day-trip ships (47%) and cabin vessels (41%), and smaller reduction values for pushed barges/tankers (29%) and cargo barges/liquid cargo barges (15%). A possible explanation for this could be the load management of the different ship classes. The fuel cell drive has its peak efficiency at about 20% load and then slightly decreases linearly until 100% load is reached, whereas conventional diesel has low efficiencies with low loads and a peak efficiency of around 40%, with high loads close to the maximum [155].

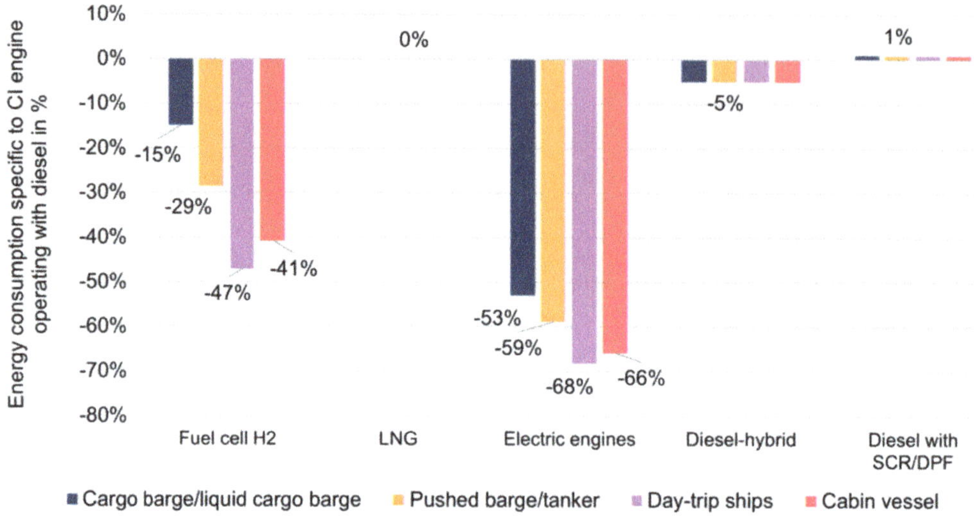

Figure 14. Energy consumption specific to conventional CI engines running with marine gas oil. Source: Own elaboration based on [109,153,155]. LNG: Liquefied natural gas, CI: Compression ignition, SCR: Selective catalytic reaction, DPF: Diesel particulate filter.

Similar reduction values are shown in Figure 14 for the energy consumption of electric engines. Day trip ships and cabin vessels exhibit the largest reductions with 66% and 68%, respectively, followed by pushed barges/tankers and cargo barges/liquid cargo barges with 59% and 53%, respectively. The assumption of a constant efficiency of 85% for battery propulsion is likely appropriate, as the efficiency–load curve of electric motors >75 hp is already close to maximum efficiency at low loads [156]. According to Otten et al. [162], ships running on LNG will probably have roughly the same energy consumption as conventional CI engines running on marine gas oil. This assumption seems fairly optimistic, as according to the studies of Bünger et al. [160], modern internal SI combustion engines in gas operation can achieve maximum efficiencies of 90–95% relative to the efficiency of a comparable diesel engine.

4.3.3. Rail Transport

In rail transport, in addition to conventional operation with catenary lines or diesel-electric drivetrains, there are three alternative drive systems: (1) Diesel–electric hybrids; (2) battery–electric hybrids; and (3) fuel cell–electric. One alternative for rail transportation is the diesel–battery hybrid drivetrain mentioned above. Alstom developed the Alstom H3 shunting locomotive with a diesel engine and an additional battery. The battery is charged during low-load operation and provides energy during high-load operation, which reduces energy consumption by up to 50% [163]. A similar technology is utilized in electric trains with additional batteries. These additional batteries enable the trains to operate in battery–electric mode on rail sections without catenary lines. The batteries are loaded on sections equipped with catenary lines. Bombardier developed the prototype, BOMBARDIER TALENT 3, which is produced in low production volumes and has an electrical mission range of 40 km [164]. 21 trains of that type are, for instance, used on three train lines in Germany [165]. With the next generation, the mission range should increase by up to 100 km [164].

Another hybrid technology is the diesel–air pressure technology, where brake energy is saved in the form of compressed nitrogen. This technology offers a lower energy consumption and less additional weight compared to the other mentioned hybrid technologies.

However, the maximum lifetime of the pressure tanks is 25 years [166]. This technology will not be investigated further in the following.

Hydrogen fuel cell trains are another promising technology for the rail transport sector. The fuel cell pilot train Alstom Coradia iLint is already operating in Lower Saxony, Germany [167]. It has a mission range of 1000 km and approximately the same weight as a conventional diesel train [167]. For comparison, the mission range of conventional diesel passenger trains, such as the Alstom Lint 54 and Alstom Lint 41, is about 1200 km [168]. Plank-Wiedenbeck et al. [168] reported that the Alstom Corodia iLint consumes about 0.18–0.25 kg_{H2}/km and a conventional equal diesel train about 1–1.8 l/km. This results in 7173 Wh/km for the hydrogen train and 13,945 Wh/km for the diesel one. The Alstom Corodia iLint is mostly comparable to the train class S-Bahn. Deutsche Bahn AG [169] reported the energy consumption of an electric and diesel S-Bahn to be 22 Wh/Seat-km and 60 Wh/Seat-km, respectively. In total, this enables a 63% energy consumption reduction for electric trains and a 49% reduction for fuel cell ones. As already noted, diesel-electric hybrid trains have the potential to reduce fuel consumption by 50%, but this only applies when using them as shunting trains. The reduction in energy demand of the different alternatives is summarized in Table 5, below.

Table 5. Reduction of energy demand in rail transport through alternative propulsion technologies.

Propulsion System	Energy Consumption Reduction
Diesel–electric hybrid	50% [163]
Battery–electric hybrid	63% based on [169]
Fuel cell–electric	49% based on [168]

Notes. Source: Own elaboration based on [163,168,169].

4.3.4. Air Transport

As a result of the low technical maturity of the alternative propulsion systems for air transport (see Roland Berger [170] and McKinsey & Company [28]), studies on energy consumption amongst aircraft are limited. Seeckt and Scholz [171] compared the combustion of conventional kerosene and hydrogen in the aircraft classes jets and turboprops. Using hydrogen in a turboprop aircraft reduced energy consumption by 5%, whereas it increased energy consumption in the jet by about 3%. According to Roland Berger [170], fuel cell propulsion systems have efficiencies about 45–50%, combined with a 55% fuel cell efficiency and 90% electric powertrain efficiency. In contrast, hydrogen combustion propulsion systems have a 40% efficiency [170]. However, other sources combine a 65% fuel cell efficiency with an 85% electric powertrain, leading to a 55% fuel cell system efficiency (see Table 4). Aircraft turbines have an efficiency of about 50% [152]. As a result of the lack of information regarding energy consumption, air transport TtW efficiency is assessed by the evaluated engine efficiencies, which are summarized in Table 6.

Table 6. Efficiencies of different propulsion systems for air transport.

Drive	Efficiency
Fuel cell–electric	45–55% [170]
Electric	82–90% [170]
Hydrogen combustion	40% [170]
Kerosene combustion	50% [152]

Notes. Source: Own elaboration based on Roland Berger [170] and Bräunling [152].

4.4. Interim Conclusion

This section shows the high compatibility of the alternative fuels FT diesel, FT kerosene, and MtG. The same accounts for FAME, HVO, and others via the FT process produced aviation fuels with already existing vehicles. It also highlights the necessity of the ASTM approval of higher blending rates for FT and MtK kerosene.

The analysis of the heating values highlights the superiority in the case of the volumetric energy density of the liquid fuels. Liquid fuels, which match with the first selection of fuels in Section 3 are FT diesel, MtG gasoline, HVO, FT kerosene, synthetic or biomass-based LNG, DME, and methanol. The volumetric energy density of the promising fuels synthetic or biomass-based SNG and H_2 is lower, while the energy density of the batteries is by far the lowest.

Furthermore, it was shown, that the drivetrain efficiencies of the battery–electric drivetrains are always the highest in the considered four sectors, followed by the fuel cell–electric drivetrain. Natural gas-based internal combustion engines lead to higher energy demand in road transport. It was also shown, that the highest energy consumption reduction can be achieved using electric drivetrains in vehicle classes, which have a high share of dynamic operation. These are motorcycles, passenger cars, light duty vehicles and buses in road transport and the passenger transport in inland waterway transport.

5. Environmental Impacts of Promising Alternative Fuels

In addition to the techno-economic assessment, alternative fuels still require environmental assessments to realize their potential and identify sustainable pathways. In Germany, the transport sector accounted for around 150 million t CO_{2eq} of emissions in 2020, despite less travel due to the COVID-19 pandemic [172]. This corresponds to a share of just under 20% of total greenhouse gas emissions in Germany. After the energy industry, the transport sector is, therefore, the largest emitter of greenhouse gases. Against the backdrop of climate change, the finite nature of fossil resources, and damaging local environmental impacts, the transport sector must therefore change significantly from its previous structure. For example, the tightening of carbon dioxide fleet limits by the EU at the beginning of 2019 has prompted numerous car manufacturers to focus more on alternative drivetrain concepts in their portfolios [173]. Not only new drive concepts but also alternative fuels are gaining importance in Germany, as well as globally, due to the outlined ecological imperatives. The global production of biofuels, therefore, reflects continuous growth since 2006 [174]. In addition to biofuels, electricity-based fuels and vehicle concepts offer the possibility of reducing the environmental impacts of the transport sector. In order to be able to make statements regarding the extent to which the substitution of fossil fuels with alternatives can reduce harmful environmental impacts, numerous analyses have been carried out in recent years. Some environmental assessments of fuel supply have been designed as so-called well-to-tank (WtT) studies. Another and very well-established method for the ecological analysis of fuels is the life cycle assessment (LCA), which is an environmental assessment procedure that is also standardized by DIN ISO 14040 [175] and 14044 [176].

A study by Moro and Helmers [177] clearly addresses the differences between WtT and well-to-wheel analyses, as well as LCAs. Therefore, WtT is understood as the energy input for the production, transportation, and distribution of fuels. Thus, these studies focus on the fuel supply pathway alone. However, emissions and the environmental impacts caused by them that relate to the construction and disposal of exploration, energy conversion, and vehicle technologies are left out of consideration. The WtT analysis is therefore characterized by the fact that it represents a relevant subset of the environmental impacts of fuels, as well as being easier to prepare and interpret. In an LCA, these other elements of the fuel supply can also be taken into account.

Literature reviews have been published focusing on both forms of environmental assessments of fuels, addressing results and trends from a wide range of earlier publications.

In the present study, the alternative fuels H_2, CNG/LNG, methanol, DME, MtG gasoline, FT diesel, and kerosene, as well as HVO, have been considered promising from a multi-layered techno-economic perspective. Thus, the key and general findings on environmental performance described below refer to these fuels.

The literature overview presented herein includes the environmental impacts of promising alternative fuels considering LCA results, as well as the WtT results with regard to the impact category climate change and the indicator of global warming potential (GWP 100a) in kg CO_{2eq}. For each result, it is clearly described what form of the two environmental assessment methods is being referred to. The focus is clearly on the consideration of LCA results. Nevertheless, some WtT studies provide interesting and valuable additional insights. For further environmental impact categories, only LCA results were taken into account. In addition to the results for climate change, the present overview of environmental impacts includes the following further impact categories and indicators: the category of acidification and the indicator of acidification potential (AP) in g SO_{2eq}/MJ; the category of eutrophication and the indicator of eutrophication potential (EP) in g PO_{4eq}/MJ; the impact category of summer smog and the indicator of photochemical ozone creation potential (POCP) in g C_2H_{4eq}/MJ; the category of ozone depletion and the indicator of ozone depletion potential (ODP) according to g CFC-11eq; and, finally, the impact category particulate matter in this case, with the corresponding indicator < 10 μm (PM10).

An overview of the WtT GWP results of different bio- and electricity-based fuels from 94 publications was presented in a study by Naumann et al. [174]. Drawing on the large number of results presented in the literature review, Figure 15 is limited to a subset of the fuels identified as promising in this study.

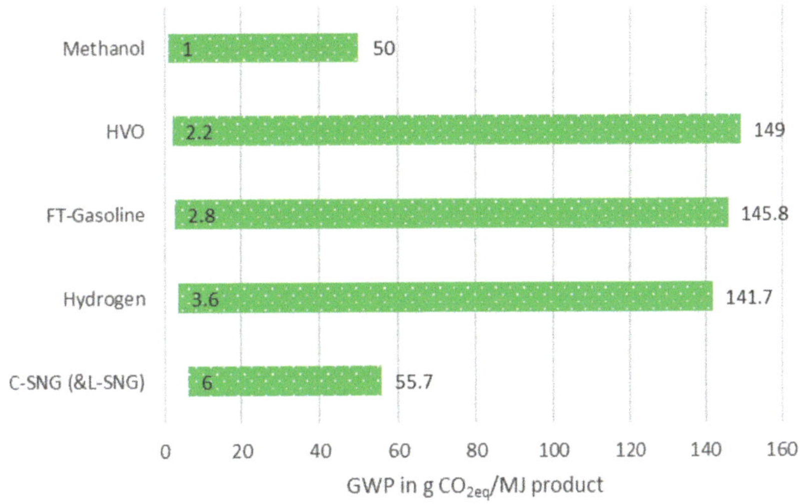

Figure 15. Ranges of GWP results of WtT assessments and promising selected fuels. Source: Own elaboration based on [174]. HVO: Hydrotreated vegetable oils, FT: Fischer–Tropsch, C-SNG: Compressed synthetic natural gas, L-SNG: Liquefied synthetic natural gas, GWP: Global warming potential, WtT: Well-to-Tank.

According to Naumann et al. [174], the wide range of environmental impacts shown in Figure 15 is due to data- and process-related factors, as well as methodological differences. However, clearly differentiated statements for the cause of the range of environmental impacts are lacking. Accordingly, on the basis of this literature analysis, it can be stated that large differences are possible depending on the design of the fuel supply and the assumptions made. However, trends regarding the possible advantages of fuels should not be directly derived from this. In order to obtain more differentiated findings in the manipulated variables of the environmental impacts of these fuels, the findings from the LCA literature reviews related to these alternative fuels were used in the next step, outlined below.

Koj et al. [178] identified 32 LCA concerning Power-to-X (PtX) fuels and pathways and evaluated them in a review study. In the study, a clear focus of earlier LCA studies of gaseous PtX fuels (i.e., H_2 and SNG) was highlighted. The origin of the electricity used for the PtX pathways and the carbon dioxide source and capture processes were identified as crucial factors in mitigating environmental impacts. Amongst others, the literature results regarding electrolytic hydrogen production showed that there is an obvious impact of electricity supply on GWP. The use of electricity generated using from RE sources was shown to have significantly lower impacts than the use of conventional hydrogen production processes. The conventional hydrogen production technologies (steam reforming from natural gas or petroleum) led to GWP results of between 90 and 120 g CO_{2eq}/MJ, whereas the results of electrolytic hydrogen production with wind energy could reach values of less than 20 g CO_{2eq}/MJ. In studies where a region- or country-specific grid electricity mix was considered, large differences in environmental effects were shown depending on the geographical conditions and different associated shares of renewable and fossil-based electricity generation. Depending on the year under consideration and further assumptions, an analysis of hydrogen production based on the EU electricity mix revealed GWP results up to nearly 270 g CO_{2eq}/MJ [179]. Therefore, from an environmental point of view, the use of electricity mixes with high specific emission factors for the production of hydrogen and fuels based on it should be avoided. In addition, the literature review includes evaluations not only for the provision of hydrogen fuel but also for the use of various fuels in passenger cars. This also includes the shares of fuel production of the electricity-based fuels SNG, hydrogen, and electricity for electric transportation in the context of total environmental impact. It also turns out that electricity inputs with the lowest possible GHG emissions should be used for fuel production. Otherwise, the contributions of fuel production to the total environmental impacts can be well over 80% and lead to no relief compared to the fossil reference [180]. For considerations up to the year 2050, it is shown that if a completely renewably-generated electricity supply is used, the contribution of fuel production to the environmental impacts per kilometer can be reduced to below 10% [181]. In addition, the literature review highlights the range and inconsistency of methodological assumptions amongst previous studies [178].

A similarly-designed review study of LCAs of PtX fuels was published by Kigle et al. [182]. The authors explicitly limited the PtX approach to consider those LCA publications that analyzed electricity-based fuels for application in the transportation sector. The literature review ultimately encompassed 23 publications. Hydrogen as a fuel was frequently analyzed in this literature review as well. The significance of the electricity input is also addressed herein. For all studies that distinguished between RE and the electricity mix as inputs in the production of electricity-based synthetic fuels, a consistent trend emerged. GHG emissions for fuels based on the electricity mix were several times higher than for fuel production based on RE. The example of methanation for SNG production was used to address the various options for accounting for the carbon dioxide required for the reaction. One of the publications considered in the review calculated a carbon dioxide credit for the carbon dioxide used in the methanation process [183]. In another study [179], the credit for carbon capture was given to the industrial plant upstream of the methanation process and operated with fossil fuels. In this instance, therefore, the environmental impact of the product of the industrial plant was reduced, but not that of the SNG. This methodological aspect is referred to as allocation and is understood as the assignment of the incoming and outgoing material and energy flows of a process or product system to the product system or technology under investigation in the case of multi-output processes [175,182].

5.1. Hydrogen

Of the promising alternative fuels considered, hydrogen production has been subject to the most environmental analysis. For example, a recent review study of the ecological effects of using hydrogen as a fuel for road transport was able to identify 72 relevant studies featuring WtT or LCA results [184]. In that literature review, a range of GWP values for WtT studies and the reference MJ could be identified. The GWP of WtT ranges from 1.6 gCO_{2eq}/MJ to 218 g CO_{2eq}/MJ, with a mean of 39.6 g CO_{2eq}/MJ, were identified. The different environmental impacts of the production of hydrogen as a fuel can be caused in particular by the different pathways and technologies used for this purpose. For instance, Wulf et al. compared ten different hydrogen pathways and identified the advantageous environmental performance of electrolyzers based on proton exchange membrane electrolysis cells (PEMECs) and alkaline electrolysis cells (AECs) [185] if operated with renewable electricity in five of the six impact assessed categories. Due to these promising ecological results and the possibility of integrating renewable electricity, the technologies, and processes of water electrolysis, in particular, are considered as a form of hydrogen production in the context of future alternative fuels. The significance of the type of electricity supply for the environmental impacts of electrolysis technologies has already been highlighted by numerous individual publications and in the preceding sections based on the findings of the cited review studies. This also means that the different power consumption of the various electrolysis technologies leads to different environmental impacts. With respect to operation, water must also be noted as an important input. However, the provision of treated water only makes a negligible contribution to the environmental impacts of the GWP indicator. In addition, different material compositions are used and developed, especially for the electrolysis cells, which can result in different environmental impacts. Another factor that influences the GHG emissions caused by component manufacturing is also the lifetime of the components used. However, the contribution of component manufacturing and plant construction on the GWP indicator tends to be almost negligible. However, other environmental impacts, such as ozone depletion potential, can be very strongly influenced by component manufacturing if, for instance, polytetrafluoroethylene (PTFE) is used in the manufacturing of seals. For example, as a result of PTFE, the manufacturing phase was responsible for over 90% of the ozone depletion potential (ODP) environmental impacts for a 6 MW alkaline electrolyzer evaluated by Koj et al. [186].

Liebich et al. [9,10] discussed additional environmental impact categories for electrolytic hydrogen production in the year 2050. With respect to acidification, potential values of between 0.043 and 0.060 g SO_{2eq}/MJ were obtained, which lie close to the value of the fossil reference 0.049 g SO_{2eq}/MJ. The main contributions were given by upstream electricity generation, due to the steel, copper, and aluminum production necessary for power plant production. Additionally, for eutrophication, upstream electricity generation and especially the underlying use of steel, copper, and aluminum are responsible for most of the impacts. The eutrophication values significantly exceed those of the fossil reference with values of around 0.023 g PO_{4eq}/MJ compared to 0.0090 g PO_{4eq}/MJ. With respect to summer smog, the electrolytic hydrogen production revealed lower values (around 40% of the fossil reference). Furthermore, for particulate matter, electrolytic hydrogen production revealed lower values of around one-third compared to the fossil reference.

Hydrogen is characterized by its capability of being stored geologically prior to fuel use, if needed. This storage can be performed, for example, in salt caverns. Koj et al. [181] found that this type of storage is only responsible for minor contributions compared to other steps in the process chains.

Germany is considered a hub of electrolysis operation in numerous LCA studies (e.g., [186–188]). A clear environmental impact reduction is also shown in these publications for electrolysis operation with renewable energy compared to operation using the electricity mix. In addition, it was shown that Germany has a location disadvantage compared to other countries with respect to electrolysis operation when certain electricity inputs are employed. When operating electrolysis with the electricity mix, Austria and Spain perform significantly better by comparison [186]. When comparing the GWP when operating with PV electricity, the GWP for Portugal is over 40% lower than the impacts of electrolysis operation using PV electricity in Germany due to superior solar radiation conditions [188].

5.2. C-SNG/L-SNG

Methanation to produce SNG, often referred to as Power-to-Methane (PtM), from hydrogen and carbon dioxide is one of the most frequently analyzed PtX technologies from an environmental perspective. The PtX LCA review of Kigle et al. [182] identified six publications with LCA considerations of SNG [179,181,183,189–191]. In the LCA review by Koj et al. [178], as many as 26 publications were identified that address the environmental impacts of SNG production.

The environmental impact results of a study by Liebich et al. [9,10] have a high relevance due to the scope on Germany, assessed variants and impact categories, as well as its actuality. Furthermore, it is one of the few studies that consider SNG liquefaction (L-SNG). The study compared 12 SNG production pathways. Furthermore, different geographical scopes (Germany, Saudi Arabia, and Morocco), as well as different electricity and carbon dioxide sources and transport options, were considered. The study found for a retrospective consideration of the year 2015 typically highest contributions of electricity generation for hydrogen production on global warming potential results [9,10]. For the retrospective assessment of the year 2015, global warming potential results reach from around 10 g CO_{2eq}/MJ to over 300 g CO_{2eq}/MJ. The lowest value is achieved with German wind onshore as a power source, while the highest impacts are provoked by a pathway based on the German electricity mix. Compared to a fossil reference (natural gas incl. upstream) with a global warming potential of 63 g CO_{2eq}/MJ, the results of the best pathway in 2015 corresponds to a reduction of almost 85%. Comparing C-SNG transported by pipeline with L-SNG and the transport by liquefied natural gas tanker revealed advantages for C-SNG. Taking methane emissions during the liquefaction and regasification into account, L-SNG transport provokes around 10 g of CO_{2eq}/MJ for the L-SNG pathway. Liquefication and regasification do not necessarily have to be considered as a component of the L-SNG transport. In the JRC WtT study [192] these process steps are considered as a conditioning and distribution step of L-SNG provision. Additionally, the JRC WtT study [192] confirms a higher global warming potential for L-SNG compared to C-SNG. An assessment of the year 2050 by Liebich et al. [9,10] shows significantly lower impacts compared to the retrospective assessment of 2015. The calculated prospective global warming potential results reach from around 6 to 17 g CO_{2eq}/MJ, with most pathways showing results in the range 7 to 15 g CO_{2eq}/MJ. Thus, a reduction of 90% would be possible compared to the fossil reference.

The prospective global warming potential obtained by the LCA of Liebich et al. [9,10] (6 to 17 g CO_{2eq}/MJ) is in the same range as the WtT values in the literature review of Naumann et al. [174] (see Figure 15).

With regard to different life cycle stages, the plant construction (electrolysis, carbon capture, and methanation plant) shows lower global warming potential contributions than the contributions related to plant operation [184,185].

Liebich et al. [9,10] assessed environmental impact categories beyond global warming potential for SNG production. Acidification potential values calculated for the year 2050 reach from 0. 075 to 0.26 g SO_{2eq}/MJ. The reference value for fossil natural gas is 0.032 g SO_{2eq}/MJ. Thus, the lowest acidification potential of the electricity-based SNG production is 130% higher than the fossil reference value. Main contributions to the acidification impacts were given by the electricity provided to electrolysis, especially steel, copper, and aluminum production for the power plants, and consumable production for electrolyzers. The eutrophication potential values vary between 0.016 and 0.073 g PO_{4eq}/MJ. Compared to the fossil reference this is in minimum 900% higher than for natural gas as reference. Main contributions to eutrophication are given by the electricity for hydrogen production and for pathways with carbon capture from the lignite-fired power plant the required energy input. Again steel, copper, and aluminum production for the upstream power plants are primarily responsible for the eutrophication potential related to the electricity used for electrolysis. The summer smog values of SNG production, calculated for the year 2050, lie between 0.018 and 0.049 g C_2H_{4eq}/MJ, corresponding to 90 and 245% of the reference value. For many paths values between 0.025 and 0.030 g C_2H_{4eq}/MJ are given. These values are slightly larger than the reference value. Main contributions to these impacts are given by electricity generation for hydrogen production. Additionally, depending on the pathways construction of the DAC plants and LNG tanker transport show contributions of around

20% to the summer smog. Particulate matter for SNG production in the year 2050 was calculated in the range from 0.06 to 0.15 g/MJ. Compared to natural gas as a reference these particulate matter impacts are 170 to 580% higher. The highest contributions to the particulate matter impacts are given for electricity generation for electrolysis, construction of the plants, and transport by LNG tanker [9,10].

Besides C-SNG/L-SNG, fossil-based CNG/LNG is currently popular but is not covered in the overview of this work. However, an analysis of the environmental impacts of fossil-based CNG/LNG is attached in Appendix B.

5.3. HVO

Various vegetable oils can be used for the production of HVO. These oils are converted into hydrocarbons by means of a catalytic reaction with the addition of hydrogen [193].

The environmental impacts are not the same for all vegetable oils and differ significantly in some instances. An LCA study by Arvidson et al. [194] compared the HVO from rapeseed, palm oil, and jatropha. The GWP for the production of HVO from palm oil turned out to be the lowest. In addition, the GWP for HVO from all three feedstocks was found to be significantly lower than for fossil-based diesel fuel. It was also shown that the nitrous oxide emissions that entail the cultivation of the plants caused about half of the GHG emissions, and thus account for a large share of the environmental impacts. However, palm oil has been particularly criticized for a number of reasons, such as the land-use changes it causes and the associated GHG emissions, and it should therefore not be considered for the production of HVO [195]. This was also addressed in the discussion of biomass-based alternative fuels above.

The review study of Bierkandt et al. [195] notes a GHG emission reduction compared to fossil fuels for HVO and the German context based on the literature. In that publication, HVO is considered overall as an admixture available in moderate quantities in the short term, and well-suited for GHG reduction [195].

The ranges for the climate change indicator shown in Figure 15 include data for used cooking and vegetable oils. For used cooking oil, a range of 2.2–16 g CO_{2eq}/MJ tends to show lower environmental impacts compared to the range for vegetable oil of 5–149 g CO_{2eq}/MJ [174]. Moreover, depending on the feedstock used, the hydrogen input also varies between about 28 and 42 kg H_2/t HVO, and thus so do the environmental impacts arising from the hydrogen supply [174].

The recent JRC WtT report compares the WtT GHG emissions of HVO pathways for rapeseed, sunflower, soy, palm oil, and waste cooking oil in the European context [192]. The range of WtT greenhouse gas emissions calculated in the report is extensive. The resulting emissions begin under 10 g and reach more than 80 g CO_{2eq}/MJ if no credits are considered. The emissions of cooking oil lie at the lower end of the range, whereas soybeans tend to exhibit the highest impacts without consideration of credits. If credits are considered, most of the plant-based HVO pathways show results in a range between 40 and 60 g CO_{2eq}/MJ. The first production step clearly dominates the impacts of plant-based HVO pathways.

5.4. Methanol

Of the promising PtL fuels considered, most environmental assessment studies focus on electricity-based methanol production (Power-to-Methanol). The PtX LCA review by Koj et al. [178] identified nine studies with LCA considerations of methanol [191,196–203]. In the LCA review by Kigle et al. [182], six LCA studies on methanol were identified [189,191,196,204–206]. A total of 13 studies with content on Power-to-Methanol were identified in both literature reviews.

With respect to electricity-based methanol production, a study by Liebich et al. [9,10] is especially noteworthy due to its geographical focus on Germany, the number of variants considered, and its actuality. In this recent study, a total of 20 methanol production pathways were compared. The study considered production in Germany and import from Saudi Arabia, Morocco, Iceland, and Sweden with a wide variety of electricity and carbon dioxide sources, as well as different transport options. Across the pathways considered, the study found that the provision of electricity for electrolysis upstream of methanol synthesis contributed the largest share of total GHG potential [9,10]. In addition, this also shows the decisive importance of the electricity source of hydrogen supply and a subordinate role for methanol transport [9,10]. For a retrospective consideration of the year 2015, a global warming potential of 20–25 g CO_{2eq}/MJ was determined for some pathways. Compared to the selected reference relating to methanol from natural gas, including upstream chains with a global warming potential of 95 g CO_{2eq}/MJ, this corresponds to a reduction of about 75%. A prospective consideration of the pathways for the year 2050 yields global warming potential results of between 8.7 and 25 g CO_{2eq}/MJ methanol, with most pathways exhibiting results in the range 10 and 15 g CO_{2eq}/MJ. Thus, a reduction of about 85–90% compared to conventional fuel would be possible. These values obtained by means of LCA are also in the range of the WtT GWP values for methanol, reaching from 1 to 50 g CO_{2eq}/MJ in the literature review of Naumann et al. [174] (see Figure 15). Plant construction contributes to the environmental impacts to a lesser extent than the use of the main inputs (hydrogen and carbon dioxide) [9,10]. From another perspective, however, the relative importance of plant construction increases due to decreasing environmental impacts from its operations. The construction of carbon capture plants mostly has higher environmental impacts than the construction of the PtX plant [9,10].

Aside from GWP, Liebich et al. [9,10] assessed further environmental impact categories for electricity-based methanol synthesis. With respect to the acidification potential, values of between 0.082 and 4.9 g SO_{2eq}/MJ were obtained. The fossil reference value was 0.052 g SO_{2eq}/MJ. As a result, the acidification potential of electricity-based methanol fuel is a minimum of 60% higher than the fossil reference. The main contributions to the acidification impacts were contributed by the electricity generation for electrolytic hydrogen production. The high contribution of the upstream electricity generation is primarily caused by steel, copper, and aluminum production for power plant construction. The eutrophication potential values vary between 0.015 and 0.097 g PO_{4eq}/MJ methanol depending on the pathways. Thus, also for these impact categories, methanol fuel production is accompanied by significantly higher values than the fossil reference. The eutrophication potential is at minimum 90%, and at maximum twelve times, higher than the fossil reference. The main contributor to the impacts of most pathways is the electricity generation for electrolysis and for some pathways the energy required for carbon capture. Again steel, copper, and aluminum production for power plant construction is particularly responsible for the eutrophication potential of this fuel. The summer smog impacts of electricity-based methanol were calculated to lie between 0.018 and 0.073 g C_2H_{4eq}/MJ. The impacts of many paths are between 0.025 and 0.050 g C_2H_{4eq}/MJ. Compared to the fossil reference of 0.037 C_2H_{4eq}/MJ, the methanol pathways provoke between 50 and 200% of the impacts. Again, electricity generation for electrolytic hydrogen production is responsible for the majority of the summer smog provoked by methanol synthesis. However, the construction of the plants and partly the energy for carbon capture, or even methanol transport, exhibited noteworthy impacts. Particulate matter concentrations for methanol synthesis were calculated in the range from 0.057 to 0.20 g/MJ. These values are 20–320% higher than the fossil reference. The main contributions were shown for the electricity input required by the electrolysis, plant construction, and methanol transport. Depending on the accounting or allocation of credit for carbon capture and its utilization, total electricity-based methanol synthesis could in fact reach near net-zero GHG emission level [9,10].

5.5. DME

With respect to electricity-based pathways for the production of DME, there have been relatively few LCA studies published to date. The review by Kigle et al. identified three LCA publications on DME [204,205].

Fernández-Dacosta et al. [204] calculated 14 g CO_{2eq}/MJ as the GWP of the fuel provision without its combustion. Bongartz et al. [189] performed an LCA on DME considering the conditions expected in 2035. The future electricity mix was calculated as the electricity source for DME production. Thereby, 18.2 g CO_{2eq}/MJ was determined to be the GWP of the fuel supply for DME production. The largest contribution (57%) to environmental impacts was found for the carbon dioxide supply. Hydrogen supply accounts for 21% and DME synthesis for 15% of the total GWP value. Transportation and distribution, including compression at fueling stations, also accounts for the smallest share (7%) for this fuel. Additionally, the LCA study by Matzen et al. [205] also confirms the largest share of carbon dioxide supply in relation to the environmental impacts of DME production.

In contrast, a recent publication by Troy et al. [207] demonstrated much smaller contributions of carbon dioxide supply to overall environmental impacts. In addition, a credit for the captured carbon dioxide is taken into account. That publication, in which the circumstances of DME production in Germany were considered, again highlights the significance of the electricity input. The analysis revealed GHG reductions of at least 75% when using wind energy instead of German grid electricity, which was calculated retrospectively with values for the year 2016. Troy et al. assessed additional environmental impact categories. For particulate matter (PM) formation, they pointed out that German grid electricity inputs produce a higher PM in comparison to wind energy. Furthermore, the construction phase of the DME synthesis provokes PM due to the upstream impacts of the steel used as a construction material in DME plants. The steel utilization also showed noteworthy effects on terrestrial acidification.

In the latest JEC WtT report [192], nearly-zero GHG emissions were calculated for a DME pathway with the utilization of renewable electricity and carbon dioxide.

As in the case of electricity-based methanol, the total GHG emissions of electricity-based DME synthesis could even reach a near net-zero level depending on the accounting or allocation of credit for the capture of carbon and its utilization.

5.6. MtG

Few environmental assessments can be found that focus on MtG. Hurtig and Yearwood [208] analyzed several fuels suitable for the incorporation of carbon dioxide. Europe was chosen as the geographic setting for the analysis. For each of the fuels, different variants were calculated, ranging from operation with the EU electricity mix to various high shares of RE in the electricity input, to scenarios that made use of decarbonization. Scenarios featuring the EU electricity mix were consistently found to have higher environmental impacts compared to a fossil-based reference. This persisted in most scenarios for MtG, even at an RE share of 30%. For an 80% share of RE and higher, clear advantages compared to the fossil reference could be observed. For the mass-based functional unit employed in the study, a maximum savings potential of more than 6 kg of CO_{2eq}/kg_{MtG} product was found compared to the fossil reference. However, the study did not contain additional information on the contributions of individual process steps or components to the overall environmental performance. Furthermore, the assessments only present the GWP results and no further environmental indicators [208].

5.7. FT Diesel and Kerosene

An increasing number of LCA publications can be found in the literature on the electricity-based FT synthesis. In the PtX fuel LCA review by Kigle et al. [182], relevant studies by Hombach et al. [209] and Alhyari et al. [210] could be identified. For FT gasoline, Naumann et al. [174] noted a broad range of 2.8 to 145.8 g CO_{2eq}/MJ as GWP. The LCA study by Liebich et al. [9,10] not only contains a broad range of assessed PtM pathways but also a variety of FT fuel supply paths. In the LCA study, 17 different fully electricity-based pathways were compared for the FT synthesis in Germany or imports to the country. However, some of these pathways are bio-based and not a form of PtL. Thus, only the results of 15 pathways that can be considered PtL options are taken into account in this study. For FT fuels, the study compared production in Germany and import from Saudi Arabia, Morocco, and Iceland, as well as different transport options in addition to a wide variety of electricity and carbon dioxide sources. Instead of pointing out results for FT diesel or gasoline, a production mix of the FT fuels was considered in the LCA study. The highest GWP results (over 350 g of CO_{2eq}/MJ) were given for a pathway in which the German electricity mix for the year 2015 is considered as electricity input. For the assessed conditions of the year 2015 for several additional fuel pathways, a GWP of 20–25 g CO_{2eq}/MJ and a reduction of around 75% compared to the conventional reference were calculated. It was also shown that FT production using offshore wind in Germany can be comparative or even advantageous to onshore wind in Morocco. Further impact reductions are expected for electricity-based FT synthesis until the year 2050. For most of the pathways, GWP values of between 10 and 15 g CO_{2eq}/MJ were calculated for FT fuel. In comparison to the conventional reference, this means a reduction of approximately 85%. The study also presents the results of further environmental impact categories. Values for additional impacts are given and described in the following section and subchapter. For acidification and LCA results for the year 2050, a range of between 0.076 and 5.1 g SO_{2eq}/MJ was calculated, whereas the impacts of the fossil reference (the average value of diesel/petrol) is lower (0.074 g SO_{2eq}/MJ). The acidification potential of FT fuel is, as discussed before for other electricity-based fuels, especially provoked by emissions from upstream steel, copper and, aluminum production used for the plant construction for the required electricity. With respect to eutrophication, the lowest value of FT pathways (0.012 g PO_{4eq}/MJ) is lower in comparison to the fossil reference (0.021 g PO_{4eq}/MJ). However, there is an FT pathway that would cause multiple impacts (0.096 g PO_{4eq}/MJ). The impacts of eutrophication are especially provoked by the upstream electricity generation of hydrogen production due to materials used for power plants. Summer smog potential can be significantly reduced for some pathways (min. 0.014 g C_2H_{4eq}/MJ) compared to 0.046 g C_2H_{4eq}/MJ. The FT pathway with the highest summer smog potential induces 0.065 g C_2H_{4eq}/MJ, which is higher than the value of the fossil reference. The main source of summer smog is also given by the electricity generation for hydrogen provision. Furthermore, the construction of the plants was revealed to have notable summer smog impacts. The particulate matter impacts of FT fuel provision in the year 2050 were analyzed to be more than 20% lower, or even three times higher than the reference value. Again, electricity generation and plant construction were shown to be primarily responsible for the environmental impacts [9,10].

5.8. Contribution Analysis for Several Alternative Fuels

As pointed out before, especially the upstream electricity production and the underlying construction of power plants are mentioned to be of high relevance for many environmental impact categories. Quantification of the contribution of electricity for electrolysis as well as from further process stages is illustrated in Figure 16.

Figure 16 depicts the median contributions on environmental for 2050 for all fully electricity-based supply paths considered in the study of Liebich et al. [9,10]. Electricity for electrolysis exhibits the largest contributions, reaching from just under 60% to 80%. The contribution caused by the fuel production plants ranges from one eighth to one third, depending on the fuel. Some materials used for the construction of these plants, especially steel, followed by aluminum, copper, and cement, provoke most of these environmental impacts.

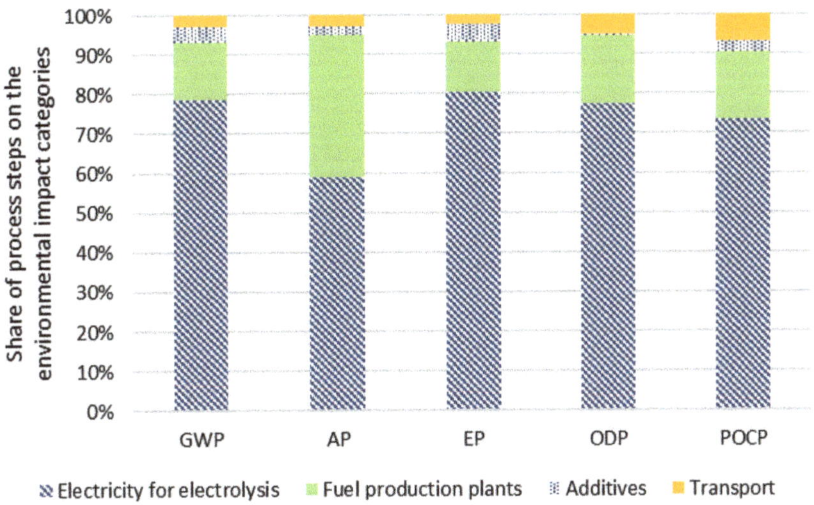

Figure 16. Contribution analysis for selected environmental impact categories for medians across electricity-based fuel supply pathways 2050. Source: Own elaboration based on [9,10]. GWP: Global warming potential; AP: Acidification potential; EP: Eutrophication potential; ODP: Ozone depletion potential; POCP: Photochemical ozone creation potential.

Transport, including its direct emissions and required infrastructure or means of transport (pipelines, ships, etc.) goes along with contributions of less than 10% to the different environmental impact categories. The materials steel, aluminum, and copper are again responsible for most of the impacts of the transport infrastructures. Contributions of additives to the environmental impacts are typically even lower and in a range between 1 and 5% of the impacts per category. These additives comprise auxiliary materials for use in synthesis and separation plants, such as catalysts, scrubbing liquids, and adsorber materials.

5.9. Interim Conclusion

Sections 3 and 4 identified the promising fuels hydrogen, CNG/LNG, HVO, methanol, DME, MtG, and FT diesel as well as FT kerosene. These promising fuels were analyzed in this sections considering their environmental impact. It was shown, that production of all these promising fuels could reach near-zero or even negative GHG emissions, bounded to mandatory preconditions. Furthermore, it was found, that the impact of long-distance transport for imports of these fuels is rather low with less than 10% for the examined impact categories. The results also show, that the materials steel, aluminum, and copper should be reduced as much as possible in alternative fuel production chains to keep environmental impacts, such as acidification and eutrophication potential, below the results of the conventional pathways.

6. Discussion

In the following, the findings of the previous sections are discussed and evaluated. First, the reviewed literature pertaining to fuel costs is discussed. Second, the application of fuels and drivetrains in road, inland waterway, rail, and air transport is discussed and compared with currently available vehicle technologies. Third, the outcome of the economic impact analysis is summarized and discussed.

6.1. Cost

Hydrogen can be transported as a gas via pipeline, liquefied, or as an LOHC by ship. The most common derivatives of hydrogen are SNG, ammonia, methanol, DME, MtG gasoline, and FT diesel [34]. At this stage, no clearly superior electricity-based fuel has been identified due to the different pros and cons of each fuel. Therefore, a clear and balanced trade-off decision between hydrogen and its derivatives has not yet been taken. The drawbacks of derivatives include lower production efficiencies and the need for a CO_2 source. The benefits include lower transportation costs compared to hydrogen due to the compatibility of the existing infrastructure [83]. H_2 transportation costs reveal insecurities due to the low TRL of LH_2 and LOHC vessels, which result in a higher spread of costs. Furthermore, the low utilization of pipelines during the ramp-up time of H_2 production is weighted differently in techno-economic analyses. In terms of capacity scaling, marine vessel transport carries advantages. However, possible unloaded return journeys of LOHC/LH_2-carrying vessels negatively impact their economic competitiveness [83,91]. Another factor that causes cost insecurity is the estimation of the demanded import quantities that influence the occurrence and expression of scaling effects [104]. The assumed economies of scale regarding electrolysis investment costs for the short-term time scope of 2030 should be treated with caution and take into account the current size of the H_2 market [83]. The extracted lower bounded cost values for cross-border prices for 2050 are 5.3 EURct/kWh_{LHV} for H_2 transported via pipeline from Morocco, 8.0 EURct/kWh_{LHV} for SNG imported from the MENA region via pipeline and PtL, and Syncrude at 8.0 EURct/kWh_{LHV} for imports from Iceland [106,110,112]. The respective upper boundaries of cross-border prices for 2050 are 11.8 EURct/kWh_{LHV} for LH_2 vessels importing from Norway or the UK [34] and 14.2 EURct/kWh_{LHV} for SNG imported from the MENA region via pipeline [112]. The upper bound value for PtL is the highest, at 19.2, for import from North Africa as determined by Schmidt et al. [108]. These snapshot values reveal optimistic and pessimistic assumptions regarding all three energy carriers and are not suitable for a conclusive assessment. Furthermore, the pessimistic cost values indicate that adequate CO_2 cost reallocations are necessary in order to ensure the market penetration of CO_2-neutral fuels, which cannot ensure cost-competitiveness in all cost calculations against their fossil equivalents [87,110].

The first studies focused on the import regions in North Africa and the Middle East due to the relatively low transportation distance and high RE FLH [89,90,110,112]. More recent studies aim to find the optimal points of high RE potential in terms of FLH and the general potential of available land, usable for RE scale up, low import costs, and politically—as well as economically—beneficial conditions [34,86,104]. This leads to a shift in the determined import regions and clearer recommendations. Suitable identified countries for production scale-up through 2030 are Iceland, Canada, Morocco, Norway, Tunisia, and Turkey. Furthermore, long-term perspectives through 2050 ascribe the highest RE potential to Egypt, Algeria, Argentina, Australia, Kazakhstan, Russia, and Saudi Arabia as favorable production countries [104]. In comparison to electricity-based fuels, biomass-based ones such as HVO and biomethane exhibit generally lower price levels and high TRLs, and should therefore be utilized in the highest quantities to ensure low ILUC risks [100].

6.2. Road Transport

C-SNG and L-SNG are promising due to the low-to-medium production costs and the market share already possessed by fossil natural gas. LNG trucks are utilized, for instance, by Volvo with HPDI diesel engines in dual-fuel operation with 5–10% diesel, or by Scania with gas engines [211,212]. The advantage of dual-fuel operation is a higher efficiency that approximately corresponds to that of conventional operation with diesel. In contrast, the gas engine offers efficiencies in the range of 75–85% and a maximum efficiency of 90–95% in relation to the conventionally-operated diesel engines [160]. A disadvantage of the diesel engine in gas operation is higher emissions, as shown by the emission factors in HBEFA 4.1 [159] or Otten et al. [162]. Methanol is used in pure form (M100) in racing [123]. Additionally, Kramer et al. [109] assessed M100 as being a possible future fuel for SI engines. Similar to natural gas, methanol is also used in diesel engines in dual-fuel operation. Dual-fuel systems for diesel engines have already been used in the automotive and marine domains. An example is the use in the ferry Stena Germanica [213], which operates between Kiel and Gothenburg. Zhai [214] reported that the utilization of methanol in heavy duty vehicles has been part of the 2012-started methanol pilot program in China. Additionally, there are the two heavy duty vehicle series Sinotruk Howo (ZZ3317N4667D1M and ZZ3257N3847D1M II) and Shacman (SX3317DR456HM and SX3317DR456HM II) that employ methanol–diesel dual-fuel engines of the manufacturers China National Heavy Duty Truck Group Corp., Ltd. and Shaanxi Heavy-duty Automobile Group Co., Ltd. [215].

However, the European Commission evaluates the use of methanol in dual-fuel operation in the diesel engines of heavy-duty vehicles as unexplored [120] despite the work of Zhai [214]. In dual-fuel operation in diesel engines, the nitrogen oxide (NO_x), PM, and soot emissions decrease, whereas the hydrocarbon ones increase [123].

In this study, passenger cars, light duty vehicles, and motorcycles were identified as vehicle classes that tend to travel over shorter distances. Furthermore, large reductions in consumption are possible for these vehicle classes through the implementation of battery–electric drivetrains (see Figure 13). Buses and rigid trucks were identified as vehicle classes with medium mission range requirements, which is why the use of a battery–electric drive also seems to be viable for these vehicle classes. However, the higher vehicle mass of such vehicles and correspondingly higher energy demand militates against the use of battery–electric drives. In the case of heavy vehicles and longer distances, the use of a battery–electric drivetrain appears to be somewhat disadvantageous due to its much lower volumetric and gravimetric energy densities (see Figure 12). Another promising technology is the use of catenary trucks in freight transport. The costs of this technology depend heavily on the length of overhead line sections, the traffic volume, and the share of vehicles using this technology [216]. As a result of this, it is not possible to compare the costs, as was done for the other energy sources above. A spatial analysis to determine the costs of overhead lines is therefore necessary. The use of catenary trucks in heavy duty transport was investigated and published in another study [217]. It concluded that catenary trucks for heavy duty transport generally lead to more technical as well as economic uncertainties than other alternative propulsion systems. Therefore, this propulsion concept is not further discussed herein. Further details can be found in Breuer et al. [217].

The utilization of C-SNG- or L-SNG-fueled vehicles is generally promising. Gas engines can be used for the vehicle classes identified as having low- and medium mission range requirements: passenger cars, light duty vehicles, buses, and rigid trucks. Of the lighter vehicle classes passenger cars and light duty vehicles, the use of hybrid concepts is viable due to the higher consumption of gas engines. For vehicles with larger mission range requirements, the use of dual-fuel diesel engines with LNG is advantageous due to the low energy density of methane and the better efficiency of dual-fuel engines. This includes articulated and trailer trucks. The use of fuel cell–electric propulsion tends to be possible and advantageous in road traffic in all vehicle classes with the exception of motorcycles. For vehicles with higher mission range requirements, larger tanks are necessary due to the lower volumetric energy density. The high gravimetric energy density and efficiency are also advantageous.

Peters et al. [218] reviewed and assessed different alternative fuels for heavy-duty transport. The investigated fuels and propulsion systems, which include diesel-like liquid fuels, hydrogen, natural gas, DME, and catenary trucks, match the results presented above.

6.3. Inland Waterway Transport

Since January 2020, the new European regulation (EU) 2016/1628 EURO V for non-road mobile machinery applies to all new engines higher than 300 kW in inland waterway transport [219]. For instance, the limit for NO_x emissions was reduced from 6 to 1.8 g/kWh. Options for attaining these new emission standards include efficiency improvements, exhaust gas after-treatment such as DPF or SCR, and alternative fuels [220]. Among the investigated Drop-In fuels, the use of synthetic diesel in inland navigation seems to be the most promising option, although this will not meet the mentioned EURO V emissions regulation without the additional use of exhaust gas after-treatment. The same applies to ships with dual-fuel LNG propulsion [162]. Only ships with LNG- or CNG-powered spark ignition engines are consistent with the new emissions regulation without exhaust gas after-treatment [162]. In addition to LNG operation, the alternative propulsion concepts of fuel cell– and battery–electric are promising due to their high efficiencies (see Figure 14). Limiting factors include the low volumetric energy densities of hydrogen and batteries, and the lower gravimetric energy density of the latter (see Figure 12). An analysis of the required mission ranges in passenger shipping was not performed. Assuming that cabin ships, ferries, day excursion vessels, and smaller boats tend to cover shorter distances, the use of battery–electric propulsion is possible and advantageous for these ship classes. The consumption reductions are also greatest for these classes (see Figure 14). The use of battery–electric drive in freight transport is not feasible due to the lower fuel consumption reductions, the lower energy densities of the batteries, and the high mission ranges required. The use of a fuel cell–electric powertrain in passenger and freight transport in inland navigation is promising and offers moderate advantages in terms of energy consumption and a high gravimetric energy density, and therefore the maximum payload is not restricted. One challenge could be the low volumetric energy density for ships with long mission ranges, such as cargo and liquid cargo barges. The literature confirms this assessment of the viability of alternative propulsion systems, and is outlined in the following.

Vessels with battery–electric propulsion are already operated in the inland navigation sector. The electric day cruiser St. Nicholas operates with a 50 kW electric drivetrain on the Rursee in Germany and the electric ferry Sankta Maria II with an 80 kW one on the Mosel in Germany [221]. Both can carry up to 250 passengers. Zero Emission Services B.V. [222] announced in a press release that it will equip a motorized cargo barge owned by the Heineken brewery with an electric drivetrain. The energy storage system used is the exchangeable battery container system developed by Zero Emission Services B.V. [222]. In the future, electric freight ships will operate on the approximately 60 km-long Zoeterwoude–Alpherium–Moerdijk route in the Netherlands. Based on the analyses of the mission range of inland freight traffic performed in this study, this technology only seems to be suitable for individual applications (see Figure 3). Moreover, according to Kasten et al. [223] battery–electric inland waterway barges are rarely considered options for inland navigation due to the low volumetric and gravimetric energy density and the resulting high weight of the batteries. LNG propulsion is already widespread in shipping, with the LNG carrier vessels the TMS Ecotank. III, ex-TMS Green Rhine, and Eiger already being in operation on the Rhine, Germany [221]. Bauen et al. [120] classify LNG technology for shipping as being at the commercial level. In Norway, five LNG ferries of the type Fjord1 have been in operation since 2007 [224].

For passenger transport, the high-speed ship Francisco was put into service in South America. Furthermore, the cruise ferries MS Stavangerfjord and MS Bergenfjord entered service in 2013 and 2014, respectively, and operate between Norway and Denmark. [224] Bauen et al. [120] report that 50 LNG-fueled ships (excluding LNG liquid cargo barges) are already in operation in the EU. At the time of the publication of the European Commission report [120], 45 more LNG-fueled ships were on order. For LNG pushers, at the time of this study, only a design study by Rolls-Royce and Canadian Robert Allan, a ship designer, is known [225]. Fossil LNG has been evaluated in numerous studies to be a good alternative for inland navigation [226], although the Federal Ministry of Transport and Digital Infrastructure in Germany [226] states that it is only a temporary solution, as fossil natural gas reserves will also run out. According to Kopyscinski et al. [227], fossil LNG could subsequently be replaced by SNG, for instance, from biomass or PtG methane.

Zerta et al. [155] provide an overview of existing fuel cell-driven vessels. Their work shows that commercial fuel cell vessels are mostly small in size, such as sport boats with engine power ranges of 4–50 kW. However, there is also a commercial ferry with 2 × 200 kW and pusher boat with an engine power of 2 × 200 kW in operation. Prototypes, pilot projects, and demonstration vessels exist with engine powers in the range of 1–2 × 1000 kW. These vessels are primarily inland passenger vessels, ferries, or small ships. [155] According to Zerta et al. [155], ships with fuel cell–electric propulsion systems of up to 10,000 kW are being investigated in theoretical studies. Amongst the German inland waterway fleet, pusher boats of up to 362 kW make up a share of 70% [228] and could presumably be covered by the already existing fuel cell-powered inland waterway vessel fleet with 2 × 200 kW. In a feasibility study by the MariGreen project [229], hydrogen is evaluated as a promising energy carrier for inland navigation and its use in combustion engines is advantageous for the ship classes cargo/liquid cargo barges and pushed barges/tankers, whereas, for ferries and cabin ships, fuel cell–electric propulsion is preferable.

More recent studies, such as that from Zerta et al. [155], focus on fuel cell–electric propulsion for all vessel classes. The Rhine Hydrogen Integration Network of Excellence (RH2INE) has set the goal of operating ten hydrogen-powered inland vessels between Rotterdam and Duisburg by 2024 [230].

Finally, the retrofit of the already mentioned Stena Germanica into methanol dual-fuel operation using marine gasoil as a pilot fuel should be mentioned [213]. As noted earlier, methanol also constitutes a promising alternative energy source and can be used in the dual-fuel operation of diesel engines.

6.4. Rail Transport

In principle, the conventional diesel fuel used in rail transport can be replaced by HVO or FT diesel. Alternatively, diesel–electric hybrids, battery–electric hybrids, or fuel cell–electric propulsion systems can be used. In contrast to other transport sectors, the choice of propulsion system for rail transport depends on the route of the respective train line. If only short, non-electrified sections of a track must be covered, the use of a battery–electric train with a catenary line connection and a battery for up to 100 km is beneficial. For longer, non-electrified distances for which electrification is not worthwhile due to low traffic volumes, hydrogen trains can be used instead.

6.5. Air Transport

As noted above, air transport requires propulsion systems with long mission ranges. Considering the low energy densities of batteries, battery–electric propulsion systems are unsuitable for commercial aviation, despite their high efficiency rates. Hydrogen as an energy carrier offers a higher gravimetric energy density. However, neither of these technologies will be operational in this context in the foreseeable future, which is why they are not further investigated herein. As will be discussed below, the literature supports this hypothesis. According to Thomson [231], the use of lithium-ion and nickel–cadmium battery systems in battery-powered aircraft is restricted by the weight and size of the battery system. Battery-powered aircraft could enter the market between 2030 and 2040, but their application is limited to small and medium-sized aircraft [232,233]. According to McKinsey & Company [28], battery–electric aircraft are applicable for commuter (<19 PAX), regional (20–80 PAX), and short-range aircraft (81–165 PAX) for mission ranges of up to 1000 km, whereas they are not suitable for larger aircraft due to their limited energy densities. A study by Roland Berger [234], however, concluded that the use of a battery-electric propulsion system in commuter aircraft would reduce mission range to 202 km. In Roland Berger [234], a battery energy density of 180 Wh/kg was assumed. Current densities for batteries in road vehicles are up to 260 Wh/kg [148]. Furthermore, according to Roland Berger [234], the efficiency would have to be increased by a reduced drag coefficient, an increased wingspan, and a reduced mass in order for battery–electric propulsion to be viable in commuter aircraft. Additionally, the battery density would need to be increased to 720 Wh/kg [234]. Eviation Aircraft has developed the nine-passenger Alice electric airplane, which has a stated mission range of 1000 km and is expected to become commercially available in 2021 [235]. The ICAO [236] has published an overview of existing electric aircraft prototypes and projects, most of them are below <19 PAX, i.e., in the commuter aircraft class, or are designed for even shorter distances, such as the Volocopter 2X with a 27 km mission range. However, only the aircraft project Airbus/Siemens/Rolls Royce E-Fan X (hybrid-electric), the Wright Electric/Easy Jet (electric), and the Boeing Sugar VOLT (hybrid-electric) are in the class of large commercial aircraft with 100 PAX, 120 PAX, and 135 PAX, respectively. Market entry is planned for 2027–2050 [236]. In another study by Roland Berger [170], existing hydrogen aircraft projects are examined and evaluated. Furthermore, fuel cell/hydrogen propulsion for aircraft is assessed. At present, the only fuel cell-driven aircraft is the DLR HY4, which has a mission range of 750–1500 km and offers space for four passengers [237]. Other prototypes under development feature up to 20 seats. Exclusively, NASA's CHEETA project is focusing on large commercial aircraft. The feasibility studies Airbus Cryoplane and NASA Concept B also investigated large commercial hydrogen aircraft powered by either fuel cells or turbines [170].

McKinsey & Company [28] investigated the use of hydrogen as an energy source in the five different classes of commuter (<19 PAX), regional (20–80 PAX), short-range (81–165 PAX), medium-range (166–250 PAX), and long-range aircraft (>250 PAX). They concluded that fuel cell propulsion is best suited for commuter and regional aircraft, whereas H_2 turbines are more suitable for medium- and long-range aircraft; hybrid engines should be used in short-range aircraft. However, costs will increase as aircraft size does. They also predict that commuter aircraft powered by hydrogen will be available in less than ten years and regional aircraft in the next 10–15 years, whereas the larger aircraft classes will not be available for more than 15 years. [28] According to McKinsey & Company [28], there is no mission range limit for hydrogen propulsion for commuter (<19 PAX), regional (20–80 PAX), and short-range aircraft (81–165 PAX), whereas for medium-range (166–250 PAX) and long-range aircraft (>250 PAX), new and more efficient aircraft concepts are needed for mission ranges above 10,000 km.

As a conclusion of the above discussion, the only short-time solution for aircraft, with small aircraft being the exception, is synthetic jet fuel. Different sustainable aviation fuels already went through the ASTM procedure and, as discussed earlier in this work, are certified with Drop-In rates up to 50% [58]. The analysis in this work showed, that higher Drop-In rates up to 100% are possible. Current demonstration projects are investigating the use of 100% unblended sustainable aviation fuels [238] and, therefore, are supporting the results of the analysis. The future should aim for certification of higher Drop-In rates of the already certified sustainable aviation fuels via the FT process and also the approval of synthetic jet fuel via the MtK pathway by the ASTM.

6.6. Environmental Impacts of Promising Alternative Fuels

Compared to conventional fuels, all promising alternative fuels showed potential for reducing GHG emissions. However, some preconditions must be specified in order to obtain superior environmental performance from these promising fuels. For the case of Germany, the use of the current, largely fossil-based electricity mix for the production of electricity-based fuels, should be avoided. In case the current electricity mix was to be used, environmental impacts would typically significantly exceed the impact of fossil fuel production. For domestic fuel production in Germany, the use of wind energy in particular for the production of the electricity-based fuels of hydrogen, SNG, DME, FT fuels, and MtG can facilitate near-zero GHG emissions. Besides the high importance of the type of electricity used for the production of these fuels, full load hours, the source of carbon, and the way in which carbon dioxide separation and utilization is implemented have a noteworthy influence on the results. In the case of full allocation of the required carbon dioxide for fuel production, the GHG emissions of SNG, DME, FT fuels, and MtG could reach a near net-zero GHG emission level. In addition to the domestic production of promising fuels, the import from countries with better production conditions can be considered from an environmental perspective. Environmental impact results show that long-distance transport goes along with contributions of typically less than 10% to the overall results of different environmental impact categories. However, the advantages of these fuels for GHG emissions in comparison to the fossil references do not necessarily show up for additional impact categories as well. Major contributions to several environmental impact categories are given by upstream impacts of the electricity used for hydrogen production. This is primarily caused by the upstream production of steel, aluminum, and copper for the power plants, e.g., the production of steel is very important for the construction of wind turbines as well as further materials like aluminum, copper and their environmental impact provoking production. To reduce the environmental impacts of many impact categories in the future, especially the upstream manufacturing of steel, aluminum, and copper must be environmentally optimized or these materials must be substituted.

The potential to reduce the GWP per MJ of fuel is also presented for biomass-based HVO fuel pathways. This is especially the case if used cooking oil is utilized. However, some plant-based pathways of HVO production do not enable environmental impact reductions compared to conventional fuels. Furthermore, the use of biomass for fuel production instead of the growing of food is discussed and is controversial, and conflicts and impacts can arise relating to land use for the production of HVO.

As previously noted with respect to biomass potential, the presented overview of the environmental impacts of fuel production indicates once again that electricity-based fuels are essential, in addition to biomass-based ones.

Taking all of the considered climate change values regarding the production of the promising identified fuels into account, hydrogen and methanol tend to exhibit the lowest impacts. Taking further impact categories into account, hydrogen reveals reductions compared to the fossil reference in further impact categories. From this perspective, electricity-based hydrogen produced by wind energy would exhibit the best environmental results of the promising fuels considered if a normalization step of different environmental impact categories were to be conducted, as stated by Koj et al. [186]. Nevertheless, as previously mentioned, the full allocation of carbon dioxide for fuel production, which depends on the conditions at the production sites, could also reduce the GHG emissions of FT fuel, SNG, methanol, DME, and MtG to a near net-zero GHG emission level.

Due to the focus area of this study and the need to set limits on the large number of publications relating to the environmental impacts of alternative fuels and vehicles, only literature results pertaining to fuel production were described and analyzed. If fuel use for transport is environmentally assessed, particularly highly efficient vehicles and fuels with low or no carbon content can lead to advantageous environmental performances.

The already-mentioned RED II also has high relevance for the present and future environmental impacts of alternative fuels in Germany. This European directive must be transposed into national legislation. In Germany, RED II is implemented by the GHG Quota (GHG Quota), an instrument that forces fuel distributors to gradually reduce and monitor the GHG emissions of their respective distributed fuels. Further information on the status and further development of the German GHG Quota is summarized in a study by Naumann et al. [239].

7. Conclusions

This study showed the importance of having a technological diversity of fuels and drivetrains available to satisfy the complex respective requirements of these. At present, the most promising are battery–and hydrogen fuel cell–electric drivetrains, DME, natural gas, methanol, and the Drop-In fuels FT-diesel, HVO, MtG, and FT-kerosene. However, this selection may change with upcoming research as well as political conditions.

It was shown that from today's perspective, electrification with battery–electric drivetrains is highly unlikely for most use cases in long-distance heavy duty transport, shipping, and aircraft transport, although there are exceptions like small short-range aircraft. It is necessary to address these vehicle classes with other emission-reducing solutions such as, if possible, fuel cell–electric propulsion or electricity-based fuels. The latter is the only short-term solution for air transport currently apparent. Among electricity-based fuels, drop-in fuels have the advantage of their compatibility with existing vehicles, which is, in the case of FT-kerosene and -diesel, as well as MtG, already legally regulated.

The results of the cost value review highlight the insecurities around the regarded cost level production costs, cross-border prices, and end user prices in the current research landscape. The extracted interval sizes of cross-border prices for 2020 are 7 EURct/kWh$_{LHV}$ for H_2, 10 EURct/kWh$_{LHV}$ for SNG, and 8 EURct/kWh$_{LHV}$ for PtL fuels. Cost insecurity increases with the predicted range and length of the value chains. At present, cost comparisons indicate that lower production costs of H_2 are nearly compensated by higher transport costs in comparison to other fuels that offer existing infrastructural compatibility. Furthermore, the domestic production of H_2 is considered cost-competitive to LH_2 imports.

The overview of environmental impacts provoked by the production of the promising alternative fuels showed influencing factors, potential reductions, but also occasional disadvantages. Significant reductions of global warming potential of alternative fuel production compared to fossil references can be achieved. Production of all promising fuels could reach near-zero or even negative GHG emissions, bounded to mandatory preconditions like the accounting approach of credits for carbon capture. Domestic fuel production in Germany should avoid fossil-based electricity and prefer wind power. Additionally, the import of these fuels from countries with promising production conditions can be an interesting alternative to produce viable fuels. The impact of long-distance transport for imports is rather low, as environmental assessments show shares of less than 10% to results of different impact categories. Along the fuel production chains, care must be taken to use as few amounts as possible from materials, such as steel, aluminum, and copper to keep further environmental impacts (e.g., acidification and eutrophication potential) below the results of the conventional pathways.

Insecurities with regard to future electricity-based fuel demand, market development, and cost structures have led to the investigation of a variety of production locations in the literature. Furthermore, electricity-based fuels strongly depend on their eligibility for CO_2 reduction targets. At present, there is no secure legal framework, which makes it difficult to determine import quantities and discourages investment, which results in increased cost insecurity in scaling effects.

Further research and demonstration of alternative fuel production plants, especially large-scale ones, and their interaction with RES, CO_2-capturing, and seawater desalination processes must be carried out in order to achieve precise cost predictions. The same applies to long-distance hydrogen transportation.

Author Contributions: Conceptualization, J.L.B. and J.S.; methodology, J.L.B. and J.S.; formal analysis, J.L.B., J.S. and J.C.K.; investigation, J.L.B., J.S. and J.C.K.; writing—original draft preparation, J.L.B., J.S. and J.C.K.; writing—review and editing, F.S., R.P., R.C.S. and M.F., visualization, J.L.B., J.S. and J.C.K., supervision, R.P., R.C.S., R.A., K.G. and D.S.; project administration, R.P.; funding acquisition, R.P. and D.S. All authors have read and agreed to the published version of the manuscript.

Funding: Funding of the center of excellence "Virtual Institute—Power to Gas and Heat" (EFRE-0400111, EFRE-0400151) by the "Operational Program for the promotion of investments in growth and employment for North Rhine-Westphalia from the European fund for regional development" (OP EFRE NRW) through the Ministry of Economic Affairs, Innovation, Digitalization and Energy of the State of North Rhine-Westphalia is gratefully acknowledged. Funded by the Deutsche Forschungsgemeinschaft (DFG, German Research Foundation)—491111487.

Acknowledgments: The authors would also like to thank Bernd Emonts from the Institute of Energy and Climate Research—Electrochemical Process Engineering (IEK-14), Forschungszentrum Jülich GmbH who triggered this original research offspring study.

Conflicts of Interest: The authors declare no conflict of interest.

Appendix A

Table A1. Technology readiness level of alternative fuel production.

Fuel	Process	Production TRL
Hydrogen	Upgraded biogas from municipal organic waste, wet manure, sewage sludge, maize, or double cropping	TRL 9 [5]
	Gasification of farmed wood	TRL 8 [5]
	Renewable electricity via alkaline electrolysis	TRL 9 [5]
	Renewable electricity via polymer electrolyte membrane electrolysis	TRL 9 [57]
	Renewable electricity via solid oxide electrolysis cell	TRL 6–7 [57]
HVO	Conventional (biomass)	TRL 9 [5,76]
FAME	Conventional (biomass)	TRL 9 [5,76]
Syndiesel	BtL, lignocellulose pyrolysis-based	TRL 6 [5,76]
	lignocellulose gasification	TRL 8 [5]
	lignocellulose hydrothermal liquefaction (HTL) and upgrading	TRL 4 [5]
	FT diesel from CO_2 and H_2	TRL 6 [4]
	Diesel via methanol from CO_2 and H_2	TRL 9 [5]
Synthetic gasoline	lignocellulose pyrolysis-based	TRL 6 [5]
	MtG	TRL 9 [4]
Methanol	Lignocellulose	TRL 8 [5]
	From CO_2 and H_2	TRL 9 [4,5]
Ethanol	Conventional (biomass)	TRL 9 [5,76]
	Lignocellulose	TRL 7 [76], TRL 8 [5]
	From CO_2 and H_2	TRL 4 [4]
Butanol(1/2)	Conventional (biomass)	TRL 7 [5]
	From CO_2 and H_2	TRL 4 [4]
DME	Lignocellulose	TRL 8 [5]
	From CO_2 and H_2	TRL 9 [4,5]
OME_1	Lignocellulose	TRL 5 [5]
	From CO_2 and H_2	TRL 5 [4]
$OME_{3\text{-}5}$	Lignocellulose	TRL 5 [5]
	From CO_2 and H_2	TRL 4–5 [4,79]
Iso-octanol	From CO_2 and H_2	TRL 4 [4]
Octanol		TRL 1 [78]
	From lignocellulose	TRL 3 based on Leitner et al. [77]

Table A1. *Cont.*

Fuel	Process	Production TRL
CNG/CBM/SNG	Biomethane/biogas from residues (e.g., biowaste, manure, stillage)	TRL 9 [5,76]
	Biomethane/synthetic natural gas (SNG) from lignocelluloses (e.g., wood and straw)	TRL 7 [76], TRL 8 [5]
	From CO_2 and H_2	TRL 9 based on [5], Deutsche Energie-Agentur [80]
SLNG/LBM	Upgraded biogas to LBM from municipal waste, wet manure, sewage sludge, maize, double cropping	TRL 9 [5]
	SNLG from gasification of lignocelluloses (e.g., waste wood and wood chips)	TRL 8 [5]
	SLNG from CO_2 and H_2	TRL 9 [5]
Synthetic jet fuel	FT-SPK from CO_2 and H_2	TRL 6 [4]
	FT-SPK from biomass via gasification	TRL 9; Fulcrum [67] and Red Rock [68] have plants under construction with 30 kt/year and 45 kt/year
	Jet fuel from MtK process	TRL 4 based on based on Tabak et al. [63] and Tabak and Yurchak [64]
	HEFA-SPK from bio-oils, animal fat, and recycled oils	TRL 9; commercial process by World Energy Paramount (former AltAir Paramounts LLC) [60] and Neste Oyj [69]
	HFS-SIP from the microbial conversion of sugars into hydrocarbons	TRL 9, commercial process by Amyris in Brazil [60]
	ATJ-SPK from agricultural waste products (stover, grasses, forestry slash, and crop straws)	TRL 9; Pilot plant by LanzaTech [70]; commercial plant by Ekobenz with 22.5 kt/year [71]. Commercial-scale plants planned by LanzaTech [72] and SWEDISH BIOFUELS AB [73]
	CHJ from triglyceride-based feedstocks (plant oils, waste oils, algal oils, soybean oil, jatropha oil, camelina oil, carinata oil, and tung oil)	TRL 6–7; demonstration plant of ARA and euglena [74]
	HHC-SPK from biologically-derived hydrocarbons such as algae	TRL 4; laboratory scale by IHI [75]

Table A2. Overview of literature studies investigating fuel production costs.

Authors	Year	Content/Investigated Regions	Identified Costs	Meta-Study
Merten et al. [83]	2020	Germany: imported vs. locally produced hydrogen	Generation Transport	yes
Kreidelmeyer et al. [92]	2020	Fuel: H_2, methane, methanol, FT-syncrude Import region: MENA Assessed Costs: production, transport, and distribution	Generation Transport Taxes/Levies	no
Robinius et al. [34]	2020	Import costs of (synfuels, SNG, and H_2) are determined through optimized technology decision-making, electricity imported from countries neighboring Germany	Generation Transport	no
Gerhardt et al. [106]	2020	Import costs from Morocco and Tunisia, energy costs, demineralization, liquefaction, and transport	Generation Transport	no
Christoph Hank et al.	2020	Import costs from Morocco for LH_2, LOHC, liquefied methane, methanol, and ammonia	Generation Transport	
Mottschall et al. [107]	2019	Difference between H_2, CH_4, and PtL import costs No own calculation	Generation Transport	yes
Schindler [89]	2019	Cost analyses of end user costs of in-Germany-produced and imported H_2, SNG, and PtL Fuels for 2020 and 2050 Primary source is unknown	Generation Transport Taxes/Levies	no
Eichhammer et al. [90]	2019	Generation costs of hydrogen and derived products from Morocco 2015, 2030, and 2050	Generation	no
Michalski et al. [105]	2019	H_2 generation costs in Germany, with electricity imported from neighboring countries	Generation	no
Jensterle et al. [104]	2019	Country-specific suitability of energy carrier export due to various soft factors	Generation Transport	no
Schemme et al. [4]	2019	Techno-economic aspects of specific synfuels: H_2, methanol, ethanol, DME, OME, MtG, FT, and butanol	Generation	no
Terlouw et al. [111]	2019	Energy system modeling for 2050, determining favorable RE locations for energy carrier production, hydrogen transport technologies' TRLs, and costs	Generation Transport	no
Perner et al. [110]	2018	Future costs of fossil fuels PtL from North and Baltic Sea/North Africa/Iceland and Germany Level: Transport + conversion losses + electricity costs without taxes and levies		
Kramer et al. [109]	2018	Min/max scenario for 2030 differed for all common alternative fuels	Generation Neglecting of transport costs	no
Hobohm et al. [112]	2018	Calculation of energy carrier demand of different sectors and end user prices	Generation transport taxes/levies	no
Pfennig et al. [6]	2017	2030/2050 generation costs of generalized PtL fuels from the North Sea and Morocco	Generation transport	no

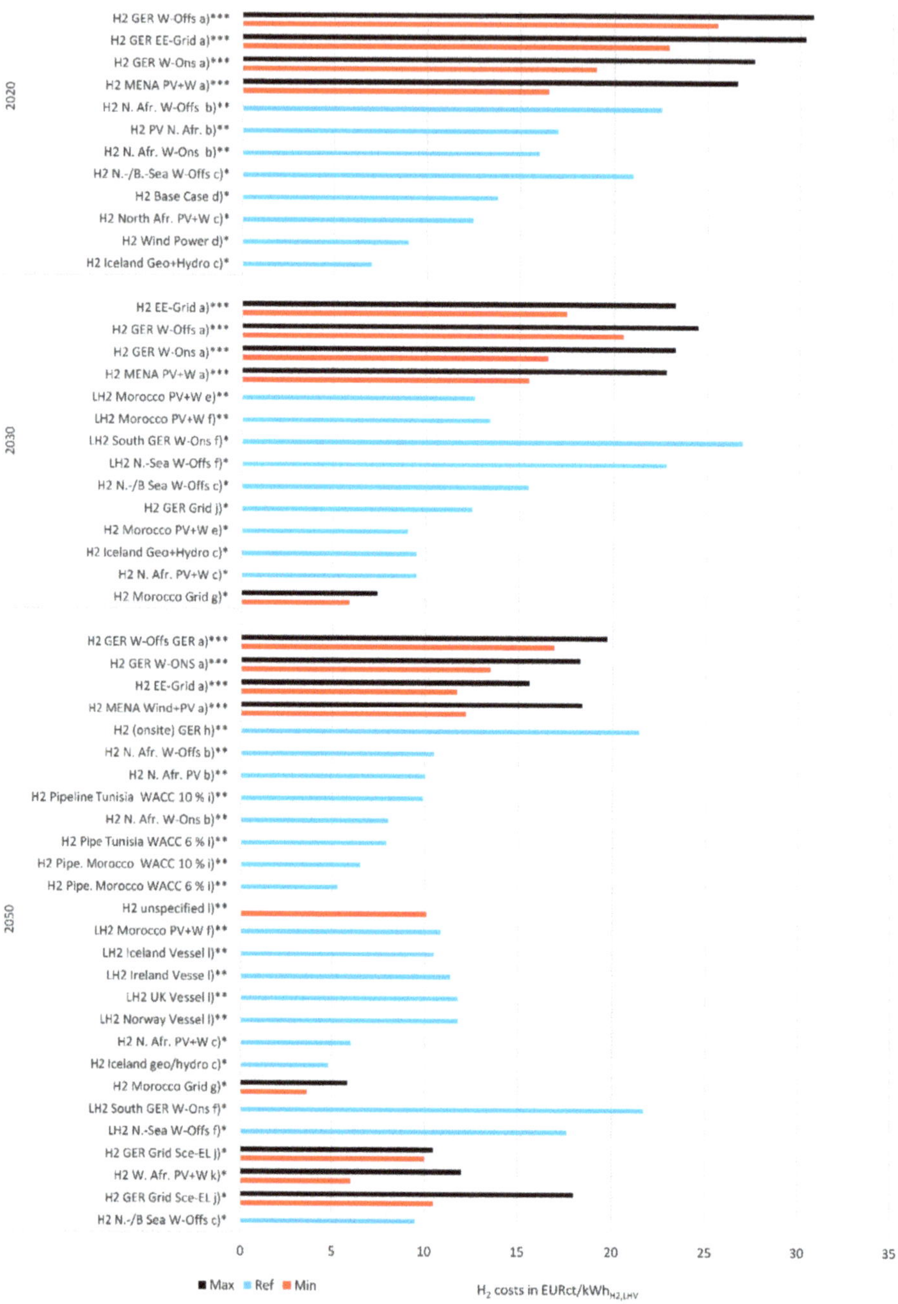

Figure A1. Cost intervals of H_2 and LH_2 production in Germany and abroad, transport costs, and taxes/levies. * production, ** +transport, *** +taxes/levies. Source: Own elaboration based on (a) [92], (b) [89], (c) [85], (d) [4], (e) [87], (f) [6], (g) [90], (h) [108], (i) [106], (j) [105], (k) [88], (l) [34].

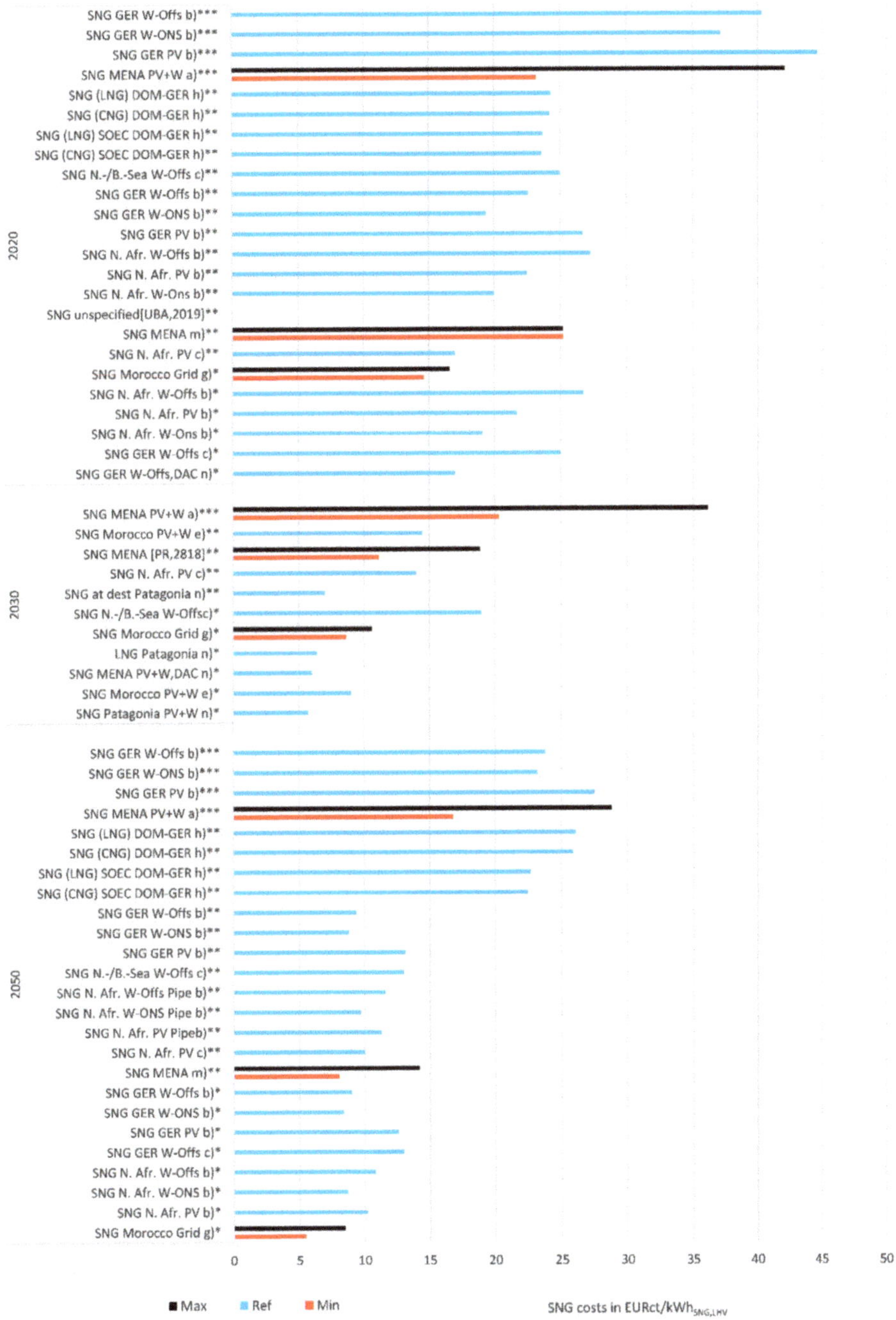

Figure A2. Cost intervals of SNG production in Germany and abroad, transport costs, and taxes/levies. * production, ** +transport, *** +taxes/levies. Source: Own elaboration based on (a) [92], (b) [89], (c) [85], (e) [87], (h) [108], (m) [112], (n) [113], (g) [109].

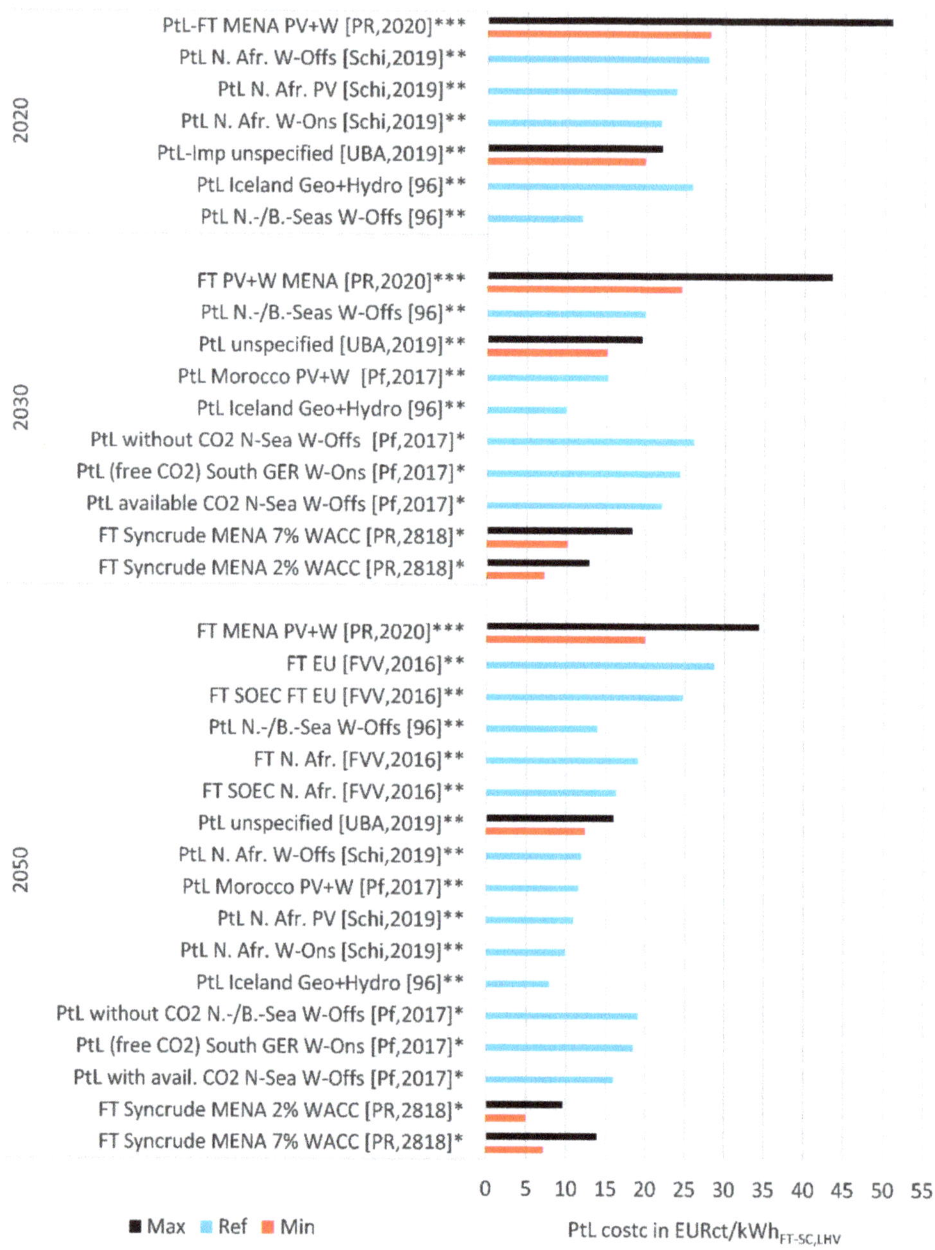

Figure A3. Cost intervals of unspecified PtL fuel production in Germany and abroad, transport costs, and taxes/levies. * production, ** +transport, *** +taxes/levies. Source: Own elaboration based on [6,85,89,92,108,112].

Appendix B

Environmental Impacts of CNG/LNG

Compared to the extensive use of petroleum-based fuels in the German transport sector today, CNG and LNG can help reduce environmental impacts. If the combustion of these gaseous fuels is taken into account, climate change impacts may be reduced due to their chemical compositions. For this reason, CNG and LNG are often referred to as bridging technologies. Thus, from an environmental perspective, vehicles that utilize natural gas-based fuels can bridge the gap between low- and zero-GHG fuels.

Due to the different states of aggregation, CNG and LNG offer different transport options. For CNG transport to the EU and Germany, transport by pipeline can be assumed. For example, an LCA publication by Heidt et al. [240] regarding the environmental effects of CNG for road transport in Germany considered a transport distance of more than 4000 km on average to the EU for the year 2013 and assumed a transport distance of 7000 km for the year 2030. Heidt et al. [240] determined the value of 17.3 g CO_{2eq}/MJ as the WtT emission factor for CNG in the case of 4000 km as the transport distance. For a transport distance of 7000 km, 20.9 g CO_{2eq}/MJ was calculated [240].

An established and well-known series of WtT studies of fuels with Europe as the geographical scope is the JEC WtT reports. In the most recent version, six CNG pathways were analyzed that differed in their assumptions along the pathways (transport distances, production, conditioning assumptions, etc.) [192]. The WtT GHG emissions for CNG vary between 11 and around 17 g CO_{2eq} depending on the respective pathway. Production and conditioning is considered to be the first step in the pathways. A fixed WtT GHG emission of 4 g CO_{2eq}/MJ for production and conditioning was assumed for all pathways. With respect to the transportation to market, as a next step of the pathways, major differences can be noted. Pathways that consider a typical natural gas EU mix with transport distances of only 1900 km to the EU border and average distances of 500 km within it demonstrate the contributions of the transport step being below WtT GHG emissions of 4 g CO_{2eq}/MJ. In a pathway with greater transport distances (4300 km to the EU border and 700 km within it), WtT GHG emissions of just under 10 g CO_{2eq}/MJ were calculated. Another essential contribution to environmental impacts was given by the step of conditioning and distribution, which also includes the compression of gas at service stations. For these steps, values of around 4 g CO_{2eq}/MJ were identified [192].

The LNG supply is also multi-stage. Gas production is followed by a processing step. This is usually then followed by gas transport (e.g., by pipeline) before gas liquefaction occurs. The LNG produced in this process is usually then transported over longer distances (especially by ship). At the end of the chain, depending on the application option, regasification may be conducted. In the most recent version of the JEC WtT report, one LNG pathway is compared to the CNG ones [192]. Around 18 g CO_{2eq}/MJ was calculated for this pathway. Compared to the CNG pathways, this one accompanies an additional noteworthy contribution. The contribution of around 4 g CO_{2eq}/MJ is given by liquefaction. For LNG, an LCA study by Wachsmuth et al. [241] offers more versatile insights in the environmental performance of LNG pathways for Germany. This study considered the environmental impacts for five different cases. A broad range of climate change impacts was featured, spanning from nearly 15 g CO_{2eq}/MJ up to around 29 g CO_{2eq}/MJ. This large range is especially due to deliveries to Germany from regions of the world whose distances vary. The supply of conventional LNG from Katar under the considered conditions carries the lowest impacts. In contrast, the supply of unconventional LNG from Australia (Queensland) exhibited the highest. In the case of unconventional LNG sources, gas production was revealed to be the main contributor to upstream GHG emissions. Furthermore, transport distances revealed a major influence on environmental impacts.

References

1. Intergovernmental Panel on Climate Change. *Climate Change 2013: The Physical Science Basis. Contribution of Working Group I to the Fifth Assessment Report of the Intergovernmental Panel on Climate Change*; Stocker, T.F., Qin, D., Plattner, G.-K., Tignor, M., Allen, S.K., Boschung, J., Nauels, A., Xia, Y., Bex, V., Midgley, P.M., Eds.; Cambridge University Press: Cambridge, UK; New York, NY, USA, 2013; p. 1535.
2. United Nations. Paris Agreement. In Proceedings of the UN Climate Change Conference (COP21), Paris, France, 30 November–12 December 2015.
3. Welke, M.; Beck, M. *Klimaschutz in Zahlen, Fakten, Trends und Impulse Deutscher Klimapolitik Ausgabe 2020*; Bundesministerium für Umwelt, Naturschutz und nukleare Sicherheit: Berlin, Germany, 2020.
4. Schemme, S.; Breuer, J.L.; Köller, M.; Meschede, S.; Walman, F.; Samsun, R.C.; Peters, R.; Stolten, D. H2-based synthetic fuels: A techno-economic comparison of alcohol, ether and hydrocarbon production. *Int. J. Hydrog. Energy* **2020**, *45*, 5395–5414. [CrossRef]
5. Prussi, M.; Yugo, M.; Prada, L.D.; Padella, M.; Edwards, R. *JEC Well-To-Wheels Report v5*; Publications Office of the European Union: Luxembourg, 2020.
6. Pfennig, M.; Gerhardt, N.; Pape, C.; Böttger, D. *Mittel-und Langfristige Potenziale von PtL-und H2-Importen aus Internationalen EE-Vorzugsregionen*; Fraunhofer-Institut für Windenergie und Energiesystemtechnik: Kassel, Germany, 2017.
7. Schorn, F.; Breuer, J.L.; Samsun, R.C.; Schnorbus, T.; Heuser, B.; Peters, R.; Stolten, D. Methanol as a renewable energy carrier: An assessment of production and transportation costs for selected global locations. *Adv. Appl. Energy* **2021**, *3*, 100050. [CrossRef]
8. Bracker, J.; Timpe, C. *An Outline of Sustainability Criteria for Synthetic Fuels Used in Transport*; Öko-Institut e.V.: Freiburg, Germany, 2017.
9. Liebich, A.; Fröhlich, T.; Münter, D.; Fehrenbach, H.; Giegrich, J.; Köppen, S.; Dünnebeil, F.; Knörr, W.; Biemann, K.; Simon, S.; et al. *Detailed Analyses of the System Comparison of Storable Energy Carriers from Renewable Energies—Final Report*; Federal Environment Agency: Dessau-Roßlau, Germany, 2021. Available online: https://inis.iaea.org/search/search.aspx?orig_q=reportnumber:%22UBA-FB--000263/ANH%22 (accessed on 9 December 2021).
10. Liebich, A.; Fröhlich, T.; Münter, D.; Fehrenbach, H.; Giegrich, J.; Köppen, S.; Dünnebeil, F.; Knörr, W.; Biemann, K.; Simon, S.; et al. *Detailed Analyses of the System Comparison of Storable Energy Carriers from Renewable Energies—Annex*; Federal Environment Agency: Dessau-Roßlau, Germany, 2021. Available online: https://www.researchgate.net/publication/350121764_Full_text_SYSEET_Detailed_analyses_of_the_system_comparison_of_storable_energy_carriers_from_renewable_energies_Annex (accessed on 9 December 2021).
11. Ziolkowska, J.R. Chapter 1—Biofuels technologies: An overview of feedstocks, processes, and technologies. In *Biofuels for a More Sustainable Future*; Elsevier: Amsterdam, The Netherlands, 2020; pp. 1–19. [CrossRef]
12. Peters, R.; Decker, M.; Eggemann, L.; Schemme, S.; Schorn, F.; Breuer, J.L.; Weiske, S.; Pasel, J.; Samsun, R.C.; Stolten, D. Thermodynamic and ecological preselection of synthetic fuel intermediates from biogas at farm sites. *Energy Sustain. Soc.* **2020**, *10*, 4. [CrossRef]
13. Peters, R.; Baltruweit, M.; Grube, T.; Samsun, R.C.; Stolten, D. A techno economic analysis of the power to gas route. *J. CO_2 Util.* **2019**, *34*, 616–634. [CrossRef]
14. Bruchof, D. Energiewirtschaftliche Verkehrsstrategie—Möglichkeiten und Grenzen Alternativer Kraftstoffe und Antriebe in Deutschland und der EU-27. Ph.D. Thesis, Universität Stuttgart, Stuttgart, Germany, 2012.
15. Mankins, J.C. *Technology Readiness Levels: A White Paper*; NASA, Office of Space Access and Technology, Advanced Concepts Office: Washington, DC, USA, 1995.
16. Assistant Secretary of Defense for Research and Engineering. *Technology Readiness Assessment (TRA) Guidance*; Department of Defense: Washington, DC, USA, 2011.
17. European Space Agency. Technology Readiness Level (TRL). Available online: http://sci.esa.int/sci-ft/50124-technology-readiness-level/ (accessed on 29 July 2019).
18. European Commission. Horizon 2020—Work Programme 2016–2017. In *20. General Annexes*; European Commission: Brussels, Belgium, 2017; p. 29.
19. *ISO 16290:2013*; Space Systems—Definition of the Technology Readiness Levels (TRLs) and Their Criteria of Assessment. International Organization for Standardization: Geneva, Switzerland, 2013.
20. Rose, A.D.; Buna, M.; Strazza, C.; Olivieri, N.; Stevens, T.; Peeters, L.; Tawil-Jamault, D. *Technology Readiness Level Guidance Principles for Renewable Energy Technologies—Final Report*; European Commission: Brussels, Belgium, 2017.
21. Australian Renewable Energy Agency. Commercial Readiness Index for Renewable Energy Sector. Available online: https://arena.gov.au/assets/2014/02/Commercial-Readiness-Index.pdf (accessed on 18 March 2021).
22. Wermuth, M.; Neef, C.; Wirth, R.; Hanitz, I.; Löhner, H.; Hautzinger, H.; Stock, W.; Pfeiffer, M.; Fuchs, M.; Lenz, B.; et al. *Kraftfahrzeugverkehr in Deutschland 2010*; Bundesministerium für Verkehr, Bau und Stadtentwicklung: Braunschweig, Germany, 2012.
23. Nobis, C.; Kuhnimhof, T. *Mobilität in Deutschland—MiD Ergebnisbericht, Studie von Infas, DLR, IVT und Infas 360 im Auftrag des Bundesministers für Verkehr und Digitale Infrastruktur*; Infas Institut für Angewandte Sozialwissenschaft GmbH: Bonn/Berlin, Germany, 2018.
24. Radke, S. *Verkehr in Zahlen 2019/2020*; Bundesministerium für Verkehr und digitale Infrastruktur: Flensburg, Germany, 2019.
25. Tesla, Inc. WLTP Reichweite Tesla Model S. Available online: https://www.tesla.com/de_de/models (accessed on 14 May 2020).

26. Breuer, J.L.; Can Samsun, R.; Peters, R.; Stolten, D. Road traffic volume map 2014 for North Rhine-Westphalia, Germany. *Harv. Dataverse* **2019**. [CrossRef]
27. Breuer, J.L.; Samsun, R.C.; Peters, R.; Stolten, D. The impact of diesel vehicles on NOx and PM10 emissions from road transport in urban morphological zones: A case study in North Rhine-Westphalia, Germany. *Sci. Total Environ.* **2020**, *727*, 138583. [CrossRef] [PubMed]
28. McKinsey & Company. *Hydrogen-Powered Aviation a Fact-Based Study of Hydrogen Technology, Economics, and Climate Impact by 2050*; Publications Office of the European Union: Luxembourg, 2020.
29. *Bundesnetzagentur für Elektrizität, Gas, Telekommunikation, Post und Eisenbahnen*; Marktuntersuchung Eisenbahnen: Bonn, Germany, 2019.
30. Deutsche Bahn Netze. Infrastrukturregister: Grundsätze Stand Juni 2019, Version 4. Available online: https://fahrweg.dbnetze.com/fahrweg-de/kunden/nutzungsbedingungen/infrastrukturregister/grundsaetze-1369462 (accessed on 5 August 2019).
31. Statistisches Bundesamt. *Güterverkehrsstatistik der Binnenschifffahrt 2016*; Statistisches Bundesamt: Wiesbaden, Germany, 2017.
32. Helgeson, B.; Peter, J. The role of electricity in decarbonizing European road transport—Development and assessment of an integrated multi-sectoral model. *Appl. Energy* **2020**, *262*, 114365. [CrossRef]
33. Köhler, J.; Kirsch, D.; Klukas, A.; Timmerberg, S.; Kaltschmitt, M. *Teilstudie "Studie über die Marktreife von Erdgasmotoren in der Binnen-und Seeschifffahrt"*; Wissenschaftliche Beratung des BMVI zur Mobilitäts-und Kraftstoffstrategie: Karlsruhe, Germany, 2018.
34. Robinius, M.; Markewitz, P.; Lopion, P.; Kullmann, F.; Heuser, P.-M.; Syranidis, K.; Cerniauskas, S.; Schöb, T.; Reuß, M.; Ryberg, S.; et al. Wege Für Die Energiewende Kosteneffiziente und klimagerechte Transformationsstrategien für das deutsche Energiesystem bis zum Jahr 2050. *Energy Environ.* **2020**, *499*, 1–141.
35. Van Wijk, A.; van der Roest, E.; Boere, J. *Solar Power to the People*; IOS Press BV: Amsterdam, The Netherlands, 2017.
36. Hydrogen Council. *Path to Hydrogen Competitiveness a Cost Perspective*; Hydrogen Council: Brussels, Belgium, 2020.
37. Hydrogen Council, McKinsey & Company. *Hydrogen Insights a Perspective on Hydrogen Investment, Market Development and Cost Competitiveness*; Hydrogen Council: Brussels, Belgium, 2021.
38. Transport & Environment. *RED II and Advanced Biofuels, Recommendations about Annex IX of the Renewable Energy Directive and Its Implementation at National Level*; Transport & Environment: Brussels, Belgium, 2020.
39. European Union. *Directive (EU) 2018/2001 of the European Parliament and of the Council of 11 December 2018 on the Promotion of the Use of Energy from Renewable Sources (Recast)*; European Union: Brussels, Belgium, 2018.
40. Gemeinsame Forschungsstelle, Institut für Energie und Verkehr; Hamje, H.; Hass, H.; Lonza, L.; Maas, H.; Reid, A.; Rose, K.D.; Venderbosch, T. *EU Renewable Energy Targets in 2020: Revised Analysis of Scenarios for Transport Fuels*; Publications Office: Brussels, Belgium, 2014.
41. European Commission. *Report from the Commission to the European Parliament, the Council, the European Economic and Social Committee and the Committee of the Regions on the Status of Production Expansion of Relevant Food and Feed Crops Worldwide*; European Commission: Brussels, Belgium, 2019.
42. European Commission. *Commission Delegated Regulation (EU) 2019/807 of 13 March 2019 Supplementing Directive (EU) 2018/2001 of the European Parliament and of the Council as Regards the Determination of High Indirect Land-Use Change-Risk Feedstock for Which a Significant Expansion of the Production Area into Land with High Carbon Stock Is Observed and the Certification of Low Indirect Land-Use Change-Risk Biofuels, Bioliquids and Biomass Fuels*; European Union: Brussels, Belgium, 2019.
43. Mineralöl Wirtschafts Verband e.V. Jahresbericht 2018, Vision 2050, Flüssige Kraftstoffe bewegen Deutschland auch 2050. Available online: https://www.mwv.de/wp-content/uploads/2016/06/180830_MWV_Jahresbericht-2018_RZ_Web_es_small.pdf (accessed on 20 February 2020).
44. Arbeitsgruppe Erneuerbare Energien-Statistik. *Erneuerbare Energien in Deutschland Daten zur Entwicklung im Jahr 2019*; Umweltbundesamt: Dessau-Roßlau, Germany, 2020.
45. Referat 221—Grundsatzangelegenheiten der Gruppe 22, Anerkennungs- und Akkreditierungsfragen, nachhaltige Biomasse. *Evaluations-und Erfahrungsbericht für das Jahr 2017—Biomassestrom-Nachhaltigkeitsverordnung—Biokraftstoff-Nachhaltigkeitsverordnung*; Bundesanstalt für Landwirtschaft und Ernährung: Bonn, Germany, 2018.
46. Fehrenbach, H. *Einsatz von Biokraftstoffen im Verkehrssektor bis 2030*; IFEU—Institut für Energie-und Umweltforschung GmbH: Heidelberg, Germany, 2019.
47. Billig, E.; Decker, M.; Benzinger, W.; Ketelsen, F.; Pfeifer, P.; Peters, R.; Stolten, D.; Thrän, D. Non-fossil CO_2 recycling—The technical potential for the present and future utilization for fuels in Germany. *J. CO_2 Util.* **2019**, *30*, 130–141. [CrossRef]
48. Brosowski, A.; Thrän, D.; Mantau, U.; Mahro, B.; Erdmann, G.; Adler, P.; Stinner, W.; Reinhold, G.; Hering, T.; Blanke, C. A review of biomass potential and current utilisation—Status quo for 93 biogenic wastes and residues in Germany. *Biomass Bioenergy* **2016**, *95*, 257–272. [CrossRef]
49. DBFZ Deutsches Biomasseforschungszentrum gemeinnützige GmbH; Brosowski, A.; Adler, P.; Erdmann, G.; Stinner, W.; Thrän, D.; Mantau, U.; Blanke, C.; Mahro, B.; Hering, T.; et al. *Biomassepotentiale von Rest-und Abfallstoffen*; Fachagentur Nachwachsende Rohstoffe e. V. (FNR): Gülzow-Prüzen, Germany, 2015.
50. Fehrenbach, H.; Giegrich, J.; Köppen, S. *BioRest: Verfügbarkeit und Nutzungsoptionen Biogener Abfall-und Reststoffe im Energiesystem (Strom-, Wärme-und Verkehrssektor)*; Umweltbundesamt TEXTE: Berlin, Germany, 2019.
51. Johansson, S. Limits to biofuels. *EPJ Web Conf.* **2013**, *54*, 01014. [CrossRef]

52. De Castro, C.; Carpintero, Ó.; Frechoso, F.; Mediavilla, M.; de Miguel, L.J. A top-down approach to assess physical and ecological limits of biofuels. *Energy* **2014**, *64*, 506–512. [CrossRef]
53. Soler, A.; Yugo, M. *Role of E-Fuels in the European Transport System—Literature Review*; No. 14/19; Concawe: Brussels, Belgium, 2020.
54. Kasten, P.; Heinemann, C. *Kein Selbstläufer: Klimaschutz und Nachhaltigkeit durch PtX.*; Öko-Institut e.V.: Berlin, Germany, 2019.
55. International Energy Agency. Technology Roadmap Biofuels for Transport. Available online: https://webstore.iea.org/download/direct/632?fileName=Biofuels_Roadmap_WEB.pdf (accessed on 18 March 2021).
56. Eurostat Statistics Explained. Land Cover Statistics. Available online: https://ec.europa.eu/eurostat/statistics-explained/index.php?title=Land_cover_statistics#Land_cover_in_the_EU (accessed on 25 November 2020).
57. Zelt, O.; Kobiela, G.; Ortiz, W.; Scholz, A.; Monnerie, N.; Rosenstiel, A.; Viebahn, P. *Multikriterielle Bewertung von Bereitstellungstechnologien Synthetischer Kraftstoffe*; Wuppertal Institut für Klima, Umwelt, Energie gGmbH: Wuppertal, Germany, 2021.
58. *ASTM D7566-20b*; Standard Specification for Aviation Turbine Fuel Containing Synthesized Hydrocarbons. ASTM International: West Conshohocken, PA, USA, 2020.
59. Wormslev, E.C.; Broberg, M.K. *Nordic Perspectives on the Use of Advanced Sustainable Jet Fuel for Aviation—Update 2019*; Nordic Energy Research: Kopenhagen, Denmark, 2020.
60. ICAO Secretariat. Estimated Prices of Aviation Alternative Fuels. In Proceedings of the Conference on Aviation and Alternative Fuels, Mexico City, Mexico, 11 October 2017.
61. Sustainable Aviation. Sustainable Aviation Fuels Road-Map. Available online: https://www.sustainableaviation.co.uk/wp-content/uploads/2020/02/SustainableAviation_FuelReport_20200231.pdf (accessed on 26 March 2020).
62. Schmidt, P.; Batteiger, V.; Roth, A.; Weindorf, W.; Raksha, T. Power-to-Liquids as Renewable Fuel Option for Aviation: A Review. *Chem. Ing. Tech.* **2018**, *90*, 127–140. [CrossRef]
63. Tabak, S.A.; Avidan, A.A.; Krambeck, F.J. *Production of Synthetic Gasoline and Diesel Fuel from Non-Petroleum Resources*; Mobil Research and Development Corp., Research Dept., Paulsboro Lab.: Paulsboro, NJ, USA, 1986; pp. 293–299.
64. Tabak, S.A.; Yurchak, S. Conversion of methanol over ZSM-5 to fuels and chemicals. *Catal. Today* **1990**, *6*, 307–327. [CrossRef]
65. Ruokonen, J.; Nieminen, H.; Dahiru, A.R.; Laari, A.; Koiranen, T.; Laaksonen, P.; Vuokila, A.; Huuhtanen, M. Modelling and Cost Estimation for Conversion of Green Methanol to Renewable Liquid Transport Fuels via Olefin Oligomerisation. *Processes* **2021**, *9*, 1046. [CrossRef]
66. Zech, K.; Naumann, K.; Müller-Langer, F.; Schmidt, P.; Weindorf, W.; Altmann, M.; Michalski, J.; Niklaß, M.; Meyer, H.; Lischke, A.; et al. *Biokerosin und EE-Kerosin für Die Luftfahrt der Zukunft—von der Theorie zu Pilotvorhaben*; Federal Ministry for Digital and Transport: Leipzig, Germany; Munich, Germany; Berlin/Heidelberg, Germany, 2011.
67. Fulcrum. Sierra BioFuels Plant. Available online: http://fulcrum-bioenergy.com/facilities/ (accessed on 3 December 2020).
68. Red Rock Biofuels. Red Rock Biofuels—Lakeview Plant. Available online: https://www.redrockbio.com/lakeview-site/ (accessed on 3 December 2020).
69. Neste Oyj. What Is Neste MY Sustainable Aviation Fuel? Available online: https://www.neste.com/products/all-products/aviation/key-benefits (accessed on 3 December 2020).
70. BETO-Funded Technology Produces Jet Fuel for Virgin Atlantic. Available online: https://www.energy.gov/eere/bioenergy/articles/beto-funded-technology-produces-jet-fuel-virgin-atlantic (accessed on 3 December 2020).
71. Ekobenz. Innovative Production of Synthetic Fuels. Available online: http://ekobenz.com/production-plant (accessed on 3 December 2020).
72. Stark, A. Lanzatech Wins Bid for World's First Large Scale ATJ Facility. Available online: www.process-worldwide.com (accessed on 21 December 2021).
73. European Comission. Production of Fully Synthetic Paraffinic Jet Fuel from Wood and Other Biomass. Available online: https://cordis.europa.eu/project/id/612763 (accessed on 3 December 2020).
74. Euglena Co. Ltd. ASTM D7566 Revised to Add a New Standard for Jet Fuel Produced through the BIC Process, the Technology Employed in Euglena Co's Demo Plant. Available online: https://www.euglena.jp/pdf/n20200131.pdf (accessed on 21 December 2021).
75. IHI. Bio-jet Fuel Manufactured from Microalgae Receives ASTM International Standard Certification-Contributing to the Reduction of CO_2 Emissions from Aircraft. Available online: https://www.ihi.co.jp/en/all_news/2020/other/1196667_2042.html (accessed on 21 December 2021).
76. Müller-Langer, F.; Majer, S.; O'Keeffe, S. Benchmarking biofuels—A comparison of technical, economic and environmental indicators. *Energy Sustain. Soc.* **2014**, *4*, 20. [CrossRef]
77. Leitner, W.; Klankermayer, J.; Pischinger, S.; Pitsch, H.; Kohse-Hoinghaus, K. Advanced Biofuels and Beyond: Chemistry Solutions for Propulsion and Production. *Angew. Chem. Int. Ed.* **2017**, *56*, 5412–5452. [CrossRef] [PubMed]
78. Schemme, S.; Breuer, J.L.; Samsun, R.C.; Peters, R.; Stolten, D. Promising catalytic synthesis pathways towards higher alcohols as suitable transport fuels based on H_2 and CO_2. *J. CO_2 Util.* **2018**, *27*, 223–237. [CrossRef]
79. Schemme, S.; Meschede, S.; Köller, M.; Samsun, R.C.; Peters, R.; Stolten, D. Property Data Estimation for Hemiformals, Methylene Glycols and Polyoxymethylene Dimethyl Ethers and Process Optimization in Formaldehyde Synthesis. *Energies* **2020**, *13*, 3401. [CrossRef]
80. Deutsche Energie-Agentur. Audi E-Gas Projekt. Available online: https://www.powertogas.info/projektkarte/audi-e-gas-projekt/ (accessed on 10 January 2020).

81. BMWi, Federal Ministry for Economic Affairs and Climate Action. *The National Hydrogen Strategy*; BMWi: Berlin, Germany, 2020.
82. Bründlinger, T.; König, J.E.; Frank, O.; Gründig, D.; Jugel, C.; Kraft, P.; Krieger, O.; Mischinger, S.; Prein, D.P.; Seidl, H.; et al. *Dena-Leitstudie Integrierte Energiewende*; Deutsche Energie-Agentur: Berlin, Germany, 2018.
83. Merten, F.; Scholz, A.; Krüger, C.; Heck, S.; Girard, Y.; Mecke, M.; Goerge, M. *Bewertung der Vor-und Nachteile von Wasserstoffimporten im Vergleich zur Heimischen Erzeugung*; Wuppertal Institut, DIW Econ: Wuppertal, Germany, 2020.
84. Dena, German Energy Agency. *Zwischenbericht, dena-Leitstudie Aufbruch Klimaneutralität, Ein Blick in die Werkstatt: Erste Erkenntnisse und Ableitungen Zentraler Handlungsfelder*; Deutsche Energie-Agentur GmbH: Berlin, Germany, 2021.
85. Perner, J.; Unteutsch, M.; Lövenich, A. *The Future Cost of Electricity-Based Synthetic Fuels*; Agora Energiewende, Frontier Economics: Berlin, Germany, 2018.
86. Fasihi, M.; Breyer, C. Baseload electricity and hydrogen supply based on hybrid PV-wind power plants. *J. Clean. Prod.* **2020**, *243*, 118466. [CrossRef]
87. Hank, C.; Sternberg, A.; Köppel, N.; Holst, M.; Smolinka, T.; Schaadt, A.; Hebling, C.; Henning, H.-M. Energy efficiency and economic assessment of imported energy carriers based on renewable electricity. *Sustain. Energy Fuels* **2020**, *4*, 2256–2273. [CrossRef]
88. Ishmam, S.; Winkler, C.; Sanchez, P.-E.; Ojieabu, N.; Heinrichs, H.; Agbo, S.; Kuckertz, P.; Khuat, H.; Patil, S.; Kotzur, L.; et al. H2ATLAS-AFRICA. Available online: https://africa.h2atlas.de/ (accessed on 21 December 2021).
89. Schindler, J. *Gestehungskosten von PtX-Produkten im Vergleich Zwischen Deutschland und Nordafrika*; EnergieAgentur.NRW: Wuppertal, Germany, 2019.
90. Eichhammer, W.; Oberle, S.; Händel, M.; Boie, I.; Gnann, T.; Wietschel, M.; Lux, B. *Study on the Opportunities of "Power-to-X" in Morocco*; Fraunhofer Institute for Systems and Innovation Research: Karlsruhe, Germany, 2019.
91. IEA, International Energy Agency. The Future of Hydrogen Seizing Today's Opportunities. Available online: https://iea.blob.core.windows.net/assets/9e3a3493-b9a6-4b7d-b499-7ca48e357561/The_Future_of_Hydrogen.pdf (accessed on 9 November 2021).
92. Kreidelmeyer, S.; Dambeck, H.; Kirchner, A.; Wünsch, M. *Kosten und Transformationspfade für Strombasierte Energieträger*; Prognos AG: Basel, Switzerland, 2020.
93. IEA, International Energy Agency. IEA G20 Hydrogen Report: Assumptions. Available online: https://iea.blob.core.windows.net/assets/29b027e5-fefc-47df-aed0-456b1bb38844/IEA-The-Future-of-Hydrogen-Assumptions-Annex_CORR.pdf (accessed on 21 December 2021).
94. Robinius, M. *Strom- und Gasmarktdesign zur Versorgung des Deutschen Straßenverkehrs mit Wasserstoff*; Forschungszentrum Jülich: Jülich, Germany, 2016; Volume 300.
95. Kawasaki. Liquefied Hydrogen Carrier with World's Highest Carrying Capacity—AiP Obtained from ClassNK. Available online: https://global.kawasaki.com/en/corp/newsroom/news/detail/?f=20210506_9983 (accessed on 15 December 2021).
96. Fraunhofer IEE. PtX-Atlas. Available online: https://maps.iee.fraunhofer.de/ptx-atlas/ (accessed on 20 December 2021).
97. Huss, A.; Maas, H.; Hass, H. *Well-to-Wheels Analysis of Future Automotive Fuels and Powertrains in the European Context TANK-TO-WHEELS (TTW) Report Version 4*; Publications Office of the European Union: Luxembourg, 2013.
98. Robinius, M.; Linßen, J.; Grube, T.; Reuß, M.; Stenzel, P.; Syranidis, K.; Kuckertz, P.; Stolten, D. Comparative Analysis of Infrastructures: Hydrogen Fueling and Electric Charging of Vehicles. *Schr. Des Forsch. Jülich Reihe Energ. Umw. Energy Environ.* **2018**, *408*, 108.
99. Union zur Förderung von Oel-und Proteinpflanzen e.V. UFOP Report on Global Market Supply 2020/2021. Available online: https://www.ufop.de/files/7216/1649/5848/UFOP_SupplyReport_2020-2021__120321.pdf (accessed on 2 December 2021).
100. Völler, K.; Reinholz, T. *Branchenbarometer Biomethan 2019*; Deutsche Energie-Agentur GmbH: Berlin, Germany, 2019.
101. Bundesministerium für Wirtschaft und Energie. Gesamtausgabe der Energiedaten—Datensammlung des BMWi. Available online: https://www.bmwi.de/Redaktion/DE/Binaer/Energiedaten/energiedaten-gesamt-xls.html (accessed on 18 February 2021).
102. Detz, R.J.; Reek, J.N.H.; van der Zwaan, B.C.C. The future of solar fuels: When could they become competitive? *Energy Environ. Sci.* **2018**, *11*, 1653–1669. [CrossRef]
103. Union zur Förderung von Oel-und Proteinpflanzen e.V. UFOP Report on Global Market Supply 2018/2019. Available online: https://www.ufop.de/files/4815/4695/8891/WEB_UFOP_Report_on_Global_Market_Supply_18-19.pdf (accessed on 18 February 2021).
104. Jensterle, M.; Narita, J.; Piria, R.; Schröder, J.; Steinbacher, K.; Wahabzada, F.; Zeller, T.; Crone, K.; Löchle, S. *Grüner Wasserstoff: Internationale Kooperations-Potenziale für Deutschland Kurzanalyse zu Ausgewählten Aspekten Potenzieller Nicht-EU-Partnerländer*; Deutsche Energie-Agentur, Deutsche Gesellschaft für Internationale Zusammenarbeit, NAVIGANT, Adelphi: Berlin, Germany, 2019.
105. Michalski, J.; Altmann, M.; Bünger, U.; Weindorf, W. *Wasserstoffstudie Nordrhein-Westfalen*; Ministerium für Wirtschaft, Innovation, Digitalisierung und Energie des Landes Nordrhein-Westfalen: Düsseldorf, Germany, 2019.
106. Gerhardt, N.; Bard, J.; Schmitz, R.; Beil, M.; Pfennig, M.; Kneiske, T. *Wasserstoff im Zukünftigen Energiesystem: Fokus Gebäudewärme*; Fraunhofer-Institut für Energiewirtschaft und Energiesystemtechnik: Kassel, Germany, 2020.
107. Mottschall, M.; Kasten, P.; Kühnel, S.; Minnich, L. *Sensitivitäten zur Bewertung der Kosten verschiedener Energieversorgungs-Optionen des Verkehrs bis zum Jahr 2050*; TEXTE 114/2019; Umweltbundesamt, Öko-Institut: Dessau-Roßlau, Germany, 2019.
108. Schmidt, P.R.; Zittel, W.; Weindorf, W.; Raksha, T. *Renewables in Transport 2050*; Forschungsvereinigung Verbrennungskraftmaschinen e.V.: Frankfurt am Main, Germany, 2016.

109. Kramer, U.; Ortloff, F.; Stollenwerk, S.; Thee, R. *Defossilisierung des Transportsektors*; Forschungsvereinigung Verbrennungskraftmaschinen e.V.: Frankfurt am Main, Germany, 2018.
110. Perner, J.; Unteutsch, M.; Lövenich, A. *Die Zukünftigen Kosten Strombasierter Synthetischer Brennstoffe*; Agora Verkehrswende, Agora Energiewende, Frontier Economics: Berlin, Germany, 2018.
111. Terlouw, W.; Peters, D.; Tilburg, J.v.; Schimmel, M.; Berg, T.; Cihlar, J.; Mir, G.U.R.; Spöttle, M.; Staats, M.; Lejaretta, A.V.; et al. *Gas for Climate. The Optimal Role for Gas in a Net-Zero Emissions Energy System*; Navigant Netherlands B.V.: Utrecht, The Netherlands, 2019.
112. Hobohm, J.; der Maur, A.A.; Dambeck, H.; Kemmler, A.; Koziel, S.; Kreidelmeyer, S.; Piégsa, A.; Wendring, P.; Meyer, B.; Apfelbacher, A.; et al. *Status und Perspektiven Flüssiger Energieträger in der Energiewende*; Fraunhofer Institut für Umwelt-, Sicherheits-und Energietechnik: Oberhausen, Germany; Deutsches Biomasseforschungszentrum gemeinnützige GmbH: Leipzig, Germany, 2018.
113. Fasihi, M.; Breyer, C. Synthetic Fuels and Chemicals: Options and Systemic Impact. In Proceedings of the Strommarkttreffen, Berlin, Germany, 29 June 2018.
114. *DIN EN 590*; Automotive Fuels—Diesel—Requirements and Test Methods. Deutsches Institut für Normung e.V.: Berlin, Germany, 2014.
115. *DIN EN 228*; Automotive Fuels—Unleaded Petrol—Requirements and Test Methods. Deutsches Institut für Normung e.V.: Berlin, Germany, 2017.
116. Schripp, T.; Herrmann, F.; Oßwald, P.; Köhler, M.; Zschocke, A.; Weigelt, D.; Mroch, M.; Werner-Spatz, C. Particle emissions of two unblended alternative jet fuels in a full scale jet engine. *Fuel* **2019**, *256*, 115903. [CrossRef]
117. Sasol AG. Sasol Takes to the Skies with the World's First Fully Synthetic Jet Fuel. Available online: https://www.sasol.com/media-centre/media-releases/sasol-takes-skies-world-s-first-fully-synthetic-jet-fuel (accessed on 21 December 2021).
118. International Air Transport Association. *IATA Guidance Material for Sustainable Aviation Fuel Management*; International Air Transport Association: Montreal, QC, Canada; Geneva, Switzerland, 2012.
119. Zech, K.; Naumann, K.; Müller-Langer, F.; Schmidt, P.; Weindorf, W.; Mátra, Z. *Drop-In-Kraftstoffe für die Luftfahrt. Studie im Rahmen des Auftrags» Wissenschaftliche Begleitung, Unterstützung und Beratung des BMVI in den Bereichen Verkehr und Mobilität mit besonderem Fokus auf Kraftstoffe und Antriebstechnologien sowie Energie und Klima für das Bundesministerium für Verkehr und digitale Infrastruktur (BMVI)*; Deutsches Zentrum Für Luft-Und Raumfahrt e. V.(DLR): Berlin, Germany, 2014.
120. Bauen, A.; Gomez, I.; OudeNijeweme, D.; Paraschiv, M. *Alternative Fuels—Expert Group Report*; European Commission: Brussels, Belgium, 2017.
121. *ASTM D4806-13a*; Standard Specification for Denatured Fuel Ethanol for Blending with Gasolines for Use as Automotive Spark-Ignition Engine Fuel. ASTM International: West Conshohocken, PA, USA, 2013. [CrossRef]
122. *ASTM D7862-13*; Standard Specification for Butanol for Blending with Gasoline for Use as Automotive Spark-Ignition Engine Fuel. ASTM International: West Conshohocken, PA, USA, 2013.
123. Institut für nachhaltige Wirtschaft und Logistik. *Potenzialanalyse Methanol als Emissionsneutraler Energieträger für Schifffahrt und Energiewirtschaft*; Maritimes Cluster Norddeutschland e. V.: Elsfleth, Germany, 2018.
124. Wei, H.; Feng, D.; Pan, J.; Shao, A.; Pan, M. Knock characteristics of SI engine fueled with n-butanol in combination with different EGR rate. *Energy* **2017**, *118*, 190–196. [CrossRef]
125. *ASTM D975-20c*; Standard Specification for Diesel Fuel. ASTM International: West Conshohocken, PA, USA, 2020. [CrossRef]
126. *DIN EN 15940*; Automotive Fuels—Paraffinic Diesel Fuel from Synthesis or Hydrotreatment—Requirements and Test Methods. Deutsches Institut für Normung e.V.: Berlin, Germany, 2019. [CrossRef]
127. *DIN EN 16709*; Automotive Fuels—High FAME Diesel Fuel (B20 and B30)—Requirements and Test Methods. Deutsches Institut für Normung e.V.: Berlin, Germany, 2019.
128. *DIN EN 14214*; Liquid Petroleum Products—Fatty Acid Methyl Esters (Fame) for Use in Diesel Engines and Heating Applications—Requirements and Test Methods. Deutsches Institut für Normung e.V.: Berlin, Germany, 2019.
129. Bohl, T.; Smallbone, A.; Tian, G.; Roskilly, A.P. Particulate number and NO trade-off comparisons between HVO and mineral diesel in HD applications. *Fuel* **2018**, *215*, 90–101. [CrossRef]
130. Kuronen, M.; Mikkonen, S.; Aakko, P.; Murtonen, T. Hydrotreated Vegetable Oil as Fuel for Heavy Duty Diesel Engines. *SAE Tech. Pap.* **2007**, *2007-01-4031*. [CrossRef]
131. Sayin, C.; Ozsezen, A.N.; Canakci, M. The influence of operating parameters on the performance and emissions of a DI diesel engine using methanol-blended-diesel fuel. *Fuel* **2010**, *89*, 1407–1414. [CrossRef]
132. Sayin, C. Engine performance and exhaust gas emissions of methanol and ethanol–diesel blends. *Fuel* **2010**, *89*, 3410–3415. [CrossRef]
133. Damyanov, A. Alcoholic Fuels in Diesel Engines. Methanol, Ethanol and Butanol. In *Zukünftige Kraftstoffe. ATZ/MTZ-Fachbuch*; Maus, W., Ed.; Springer: Berlin/Heidelberg, Germany, 2019. [CrossRef]
134. Rajesh Kumar, B.; Saravanan, S.; Rana, D.; Anish, V.; Nagendran, A. Effect of a sustainable biofuel—n-octanol—On the combustion, performance and emissions of a DI diesel engine under naturally aspirated and exhaust gas recirculation (EGR) modes. *Energy Convers. Manag.* **2016**, *118*, 275–286. [CrossRef]

135. Continental. Klimaneutral Fahren: Continental Testet Erfolgreich Synthetischen Diesel-Ersatzkraftstoff OME. Available online: https://www.continental.com/de/presse/pressemitteilungen/2017-08-01-klimaneutral-fahren-92062 (accessed on 21 December 2021).
136. Beidl, C.; Münz, M.; Mokros, A. Synthetische Kraftstoffe. In *Zukünftige Kraftstoffe. ATZ/MTZ-Fachbuch*; Maus, W., Ed.; Springer Vieweg: Berlin/Heidelberg, Germany, 2019; pp. 814–849. [CrossRef]
137. Hemighaus, R.; Boval, T.; Bacha, J.; Barnes, F.; Franklin, M.; Gibbs, L.; Hogue, N.; Jones, J.; Lesnini, D.; Lind, J.; et al. Aviation Fuels Technical Review. Available online: https://skybrary.aero/bookshelf/books/2478.pdf (accessed on 28 June 2021).
138. *ASTM D7467-20a*; Standard Specification for Diesel Fuel Oil, Biodiesel Blend (B6 to B20). ASTM International: West Conshohocken, PA, USA, 2020. [CrossRef]
139. *DIN EN 16734*; Automotive Fuels—Automotive B10 Diesel Fuel—Requirements and Test Methods. Deutsches Institut für Normung e.V.: Berlin, Germany, 2016.
140. Mignard, D.; Pritchard, C. On the use of electrolytic hydrogen from variable renewable energies for the enhanced conversion of biomass to fuels. *Chem. Eng. Res. Des.* **2008**, *86*, 473–487. [CrossRef]
141. Eichlseder, H.; Klell, M. Speicherung und Transport. In *Wasserstoff in der Fahrzeugtechnik: Erzeugung, Speicherung, Anwendung*; Vieweg+Teubner: Wiesbaden, Germany, 2010; pp. 95–150.
142. *Aspen Plus® V10 [Computer software]*; Aspen Technology Inc.: Delaware, DE, USA, 2017.
143. ExxonMobil. ExxonMobil Jet Fuel. Available online: https://www.exxonmobil.com/en-FI/Commercial-Fuel/pds/GL-XX-JetFuel-Series (accessed on 26 May 2020).
144. Hahne, E. *Technische Thermodynamik*; Springer: Oldenbourg, Germany, 2011.
145. National Institute of Standards and Technology. *NIST Standard Reference Simulation Website*; NIST Standard Reference Database Number: Gaithersburg, MD, USA, 2019; Volume 173.
146. Kim, S.; Thiessen, P.A.; Bolton, E.E.; Chen, J.; Fu, G.; Gindulyte, A.; Han, L.; He, J.; He, S.; Shoemaker, B.A.; et al. PubChem Substance and Compound databases. *Nucleic Acids Res.* **2016**, *44*, D1202–D1213. [CrossRef] [PubMed]
147. Liu, H.; Wang, Z.; Wang, J.; He, X.; Zheng, Y.; Tang, Q.; Wang, J. Performance, combustion and emission characteristics of a diesel engine fueled with polyoxymethylene dimethyl ethers (PODE3-4)/diesel blends. *Energy* **2015**, *88*, 793–800. [CrossRef]
148. Ulvestad, A. A brief Review of Current Lithium Ion Battery Technology and Potential Solid State Battery Technologies. *arXiv* **2018**, arXiv:1803.04317.
149. Institut für Arbeitsschutz der Deutschen Gesetzlichen Unfallversicherung. GESTIS-Stoffdatenbank Methan. Available online: http://gestis.itrust.de/nxt/gateway.dll/gestis_de/000000.xml?f=templates$fn=default.htm$vid=gestisdeu:sdbdeu$3.0 (accessed on 26 May 2020).
150. BP P.L.C. Safety data sheet Residual Fuel Oil (Flashpoint > 100 °C)—Blended. Available online: https://www.bp.com/content/dam/bp/business-sites/en/global/bp-global-energy-trading/documents/what-we-do/marine/ara/belguim-rma-rmk-terminal-high-flash.pdf (accessed on 18 February 2021).
151. Ernst, C.-S. Energetische, Ökologische und Ökonomische Lebenszyklusanalyseelektrifizierter Antriebsstrangkonzepte. Ph.D. Thesis, RWTH Aachen University, Aachen, Germany, 2015.
152. Bräunling, W.J.G. Reale Triebwerkskreisprozesse. In *Flugzeugtriebwerke: Grundlagen, Aero-Thermodynamik, Ideale und Reale Kreisprozesse, Thermische Turbomaschinen, Komponenten, Emissionen und Systeme*; Springer: Berlin/Heidelberg, Germany, 2009; pp. 1141–1251.
153. Van Basshuysen, R.; Schäfer, F. (Eds.) *Handbuch Verbrennungsmotor: Grundlagen, Komponenten, Systeme, Perspektiven. 8. Überarbeitete Auflage ed.*; Springer Vieweg: Wiesbaden, Germany, 2017; p. 1841.
154. Füßel, A. Vergleich der aktuellen BEV-und ICE-Technologie. In *Technische Potenzialanalyse der Elektromobilität*; Springer Vieweg: Wiesbaden, Germany, 2017; pp. 15–33.
155. Zerta, M.; Schmidt, P.; Weindorf, W.; Bünger, U.; Langfeldt, L.; Scholz, B.; Giebel, L.-V.; Klemm, P.; Sattler, G. *SHIPFUEL–Strombasierte Kraftstoffe für Brennstoffzellen in der Binnenschifffahrt–Hintergrundbericht; Studie im Auftrag der Nationalen Organisation Wasserstoff-und Brennstoffzellentechnologie (NOW) für das Bundesministerium für Verkehr und Digitale Infrastruktur (BMVI)*; Nationale Organisation Wasserstoff- und Brennstoffzellentechnologie: München, Germany; Berlin/Hamburg; Germany, 2019.
156. United States Department of Energy. *Determining Electric Motor Load and Efficiency*; EERE Publication and Product Library: Washington, DC, USA, 1997.
157. Breeze, P. The Proton Exchange Membrane Fuel Cell. In *Fuel Cells*; Elsevier: Amsterdam, The Netherlands, 2017; pp. 33–43.
158. Ahluwalia, R.K.; Wang, X. Direct hydrogen fuel cell systems for hybrid vehicles. *J. Power Sources* **2005**, *139*, 152–164. [CrossRef]
159. INFRAS. Handbuch für Emissionsfaktoren des Straßenverkehrs (HBEFA). Available online: https://www.hbefa.net/d/ (accessed on 25 January 2021).
160. Bünger, U.; Landinger, H.; Weindorf, W.; Wurster, R.; Zerhusen, J.; Zittel, W. *Vergleich von CNG und LNG zum Einsatz in LKW im Fernverkehr, Eine Expertise für die Open Grid Europe GmbH | Abschlussbericht*; Ludwig-Bölkow-Systemtechnik GmbH: Ottobrunn, Germany, 2016.
161. Weiss, M.; Cloos, K.C.; Helmers, E. Energy efficiency trade-offs in small to large electric vehicles. *Environ. Sci. Eur.* **2020**, *32*, 46. [CrossRef]
162. Otten, M.; 'T Hoen, M.; den Boer, E. *STREAM Freight Transport 2016: Emissions of Freight Transport Modes*; CE Delft: Delft, The Netherlands, 2017.

163. CE Cideon Engineering GmbH Co. KG. Produktblatt Hybrid Lokomotive. Available online: https://www.cideon-engineering.com/de/wp-content/uploads/sites/2/2019/02/CE_Produktblatt_HybidLokomotive.pdf (accessed on 18 February 2021).
164. Bombardier. World Premiere: Bombardier Transportation Presents a New Battery-Operated Train and Sets Standards for Sustainable Mobility. Available online: https://rail.bombardier.com/en/about-us/worldwide-presence/germany/de.html/bombardier/news/2018/bt-20180912-world-premiere--bombardier-transportation-presents-a/de (accessed on 21 December 2021).
165. Vlexx GmbH. Bombardier TALENT 3 Datenblatt. Available online: https://www.vlexx.de/media/vlexx/downloads/ens/Datenblatt_TALENT-3_lr.pdf (accessed on 15 October 2021).
166. Betz, P. Neue Diesel-Hybrid-Technik von Voith Turbo für Schienenfahrzeuge. Available online: https://www.openpr.de/news/194991/Neue-Diesel-Hybrid-Technik-von-Voith-Turbo-fuer-Schienenfahrzeuge.html (accessed on 21 December 2021).
167. Alstom. World Premiere: Alstom's Hydrogen Trains Enter Passenger Service in Lower Saxony. Available online: https://www.alstom.com/press-releases-news/2018/9/world-premiere-alstoms-hydrogen-trains-enter-passenger-service-lower (accessed on 21 December 2021).
168. Plank-Wiedenbeck, U.; Jentsch, M.; Lademann, F.; Büttner, S.; Meyer, N.; Ivanov, A. *Schlussbericht Machbarkeitsstudie Pilotprojekt Einsatz von H2BZ-Triebwagen in Thüringen*; Freistaat Thüringen Ministerium für Umwelt, Energie und Naturschutz: Weimar, Germany, 2019.
169. Deutsche Bahn AG. Grundlagenbericht UmweltMobilCheck, Stand 01.04.2016. Available online: https://www.deutschebahn.com/resource/blob/265408/a54aed9e6a0513294c3b9607986fb38f/umc__grundlagenbericht__2016-data.pdf (accessed on 22 June 2019).
170. Roland Berger. Hydrogen—A Future Fuel for Aviation? Available online: https://www.rolandberger.com/publications/publication_pdf/roland_berger_hydrogen_the_future_fuel_for_aviation.pdf (accessed on 21 December 2021).
171. Seeckt, K.; Scholz, D. Jet versus Prop, Hyrogen versus Kerosene for a Regional Freighter Aircraft. In Proceedings of the Deutscher Luft- und Raumfahrtkongress. 2009. Available online: https://www.fzt.haw-hamburg.de/pers/Scholz/GF/GF_Paper_DLRK_09-09-08.pdf (accessed on 21 December 2021).
172. Umweltbundesamt. *Jährliche Treibhausgas-Emissionen in Deutschland/Annual Greenhouse Gas Emissions in Germany—Nationales Treibhausgasinventar 2021, 12/2020*; Umweltbundesamt: Dessau-Roßlau, Germany, 2021.
173. Meyer, K.; Helms, H.; Fehrenbach, H.; Biemann, K.; Kämper, C.; Lambrecht, U.; Jöhrens, J. *Klimabilanz von Strombasierten Antrieben und Kraftstoffen*; Agora Verkehrswende: Berlin, Germany, 2019; p. 56.
174. Naumann, K.; Schröder, J.; Oehmichen, K.; Etzold, H.; Müller-Langer, F.; Remmele, E.; Thuneke, K.; Raksha, T.; Schmidt, P. *Monitoring Biokraftstoffsektor*; Deutsches Biomasseforschugnszentrum gemeinnützige GmbH: Leipzig, Germany, 2019.
175. *ISO 14040*; Environmental Management—Life Cycle Assessment—Principles and Framework. International Organization for Standardization (ISO): Geneva, Switzerland, 2006.
176. *ISO 14044*; Environmental Management—Life Cycle Assessment—Requirements and Guidelines. International Organization for Standardization (ISO): Geneva, Switzerland, 2006.
177. Moro, A.; Helmers, E. A new hybrid method for reducing the gap between WTW and LCA in the carbon footprint assessment of electric vehicles. *Int. J. Life Cycle Assess.* **2017**, *22*, 4–14. [CrossRef]
178. Koj, J.C.; Wulf, C.; Zapp, P. Environmental impacts of power-to-X systems—A review of technological and methodological choices in Life Cycle Assessments. *Renew. Sustain. Energy Rev.* **2019**, *112*, 865–879. [CrossRef]
179. Zhang, X.; Bauer, C.; Mutel, C.L.; Volkart, K. Life Cycle Assessment of Power-to-Gas: Approaches, system variations and their environmental implications. *Appl. Energy* **2017**, *190*, 326–338. [CrossRef]
180. Walker, S.B.; Fowler, M.; Ahmadi, L. Comparative life cycle assessment of power-to-gas generation of hydrogen with a dynamic emissions factor for fuel cell vehicles. *J. Energy Storage* **2015**, *4*, 62–73. [CrossRef]
181. Koj, J.C.; Wulf, C.; Linssen, J.; Schreiber, A.; Zapp, P. Utilisation of excess electricity in different Power-to-Transport chains and their environmental assessment. *Transp. Res. Part D Transp. Environ.* **2018**, *64*, 23–35. [CrossRef]
182. Kigle, S.; Pichlmaier, S.; Huber, J.; Regett, A. Ökobilanz strombasierter synthetischer Kraftstoffe: Worauf es ankommt. *EW* **2019**, *2019*, 18–23.
183. Wettstein, S.; Itten, R.; Stucki, M. *Life Cycle Assessment of Renewable Methane for Transport and Mobility*; Swiss National Science Foundation: Zürich, Switzerland, 2018; pp. 1–35.
184. Rinawati, D.I.; Keeley, A.R.; Takeda, S.; Managi, S. A systematic review of life cycle assessment of hydrogen for road transport use. *Prog. Energy* **2021**, *4*, 012001. [CrossRef]
185. Wulf, C.; Kaltschmitt, M. Hydrogen Supply Chains for Mobility—Environmental and Economic Assessment. *Sustainability* **2018**, *10*, 1699. [CrossRef]
186. Koj, J.C.; Wulf, C.; Schreiber, A.; Zapp, P. Site-dependent environmental impacts of industrial hydrogen production by alkalinewater electrolysis. *Energies* **2017**, *10*, 860. [CrossRef]
187. Bareiß, K.; de la Rua, C.; Möckl, M.; Hamacher, T. Life cycle assessment of hydrogen from proton exchange membrane water electrolysis in future energy systems. *Appl. Energy* **2019**, *237*, 862–872. [CrossRef]
188. Kanz, O.; Bittkau, K.; Ding, K.; Rau, U.; Reinders, A. Review and Harmonization of the Life-Cycle Global Warming Impact of PV-Powered Hydrogen Production by Electrolysis. *Front. Electron.* **2021**, *2*, 711103. [CrossRef]

189. Bongartz, D.; Doré, L.; Eichler, K.; Grube, T.; Heuser, B.; Hombach, L.E.; Robinius, M.; Pischinger, S.; Stolten, D.; Walther, G.; et al. Comparison of light-duty transportation fuels produced from renewable hydrogen and green carbon dioxide. *Appl. Energy* **2018**, *231*, 757–767. [CrossRef]
190. Reiter, G.; Lindorfer, J. Global warming potential of hydrogen and methane production from renewable electricity via power-to-gas technology. *Int. J. Life Cycle Assess.* **2015**, *20*, 477–489. [CrossRef]
191. Uusitalo, V.; Väisänen, S.; Inkeri, E.; Soukka, R. Potential for greenhouse gas emission reductions using surplus electricity in hydrogen, methane and methanol production via electrolysis. *Energy Convers. Manag.* **2017**, *134*, 125–134. [CrossRef]
192. Prussi, M.; Yugo, M.; De Prada, L.; Padella, M.; Edwards, R.; Lonza, L. *JEC Well-to-Tank report v5—Well-to-Wheels Analysis of Future Automotive Fuels and Powertrains in the European Context*; Publications Office of the European Union: Luxembourg, 2020.
193. Fachagentur Nachwachsende Rohstoffe e.V. *Biokraftstoffe*; Biokraftstoffe: Gülzow-Prüzen, Germany, 2014.
194. Arvidsson, R.; Persson, S.; Fröling, M.; Svanström, M. Life cycle assessment of hydrotreated vegetable oil from rape, oil palm and Jatropha. *J. Clean. Prod.* **2011**, *19*, 129–137. [CrossRef]
195. Bierkandt, T.; Severin, M.; Ehrenberger, S.; Köhler, M. *Klimaneutrale Synthetische Kraftstoffe im Verkehr—Potenziale und Handlungsempfehlungen*; DLR: Stuttgart, Germany, 2018.
196. Biernacki, P.; Röther, T.; Paul, W.; Werner, P.; Steinigeweg, S. Environmental impact of the excess electricity conversion into methanol. *J. Clean. Prod.* **2018**, *191*, 87–98. [CrossRef]
197. Deutz, S.; Bongartz, D.; Heuser, B.; Kätelhön, A.; Schulze Langenhorst, L.; Omari, A.; Walters, M.; Klankermayer, J.; Leitner, W.; Mitsos, A.; et al. Cleaner production of cleaner fuels: Wind-to-wheel—environmental assessment of CO_2-based oxymethylene ether as a drop-in fuel. *Energy Environ. Sci.* **2018**, *11*, 331–343. [CrossRef]
198. Hoppe, W.; Bringezu, S. Vergleichende Ökobilanz der CO_2-basierten und konventionellen Methan- und Methanolproduktion. *Uwf Umw.* **2016**, *24*, 43–47. [CrossRef]
199. Hoppe, W.; Thonemann, N.; Bringezu, S. Life Cycle Assessment of Carbon Dioxide-Based Production of methane and Methanol and Derived Polymers. *J. Ind. Ecol.* **2017**, *22*, 327–340. [CrossRef]
200. Schmidt, P.; Weindorf, W. *Power-to-Liquids: Potentials and Perspectives for the Future Supply of Renewable Aviation Fuel*; German Environment Agency: Dessau-Roßlau, Germany, 2016.
201. Spielmann, M.; Ruiz, S.; Zah, R. *Analyse der Umwelt-Hotspots von Strombasierten Treibstoffen*; BAFU: Zürich, Switzerlands, 2015; pp. 1–68.
202. Sternberg, A.; Bardow, A. Power-to-What?—Environmental assessment of energy storage systems. *Energy Environ. Sci.* **2015**, *8*, 389–400.
203. Sternberg, A.; Teichgräber, H.; Voll, P.; Bardow, A. CO_2 vs. Biomass: Identification of Environmentally Beneficial Processes for Platform Chemicals from Renewable Carbon Sources. *Comput. Aided Chem. Eng.* **2015**, *37*, 1361–1366.
204. Fernández-Dacosta, C.; Shen, L.; Schakel, W.; Ramirez, A.; Kramer, G.J. Potential and challenges of low-carbon energy options: Comparative assessment of alternative fuels for the transport sector. *Appl. Energy* **2019**, *236*, 590–606. [CrossRef]
205. Matzen, M.; Demirel, Y. Methanol and dimethyl ether from renewable hydrogen and carbon dioxide: Alternative fuels production and life-cycle assessment. *J. Clean. Prod.* **2016**, *139*, 1068–1077. [CrossRef]
206. Wagner, U.; Eckl, R.; Tzscheutschler, P. Energetic life cycle assessment of fuel cell powertrain systems and alternative fuels in Germany. *Energy* **2006**, *31*, 3062–3075. [CrossRef]
207. Troy, S.; Zapp, P.; Kuckshinrichs, W.; Weiske, S.; Peters, R.; Moser, P.; Stahl, K. Life Cycle Assessment for full chain CCU demonstration in the ALIGN-CCUS project—dimethyl ether and polyoxymethylen dimethyl ethers production from CO_2 and its usages in the mobility and electricity sectors. In Proceedings of the 15th International Conference on Greenhouse Gas Control Technologies, GHGT-15, Abu Dhabi, United Arab Emirates, 15–18 March 2021.
208. Hurtig, O.; Yearwood, J. Deliverable 4.3: Economic and Environmental Impacts of Most Promising CCU Pathways; Trinomics, Dechema: 2018. Available online: http://carbonnext.eu/Deliverables/_/D4.3%20Economic%20and%20environmental%20impacts%20of%20most%20promising%20CCU%20pathways.pdf (accessed on 21 December 2021).
209. Hombach, L.E.; Doré, L.; Heidgen, K.; Maas, H.; Wallington, T.J.; Walther, G. Economic and environmental assessment of current (2015) and future (2030) use of E-fuels in light-duty vehicles in Germany. *J. Clean. Prod.* **2019**, *207*, 153–162. [CrossRef]
210. Alhyari, M.M.; Al-Salaymeh, A.; Irshidat, M.; Kaltschmitt, M.; Neuling, U. The Impact of Energy Source on the Life-Cycle Assessment of Power-to-Liquid Fuels. *J. Ecol. Eng.* **2019**, *20*, 239–244. [CrossRef]
211. Volvo. LKW MIT LNG-ANTRIEB. Available online: https://www.volvotrucks.de/de-de/trucks/alternative-antriebe/lng-lkw.html (accessed on 19 December 2020).
212. Scania. Praxisinfos zu Gas-Lkw. Available online: https://www.scania.com/de/de/home/products-and-services/articles/vorteile-erdgas-lkw.html (accessed on 19 December 2020).
213. Stojcevski, T. METHANOL—As Engine Fuel, Status Stena Germanica and Market Overview. Available online: https://www.zerovisiontool.com/sites/www.zerovisiontool.com/files/attachments/pilo (accessed on 19 August 2019).
214. Zhai, K. *A Brief Review of China's Methnaol Vehicle Pilot and Policy*; Methanol Institute: Beijing, China, 2019.
215. AUTO-CHE.COM. Methanol/Diesel Dual Fuel Dump Truck. Available online: http://auto-che.com/t/methanol-diesel-dump-truck.html (accessed on 21 August 2019).

216. Wietschel, M.; Gnann, T.; Kühn, A.; Plötz, P.; Moll, C.; Speth, D.; Buch, J.; Boßmann, T.; Stütz, S.; Schellert, M.; et al. *Machbarkeitsstudie zur Ermittlung der Potentiale des Hybrid-Oberleitungs-Lkw*; Fraunhofer Institut für System und Innovationsforschung (ISI): Karlsruhe, Germany, 2017.
217. Breuer, J.L.; Samsun, R.C.; Stolten, D.; Peters, R. How to reduce the greenhouse gas emissions and air pollution caused by light and heavy duty vehicles with battery-electric, fuel cell-electric and catenary trucks. *Environ. Int.* **2021**, *152*, 106474. [CrossRef] [PubMed]
218. Peters, R.; Breuer, J.L.; Decker, M.; Grube, T.; Robinius, M.; Samsun, R.C.; Stolten, D. Future Power Train Solutions for Long-Haul Trucks. *Sustainability* **2021**, *13*, 2225. [CrossRef]
219. European Union. *Regulation (EU) 2016/1628 of the European Parliament and of the Council of 14 September 2016 on Requirements Relating to Gaseous and Particulate Pollutant Emission Limits and Type-Approval for Internal Combustion Engines for Non-Road Mobile Machinery, Amending Regulations (EU) No 1024/2012 and (EU) No 167/2013, and Amending and Repealing Directive 97/68/EC*; European Union: Brussels, Belgium, 2016.
220. Quispel, M. Reducing Air Pollutant Emissions of Inland Waterway Transport in Europe. Available online: https://www.ccr-zkr.org/files/documents/workshops/wrshp081013/1_MQuispel_en.pdf (accessed on 16 February 2021).
221. Dahlke, F. Binnenschifffahrt—Innovative Antriebskonzepte. In Proceedings of the Jahrestagung Mobilität 2019, Düsseldorf, Germany, 27 June 2019.
222. Zero Emission Services B.V. Schlüsselakteure Entwickeln ein Emissionsloses Antriebssystem für die Binnenschifffahrt. Available online: https://zeroemissionservices.nl/de/schlusselakteure-entwickeln-ein-emissionsloses-antriebssystem-fur-die-binnenschifffahrt/ (accessed on 21 December 2021).
223. Kasten, P.; Mottschall, M.; Köppel, W.; Degünther, C.; Schmied, M.; Wüthrich, P. Erarbeitung einer Fachlichen Strategie zur Energieversorgung des Verkehrs bis zum Jahr 2050. *TEXTE* **2016**, *72/2016*. Available online: https://www.umweltbundesamt.de/sites/default/files/medien/377/publikationen/2016-11-10_endbericht_energieversorgung_des_verkehrs_2050_final.pdf (accessed on 19 December 2020).
224. DNV GL. Highlight Projects in the LNG as Fuel History. Available online: https://brandcentral.dnvgl.com/download/DownloadGateway.dll?h=BE1B38BB718539CC0AB58A5FF2EA7A8335EF9A5BC87A42E8FE938B52042522100D70F7679F884F56637AC7C6F561AD0F (accessed on 7 April 2021).
225. Rolls-Royce. Rolls-Royce and Robert Allan Present the World's First LNG-Powered Shallow-Water Push Boat. Available online: https://www.rolls-royce.com/media/press-releases/2019/06-12-2019-rr (accessed on 21 December 2021).
226. Bundesministerium für Verkehr, Bau und Stadtentwicklung. *Die Mobilitäts-und Kraftstoffstrategie der Bundesregierung (MKS)*; Bundesministerium für Verkehr, Bau und Stadtentwicklung: Berlin, Germany, 2013.
227. Kopyscinski, J.; Schildhauer, T.; Biollaz, S. Production of synthetic natural gas (SNG) from coal and dry biomass–A technology review from 1950 to 2009. *Fuel* **2010**, *89*, 1763–1783. [CrossRef]
228. Generaldirektion Wasserstraßen und Schifffahrt. Veränderungen des Schiffsbestandes der Deutschen Binnenflotte im Jahr 2017—Zentrale Binnenschiffsbestandsdatei. Available online: https://www.gdws.wsv.bund.de/SharedDocs/Downloads/DE/Binnenschiffsbestandsdatei/2017.pdf?__blob=publicationFile&v=3 (accessed on 20 February 2021).
229. MariGreen project. *Perspectives for the Use of Hydrogen as Fuelin Inland Shipping*; Institute for Combustion Engines VKA: Leer, Germany; Mariko GmbH: Aachen, Germany; FME: Groningen, The Netherlands, 2018.
230. RH2INE Awarded Co-Funding of the EU Connecting Europe Facility! Available online: https://www.rh2ine.eu/latest-news/253/ (accessed on 12 December 2020).
231. Thomson, R. Hydrogen Has Emerged as a Key Contender in the Battle to Secure a Sustainable Future for Aviation. Available online: https://www.rolandberger.com/en/Publications/Hydrogen-A-future-fuel-of-aviation.html (accessed on 20 August 2020).
232. Schmied, M.; Wüthrich, P.; Zah, R.; Friedl, C. Postfossile Energieversorgungsoptionen für einen Treibhausgasneutralen Verkehr im Jahr 2050: Eine Verkehrsträgerübergreifende Bewertung. *TEXTE* **2015**, *30/2015*. Available online: https://www.umweltbundesamt.de/publikationen/postfossile-energieversorgungsoptionen-fuer-einen (accessed on 6 July 2020).
233. International Civil Aviation Organization. Electric and Hybrid Aircraft Platform for Innovation (E-HAPI). Available online: https://www.icao.int/environmental-protection/Pages/electric-aircraft.aspx (accessed on 6 July 2020).
234. Roland Berger. *Aircraft Electrical Propulsion—The Next Chapter of Aviation?* 2017. Available online: http://docplayer.net/59715725-Aircraft-electrical-propulsion-the-next-chapter-of-aviation.html (accessed on 6 July 2020).
235. Eviation Aircraft. ALICE Specification. Available online: https://www.eviation.co/aircraft/#4 (accessed on 22 September 2020).
236. ICAO Secretariat. Electric, Hybrid, and Hydrogen Aircraft—State of Play. In *ICAO Environmental Report 2019*; ICAO: Montreal, QC, Canada, 2019; pp. 124–130.
237. Deutsches Zentrum für Luft-und Raumfahrt. HY4 Emissionsfreie Passagierflugzeuge. Available online: https://www.dlr.de/content/de/downloads/2015/dlr-flyer-hy4_1892.pdf?__blob=publicationFile&v=16 (accessed on 18 February 2021).
238. AIRBUS. This Chase Aircraft Is Tracking 100% SAF's Emissions Performance. Available online: https://www.airbus.com/en/newsroom/stories/2021-11-this-chase-aircraft-is-tracking-100-safs-emissions-performance (accessed on 21 December 2021).
239. Naumann, K.; Müller-Langer, F.; Meisel, K.; Majer, S.; Schröder, J.; Schmieder, U. *Further Development of the German Greenhouse Gas Reduction Quota, Background Paper*; Deutsches Biomasseforschungszentrum gemeinnützige GmbH: Leipzig, Germany, 2021.

240. Heidt, C.; Lambrecht, U.; Hardinghaus, M.; Knitschky, G.; Schmidt, P.; Weindorf, W.; Naumann, K.; Majer, S.; Müller-Langer, F.; Seiffert, M. *CNG und LPG—Potenziale dieser Energieträger auf dem Weg zu einer Nachhaltigeren Energiever-Sorgung des Straßenverkehrs;* DLR: Ottobrunn, Germany; Leipzig, Germany; Berlin/Heidelberg, Germany, 2013.
241. Wachsmuth, J.; Oberle, S.; Zubair, A.; Köppel, W. *Wie klimafreundlich ist LNG?—Kurzstudie zur Bewertung der Vorkettenemissionen bei Nutzung von Verflüssigtem Erdgas (LNG);* CLIMATE CHANGE 21/2019; Umweltbundesamt: Dessau-Roßlau, Germany, 2019.

energies

Article

Mapping Bio-CO_2 and Wind Resources for Decarbonized Steel, E-Methanol and District Heat Production in the Bothnian Bay

Hannu Karjunen *, Eero Inkeri and Tero Tynjälä

LUT School of Energy Systems, Lappeenranta-Lahti University of Technology LUT, 53851 Lappeenranta, Finland; eero.inkeri@lut.fi (E.I.); tero.tynjala@lut.fi (T.T.)
* Correspondence: hannu.karjunen@lut.fi

Abstract: Hydrogen is a versatile feedstock for various chemical and industrial processes, as well as an energy carrier. Dedicated hydrogen infrastructure is envisioned to conceptualize in hydrogen valleys, which link together the suppliers and consumers of hydrogen, heat, oxygen, and electricity. One potential hydrogen valley is the Bay of Bothnia, located in the northern part of the Baltic Sea between Finland and Sweden. The region is characterized as having excellent wind power potential, a strong forest cluster with numerous pulp and paper mills, and significant iron ore and steel production. The study investigates the hydrogen-related opportunities in the region, focusing on infrastructural requirements, flexibility, and co-operation of different sectors. The study found that local wind power capacity is rapidly increasing and will eventually enable the decarbonization of the steel sector in the area, along with moderate Power-to-X implementation. In such case, the heat obtained as a by-product from the electrolysis of hydrogen would greatly exceed the combined district heat demand of the major cities in the area. To completely fulfil its district heat demand, the city of Oulu was simulated to require 0.5–1.2 GW of electrolyser capacity, supported by heat pumps and optionally with heat storages.

Keywords: hydrogen; electrolysis; steel; Power-to-X; wind power; thermal energy storages

Citation: Karjunen, H.; Inkeri, E.; Tynjälä, T. Mapping Bio-CO_2 and Wind Resources for Decarbonized Steel, E-Methanol and District Heat Production in the Bothnian Bay. *Energies* **2021**, *14*, 8518. https://doi.org/10.3390/en14248518

Academic Editor: Wei-Hsin Chen

Received: 30 November 2021
Accepted: 15 December 2021
Published: 17 December 2021

1. Introduction

Industrial clusters are locations where various companies perform their individual activities, while also co-operating and sharing risks and resources between each other, thus increasing the operation efficiency and profitability. Decarbonization of these industrial clusters leads to a host of new technological challenges, but also a broad field of opportunities for renewable power generation, widespread use of hydrogen (H_2), Power-to-X products, intelligent heating, and electrification.

Hydrogen initiatives have recently been gathering interest. The global hydrogen valley platform currently lists 36 distinct hydrogen valley projects [1], including the Basque Hydrogen Corridor (BH2C) project that aims to invest 2.9 BEUR between 2020 and 2030 to create a hydrogen ecosystem in the Basque Country in Spain [2]. Another 1 BEUR H_2 investment project is envisioned for the Zuid-Holland/Rotterdam region in the Netherlands. The Rotterdam port has also been associated with carbon capture use and storage (CCUS) projects [3]. Yet, another Netherlandic project, NorthH2, aims to produce 4 GW of offshore wind power by 2030 that would be used for producing green hydrogen, and upscaling that to 10 GW by 2040 [4].

The Bay of Bothnia is another interesting industrial cluster location with a strong presence from the pulp and paper industry and steel, cement, and traditional power and heat production. The majority of industrial sectors and population centres from the region are concentrated on the coastal regions as illustrated in Figure 1. The initial motivation for this research was inspired by the interest of local industrial actors in the area, which is also reflected in the ongoing activities.

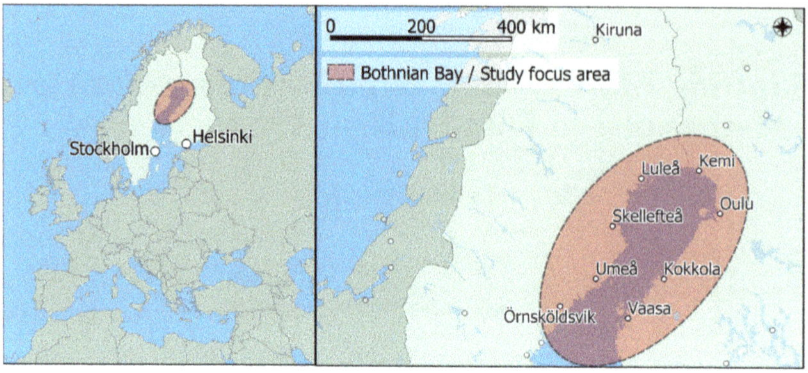

Figure 1. Location of the Bothnia Bay and the study focus area.

The available volume of biogenic CO_2 emissions from the pulp and paper sector sets the region apart from other similar industrial clusters. Thus, the application of BECCUS (bioenergy with carbon capture and utilization/storage) would offer tremendous opportunities for decreasing emission levels. Moreover, pulp mills also consume a variety of chemicals, which could be linked with Power-to-X. With the steel industry working as another driver for Power-to-X implementation, the Bothnian Bay could have a major role in pioneering upcoming technologies and cross-industry cooperation.

Recently, the BotH_2nia initiative was set in motion, facilitating communication and cooperation with different actors and projects in the area [5]. Additionally, the current status and plans related to hydrogen activities were mapped in a questionnaire study which included 37 companies from the Bothnian Bay area [6]. According to the study, hydrogen adoption is still considered to be in a very early stage, although some first concrete actions are planned for demonstration plants related to steel production and synthetic fuel manufacturing. The availability of hydrogen and its high cost is considered to be a major obstacle according to potential users.

The cost gap between fossil and electrolytic hydrogen could partly be narrowed by a reduction in electrolyser capital expenditures, which is expected to occur in phases, with increased deployment of the technology and more efficient manufacturing techniques. Still, low-cost electricity remains an essential ingredient [7]. During the last decade, solar and wind power generation costs have decreased to a level where they can outperform fossil generation [8]. In the Bothnian Bay region, low-cost wind power could become the dominant power production method in a relatively short timeframe.

The cost-competitiveness of electrolytic hydrogen can be improved even further by efficiently utilizing the by-products of electrolysers [9,10]. For instance, the integration of electrolysis with low-grade heat utilization has been demonstrated in small pilot projects in Germany and Denmark [11]. On a more theoretical scale, Böhm et al. [12] analysed the sector coupling potential of hydrogen generation and district heating by performing a survey on the technological status and development of electrolysers and district heat systems. Additionally, the study summarized the findings from an expert survey into a SWOT (strengths, weaknesses, opportunities and threats) analysis. One conclusion of the paper was that there are many potential synergies with coupling district heat and hydrogen generation, but the application of electrolysers as a heat source is not typically taken into account by the district heat industry. The study also highlighted the need for developing effective planning tools for mapping the demand for hydrogen, heat, and oxygen–which coincides exceptionally well with the topic of this work.

A related but different heat integration concept is to utilize high-temperature electrolysers (i.e., solid oxide electrolyser cells) that have been proposed to be linked with an external heat source, which provides heat at 700–900 °C and thus increases the efficiency of the electrolyser [13–15].

For synthesis products beyond hydrogen, Ikäheimo et al. [16] performed a simulation of the power and district heat system of Northern European countries, concluding that excess heat from power-to-ammonia systems could significantly contribute to district heat generation. Residual heat from electrolysers was not investigated, although its potential in future low-temperature district heat networks (i.e., 4th generation networks [17]) was acknowledged. Regional imbalance was also noted to exist between the district heat demand and potential locations of ammonia production plants. In other studies, the operation principle and optimal application of hydrogen and CO_2 storages have also been found to be important for profitability of Power-to-X [18,19].

Electrolysers also produce oxygen as a by-product, which is widely used in the pulp and paper industry [20] and in healthcare applications [21]. Saxe and Alvfors [22] already evaluated the link between pulp and paper mills and electrolysers back in 2007 and concluded that oxygen utilization could potentially have an even bigger effect for profitability than mere heat utilization.

GIS (geographic information system) framework has been widely used before in energy-related research topics [23], such as mapping and identification of district heat grid expansion sites [24,25]. Further examples include utilization of CO_2 emission data to identify industrial excess heat sources [26]. Welder et al. [27] combined GIS tools with mixed integer linear programming to optimize hydrogen-based energy infrastructure containing cavern storages, wind turbines and hydrogen pipelines. Different implementation strategies for CO_2-related transport infrastructure have also been studied [28].

Although the potential for coupling hydrogen production with district heat has been acknowledged in previous studies, few have delved into energy balance investigations and regional matching of electricity production capacity, CO_2 resources, and hydrogen demand. Intelligent placement of electrolysers could therefore have an impact on the cost-efficiency of hydrogen production. Pilot projects and demonstrations in promising sites are an essential element of commercializing sustainable hydrogen production.

The aim of the work is to perform a regional decarbonization assessment on the Bothnian Bay area, highlighting the key industries at the core of the transition. Specifically, the focus is on the conceptualization of hydrogen supply and demand, as well as its potential links to the heating sector. The study includes two detail levels: a high-level overview of the area, which focuses on electricity generation and its use, and a more detailed region-specific study concentrating on the heat sector. Three configurations are presented for integrating electrolyser systems within the existing energy infrastructure. This study illustrates the massive scale of transition that is about to take place in wind power generation and local consumption of electricity for the decarbonization of steel and implementation of Power-to-X. Proper understanding of the links and scales between these resources, demands, and potentials is a vital first step for their implementation. Moreover, the methodology developed for this analysis will also serve in future studies in other regions.

2. Materials and Methods

This study's methodology combines various databases and statistics, GIS tools, and computer simulations. The details are outlined in the following sections, which have been categorized into hydrogen-related aspects, power generation, and heat systems.

2.1. Hydrogen Production and Use

In this work, the electricity demand of electrolysis is assumed to be 53 MWh_{el} per ton of hydrogen, which corresponds to about 62% efficiency when defined for lower heating value of hydrogen. Two primary uses for hydrogen are included in the analysis: steel industry and Power-to-X. Although Power-to-X could be used to produce a variety of different products, this work uses a simplified approach and considers e-methanol as the only final product. Methanol (MeOH) could be used directly as a fuel, but also as a feedstock in numerous products, such as olefins, resins, or gasoline [29].

2.1.1. Steel Industry

Decarbonization of steel manufacturing can be achieved with the commercial HYBRIT-process, which utilizes direct hydrogen reduction instead of coal as the reducing agent. Total electricity demand of the process is estimated to be 3.5 MWh for one tonne of steel, with hydrogen production being responsible for about 75% of the total demand [30]. Two alternative values are presented for the total electricity demand of steel. Low demand is associated with converting existing steel mills into decarbonized processes, whereas the high demand scenario assumes that all iron ore obtained from the Kiruna mine in Sweden is processed to sponge iron.

Currently, the studied region has active steel mills in Tornio, Luleå, and Raahe. Additionally, a new steel mill is being planned for Boden. The Tornio mill uses an electric arc furnace and requires about 3.5 TWh of electricity annually [31], whereas the other existing mills are traditional blast furnace mills. Conversion of the Raahe mill to the HYBRIT process is expected to require about 9 TWh of electricity, and Luleå about 8 TWh, based on their respective crude steel production capacities [32]. Together, the low demand scenario amounts to about 21 TWh of annual electricity demand, consisting of the demand of existing facilities.

The regional electricity demand would be much higher if all of the iron ore mined in Kiruna were to be refined into sponge iron. The Swedish government-owned mining company LKAB has estimated that 55 TWh of renewable electricity would be required annually once the six planned sponge iron refining sites are online in 2045. Existing iron ore pelletization capacity would be dismantled in phase with the deployment of the new sponge iron processing facilities, the first of which is scheduled to be in operation already in 2030 [33,34]. The planned actions would radically increase the electricity consumption of the region, requiring significant investments in power production and transmission capacity.

2.1.2. Power-to-X and Carbon Sources

Power-to-X can produce fuels, products, and chemicals. Thus, the decarbonization of aviation, road transport, and numerous manufacturing sectors could be possible by using Power-to-X. One strong product candidate for Power-to-X is methanol, which is currently used widely in chemical manufacturing.

Two estimates are presented for the possible extent of e-methanol production in this work, both of which are based on the availability of CO_2 from industrial point sources the region. The available CO_2 quantities from the region are identified from the European pollutant release and transfer register [35], which has been complemented by manual additions based on recent news and company press releases. Figure 2 shows the projected CO_2 emissions from various sectors in the region. Facilities close to each other have been aggregated into a single symbol in the illustration.

The viable sectors for CO_2 capture were assumed to be the pulp and paper sector, and the cement industry. These sectors were considered to be relatively stable in the foreseeable future, preserving their current CO_2 volumes. The power sector, which also includes combined heat and power, was excluded as a potential CO_2 source because it was assumed to have lower full load utilization hours compared to industrial units, thus being less desirable for CO_2 capture.

The high estimate for e-methanol potential is simply based on capturing all current emissions from the accepted sectors, whereas the lower potential assumes that only about 35% of the available emissions would be captured. For reference, the low e-methanol potential approximately corresponds to capturing the flue gas emissions from the largest pulp mill in the area. Alternatively, the same CO_2 quantity could also be obtained from about four mid-sized pulp mills from the region. By contrast, the high estimate for methanol production is unlikely to be realized completely, as it would require equipping all local pulp and paper sites with CO_2 capture devices. Furthermore, the emissions from pulp mills are typically divided into different process sections, e.g., the power boiler, bark recovery boiler, and lime kiln.

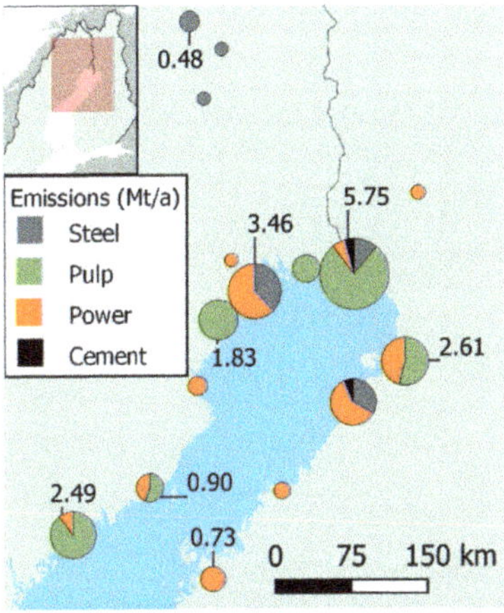

Figure 2. Distribution of the largest emitters of CO_2 from the region (data from [35]).

This work uses stoichiometric ratios to estimate the product yield and required inputs for e-methanol production. The methanol synthesis input mass ratio of hydrogen to carbon dioxide is assumed to be 0.137, and methanol yield is estimated as 0.728 kgMeOH/kgCO_2. CO_2 capture is used to obtain carbon for the methanol synthesis with an assumed CO_2 capture efficiency of 90%. The process requires heat, about 3.0–6.5 MJ/kgCO_2 [36,37]. In this paper, an optimistic value of 3.6 MJ/kgCO_2 is used. Auxiliary electricity consumption for CO_2 capture is assumed to be 250 kWh/tCO_2 [37]. Methanol synthesis is assumed to generate 430 kWh/tMeOH of additional heat, while also requiring 170 kWh/tMeOH electricity input [38,39].

2.2. Wind Power Generation Potential

In practice, the availability of electricity is another likely bottleneck for hydrogen and e-methanol production. For this reason, the regional potential of wind energy is identified by investigating the declared wind park power capacities from the Finnish [40] and the Swedish [41] perspective. Only onshore wind projects were included in the investigation, although the area has potential for offshore deployment.

The wind production data is geographically tracked in the most accurate GADM (Database of Global Administrative Areas) level available, corresponding to municipal accuracy for Sweden (GADM 2) and Finland (GADM 4). The municipalities in the studied region were allocated into twelve different groups, as illustrated in Figure 3.

The annual production of wind turbines is estimated by assuming 2600 annual full load hours, equivalent to a capacity factor of 29.7%. In reality, both the local wind conditions and the specifications of the wind turbine affect the capacity factor.

This work presents the wind data projections in three different time frames:

1. Current active turbines that have already been commissioned;
2. Short-term capacity additions (occurring roughly during the next 5 years);
3. Long-term capacity (coming online over the next 5–15 years).

The original wind power data from Finland is classified into six different groups: already commissioned wind turbines (class 6), but also projections about future wind power capacity based on turbines that are under construction, permitted sites, sites currently in various permitting and planning phases (classes 1–5), and identified projects that are in the early development phase (class 0). In this work, class 6 was classified into current capacity, classes 1–5 to short-term capacity, and class 6 into long-term capacity.

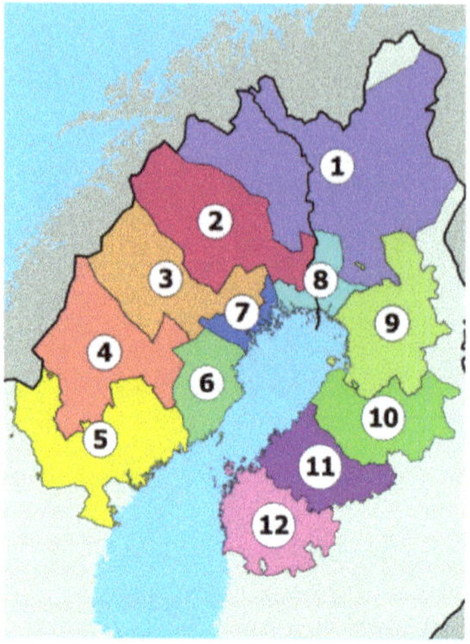

Figure 3. Allocation of municipalities into twelve calculation groups.

For Sweden, the original data is similarly divided into different classes. Commissioned turbines were included in current capacity, whereas short-term capacity included projects which were classified either as already permitted and approved, or as having an ongoing permitting process. The long-term capacity was assumed to include the class which was defined as "postponed or cancelled". Although not all postponed sites are guaranteed to realize exactly as envisioned, the data serves as an initial estimate for the long-term capacity. The data for Sweden also contains projects which have been denied a permit, or which have an ongoing appeal process, but these were not included in the analysis of this work.

The long-term capacity will likely also include sites which have not yet been identified. Furthermore, the production of existing sites could also be increased by adding new turbines or upgrading them. Thus, the long-term capacity was increased by extrapolating the cumulative capacity of existing and short-term cases. The amount of extrapolation was estimated using a parameter termed in this work as the wind power density (WPD). It is calculated by dividing the installed wind power capacity in kilowatts and the total area in square kilometres. Denmark can be calculated to have a WPD value of about 144 kW/km^2, and Germany reaches 175 kW/km^2. A recent report on the Finnish energy system presents a maximum wind power potential of 54 GW in Finland [42], corresponding to a WPD value of 160 kW/km^2. The long-term scenario was assumed to have a conservative WPD of 100 kW/km^2 for the whole Bothnian Bay region. The long-term capacity of all nodes was increased stepwise by a fixed amount until one of the two conditions is fulfilled:

- The wind power density of an individual node exceeds 150 kW/km^2, after which the node is excluded from further capacity additions;
- The total wind power generation in the region reaches 39 GW, which corresponds to average WPD of 100 kW/km^2 for the whole studied domain.

Overall, the extrapolation procedure was responsible for about 87% of the additional capacity, with the remaining 13% coming from data of already identified and potential wind projects. The resulting capacity estimates are highly dependent on the two WPD limits assumed. The WPD value of node 7 already exceeds 500 in the short-term case, so the local limit of WPD could potentially be adjusted much higher than what was assumed in this work. The current methodology does not account for differences in wind conditions between the regions, while also neglecting other land use such as farming, housing, recreational use, or reindeer herding. The methodology also favours regions which are less saturated due to the relatively low WPD limit. These aspects could lead to significant deviations between the predicted and actual capacity distributions. A more realistic estimate would require customization of local WPD limits for each region individually. Additionally, the estimate for the total wind power capacity in the region should also be verified by alternative methods.

2.3. Heat Supply and Demand

The Bothnian Bay area is located partially inside the Artic Circle, so considerable heating is required during the winter, while there is less demand during the summer. As there is significant variation in heat demand over the year, it is investigated how the potential heat production by electrolysis matches to the district heat demand in the area. Methanol synthesis also results in the generation of additional heat, while CO_2 capture systems require heat. These systems are studied on an annual level, but not with the hourly accuracy as with electrolysis heat.

As in [43], it is assumed that all losses of electrolysis are converted to heat, and that the electrolyser is cooled with a water stream which reaches an outlet temperature of 50 °C. This residual heat is then primed to higher temperatures by using a heat pump.

As an example configuration, district heat demand of the city of Oulu in Finland is studied. Oulu is one of the most populous cities in the region, which makes it highly interesting for district heat coupling with electrolysers. The overall annual heat demand, 1541 GWh, is obtained from the data of the local district heat companies [44]. The supply temperature of district heat is obtained from [45] as a function ambient temperature, as presented in Figure 4.

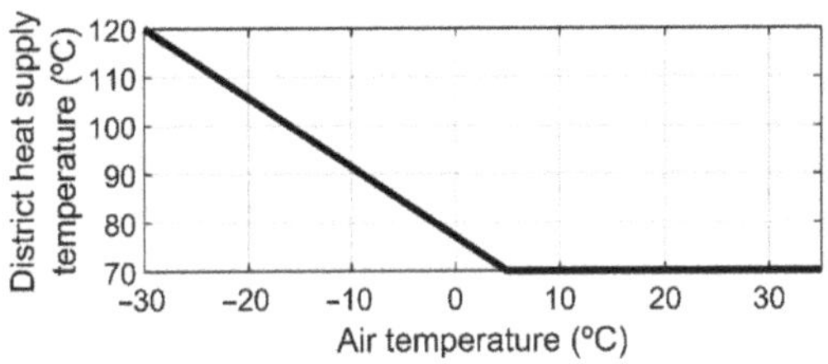

Figure 4. Supply temperature of district heat varies as a function of ambient temperature. Reproduced from [45].

The required heat power is also assumed to be a function of ambient temperature. The duration curve in Figure 5 is simplified from [46], and it is distributed on an hourly basis by the ambient temperature. The basic form of the duration curve is fixed by setting the 6000-h heat power to 10% of the maximum, the 700 h mark to 70%, and assuming linear increase between the fixed points. The peak power is then iteratively adjusted to match the annual heat energy demand.

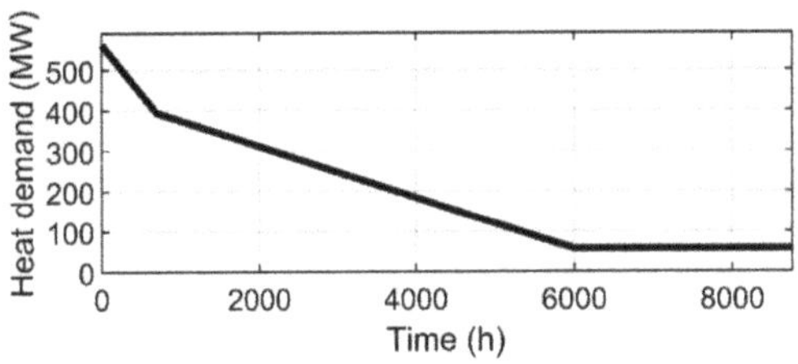

Figure 5. Duration curve of the district heat demand. Data simplified from [46].

Waste heat from electrolysis is used as a heat source for heat pump, which produces district heat. The coefficient of performance (COP) of the heat pump is defined by first calculating the idealized Carnot heat pump COP (COP_{Carnot}) from the condensing (T_{cond}) and evaporation (T_{evap}) temperatures of the heat pump

$$COP_{Carnot} = \frac{T_{cond}}{T_{Cond} - T_{evap}}, \tag{1}$$

which is then multiplied by a constant efficiency factor η to obtain the real COP

$$COP = \eta COP_{HP,Carnot}. \tag{2}$$

This simplified approach is chosen as detailed mass flow rate and inlet temperature for return flow of the district heat are not available, but the model is still capable to consider variations of ambient and district heat temperatures. The duration curve of air temperature is presented in Figure 6. An efficiency factor (η) of 50% is used, as it has been shown accessible in large scale [47].

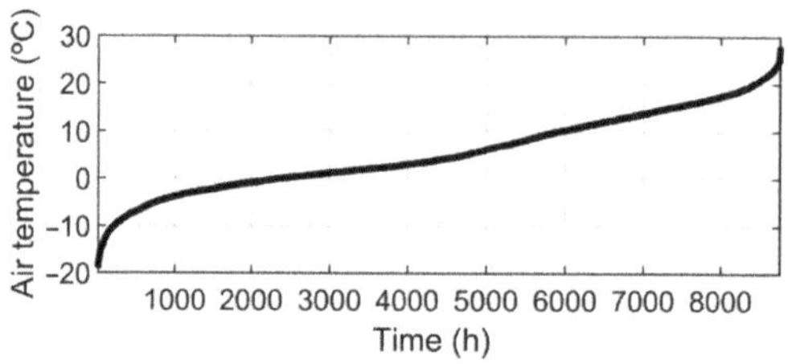

Figure 6. Duration curve of air temperature in Oulu 2020 [48].

As the heat demand varies heavily in time, there might be a demand for thermal energy storage (TES) to balance heat production from the electrolyser and heat pump, depending on the system configuration. TES is modelled with charge and discharge efficiencies of 90%, and an hourly heat loss that is proportional to stored energy. The value of the heat loss, 0.005%, is selected such that the annual efficiency of TES is approximately 80%, which is comparable to values in the review by [49]. The required TES capacity is obtained as a result of the simulation.

Although the Oulu region was in focus in this work, an estimate for the total district heat demand for all the major cities is also obtained by using the per capita district heat demand from Oulu as a benchmark (7.4 MWh/capita). The population centres that were included in the analysis are presented Figure 7.

System Configurations

Several scenarios are used to distinguish the different possible heating system design principles.

1. ElVar: Electrolyser and heat pump capacity at Oulu is matched with peak thermal power demand, so that district heat demand can be fulfilled solely with an electrolyser and a heat pump. Electrolyser operation is adjusted down during low heat demand. TES is not needed.
2. ElFix: Annual heat demand is to be met by electrolyser and a TES. The electrolyser operates at its nominal power at all times, and excess heat is delivered to TES. During high heat demand, TES is discharged.
3. ElWind: Annual heat demand is to be met by electrolyser and a TES. The electrolyser runs on wind power, and excess heat is delivered to TES.

This is not an exhaustive list covering all options. Rather, it illustrates some of the ultimate options for how to operate the electrolyser and provide heat, and also the upper and lower end of required TES. A block diagram of the simulation procedure is presented in Figure 8.

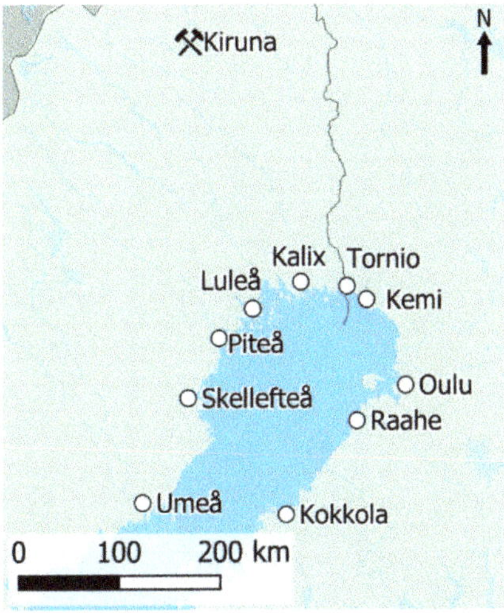

Figure 7. Major cities in the region that were included in the district heat estimation.

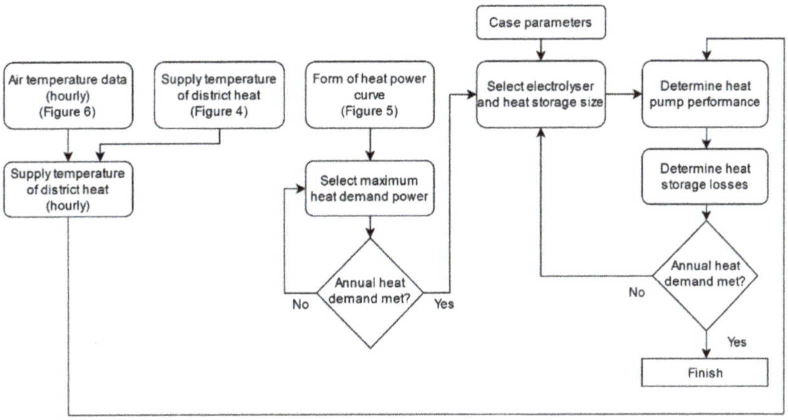

Figure 8. Block diagram of the heat balance simulation.

3. Results

The study first evaluates the renewable electricity generation potential in the region, which is then reflected against the demand side. Lastly, the heat-related investigations are presented.

3.1. Wind Power Generation Potential

The amount of renewable electricity available in the region is related to the maximum potential for Power-to-X. Estimation of the total cumulative installed capacity and annual production is given in Table 1, whereas the regional distribution of the capacity is shown in Figure 9. The estimated growth in wind power potential is vast, with capacity tripling in the short term and sextupling in the long term. From a statistical perspective, the long-term capacity could be even higher, but practical reasons could also limit the potential to lower amounts.

Table 1. Estimated wind power capacities in the Bothnian Bay region.

Time Frame	Cumulative Capacity (GW)	Annual Production (TWh)
Current	6	16
Short term	19	50
Long term	39	102

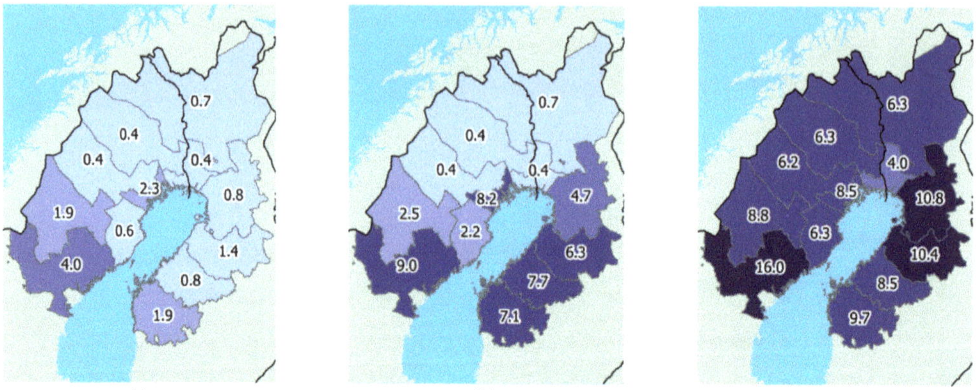

(a) (b) (c)

Figure 9. (**a**) Annual wind power production (TWh) from current turbines; (**b**) corresponding production when the *short-term* capacity additions are included; (**c**) production with long-term assumptions.

3.2. Electricity Demand Development

The electricity demand for steel decarbonization in the low and high estimates are presented in Figure 10a, whereas the projected electricity demand for methanol production in the high estimate is visible in Figure 10b. As there are numerous viable configurations for reaching the low estimate for methanol production, it is not illustrated here. The low estimate for Power-to-Methanol would essentially require selecting some sites in Figure 10b so that the total amounts to 30 TWh.

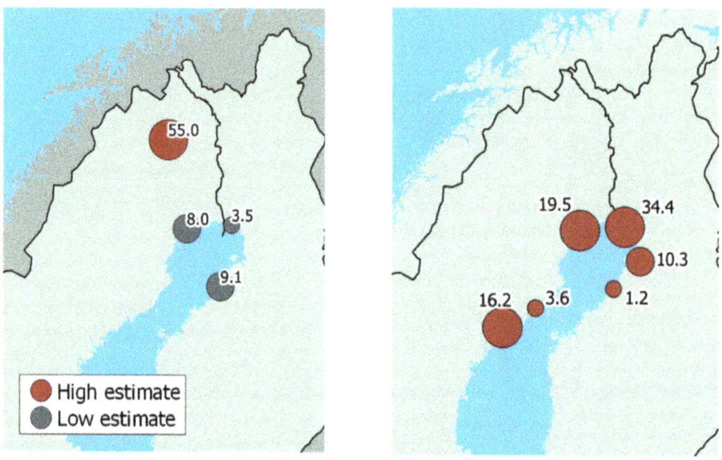

(a) (b)

Figure 10. (**a**) Steel electricity demand (TWh); (**b**) Power-to-Methanol electricity demand (TWh) if all selected carbon sources were utilized.

The resulting electricity and heat balances for the whole Bothnian Bay region are shown in Table 2. The estimated electricity production in the long-term scenario would not be enough for both high steel and high Power-to-X demand, but other combinations would be possible (e.g., high steel demand with low Power-to-X demand, or vice versa). The resulting excess heat would greatly exceed its demand in all examined cases. The additional electricity consumption required for application of heat pumps would be quite modest compared to overall electricity demand for hydrogen generation.

Table 2. Summary of electricity and heat balances in the Bothnian Bay area.

Electricity Demand Scenario **Electricity Production Scenario**	**High** **Long Term**	**Low** **Short Term**
Electricity		
Regional wind power generation (TWh)	102	50
Steel electricity demand (TWh)	55	21
e-Methanol electricity demand (TWh)	85	30
Heat pump compressor demand (TWh)	7	3
CO_2 capture demand (TWh)	3	1
Synthesis demand (TWh)	1	0.4
Electricity balance (TWh)	−50	−5
Heat		
Heat from heat pumps (TWh)	54	20
Heat from methanol synthesis (TWh)	3	1
CO_2 capture heat demand (TWh)	11	4
Regional district heat demand (TWh)	5	5
Heat balance (TWh)	+41	+12

If the total district heat demand of the major cities in the region were to be covered completely by electrolysers supported by heat pumps, about 11.5 TWh of electricity would be converted into hydrogen. This represents roughly half of the steel industry's low demand.

3.3. *Regional Insights*

Regional results focus on the city of Oulu in Finland. The necessary capacities of electrolysers, heat pumps and thermal energy storages are listed in Table 3. These configurations could completely and independently fulfil the local district heat demand of the region. Even though electrolysers would not likely be practical as the only source of heat, it illustrates the requirements and performance of the configuration. The obtained results are based on simulations for one year, and different weather conditions could lead to slightly different outcomes.

Table 3. Required capacities for producing district heat for Oulu.

Case	Electrolyser Capacity (MW)	Compressor Size (MW)	Max. Charge (MW)	Max. Discharge (MW)	TES Capacity (GWh)
ElVar	1078	162	-	-	-
ElFix	494	76	127	352	446
ElWind	1232	128	386	595	291

In contrast to electrolyser capacity, the highest capacity for the heat pump compressor is required in ElVar, 162 MW. Even though the maximum electrolyser capacity is higher in ElWind, a compressor capacity of 128 MW is already adequate. This might be due to better COP when the wind power production is peaking, during which heat is stored to TES. With the fixed electrolyser operation in ElVar, compressor capacity of 76 MW is needed.

The charging and discharging behaviour of ElFix and ElWind differ clearly, as ElWind requires higher charging and discharging rates, 386 MW and 595 MW at the maximum, respectively, while ElFix can manage with 127 MW and 352 MW, respectively.

In the case of ElVar, no TES is needed. ElWind requires 291 GWh of storage to supply heat throughout the year. ElFix needs 53% larger TES, 446 GWh. The capacity of the storage is fivefold to the world's largest water-based TES planned to Vantaa, which will have a capacity of 90 GWh [50].

The main values regarding the annual performance of the system are presented in Table 4. With the current assumptions for charge and discharge efficiency, and heat loss, the annual efficiency of TES is 79.0% and 82.2% for cases ElFix and ElWind. TES share describes the amount of heat delivered to the district heat network via TES. The rest is supplied directly with the heat pump. TES share is the highest in ElWind, 35.8%, and 29.5% for ElFix.

Table 4. Annual performance of the heating system.

Case	TES Efficiency (%)	TES Share (%)	H_2 Production (TWh)	Average COP (-)	Electricity Demand (TWh)
ElVar	-	-	2.30	7.4	3.98
ElFix	79.0	29.5	2.68	7.4	4.60
ElWind	82.2	35.8	2.72	7.4	4.66

The annual production of hydrogen is smallest in ElVar, 2.30 TWh. In ElFix and ElWind, 2.68 TWh and 2.72 TWh is produced, respectively. The difference between cases is probably due to heat losses of TES, which requires higher total heat generation by heat pump and electrolyser, which increases the required electrolyser power and consequently hydrogen production. The 18% difference in hydrogen production between ElVar and ElWind can be considered significant. However, the total H_2 production quantities are still quite modest compared to the demand of the Raahe steel mill, for instance. Raahe mill would require about 6.75 TWh electricity for hydrogen production or about 4.2 TWh of hydrogen if the estimations presented in this work are correct. Thus, the demand of H_2 could greatly exceed the production in these heat-coupled electrolyser scenarios.

Another important aspect is to compare the local electricity consumption to the available wind power in Oulu region. As seen in Table 5, the current annual production of 2.2 TWh is not sufficient, but the short-term estimate is already over double the demand, also leaving room for other electricity uses. For the local wind power potential estimation, node 9 and the neighbouring node 10 were both included (cf. Figure 3). The local growth of wind power potential in Oulu region is greater than the average for the whole Bothnian Bay.

Table 5. Estimated wind power capacities in the Oulu region.

Time	Cumulative Capacity (GW)	Annual Production (TWh)
Current	0.9	2.2
Short term	4.2	11.0
Long term	8.2	21.2

To summarize, matching the local heat demand with hydrogen production results in small and insufficient supply of hydrogen for other sectors. Conversely, if the hydrogen production is dimensioned according to available CO_2 for e-methanol production, an oversupply of 2 TWh of heat occurs. Similarly, using all projected wind electricity would result in 5–7 TWh of excess heat in Oulu.

The dynamical behaviour of TES is presented in Figure 11 for ElFiX and ElWind. Due to variation of both ambient temperature and wind production, there is more variation from charge to discharge in case ElWind. In addition, there seems to be more wind during the winter than during the summer, as the demand for seasonal storage is smaller than in ElFix. In ElFix, the strong seasonal variation is much clearer, and there are less occasions when TES operation is switched from charge to discharge or the other way round.

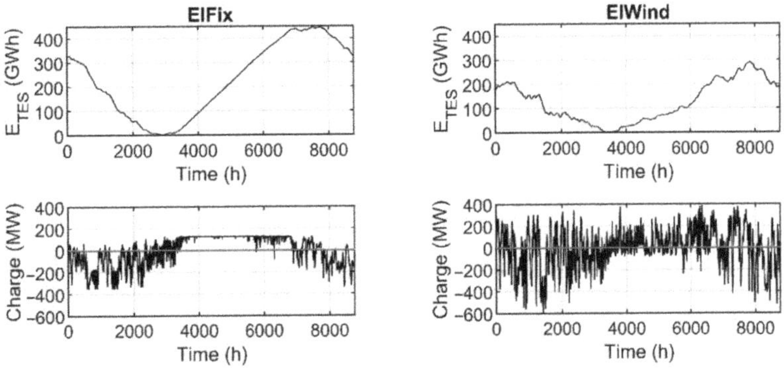

Figure 11. State of charge of TES and charge/discharge rates in cases ElFix and ElWind.

4. Discussion

Wind capacity is growing strongly in the short term in the Bothnian Bay region: ongoing projects in various stages of development could increase the cumulative capacity to 19 GW from the current capacity of 6 GW. Long-term wind capacity could in turn increase to nearly 40 GW, but the results should be verified and amended in future studies by using more sophisticated evaluation methodologies. The drastic increase in wind power is likely reflected in development needs for energy storage and transmission infrastructure, which would need to be developed and extended for electricity, hydrogen, or both.

Steel appears to be an important driver for implementing electrolytic hydrogen generation in the region. Nearly 55 TWh of renewable electricity would be required annually if all locally mined iron ore is processed to sponge iron according to existing company visions. Obtaining such amounts of electricity would necessitate harnessing a large portion of future wind power potential for hydrogen production. Sweden is, on a global scale, a minor player in iron mining or steel production, and yet hydrogen demand from the steel industry plays a dominant role in this regional study. Thus, it seems probable that a large portion of future global steel-related hydrogen production cannot be reliably linked with heat utilization, simply because there is not enough local demand for heat.

The potential volume for applying BECCUS in the region is vast, but CO_2 utilization aspects are likely limited by available electricity—particularly in the upper end of utilization visions. Biogenic CO_2 emissions from the pulp and paper sector currently amount to 10 Mt annually, which would require about 85 TWh of electricity if all were to be converted to methanol. In practice, it is not likely that all available capacities would be implemented. As both steel decarbonization and Power-to-X systems compete for the same electricity resources, some local compromises in implementation scale and timeline are likely. The availability of electricity and local wind power resources can therefore be considered a bottleneck, especially since the demand for electricity also increases in other applications, such as electric vehicles. The market demand for Power-to-X products was not evaluated in depth in this work, but it is likely not a limiting factor for manufacturing. For instance, the estimated methanol production potential in the Bothnian Bay area corresponds to 100–200% of the combined biofuel use in Sweden and Finland when compared in terms of thermal energy content of the liquids. As the potential customer sectors for methanol and other Power-to-X products would not be limited only to blending in road transport, market uptake can be expected to be much larger.

Ideally, hydrogen production should be integrated with district heat or similar uses for maximal energy efficiency. However, local heat demand can be satisfied already with a moderately sized electrolyser, which would then not be adequate for producing the necessary hydrogen volumes for decarbonizing other sectors. As a case example, if the district heat demand of the city of Oulu was produced by heat pump and electrolyser, the hydrogen production would cover only roughly a third of the demand of Raahe steel mill if it were converted to the carbon neutral HYBRIT process. From another perspective, decentralizing hydrogen production to the major cities in the region would enable all of those cities to meet their district heat demand primarily on electrolyser residual heat which would otherwise have no apparent use. Excess heat would likely be formed eventually as the hydrogen production volumes increase, but the first crucial pilots could be implemented with heat integration, increasing system efficiency and profitability. Another important aspect is the utilization of the oxygen that is formed as a by-product.

One open question is how well the operation profile of the electrolyser compares to wind power, heat demand and hydrogen demand. Heat demand and wind power vary significantly, while the demand for hydrogen might be rather stable. Therefore, some storage or balance mechanism is likely required for either hydrogen or heat, or both. In this study, the modelled heat scenarios are somewhat ultimate examples, as no other production methods for heat or hydrogen were considered. Hydrogen pipe network could balance the feed even though some of the electrolysers were to operate flexibly according to the heat demand. Similarly, some other stable heat sources would likely be available, which would decrease the required TES capacity.

Future studies should therefore focus on a more holistic overview of the dynamic performance of the system, including aspects of both electricity and heat availability and demand. It is likely this would require a more detailed analysis of the existing and future energy infrastructure, which is also related to energy transmission between neighbouring countries. The determination of oxygen market volumes could also be worthwhile, given the large number of pulp and paper mills in the area.

Author Contributions: Conceptualization, H.K., E.I. and T.T.; Data curation, H.K. and E.I.; Investigation, H.K. and E.I.; Methodology, H.K., E.I. and T.T.; Project administration, H.K.; Supervision, T.T.; Writing—original draft, H.K. and E.I.; Writing—review and editing, H.K. and T.T. All authors have read and agreed to the published version of the manuscript.

Funding: This research was funded by P2XEnable project (p2xenable.fi) provided by Business Finland grant number 8797/31/2019. The APC was funded by LUT University.

Institutional Review Board Statement: Not applicable.

Informed Consent Statement: Not applicable.

Data Availability Statement: Not applicable.

Conflicts of Interest: The authors declare no conflict of interest.

References

1. Mission Innovation Hydrogen Valley Platform. Available online: https://www.h2v.eu/ (accessed on 17 November 2021).
2. Basque Hydrogen Corridor. Available online: https://bh2c.es/en/ (accessed on 17 November 2021).
3. Porthos–CO_2 Reduction through Storage Beneath the North Sea. Available online: https://www.porthosCO2.nl/en/ (accessed on 17 November 2021).
4. NortH2–Kickstarting the Green Hydrogen Economy. Available online: https://www.north2.eu/en/ (accessed on 17 November 2021).
5. BotH2nia–CLIC Innovation. Available online: https://clicinnovation.fi/project/both2nia/ (accessed on 17 November 2021).
6. Sulasalmi, P.; Kärkkäinen, M.-L.; Fabritius, T. Hydrogen Study of the Bay of Bothnia. 2021. University of Oulu. Available online: https://www.businessoulu.com/media/2019/vetywebinaari/hydrogen-study-of-the-bay-of-bothnia-final-report.pdf (accessed on 17 November 2021).
7. IRENA. *Green Hydrogen Cost Reduction: Scaling up Electrolysers to Meet the 1.5 °C Climate Goal. International Renewable Energy Agency*; International Renewable Energy Agency: Abu Dhabi, United Arab Emirates, 2020.
8. IRENA. *Renewable Power Generation Costs in 2019. International Renewable Energy Agency*; International Renewable Energy Agency: Abu Dhabi, United Arab Emirates, 2020.
9. Parra, D.; Patel, M.K. Techno-Economic Implications of the Electrolyser Technology and Size for Power-to-Gas Systems. *Int. J. Hydrogen Energy* **2016**, *41*, 3748–3761. [CrossRef]

10. Tsupari, E.; Kärki, J.; Vakkilainen, E. Economic Feasibility of Power-to-Gas Integrated with Biomass Fired CHP Plant. *J. Energy Storage* **2016**, *5*, 62–69. [CrossRef]
11. Buttler, A.; Spliethoff, H. Current Status of Water Electrolysis for Energy Storage, Grid Balancing and Sector Coupling via Power-to-Gas and Power-to-Liquids: A Review. *Renew. Sustain. Energy Rev.* **2018**, *82*, 2440–2454. [CrossRef]
12. Böhm, H.; Moser, S.; Puschnigg, S.; Zauner, A. Power-to-Hydrogen & District Heating: Technology-Based and Infrastructure-Oriented Analysis of (Future) Sector Coupling Potentials. *Int. J. Hydrogen Energy* **2021**, *46*, 31938–31951. [CrossRef]
13. Buttler, A.; Koltun, R.; Wolf, R.; Spliethoff, H. A Detailed Techno-Economic Analysis of Heat Integration in High Temperature Electrolysis for Efficient Hydrogen Production. *Int. J. Hydrogen Energy* **2015**, *40*, 38–50. [CrossRef]
14. Min, G.; Park, Y.J.; Hong, J. Thermodynamic Analysis of a Solid Oxide Co-Electrolysis Cell System for Its Optimal Thermal Integration with External Heat Supply. *Energy Convers. Manag.* **2020**, *225*, 113381. [CrossRef]
15. Gruber, M.; Weinbrecht, P.; Biffar, L.; Harth, S.; Trimis, D.; Brabandt, J.; Posdziech, O.; Blumentritt, R. Power-to-Gas through Thermal Integration of High-Temperature Steam Electrolysis and Carbon Dioxide Methanation–Experimental Results. *Fuel Process. Technol.* **2018**, *181*, 61–74. [CrossRef]
16. Ikäheimo, J.; Kiviluoma, J.; Weiss, R.; Holttinen, H. Power-to-Ammonia in Future North European 100% Renewable Power and Heat System. *Int. J. Hydrogen Energy* **2018**, *43*, 17295–17308. [CrossRef]
17. Lund, H.; Werner, S.; Wiltshire, R.; Svendsen, S.; Thorsen, J.E.; Hvelplund, F.; Mathiesen, B.V. 4th Generation District Heating (4GDH). *Energy* **2014**, *68*, 1–11. [CrossRef]
18. Gorre, J.; Ruoss, F.; Karjunen, H.; Schaffert, J.; Tynjälä, T. Cost Benefits of Optimizing Hydrogen Storage and Methanation Capacities for Power-to-Gas Plants in Dynamic Operation. *Appl. Energy* **2020**, *257*, 113967. [CrossRef]
19. Astiaso Garcia, D.; Barbanera, F.; Cumo, F.; Di Matteo, U.; Nastasi, B. Expert Opinion Analysis on Renewable Hydrogen Storage Systems Potential in Europe. *Energies* **2016**, *9*, 963. [CrossRef]
20. Kuparinen, K.; Vakkilainen, E.; Ryder, P. Integration of Electrolysis to Produce Hydrogen and Oxygen in a Pulp Mill Process. *Appita* **2016**, *69*, 81–88.
21. Maggio, G.; Squadrito, G.; Nicita, A. Hydrogen and Medical Oxygen by Renewable Energy Based Electrolysis: A Green and Economically Viable Route. *Appl. Energy* **2022**, *306*, 117993. [CrossRef]
22. Saxe, M.; Alvfors, P. Advantages of Integration with Industry for Electrolytic Hydrogen Production. *Energy* **2007**, *32*, 42–50. [CrossRef]
23. Resch, B.; Sagl, G.; Törnros, T.; Bachmaier, A.; Eggers, J.-B.; Herkel, S.; Narmsara, S.; Gündra, H. GIS-Based Planning and Modeling for Renewable Energy: Challenges and Future Research Avenues. *IJGI* **2014**, *3*, 662–692. [CrossRef]
24. Petrovic, S.N.; Karlsson, K.B. Danish Heat Atlas as a Support Tool for Energy System Models. *Energy Convers. Manag.* **2014**, *87*, 1063–1076. [CrossRef]
25. Nielsen, S.; Möller, B. GIS Based Analysis of Future District Heating Potential in Denmark. *Energy* **2013**, *57*, 458–468. [CrossRef]
26. Persson, U.; Möller, B.; Werner, S. Heat Roadmap Europe: Identifying Strategic Heat Synergy Regions. *Energy Policy* **2014**, *74*, 663–681. [CrossRef]
27. Welder, L.; Ryberg, D.S.; Kotzur, L.; Grube, T.; Robinius, M.; Stolten, D. Spatio-Temporal Optimization of a Future Energy System for Power-to-Hydrogen Applications in Germany. *Energy* **2018**, *158*, 1130–1149. [CrossRef]
28. Karjunen, H.; Tynjälä, T.; Hyppänen, T. A Method for Assessing Infrastructure for CO 2 Utilization: A Case Study of Finland. *Appl. Energy* **2017**, *205*, 33–43. [CrossRef]
29. Simon Araya, S.; Liso, V.; Cui, X.; Li, N.; Zhu, J.; Sahlin, S.L.; Jensen, S.H.; Nielsen, M.P.; Kær, S.K. A Review of The Methanol Economy: The Fuel Cell Route. *Energies* **2020**, *13*, 596. [CrossRef]
30. Vogl, V.; Åhman, M.; Nilsson, L.J. Assessment of Hydrogen Direct Reduction for Fossil-Free Steelmaking. *J. Clean. Prod.* **2018**, *203*, 736–745. [CrossRef]
31. Tekniikka ja Talous. Tämä Sähkösyöppö Kuluttaa 3,5 TWh/v-4 % Koko Suomen Sähkönkulutuksesta. Available online: https://www.tekniikkatalous.fi/uutiset/tama-sahkosyoppo-kuluttaa-3-5-twh-v-4-koko-suomen-sahkonkulutuksesta/d30cc904-3c00-37d9-8797-bb01f23da998 (accessed on 28 November 2021).
32. SSAB. Annual Report 2020. Available online: https://mb.cision.com/Main/980/3309155/1388946.pdf (accessed on 28 November 2021).
33. LKAB. HYBRIT: SSAB, LKAB and Vattenfall First in the World with Hydrogen-Reduced Sponge Iron. Available online: https://www.lkab.com/en/news-room/press-releases/hybrit-ssab-lkab-and-vattenfall-first-in-the-world-with-hydrogen-reduced-sponge-iron/?aid=16447 (accessed on 28 November 2021).
34. LKAB. Annual and Sustainability Report 2020. Available online: https://www.lkab.com/en/SysSiteAssets/documents/finansiell-information/en/annual-reports/lkab_2020_annual_and_sustainability_report.pdf (accessed on 28 November 2021).
35. European Pollutant Release and Transfer Register (E-PRTR). Database Version v17. 2019. Available online: https://www.eea.europa.eu/ds_resolveuid/4fd0ddc476b94f2f99ecc79a81a2abdc (accessed on 8 January 2020).
36. Rabensteiner, M.; Kinger, G.; Koller, M.; Hochenauer, C. Pilot Plant Study of Aqueous Solution of Piperazine Activated 2-Amino-2-Methyl-1-Propanol for Post Combustion Carbon Dioxide Capture. *Int. J. Greenh. Gas Control.* **2016**, *51*, 106–117. [CrossRef]
37. Hwang, H.; Han, J.-H.; Lee, I.-B. Technoeconomic Feasibility Study of Monoethanolamine-Based CO_2 Capture System Deployment to Be Retrofitted to an Existing Utility System in a Chemical Plant. *Ind. Eng. Chem. Res.* **2013**, *52*, 18334–18344. [CrossRef]

38. Szima, S.; Cormos, C.-C. Improving Methanol Synthesis from Carbon-Free H2 and Captured CO2: A Techno-Economic and Environmental Evaluation. *J. CO_2 Util.* **2018**, *24*, 555–563. [CrossRef]
39. Pérez-Fortes, M.; Schöneberger, J.C.; Boulamanti, A.; Tzimas, E. Methanol Synthesis Using Captured CO_2 as Raw Material: Techno-Economic and Environmental Assessment. *Appl. Energy* **2016**, *161*, 718–732. [CrossRef]
40. Finnish Wind Power Association. FWPA Project List. 2021. Available online: https://tuulivoimayhdistys.fi/en/wind-power-in-finland/projects-under-planning (accessed on 17 November 2021).
41. VBK. County Administrative Board of Västra Götaland County. *Vindbrukskollen*. Available online: https://vbk.lansstyrelsen.se/ (accessed on 9 November 2021).
42. Roques, F.; Le Thieis, Y.; Aue, G.; Spodniak, P.; Pugliese, G.; Cail, S.; Peffen, A.; Honkapuro, S.; Sihvonen, V. Enabling Cost-Efficient Electrification in Finland. In *Sitran Studies 194*; Sitra: Helsinki, Finland, 2021.
43. Michailos, S.; Walker, M.; Moody, A.; Poggio, D.; Pourkashanian, M. A Techno-Economic Assessment of Implementing Power-to-Gas Systems Based on Biomethanation in an Operating Waste Water Treatment Plant. *J. Environ. Chem. Eng.* **2021**, *9*, 104735. [CrossRef]
44. Energiateollisuus. Kaukolämpötilasto. 2019.
45. Hiltunen, P.; Syri, S. Highly Renewable District Heat for Espoo Utilizing Waste Heat Sources. *Energies* **2020**, *13*, 3551. [CrossRef]
46. Flores, J.F.C.; Lacarrière, B.; Chiu, J.N.W.; Martin, V. Assessing the Techno-Economic Impact of Low-Temperature Subnets in Conventional District Heating Networks. *Energy Procedia* **2017**, *116*, 260–272. [CrossRef]
47. Ahrens, M.U.; Foslie, S.S.; Moen, O.M.; Bantle, M.; Eikevik, T.M. Integrated High Temperature Heat Pumps and Thermal Storage Tanks for Combined Heating and Cooling in the Industry. *Appl. Therm. Eng.* **2021**, *189*, 116731. [CrossRef]
48. Finnish Meteorological Institute. Observation Data. Available online: https://www.ilmatieteenlaitos.fi/havaintojen-lataus (accessed on 9 November 2021).
49. Yang, T.; Liu, W.; Kramer, G.J.; Sun, Q. Seasonal Thermal Energy Storage: A Techno-Economic Literature Review. *Renew. Sustain. Energy Rev.* **2021**, *139*, 110732. [CrossRef]
50. Vantaan Energia. Maailman Suurin Lämmön Kausivaraso Vantaalle. Available online: https://www.vantaanenergia.fi/fossiiliton-2026/maailman-suurin-lammon-kausivarasto-vantaalle/ (accessed on 30 November 2021).

energies

Article

Integrating System and Operator Perspectives for the Evaluation of Power-to-Gas Plants in the Future German Energy System

Johannes Schaffert [1,*], Hans Christian Gils [2,*], Max Fette [3,*], Hedda Gardian [2], Christine Brandstätt [3], Thomas Pregger [2], Nils Brücken [1], Eren Tali [1], Marc Fiebrandt [1], Rolf Albus [1] and Frank Burmeister [1]

1 Gas-und Wärme-Institut Essen e.V. (GWI), 45356 Essen, Germany; nils.bruecken@gwi-essen.de (N.B.); eren.tali@gwi-essen.de (E.T.); marc.fiebrandt@gwi-essen.de (M.F.); rolf.albus@gwi-essen.de (R.A.); frank.burmeister@gwi-essen.de (F.B.)

2 German Aerospace Center (DLR), Institute of Networked Energy Systems, 70563 Stuttgart, Germany; Hedda.Gardian@dlr.de (H.G.); Thomas.Pregger@dlr.de (T.P.)

3 Fraunhofer Institute for Manufacturing Technology and Advanced Materials IFAM, 28359 Bremen, Germany; cbr.eco@cbs.dk

* Correspondence: schaffert@gwi-essen.de (J.S.); Hans-Christian.Gils@dlr.de (H.C.G.); max.fette@ifam.fraunhofer.de (M.F.); Tel.: +49-201-3618235 (J.S.)

Abstract: In which way, and in which sectors, will renewable energy be integrated in the German Energy System by 2030, 2040, and 2050? How can the resulting energy system be characterised following a −95% greenhouse gas emission reduction scenario? Which role will hydrogen play? To address these research questions, techno-economic energy system modelling was performed. Evaluation of the resulting operation of energy technologies was carried out from a system and a business point of view. Special consideration of gas technologies, such as hydrogen production, transport, and storage, was taken as a large-scale and long-term energy storage option and key enabler for the decarbonisation of the non-electric sectors. The broad set of results gives insight into the entangled interactions of the future energy technology portfolio and its operation within a coupled energy system. Amongst other energy demands, CO_2 emissions, hydrogen production, and future power plant capacities are presented. One main conclusion is that integrating the first elements of a large-scale hydrogen infrastructure into the German energy system, already, by 2030 is necessary for ensuring the supply of upscaling demands across all sectors. Within the regulatory regime of 2020, it seems that this decision may come too late, which jeopardises the achievement of transition targets within the horizon 2050.

Keywords: energy transition; power-to-gas; PtG; hydrogen; H_2; energy system; energy modelling; energy system optimisation; system analysis

Citation: Schaffert, J.; Gils, H.C.; Fette, M.; Gardian, H.; Brandstätt, C.; Pregger, T.; Brücken, N.; Tali, E.; Fiebrandt, M.; Albus, R.; et al. Integrating System and Operator Perspectives for the Evaluation of Power-to-Gas Plants in the Future German Energy System. *Energies* **2022**, *15*, 1174. https://doi.org/10.3390/en15031174

Academic Editor: Abdelali El Aroudi

Received: 21 December 2021
Accepted: 2 February 2022
Published: 5 February 2022

1. Introduction

1.1. Background

The energy transition towards a renewable energy system that serves the demands of the electricity, gas, heat, and transport sectors is one of the most complex societal projects of our time. The green transformation of all energy-dependent activities touches all individuals, all economic activities, and administrations worldwide. While the first steps have been taken, the local, regional, and national roadmaps for the future energy system, e.g., in 2050, remain a constant challenge and need permanent scientific assessments, course corrections, and refinements.

1.2. State of Research

High temporal and spatial resolution energy system models have been limited to the electricity sector in previous analyses. These focused, for example, on the grid, storage, and power plant capacities needed to balance electricity generation from variable renewable energy (VRE) [1,2]. Continuous development has successively added coupling to other sectors, such as heating and electric mobility, to these analyses [3,4]. In parallel, models of the gas market and the gas system have been further developed to analyse future scenarios [5,6]. Against the background of the political goals of reducing CO_2 emissions, the integration of power-to-gas plants for the generation of synthetic gas also received increasing attention [7]. Recently, the energy science community has made strong progress in integrating electricity system focused models with natural gas system focused models [8]. This significantly improves the capability to analyse energy systems that are integrated across different sectors [9,10].

One continuing challenge in interdisciplinary energy system research is the coupling of models [11]. Additionally, the identification of business models for power-to-gas plant operators remains challenging [9,12,13]. Besides these aspects, in many studies, the techno-economical level of detail during optimisation of energy systems remains shallow, as the representation of gas infrastructures, for example, suffers strong simplifications, and the decision-making by individual stakeholders, such as plant operators, is not integrated.

1.3. Contribution of This Paper

This analysis is dedicated to cost-minimising strategies for the construction and operation of power-to-gas plants along the transformation of the German energy system to a climate-neutral supply. This is done from two perspectives: that of the macro-economic planner and that of the plant operator. The focus is on the incorporation of power-to-gas into an energy system that is integrated across all sectors. In addition to the electricity sector and the heating sector, the interfaces to the transport sector, via electro mobility and hydrogen vehicles, are considered. This allows the evaluation of the contribution of flexible operation of power-to-gas plants, as well as other electrical equipment in the gas system, to balance the fluctuating power generation from VRE. In addition, the regional distribution of gas and hydrogen infrastructures in Germany is considered. The methodological basis is the adequate representation of the gas system in two energy system models and their coupling via a data interface. This coupling makes it possible to analyse which adjustments to the regulatory framework are needed to make power-to-gas plants economically attractive.

2. Materials and Methods

The study relies on the enhancement and application of two models providing different perspectives on the energy system. While the plant capacities and their hourly dispatch in the *REMix* model (Section 2.1.1) result from the minimisation of economic costs on a macro-economic scale, *MuGriFlex* (Section 2.1.2) aims at the profit maximisation of the operator of one or more individual plants. These models are parametrised and applied in a harmonised and partially coupled manner (Section 2.1.3). The case study, presented here, analyses the future energy system in Germany and its neighbouring countries (Section 2.2). It relies on a detailed normative scenario for the achievement of emission reduction goals (Section 2.3). Furthermore, it is based on extensive data research of the plant inventory and possible technology development paths, especially in the gas sector (Section 2.4), as well as the other sectors (Section 2.5). Finally, we present the regulatory framework in Germany that we considered in the modelling (Section 2.6). For clarity, structure of this work is depicted in the graphical abstract Figure 1. Assumptions have been published online [14].

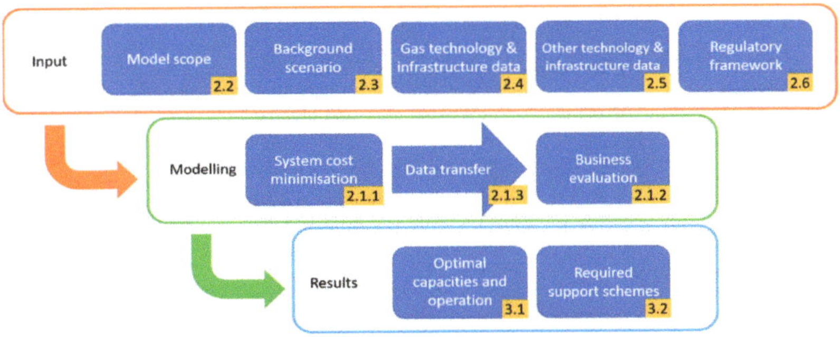

Figure 1. Overview of the modelling procedure and indication of respective sections in the paper.

2.1. Modelling Approach

The analysis relies on the combined application of the energy system models *REMix* and *MuGriFlex*, which are introduced in the following.

2.1.1. REMix

The optimisation framework *REMix* was designed for the analysis of future integrated energy systems in high spatial and temporal resolution [15]. It relies on a linear programming approach, which is typically used to minimise costs, from a central system operator's perspective, under multiple technological and economic boundary conditions. Originally limited to the power sector, it has been continuously enhanced to also include electric mobility [16], the heating sector [17], as well as hydrogen production, storage, and consumption [18]. For the case study presented here, it has been further enhanced to include the gas sector [19]. The model is designed to optimise capacities and hourly operation of all technologies in a multi-node approach and with perfect foresight over one year. Depending on the use case, many hundreds of nodes, or up to one hundred technologies, can be considered. In addition to the objective function of the system costs to be minimised, the energy carrier-specific balances are the central equations of the model. These ensure that the demand and supply of energy are balanced for each region and hour. This is achieved by using different technologies for the conversion, storage, and transport of energy, depending on the scope of the model. These technologies are limited in their use by the sum of exogenously given and, if applicable, endogenously added capacities. The mathematical framework of the model has been documented in [15–19], Figure 2 provides an overview of the framework. Details on the model scope and utilized input data considered here are provided in Sections 2.2–2.5.

2.1.2. MuGriFlex

The *MuGriFlex* model serves to analyse individual energy systems for profitability, optimal investment, and operation of the systems' components. It considers interrelated technical assets, generating, using, and storing electricity, heat, and gas, their cost, and the relevant regulatory framework [20,21]. Thereby, it adds a business perspective on the feasibility of the scenarios modelled with *REMix* [22]. Based on plant parameters, time series for energy demand, weather, and energy prices, as well as surcharges and tariffs, *MuGriFlex* simulates the operation of a combination of technical assets in hourly resolution. Thereby, it enables the assessment of the economic feasibility for defined individual energy systems, or it optimises the design and dimensioning of such energy systems within a specified regulatory framework.

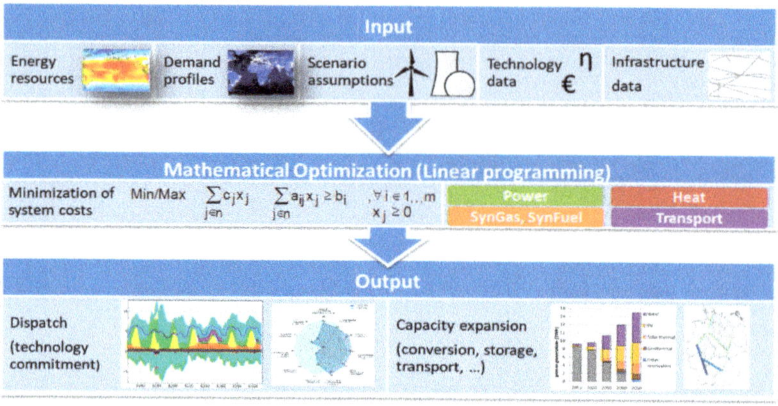

Figure 2. Overview of inputs, method, and outputs of the REMix energy system modelling framework.

2.1.3. Model Coupling

For an integrated analysis, the overall optimised energy system is looked at from the business perspective within the given regulatory framework. Hourly time series for plant operation and electricity cost, as well as the optimal gas mix per year, are central to the coupling between the two models. Outputs of *REMix* are fed into *MuGriFlex* in order to determine whether the regulatory framework is suitable to implement the desirable overall system development and its operation.

These outputs include the following values:

- Plant sizes (expressed as rated thermal output relative to peak requirement of the local energy system) for combined heat and power (CHP) plants, gas- and electric boilers, heat pumps (HP), thermal energy storage, etc.
- Operation of plants: full load hours per year
- Hourly time-series of power generation costs: These are assumed to be the electricity cost of the power plant running at the margin. To receive electricity prices, the surcharges, to be paid by the respective use case, are added.
- Time-series of produced synthetic gas to establish the gas production costs, taking into account the electricity cost at the given time

If a given framework promotes investment and operation of plants that deviates from the techno-economic optimum, *MuGriFlex* enables the exploration of alternative frameworks (see Section 3.2).

2.2. Set-Up of the Case Study

The transformation of the German energy system is the focus of this analysis. To consider the balancing effects of the European power grid, the neighbouring countries, as well as Italy, Sweden, and Norway, are also modelled in *REMix*. However, a detailed analysis of the flexible sector coupling and the gas transport is carried out only for Germany. To be able to show regional effects and to evaluate the expansion of electricity and hydrogen grid capacities, Germany is divided into 10 regions in the model. These result from partial aggregation of the federal states, according to Figure 3. To be able to describe the transformation path of the system, the scenario years 2020, 2030, 2040, and 2050 are modelled in *REMix*. The model is applied myopically, i.e., the investment decisions are carried over into the later years until the plant lifetime is reached.

To evaluate the interaction of power-to-gas plants in an integrated overall system, *REMix* includes a wide range of technologies, especially with regard to flexible sector coupling. For Germany, the model includes almost 100 technologies in the electricity, heat, gas and transport sectors. In particular, the electricity and heat supply are modelled with a high degree of granularity. Photovoltaics (PV), concentrating solar power (CSP), reservoir and run-of-the-river hydro power, onshore and offshore wind, geothermal, and biomass are being considered for electricity generation from renewable sources. An endogenous capacity expansion is considered for wind, solar, and biomass power plants. Furthermore, it is assumed that the existing wind, PV, and hydro power plants will be replaced at the end of their service life. This prevents extreme characteristics in the spatial distribution of the plants.

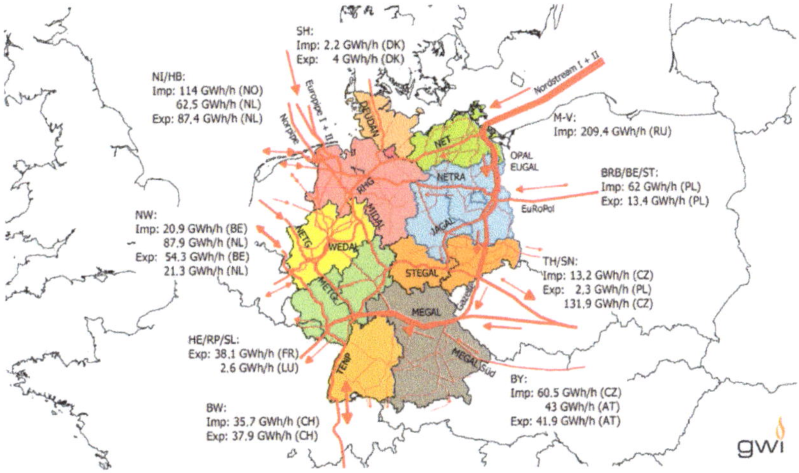

Figure 3. Simplified representation of the gas transportation network in the ten investigated model regions and assumed interconnection capacities to the neighbouring countries and regions used as start values for the *REMix* model calculations.

Conventional power generation is possible with nuclear, coal, oil, and gas power plants. The existing power plant fleet will be successively decommissioned. The exogenously assumed plant capacities and their future development are listed in [14]. While coal and nuclear power plants cannot be replaced at the end of their service life, an endogenous addition of gas-fired power plants is possible. This applies throughout the study area and equally to condensing power plants and CHP plants. For cogeneration of electricity and heat in CHP systems, 15 technologies are considered, which differ in heat consumers, plant size, and fuel. All CHP plants also have a peak load boiler, and some can be supplemented by the model with thermal storage, electric boilers, heat pumps, and solar thermal systems. Energy transport can be realised via direct current (DC) and alternating current (AC) power lines, gas pipelines, and hydrogen pipelines. For power and gas pipelines, the existing capacities, as well as the planned expansion, are taken into account. An endogenous expansion of power lines is possible from 2040, but it is limited to 5 GW per line and decade. Hydrogen pipelines within Germany can be built from the scenario year 2030. Other energy storage in the system includes underground gas storage, hydrogen cavern and tank storage, stationary battery storage, and pumped storage. Battery storage and hydrogen storage are optimised in their capacity. Flexibility can also be provided by battery electric vehicles (BEV) with bidirectional charging, decentralised heat pumps with thermal storage, and load management in industry and commerce. As described below, the production of hydrogen and methane is also optimised endogenously in the model.

In other European countries, flexible sector coupling is only considered to a very limited extent. For example, consideration of the heat sector is limited to electric heat generation, which is inflexible, as is BEV charging. The decentralised generation of hydrogen, on the other hand, is partially made more flexible via the consideration of tank storage. Pipelines and underground storage facilities for hydrogen and natural gas are only considered for Germany. While natural gas can be imported without limit at national borders, hydrogen must be produced domestically. Net electricity import, on the other hand, are possible, but limited to 20% of demand, including system losses.

2.3. Main Assumptions about the Energy Future

For the parameterization of the models and the consistency of the model coupling, quantitative scenario frameworks are an essential basis. There is also the need to document the overall energy future considered, for which the calculated results and conclusions derived from them are valid.

2.3.1. General Assumptions

The framework scenario was defined exogenously, from which further assumptions were made regarding the technology paths for the model parameterization. It is based on a socio-economic context framework similar to [23] that follows the narratives of a long-term decrease in the population in Germany from 81 to below 75 million, moderate economic growth at 1.2% per year, a further slight increase in heated building areas and vehicles in passenger transport (with 10% lower mileage by 2050), and a continuous increase in freight transport of about 1% per year. For the European countries, similar socio-economic paths are assumed according to the European project e-Highway 2050 [24], for which a decrease in the European population by 10% was assumed in the scenario variant "Small & Local", as well as a similarly moderate economic growth, with a 1.3% increase in gross domestic product (GDP) per year.

The scenario assumes a slight increase in fossil fuel prices in the future based on the national transformation scenario of [23] (Table 1). The prices for solid biomass and biogas, on the other hand, are assumed to remain constant, as biomass is only used to a limited extent in the scenario within the limits of sustainable potentials. The incineration of waste, as well as the use of geothermal heat, is not associated with any energy carrier costs. Nevertheless, it is associated with variable costs of plant operation.

Table 1. Assumed fuel costs in €/MWh in the scenario.

Fuel	2020	2030	2040	2050
Natural gas	38.4	41.0	43.2	42.1
Hard coal	15.1	16.2	17.3	20.5
Lignite	4.1	4.1	4.1	4.1
Uranium	3.2	3.2	3.2	3.2
Oil	58.3	60.5	65.9	71.3
Biogas	28.1	28.1	28.1	28.1
Solid biomass	26.9	26.9	26.9	26.9

In the scenario, it is also assumed that the emission of CO_2 is subject to costs via certificate trading. The values assumed for this were assumed to increase sharply, in line with the targets. The values used there were adjusted to the base year of the cost data (2015), taking inflation into account (see Table 2).

Table 2. Assumed emission certificate costs in €/t CO_2 in the scenario.

Scenario	2020	2030	2040	2050
Emission cost in €/t CO_2	32	94	154	216

2.3.2. Energy Demand Scenario for Germany and Europe

The scenario was developed with the aim of illustrating an exemplary development path for Germany, with regard to the large reduction in CO_2 emissions in the energy system and the resulting demand for electricity and green synthetic gas, while remaining within the range of possibilities that seem plausible from today's perspective for transformation processes in the sectors. The scenario (called THG95) implements the goal of climate neutrality of the energy system and maximum shares of renewable energies, in line with the goal of a 95% reduction in total greenhouse gas (GHG) emissions by 2050. Strong efficiency developments, in all sectors, are envisaged according to the goals of the German government's 2010 energy concept [25]. This leads to a strong use of electricity for the direct electrification of heat generation and vehicles in transport, with complementary use of hydrogen via fuel cell vehicles and, if necessary, for the storage and reconversion of hydrogen in the future energy system. The complete substitution of fossil energy carriers, including gas for backup power plants, results in a high demand for synthetic energy carriers with corresponding conversion losses.

For the neighbouring countries the developments are based on the 100% Renewable Energy Scenario (RES) of the European e-Highway 2050 project [24]. The increase in the total electricity demand in the neighbouring countries is lower compared to Germany, especially for H_2 generation, which plays a smaller role in the e-Highway 2050 scenarios. Deviating from this, the developments for electric mobility were projected in the same way in all countries to increase comparability. The resulting assumptions for the exogenously specified electricity demand are shown in the following Table 3. Further information can be found in [22].

Table 3. Electricity demand scenario for Europe in TWh per year.

Country	2020 Conv.	2050 Conv.	2050 BEV	2050 H_2	2050 HP	2050 E-H
Germany	428	344	145	423	70	159
Austria	72	47	12	10	4	3
Belgium	91	67	16	15	9	5
Czech Republic	67	41	10	10	4	4
Denmark (East)	14	8	3	3	1	0.6
Denmark (West)	23	13	5	5	2	1
France	486	380	99	90	36	6
Italy	325	284	84	77	17	12
Luxembourg	7	4	1	0.5	0.3	0.2
Netherlands	115	93	19	17	11	7
Norway	131	84	8	7	2	0.6
Poland	161	79	34	29	9	7
Sweden	146	91	16	15	6	5
Switzerland	64	49	10	10	4	2
Total	2129	1582	463	709	174	212

Conv: Conventional electricity demand of consumers; BEV: Electricity for electro-mobility; H_2: Electricity for hydrogen production; HP: Electricity for heat pumps; E-H: Electricity for electric heaters.

2.4. *Fundamentals and Modelling Assumptions for the Natural Gas and Hydrogen Sector*

The complementary consideration of the gas system in *REMix* requires extensive parameterization with infrastructure inventory data and techno-economic parameters. The procedure and data sources used for this are presented in the following.

2.4.1. Natural Gas Transportation Grids and Hydrogen Transport Option

The natural gas networks can be classified into the long-distance transport system and the finer-meshed distribution system. Within this project, the distribution level is not modelled. Instead, an ideal distribution within a model region is assumed. The intra-regional transport, via the transport system, is represented by a balance-sheet approach that is based on the physical cross-border pipeline interconnections represented in Figure 3. Following the trend of increasingly fluctuating gas flows, and anticipating a trend towards technical retrofitting for bidirectional gas flow, we allow the model to expand to all pipelines in the scenario years, in both directions, at zero additional investment cost. As a further simplification, we assume that only one natural gas quality is distributed, anticipating the discontinuation of Dutch low calorific natural gas exports to Germany planned for 2029 [26]. The model was allowed to expand the gas transport networks at a cost of 1.880 M€/km, a value which was deduced from the national natural gas grid expansion plan 2016 [27].

Gas transport capacities per hour were deduced for each border between neighbouring model regions using the above mentioned simplifications. Publicly available information from the European Network of Transmission System Operators for Gas (ENTSOG) [28] were used. It was assumed that the hourly maximum of transmission capacity is 60% higher than the reported daily capacity. Additional pipelines, which were under construction in the ENTSOG data (e.g., Nord Stream 2) were taken into account as well.

For the future scenario years, *REMix* was allowed to build an additional infrastructure, dedicated for hydrogen transport, at an estimated investment cost of 2.162 M€/km, i.e., at 15% higher cost compared to the natural gas infrastructure.

Import options other than pipeline-bound gas imports were not modelled. Liquefied imports of natural gas or hydrogen were not allowed for the *REMix* model.

2.4.2. Natural Gas Storage and Hydrogen Storage Option

An essential technical element of the German energy system is the availability of large underground gas storage facilities (Figure 4), which allow a temporal decoupling of purchase and sale of natural gas. With regard to renewable hydrogen, the storage capacities offer the temporal decoupling of production and use.

In general, hydrogen can be stored in analogy to the existing natural gas storage facilities. However, two main storage categories have to be distinguished.

Cavern storage facilities are man-made structures washed out from geological salt deposits. The salt deposit surrounding the resulting salt dome reliably seals the cavern. Due to the necessary geological structures, cavern storage facilities can only be found in certain regions. Within Germany, cavern storage potentials are found in the northern part of the territory, while in the southern part, pore storages are operated (Figure 4). In Europe, and in Germany specifically, extraordinary cavern storage potentials exist, exceeding today's storage capacities by far [29]. Salt cavern storage is suitable for hydrogen storage. For porous rock storage (depleted oil or gas fields or aquifers) the same is thought to be true in general [30,31]. However, due to uncertainties concerning underground microbiological processes and ongoing research [31], porous rock hydrogen storage was excluded for the case study presented here.

The cavern storage facilities were assigned to the respective model regions, and for the future scenario years, the model was allowed to build hydrogen caverns at an assumed cost of 220 €/MWh of hydrogen (LHV) within the same model regions, which already exhibited one or more storage facilities in 2019. The assumption implies that several additional caverns can be added to the existing cavern fields, taking advantage of the existing infrastructures. At the same time, model regions that lack cavern storage options due to disadvantageous geological conditions cannot be chosen for newly-built caverns by *REMix*.

2.4.3. Renewable Gas Production: Electrolysis and Methanation

From the portfolio of power-to-gas technologies [32], one exemplary electrolysis and one methanation technology were chosen for energy system modelling: the proton exchange membrane electrolysis (PEM) and the technical methanation.

The PEM electrolysis is assumed to operate at an efficiency of 69.1% in 2020, referring to the higher heating value of hydrogen and including grid injection (73.7% in 2030, 77.4% in 2040, and 80.4% in 2050). Investment costs of 900 €/kW electrical capacity are assumed for 2020 (550 €/kW in 2030, 450 €/kW in 2040, 350 €/kW in 2050). Fixed operating costs are estimated as 2% of the investment costs per year, and variable operating costs are estimated 0.001 €/kWh of consumed electricity.

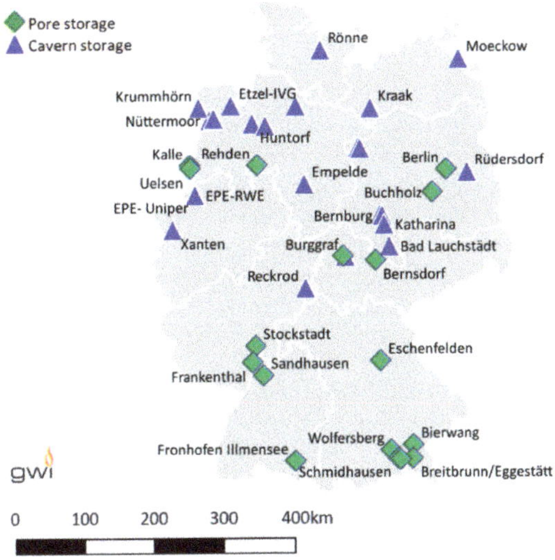

Figure 4. Underground gas storage facilities in the ten model regions in Germany.

The technical methanation is parametrised with efficiencies of 74.6% in 2020 (79.6% in 2030, 84.6% in 2040, 89.6% in 2050) including grid injection. The investment costs are assumed to be 1500 €/kWh, with respect to the higher heating value of methane in 2020 (1000 €/kWh, 900 €/kWh, 800 €/kWh). Fixed operating costs are estimated as 2.5% of the investment costs per year, variable operating costs are estimated 0.001 €/kWh of consumed electricity, and additional costs for load change of 0.001 €/kW_CH_4 were applied.

The thermal coupling of methanation (exothermal reaction) and the electrolysis process [33], as well as reversible electrolysers/fuel cells, biological methanation, and other carbon capture and usage technologies, were not taken into account.

2.4.4. Injection of Hydrogen and Biomethane into the Existing Natural Gas Grids

The injection of hydrogen to existing gas grids is one technical option for the integration of hydrogen into existing energy supply systems. Today, hydrogen is already being fed in at the gas transmission network level and at the gas distribution network level—but, to date, only on a small scale, typically at demonstration plants.

Due to modelling constraints, the admixture of hydrogen is only considered at the distribution grid level. For the scenario years, a continuous increase in the permitted maximum volumetric share of hydrogen in the natural gas infrastructure is considered, starting from 10% in 2020 to 15% in 2030, 20% in 2040, and 25% in 2050. The gradual introduction of higher hydrogen concentrations ensures that the hydrogen tolerance of the natural gas infrastructure, with all of its downstream end-use technologies, can be achieved.

The injection of biogas into the natural gas grids is modelled on the premise that the fuel quality has been upgraded to that of natural gas (biomethane) through previous processing. This corresponds to the state of the art for biomethane feed-in plants in Germany. In *REMix*, biomethane is, therefore, treated equivalently to natural gas, and blending is not limited. However, a maximum potential is specified. The domestic biomethane production potential was assumed to be 32 TWh, based on the medium scenario for manure and sewage sludge from [34].

The potentials for the specific countries and model regions considered are available in [14].

2.4.5. Gas Compression

In *REMix*, electric, as well as gas-powered, gas compressor units are considered for the transport and storage of gas. The existing compressor stations in Germany are considered as a model input, and in addition, an endogenous expansion is made possible in the model. Typical turbo compressors are assumed, for which electrification of the drive is made possible. Waste heat losses are not taken into account.

In the case of an endogenously built hydrogen infrastructure, the compression demand for transport of pure hydrogen is only covered by electric driven compressor units. Assumptions are published in [14].

2.4.6. Pre-Heating of Natural Gas for Decompression

For its distribution to the end customers, natural gas is transferred from the transport network, which is operated at high pressure, to the regional distribution networks at pressure regulation stations. In the distribution networks, it is first transported under high or medium pressure and then expanded into the low-pressure range (≤100 mbar) for the purpose of fine-mesh distribution. With each expansion, natural gas cools down due to the Joule–Thomson effect. In order to avoid condensation inside pipelines and in the pressure regulation stations and ice formation that might render the armatures inoperable, gas preheating is necessary before the gas is expanded. The heat demand for gas preheating in Germany is taken into account as a model heat sink that can be equipped with bivalent technology. The choice of technologies is the result of optimization. The model can use electric boilers, gas condensing boilers, heat storage, and gas-fired CHP plants. In order to minimise the number of model variables for the small heat demand compared to the industrial or household sector, a regional breakdown of the gas preheating demand in *REMix* was dispensed with, and the demand for gas preheating in the gas grids was aggregated and assigned to the model region North Rhine–Westphalia. For this purpose, the total demand for thermal energy is distributed over the hours of the year, using a representative demand profile for gas preheating. The annual heat demand of preheating amounts to 253 GWh in 2020, 179 GWh in 2030, 104 GWh in 2040, and 38 GWh in 2050.

2.5. Further Model Input Assumptions

Like the technologies in the gas sector, those in the other sectors are described by extensive techno-economic data sets. These include, in particular, the investment and operating costs of the plants, as well as their efficiencies and other technical parameters. The model assumptions are available in [14]. Of particular importance, to the desired transformation of the energy system, is the assumed CO_2 price that accrues system-wide on all emissions (Table 2). In Germany, no CO_2 emissions, at all, will be permitted in 2050, meaning that only renewable gases can be used in the model.

For spatially and temporally resolved modelling, the demand data, as well as the VRE potentials, must be disaggregated accordingly. For the latter, results of the *EnDAT* model [35] are used, and historical data of the weather year 2006 are applied. The procedure for the spatial distribution of the demands and the determination of the load profiles is described in detail in [19].

2.6. Legal and Regulatory Framework in Germany

The electricity sector is highly regulated, and hence, the cost of electricity consumption is at the centre of regulatory influence on investment decisions and feasibility. In Germany in 2020, for small industrial customers (50 MWh/a), cost per kWh was comprised by roughly one quarter of actual energy cost, by 15% of network charges, and by 40% of a surcharge for renewable energy support [36]. The rest were other taxes and levies. For the consumer categories most relevant to this analysis, there are exemptions and rebates. A representative power-to-heat application, like any small industry customer, is able to purchase electricity at lower cost than household customers. In contrast to other industries and power-to-heat, power-to-gas plants are additionally exempted from electricity tax of roughly 0.02 €/kWh [37]. In the meantime, since the modelling took place in 2020, an exemption from the renewable energy surcharge was granted as well, albeit just under certain conditions.

Projections on future electricity cost, as they enter into the evaluation of economic feasibility of the required investments with *MuGriFlex*, are based on a number of assumptions. Hourly electricity costs are an output of techno-economic modelling with *REMix* in the respective scenario as presented above. In line with political decisions and current discussions in Germany, we assume that the renewable energy surcharge will phase out, as future investments into wind and solar power will receive less and eventually no support and past subsidy commitments are already phasing out. Network charges, on the other hand, are likely to rise with grid expansion to integrate renewable electricity. In line with projected investments in the electricity grid, network charges, and other levies and surcharges, drop from roughly 0.12 €/kWh to around roughly 0.08 €/kWh in 2050.

Given the limited economic feasibility, additional support policies are in place or under consideration for certain relevant technologies. By and large, support occurs in the form of investment support or operational subsidies. In 2020, investment support was administered to district heating pipes and thermal energy storage, as well as, under specific circumstances, to electric boilers and to power-to-gas demonstration plants. Operational subsidies for a representative CHP plant were between 0.03 and up to 0.11 €/kWh [36,38]. Operational support of electrolysis happens only in the form of reduced taxes and surcharges, as discussed above. The scale of additional support that might be needed to achieve the investment levels and operation schedules, found optimal in the overall system modelling, is discussed in Section 3.2.

3. Results

3.1. Results of the Energy System Optimisation

3.1.1. Development of Energy Demand

The goal of a drastic CO_2 emission reduction requires a fundamental shift in energy demand driven by sector coupling. For the system considered in *REMix*, this mostly concerns a significant decrease in gas demand and a strong increase in power demand (Figure 5). These demands are partially exogenously defined, and partially model output. The endogenous power demand includes, most dominantly, the electrolytic production of hydrogen and the usage of electrical heat generation in district heating systems, as well as industry. Regarding the gas demand, including both hydrogen and pipeline gas, the usage in power plants and boilers are a model output.

3.1.2. Development of Power Supply and Flexibility Provision

The supply of the increasing power demand, and the substitution of the conventional power plant park, requires a substantial increase in the installed renewable power generation capacity (Figure 6). Already, until 2030, PV and wind capacities are more than doubled, compared to 2020, to enable the phase-out of nuclear and coal power plants. Further capacity installations are required along the transformation towards an integrated energy system. The sharp increase in hydrogen production between 2040 and 2050, especially, drives the installation of additional offshore wind turbines and PV systems. To ensure security of supply, dispatchable generation capacities will be required until 2050. For that, *REMix* mostly chooses gas CHP units in district heating systems, which also contribute to the heat supply. While these are, at first, operated using natural gas, they only have biomethane and synthetic methane available in 2050. Based on these installations, the power generation structure sees a major shift to emission-free technologies. Driven by the CO_2 price assumed, coal power plants are almost not used anymore already in 2030. Instead, onshore wind power provides almost half of the power supply. In 2040, also gas power plants are reduced to a minor share in power supply, whereas additional electricity imports become significant. In the zero-emission system of 2050, PV takes over the role as the most important source of electricity, followed by onshore and offshore wind power. Other technologies contribute less than 10% of the overall supply, while the imports reach the exogenously defined limit of 20%.

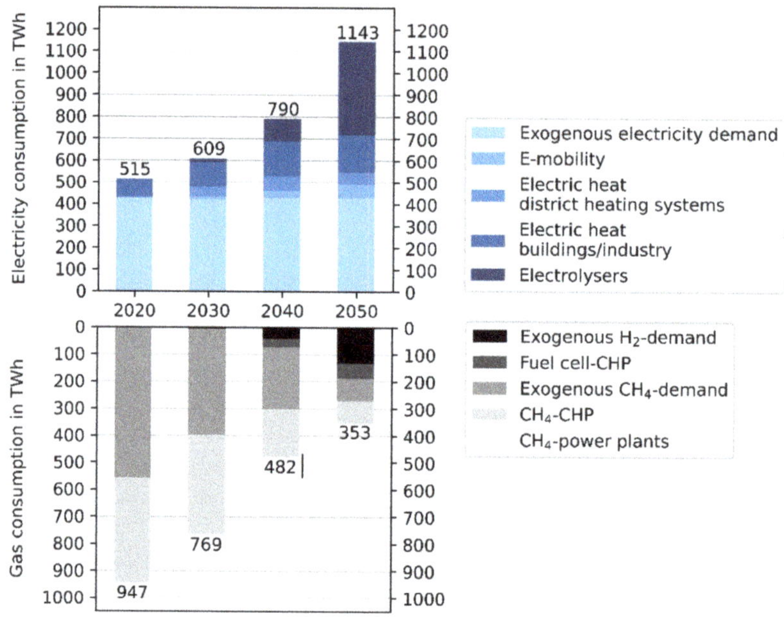

Figure 5. Development of power demand (upper diagram) and gas demand (lower diagram) along the transformation pathway until the year 2050.

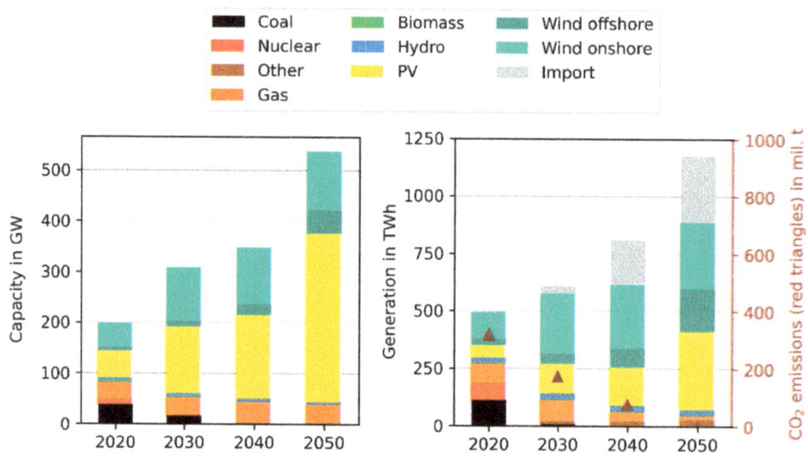

Figure 6. Development of the power supply in Germany. The left figure shows the installed power generation capacities. The right figure shows the annual power generation and imports (**left axis**), as well as the CO_2 emissions (**right axis**, triangle symbols). The technology "others" subsumes waste incineration, oil, and geothermal energy.

The strong increase in renewable electricity generation is accompanied by an increase in the required flexibility demand. This is covered by numerous technologies, the suitable combination of which makes it possible to limit the share of VRE curtailment to less than 1.5% of potential electricity generation. At 0.7%, the maximum value achieved in Germany is even lower. Due to the change in energy demand and the power plant fleet, the use of the various flexibility options shows different trends over the course of the scenario years (Figure 7). Thus, the power generation in controllable power plants already decreases significantly until 2030. In contrast, there is an increase in the use of all types of energy storage. In addition to electricity storage, these also include heat storage, which serves to make CHP plants and heat pumps more flexible, as well as hydrogen storage. The latter allow electrolyser operation to be adapted to VRE availability. Stationary energy storage is complemented by flexible and bidirectional charging of battery vehicles and demand response in industry and commerce. Extensive shifts in the use of transportation networks for energy are also evident over the course of the transformation. For example, due to the decline in demand, the volume of gas transported across regional borders in Germany falls from just under 700 TWh in 2020 to about 200 TWh in 2050. This is partially compensated for by the construction and use of a hydrogen network, which, in 2050, will transport an energy volume of about 200 TWh across regional borders. The power grid also shows an increase in transported energy, from just under 90 TWh in 2020 to 200 TWh in 2050. The investments required for the transformation of the gas sector are described in more detail below, and technology-specific values are provided in [19].

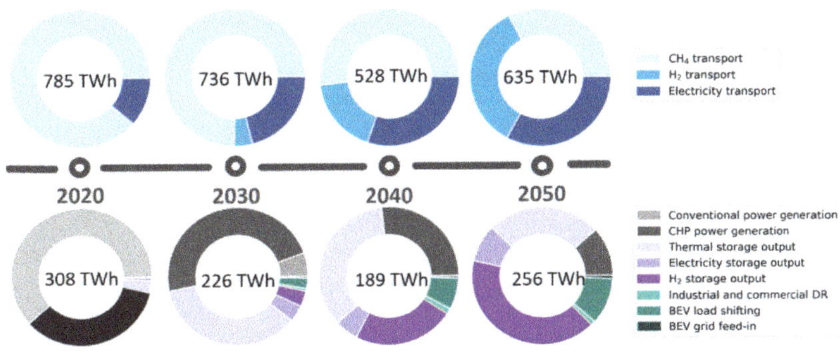

Figure 7. Development of the load balancing technology usage in Germany. The upper diagrams show the grid-bound technology use, while the lower graphs depict the need for local balancing options. The total annual numbers are embedded in the centre of each graph.

3.1.3. Deployment and Operation of Gas Infrastructures in Germany

REMix features an aggregated, but explicit, consideration of the infrastructures in the gas system. This allows for an analysis of the capacities and operation of this equipment and its development along the transformation process. In the course of implementing sector coupling, the expansion of hydrogen infrastructures plays a central role. Thus, with the increase in demand, there is a continuous growth in the capacities of gas generation plants, storage facilities, transport pipelines, and compressors (Figure 8). Compression for gas transport and storage is assumed to be electricity-based only for hydrogen, but both gas- and electricity-based for natural gas and synthetic methane. Based on the compressor capacities available today, *REMix* can invest endogenously in both technologies. The results show that, in the case of gas storage, the only investment is in electric compressors, and these also do all the compression work. It follows that storage injection of hydrogen and natural gas/methane occurs especially at times of high VRE generation. In the gas grid, on the other hand, a mixture of both technologies is used, mainly using the compressor capacities already available today. However, the share of compression work provided by gas-based compressors decreases from 55% in 2020 to 15% in 2050.

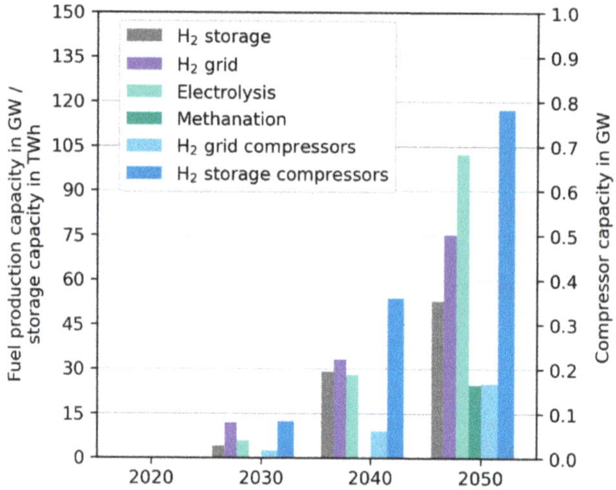

Figure 8. Development of the overall capacities of hydrogen and methane production, hydrogen storage, hydrogen pipeline (**left axis**), and compression (**right axis**) capacities in Germany. All these are endogenously optimised by *REMix*.

In its final state in 2050, the hydrogen transport infrastructure, added endogenously by *REMix*, connects the west of the country with the northwest and the south. Due to the underground caverns only available there (Figure 4), hydrogen storage facilities will be built especially in the north of the country. For the methanation plants, installation close to the storage facilities is preferred. Instead, the electrolysers are distributed evenly across the country. This is reflected in the quantities of hydrogen produced, stored, and transported (Figure 9).

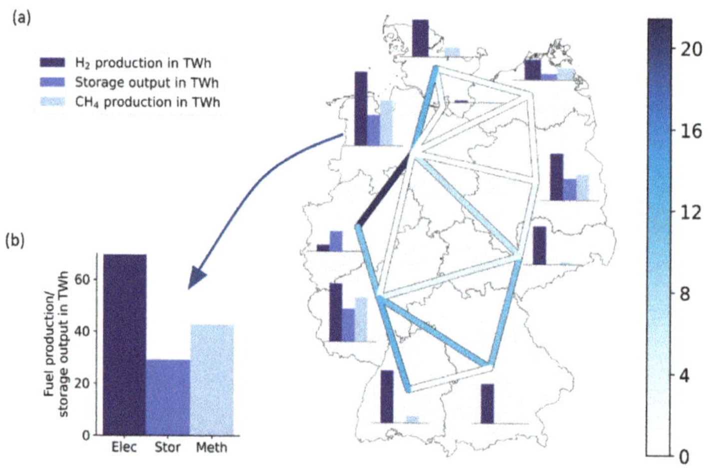

Figure 9. Production of hydrogen, hydrogen storage output, and methane production in the ten model regions in Germany (bar charts of figure (**a**)) and grid-bound hydrogen transport between the model regions (network plot and right hand colour scale). Exemplary bar chart of Lower Saxony's gas production and storage usage (**b**).

To evaluate the use of flexibility in the gas system, for integrating VRE generation, analysis of hourly plant dispatch is helpful. The hourly dispatch shows that the compressors respond to the VRE availability and thus, contribute to load balancing (Figure 10). This mainly concerns phases of very low VRE generation in winter, as can be seen, e.g., in the area of hour 770. There, it can also be seen that the demand for compression energy in the gas grid is mainly driven by the operation of the methanation plants. These operate at different times than the electrolysers, at least in winter, and are driven by methane demand, which is particularly high during periods of low VRE power generation. Compression for gas storage correlates, primarily, with the times of electrolysis operation, which generally coincides with high VRE generation (Figure 11).

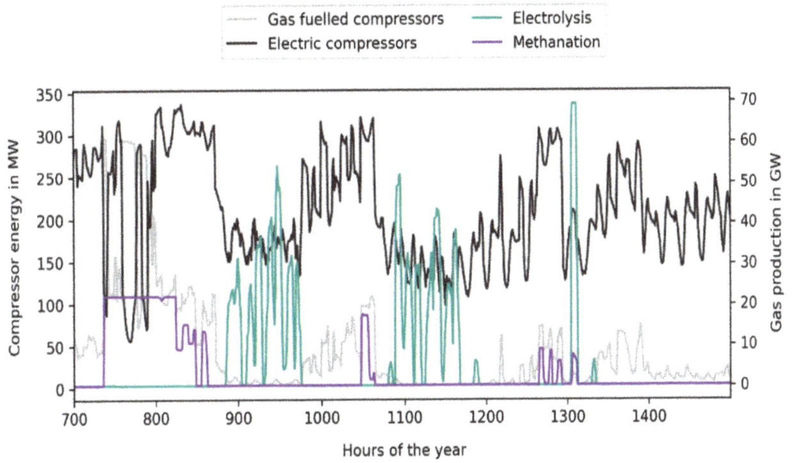

Figure 10. Provision of compression energy in the gas network in February 2050.

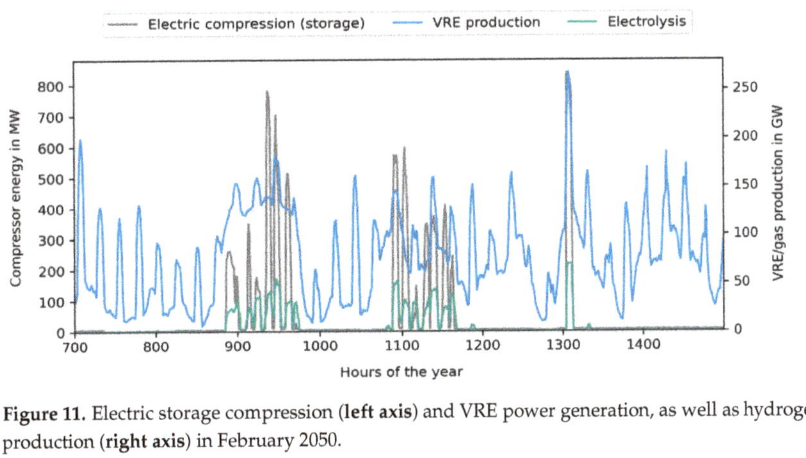

Figure 11. Electric storage compression (**left axis**) and VRE power generation, as well as hydrogen production (**right axis**) in February 2050.

The annual electricity demand for compression in gas transport is about 1 TWh regardless of the scenario year, with the share of the hydrogen network exceeding that of the natural gas network only in 2050. While the electricity demand of compression in gas storage facilities is significantly lower than that of transmission pipelines in the early years of the scenario, it exceeds it in 2050 due to the strong increase in the use of hydrogen storage facilities, whose annual electricity demand in 2050 rises to 3.5 TWh. Due to these orders of magnitude, even the flexible compression of gas does not make a significant contribution to VRE integration, as the comparison of Figures 10 and 11 shows. At least in the case of compressors in gas storage, there is a clear correlation of operation with VRE generation, which is mainly caused by the simultaneous electrolysis operation.

The model results show that gas preheating is increasingly electrified in the course of the scenario years (see Figure 12). The share of gas boilers in demand coverage decreases from the assumed 100% in 2020 to 50% in 2030 and about 20% in subsequent years. CHP plants are added endogenously in 2030, and then supply a quarter of the required heat; however, thereafter, their supply share drops to 14% and 9%, respectively. In turn, the share of heat generated by electric boilers increases from about a quarter in 2030 to above 70% in 2050. To make the CHP plants and electric boilers more flexible, heat storage facilities are added in 2030 that can absorb about 3 h of the aggregated thermal generation capacity. However, as this is only about 100 MW in total, and it drops to less than 30 MW by 2050, these plants do not contribute substantially to the integration of VRE generation in Germany. This is also reflected in the magnitude of the heat generated, which decreases from 265 GWh to 150 GWh over the course of the scenario years.

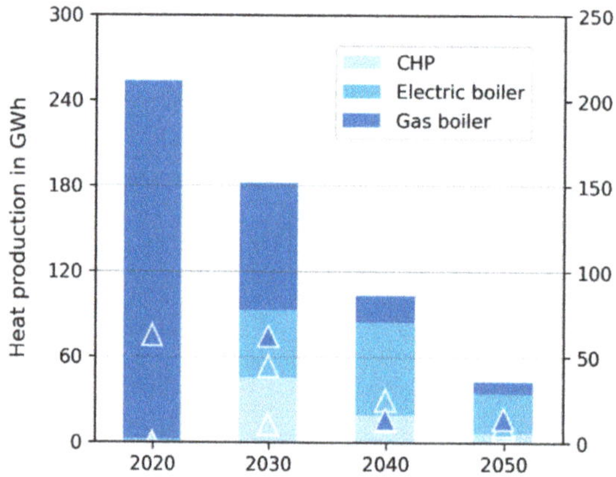

Figure 12. Heat production for gas preheating per year (**left axis**, bar charts) and capacities of gas preheating technologies (**right axis**, data points).

3.2. Business Perspective of Power-to-Gas-Plants

Based on model calculations with *MuGriFlex*, the economic efficiency of water electrolysis is examined. Figure 13 shows the yearly utilisation of power-to-gas plants (in full load hours) resulting from *REMix*. Since these vary greatly from region to region, they are shown as a range. Regions with the lowest full load hours are usually interior regions, and those with the highest utilization usually coastal regions. An additional marker shows the average (weighted by produced methane per year) of all power-to-gas plants in all regions:

The application of *MuGriFlex* aims at showing how the plants would be utilised if only the operating costs are taken into account. The number of feasible operation hours are hours in which gas can be produced at the same, or lower, costs than the reference gas price, based on the input cost of electricity, including projected surcharges.

The result is that the operation is economically feasible without bonus payments only in the scenario year 2050 (but then, for as many as 7500 h/a). In all other scenario years, synthetic gas cannot be produced economically in any hour of the year, even without considering the investment and fixed operating costs (Figure 14; intersection of the coloured lines with the *Y*-axis). This implies that a plant operator who receives full funding for the electrolyser would still incur a loss every hour in which the system operates.

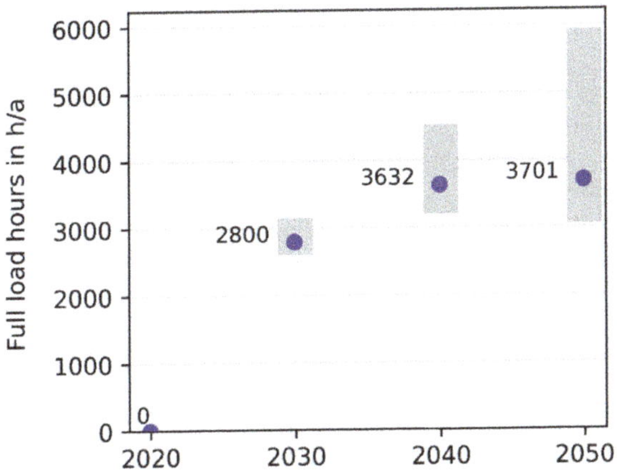

Figure 13. Yearly utilisation of electrolysers: the blue dots represent the weighted average value of all model regions (weighted by produced methane per year). The grey bars indicate the bandwidth of the individual model region values.

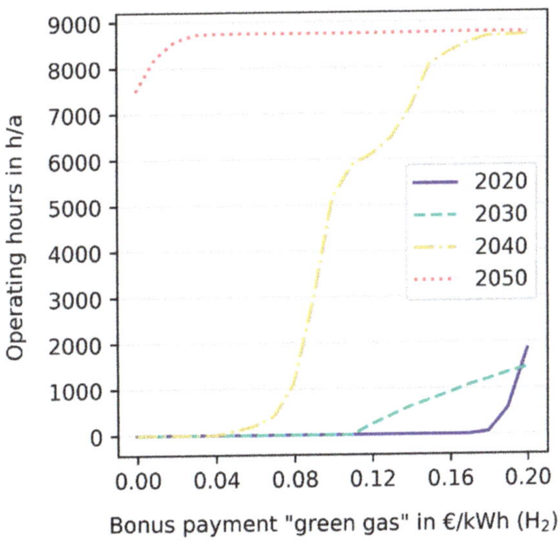

Figure 14. Economically feasible operating hours of electrolysers, depending on a premium paid on the output.

Therefore, the second step is to analyse what incentive is necessary to allow plants to operate in more hours per year under the assumed framework conditions. The assumed incentive is in the form of a "premium for green gas" in addition to the revenue from the competitive gas price.

The feasible operating hours with this additional support are shown in Figure 14, for the four scenario years and with the premium increasing from zero to up to 0.2 €/kWh H_2; still without taking into account investment and fixed operating costs.

The results show that in the year 2050, a premium of just 0.02 €/kWh H_2 would lead to an increase in operating hours to more than 8500 h/a, which would correspond to a utilization factor of 97%. In all other years, considerable additional revenues, e.g., in the form of the above-mentioned premium on the gas produced, are necessary to reach noteworthy operating hours. Note, however, that similar effects to a premium payment on the output are achieved by correspondingly subsidizing the input. Thus, a reduction in electricity cost and, particularly, the respective surcharges represent an alternative option to the premium payment, analysed here, for improving the feasibility of electrolysis.

Finally, adding to the insights on feasibility of operation, it was determined whether, with these assumptions, investments would also be profitable from a business perspective. This is expressed by the profit that can be expected yearly on plant capacity (the basis is the electrical input capacity of the electrolyser). With the plant operation and electricity cost outlined above, fixed operating costs and annuity of the plant investment are deducted from the revenues of the gas trade.

The result of this evaluation is shown in Figure 15. It depicts the profit to be achieved with the plants, depending on the additional premium paid. If this value is negative, the plants may be running because they would generate a profit in some hours of the year, but the sum of the annual profit does not exceed the annual fixed operating costs and the annuity of the plant investments to be recovered.

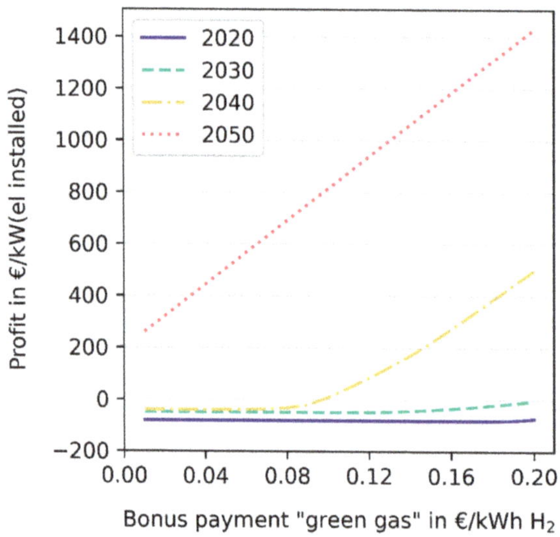

Figure 15. Annual profit per installed electrolysis capacity, depending on an additional premium paid in addition to gas.

Here, too, the year 2050 is the exception, in which an annual profit of around 200 €/kW of plant capacity can be generated even without an additional premium. However, this scenario year is also the only year in which complete defossilisation of fuels was required in *REMix*, which, in turn, leads to high gas prices that are advantageous from the business perspective taken here.

In 2040, premiums between about 0.08–0.12 €/kWh H_2 are necessary to make a profit. As there is a significant demand for hydrogen in these years, according to the *REMix* results (up to about 9% energetic share in the gas mix), it can be concluded that considerable support will be necessary to operate this technology during this period of time.

In 2020 and 2030, even premiums of 0.20 €/kWh H_2 would not be sufficient to ensure an economic operation of the electrolysers.

4. Discussion

The modelling is based on an exemplary long-term scenario of the development of the energy system until the year 2050, according to the goals of long-term climate neutrality. The resulting energy demands and calculated supply structures are conditioned by this narrative. Here, both of the other possible societal, political, and economic context scenarios, as well as different implementation paths of alternative technical options are not explicitly considered. In this respect, the results are exemplary and have only limited robustness with regard to their conclusions on options for action and the further design of the energy transition. In addition, it must be taken into account that the development path of the German and European energy system assumed here is not in line with the 1.5 °C temperature target formulated in Paris, but it only corresponds to the 95% reduction in GHG emissions by 2050 (relative to 1990) formulated by the German Federal Government in 2010 [25].

The combined techno-economic and business perspective energy system analysis presented here clearly has some shortcomings, which are briefly summarised in the following. The regional detail was restricted to ten model regions representing German federal states, some of which combined to reduce modelling effort. Results can, therefore, not be differentiated for each federal state and the European perspective is missing. As a consequence, the gas transport system for natural gas and hydrogen was only modelled among the ten German model regions with neighbouring countries treated in term of fixed boundary conditions. Gas transits across Europe, and the resulting international dependencies, were not part of the study. From this follows that the statements on the operation of the methane or natural gas grid in the earlier scenario years cannot be compared with today's reality. Storage facilities and transport pipelines are used to a much lesser extent than is the case today. This is due to the restriction to Germany in the simplified representation of the gas system chosen here, which does not take transit flows into account, and to the neglect of aspects of gas trading and reserve stockpiling. These ensure that natural gas is imported according to demand without intermediate storage, which, in turn, is favoured by the model-related neglect of transport times for gases. Our results are additionally affected by not explicitly modelling CO_2 costs and availability, which have an influence on siting and operation of methanation plants.

The system modelling with *REMix* comprises a broad range of technologies across all sectors. This allows for an improved understanding of interactions between different balancing technologies. However, for the countries outside Germany, a reduced technology scope is considered, which may have an impact on the observed cross-border power flows and the required balancing technologies. Compared to previous works [39,40], the *REMix* results show a different allocation of electrolysers, which is caused by a different technology scope, deviation assumptions regarding VRE potentials, and the neglect of hydrogen imports. This also drives differences in the corresponding storage and pipeline infrastructure. Still, the exploitation of hydrogen cavern storage in northern and central Germany, as well as the installation of a pipeline infrastructure from northern to western and southern Germany, is a robust result across this and previous analyses. The sensitivity of the infrastructure design to changes, in decisive input parameters or our scenario, have been assessed in [19].

Despite the broad range of technologies, the characteristics of the future gas system are not fully captured in our modelling. For example, energy import options for liquefied natural gas, as well as liquefied biogas and liquefied hydrogen, or any type of liquid organic hydrogen carrier (LOHC) on the world market have not been considered. Concerning hydrogen production, for the reason of limited modelling capacity, alternative technologies of alkaline electrolysis, and solid oxide electrolysers were not modelled, and pyrolysis was excluded as well, due to low technology readiness level. Additionally, the storage aspect was covered by cavern storages (salt dome), while potential future hydrogen storage options in porous rock formations were excluded due to their low technology readiness level as of today. Future assessments should comprise these technologies to evaluate further relevant dimensions of an emerging coupled energy system based on renewable electricity and renewable gases.

The aggregated modelling approach in *REMix* completely neglects the energy distribution within the model regions. This may cause an overestimation of the spatial balancing linked to an underestimation of temporal balancing, e.g., via stationary battery storage.

By integrating the *REMix* results into *MuGriFlex*, the business case for the determined least-cost infrastructures can be evaluated. The unprofitability of Power-to-Gas, under current market conditions, with a need for lower cost and higher revenues was similarly found by van Leeuwen and Mulder [41] and with a focus on flexibility provision by Li and Mulder [13]. Still, the combination of the two modelling approaches confirms the need for adjustments to the regulatory framework required to create a favourable investment environment for Power-to-Gas plants.

5. Conclusions

Our modelling results describe the path to the economic integration of power-to-gas plants into the German energy system of the future. This is based on an enhancement and coupling of the *REMix* and *MuGriFlex* models, the conception of a framework scenario of the energy system transformation in Germany and its neighbouring countries, and the research and integration of extensive data sets on gas system technologies. The results underline the significant role of flexible hydrogen and methane production for the integration of VRE power generation. In particular, flexible electrolysis, together with other sector coupling technologies, contributes to the result that, even in the case of zero-emission electricity generation, hardly any VRE curtailment is needed.

Our results show that, especially the operation of electrolysers, but also that of electric compressors in transmission networks and storage facilities, is concentrated in periods of high VRE availability. Complementarily, our analyses show that in the course of the system transformation by 2050, electrification of the equipment in the gas system occurs not only in the compressors but also in the gas preheating. A flexible operation is here realised by the installation of hybrid heat supply options and thermal energy storage. However, the amounts of electricity used for gas preheating are negligible compared to the overall system.

While the installation and use of electrolysers in the overall system optimization proves to be economically viable as early as 2030, the business analysis shows a different picture. Our results show that investments in power–to-gas plants are not economically feasible in the scenario considered using todays and the projected market conditions. Due to the high level of surcharges to be paid on the electricity needed, an operator would not run the plants at any hour of the year except in the scenario year 2050, even when investment and fixed operating costs are not taken into account. Additional bonus payments between 0.08 and 0.20 €/kWh of produced hydrogen are necessary to incentivise more than 2000 h of operation per year. To generate an overall profit, i.e., to receive an overall income higher than the total costs (including fixed operating costs and capital), bonus payments of more than 0.09 €/kWh (2040) or more than 0.20 €/kWh of hydrogen (2030) would be required. However, these incentives would require very small-scale calibration and a significant financial outlay, given the large role those synthetic gases would play from an overall system perspective. Under the framework conditions assumed, the electrolysers required from an overall system point of view could, at best, be realised in spatial proximity to, e.g., large wind farms and with consequently reduced surcharges on the electricity price.

In order to enable the rapid expansion of hydrogen infrastructures, which is both needed for the transformation of the energy system and to be economically attractive, ways must be found to close this financing gap.

Author Contributions: Conceptualization, H.C.G., M.F. (Max Fette) and J.S.; methodology, H.C.G., M.F. (Max Fette), H.G., C.B., J.S. and T.P.; software, H.C.G., H.G. and M.F. (Max Fette); validation, H.C.G., M.F. (Max Fette), C.B., H.G. and J.S.; formal analysis, H.C.G., M.F. (Max Fette), H.G. and J.S.; investigation, J.S., H.C.G., M.F. (Max Fette), H.G., C.B., T.P., N.B., E.T. and M.F. (Marc Fiebrandt); resources, J.S., H.C.G., M.F. (Max Fette), R.A. and F.B.; data curation, H.G., H.C.G., M.F. (Max Fette), J.S., E.T., N.B., M.F. (Marc Fiebrandt) and T.P.; writing—original draft preparation, J.S., H.C.G., M.F. (Max Fette), T.P., H.G. and C.B.; writing—review and editing, J.S., H.C.G., M.F. (Max Fette), H.G., C.B. and T.P.; visualization, H.G.; supervision, H.C.G., M.F. (Max Fette), F.B. and R.A.; project administration, M.F. (Max Fette), H.C.G. and J.S.; funding acquisition, M.F. (Max Fette), C.B., H.C.G., T.P. and J.S. All authors have read and agreed to the published version of the manuscript.

Funding: This research was funded by the German Federal Ministry for Economic Affairs and Energy, grant numbers 03ET4038 (Project 'MuSeKo') and 03EI1030 (Project 'Fahrplan Gaswende').

Data Availability Statement: Input data for the REMix model of this research project are publicly available at https://zenodo.org/record/5705414#.YZTaTmDMKUk (accessed on 16 November 2021).

Acknowledgments: The authors want to thank the following colleagues and students for their help during the project: Leander Kimmer, Christopher Rickert and Isabelle Roller; Felix Cebulla and Eileen Meyer, Klaus Peters, Philipp Fischer, Manfred Lange, Markus Köppke, Norman Dünne, Sophia von Berg, Enado Pineti, Aml Mahmoud, Alexandra Botor. All external feedback received during the workshops and expert interviews of the *MuSeKo* project is highly appreciated.

Conflicts of Interest: The authors declare no conflict of interest. The funders had no role in the design of the study; in the collection, analyses, or interpretation of data; in the writing of the manuscript, or in the decision to publish the results.

References

1. Heide, D.; Von Bremen, L.; Greiner, M.; Hoffmann, C.; Speckmann, M.; Bofinger, S. Heide Seasonal Optimal Mix of Wind and Solar Power in a Future, Highly Renewable Europe. *Renew. Energy* **2010**, *35*, 2483–2489. [CrossRef]
2. Heide, D.; Greiner, M.; von Bremen, L.; Hoffmann, C. Reduced Storage and Balancing Needs in a Fully Renewable European Power System with Excess Wind and Solar Power Generation. *Renew. Energy* **2011**, *36*, 2515–2523. [CrossRef]
3. Gils, H.C.; Simon, S. Carbon Neutral Archipelago-100% Renewable Energy Supply for the Canary Islands. *Appl. Energy* **2017**, *188*, 342–355. [CrossRef]
4. Brown, T.; Schlachtberger, D.; Kies, A.; Schramm, S.; Greiner, M. Synergies of Sector Coupling and Transmission Reinforcement in a Cost-Optimised, Highly Renewable European Energy System. *Energy* **2018**, *160*, 720–739. [CrossRef]
5. Hauser, P.; Heinrichs, H.U.; Gillessen, B.; Müller, T. Implications of Diversification Strategies in the European Natural Gas Market for the German Energy System. *Energy* **2018**, *151*, 442–454. [CrossRef]
6. Kabirian, A.; Hemmati, M.R. A Strategic Planning Model for Natural Gas Transmission Networks. *Energy Policy* **2007**, *35*, 5656–5670. [CrossRef]
7. Wulf, C.; Linßen, J.; Zapp, P. Review of Power-to-Gas Projects in Europe. *Energy Procedia* **2018**, *155*, 367–378. [CrossRef]
8. Farrokhifar, M.; Nie, Y.; Pozo, D. Energy Systems Planning: A Survey on Models for Integrated Power and Natural Gas Networks Coordination. *Appl. Energy* **2020**, *262*, 114567. [CrossRef]
9. Breyer, C.; Tsupari, E.; Tikka, V.; Vainikka, P. Power-to-Gas as an Emerging Profitable Business Through Creating an Integrated Value Chain. *Energy Procedia* **2015**, *73*, 182–189. [CrossRef]
10. Caglayan, D.; Heinrichs, H.U.; Robinius, M.; Stolten, D. Robust Design of a Future 100% Renewable European Energy Supply System with Hydrogen Infrastructure. *Int. J. Hydrog. Energy* **2021**, *46*, 29376–29390. [CrossRef]
11. Torralba-Díaz, L.; Schimeczek, C.; Reeg, M.; Savvidis, G.; Deissenroth-Uhrig, M.; Guthoff, F.; Fleischer, B.; Hufendiek, K. Identification of the Efficiency Gap by Coupling a Fundamental Electricity Market Model and an Agent-Based Simulation Model. *Energies* **2020**, *13*, 3920. [CrossRef]
12. van Leeuwen, C.; Mulder, M. Power-to-Gas in Electricity Markets Dominated by Renewables. *Appl. Energy* **2018**, *232*, 258–272. [CrossRef]
13. Li, X.; Mulder, M. Value of Power-to-Gas as a Flexibility Option in Integrated Electricity and Hydrogen Markets. *Appl. Energy* **2021**, *304*, 117863. [CrossRef]
14. Gils, H.C.; Gardian, H.; Pregger, T.; Schaffert, J.; Tali, E.; Fette, M.; Brandstätt, C. REMix Model Input Data for the THG95/GHG95 Scenario Analysed within the MuSeKo Project. Available online: https://zenodo.org/record/5705414 (accessed on 16 November 2021).
15. Gils, H.C.; Scholz, Y.; Pregger, T.; de Tena, D.L.; Heide, D. Integrated Modelling of Variable Renewable Energy-Based Power Supply in Europe. *Energy* **2017**, *123*, 173–188. [CrossRef]
16. Luca de Tena, D.; Pregger, T. Impact of Electric Vehicles on a Future Renewable Energy-Based Power System in Europe with a Focus on Germany. *Int. J. Energy Res.* **2018**, *42*, 2670–2685. [CrossRef]
17. Gils, H.C. Balancing of Intermittent Renewable Power Generation by Demand Response and Thermal Energy Storage. Ph.D. Thesis, Uniersität Stuttgart, Stuttgart, Germany, 2015.
18. Michalski, J.; Bünger, U.; Crotogino, F.; Donadei, S.; Schneider, G.-S.; Pregger, T.; Cao, K.-K.; Heide, D. Hydrogen Generation by Electrolysis and Storage in Salt Caverns: Potentials, Economics and Systems Aspects with Regard to the German Energy Transition. *Int. J. Hydrog. Energy* **2017**, *42*, 13427–13443. [CrossRef]
19. Gils, H.C.; Gardian, H.; Schmugge, J. Interaction of Hydrogen Infrastructures with Other Sector Coupling Options towards a Zero-Emission Energy System in Germany. *Renew. Energy* **2021**, *180*, 140–156. [CrossRef]

20. Brandstätt, C.; Fette, M. Feasibility of Balancing Renewable Electricity through Heat and Gas Based "Storage Chains". In Proceedings of the International Renewable Energy Storage Conference (IRES), Düsseldorf, Germany, 9–11 March 2015.
21. Brandstätt, C.; Fette, M.; Meyer, S. Multi-Grid-Storage Flexibilität für die Stromversorgung aus Gas- und Wärmenetzen. Final Report. 2015. Available online: https://www.ifam.fraunhofer.de/de/Institutsprofil/Standorte/Bremen/Formgebung_Funktionswerkstoffe/Energiesystemanalyse/Projektdetails/MuGriSto.html (accessed on 5 July 2021).
22. Brandstätt, C.; Fette, M.; Gils, H.C.; Gardian, H.; Pregger, T.; Schaffert, J.; Brücken, N.; Tali, E. *Final Report: Multi-Sektor-Kopplung (MuSeKo)—Modellbasierte Analyse der Integration Erneuerbarer Stromerzeugung Durch Die Kopplung der Stromversorgung Mit Dem Wärme, Gas- und Verkehrssektor; BMWi; Grant Agreement No, 03ET4038*; Fraunhofer-Institut für Fertigungstechnik und Angewandte Materialforschung (IFAM): Bremen, Germany; Deutsches Zentrum für Luft- und Raumfahrt (DLR): Stuttgart, Germany; Gas- und Wärme-Institut Essen e.V. (GWI): Essen, Germany, 2020. [CrossRef]
23. Pfluger, B.; Tersteegen, B.; Franke, B.; Bernath, C.; Boßmann, T.; Deac, G.; Elsland, R.; Fleiter, T.; Kühn, A.; Ragwitz, M.; et al. *Langfristszenarien Fur Die Transformation Des Energiesystems in Deutschland—Modul 1: Hintergrund, Szenarioarchitektur Und Übergeordnete Rahmenparameter*; Fraunhofer ISI, Consentec GmbH, ifeu, Bundesministerium für Wirtschaft und Energie: Berlin, Germany, 2017; p. 45.
24. Project E-Highway2050—Modular Development Plan of the Pan-European Transmission System 2050. Available online: https://docs.entsoe.eu/baltic-conf/bites/www.e-highway2050.eu/results/ (accessed on 17 December 2021).
25. Bundesministerium für Wirtschaft und Technologie (BMWi). *Bundesministerium für Umwelt, Naturschutz und Reaktorsicherheit Energiekonzept Für Eine Umweltschonende, Zuverlässige Und Bezahlbare Energieversorgung*; BMWi: Berlin, Germany, 2010; p. 34.
26. International Energy Agency (IEA); European Network of Transmission System Operators for Gas (ENTSO-G). Netherlands Ministry of Economic Affairs and Climate Policy L-Gas Market Conversion Review—Winter Report 2021 Task Force Monitoring L-Gas Market Conversion 2021. Available online: https://groenevinger.files.wordpress.com/2021/02/bijlage-2-winter-report.pdf (accessed on 17 December 2021).
27. FNB Gas—Die Fernleitungsnetzbetreiber. *German Gas Network Development Plan (Netzentwicklungsplan Gas) 2016–2026*; FNB Gas: Berlin, Germany, 2017. Available online: https://www.bundesnetzagentur.de/DE/Sachgebiete/ElektrizitaetundGas/Unternehmen_Institutionen/NetzentwicklungSmartGrid/Gas/fruehereNEP/NEP_Gas2016/NEP_Gas2016_node.html (accessed on 31 January 2022).
28. European Network of Transmission System Operators for Gas (entsog): The European Natural Gas Network 2017. Available online: https://www.entsog.eu/sites/default/files/2018-09/ENTSOG_CAP_2017_A0_1189x841_FULL_064.pdf (accessed on 3 January 2019).
29. Caglayan, D.; Weber, N.; Heinrichs, H.; Linssen, J.; Robinius, M.; Kukla, P.; Stolten, D. Technical Potential of Salt Caverns for Hydrogen Storage in Europe. *Int. J. Hydrog. Energy* **2020**, *45*, 6793–6805. [CrossRef]
30. Stolzenburg, K.; Hamelmann, R.; Wietschel, M.; Genoese, F.; Michaelis, J.; Lehmann, J.; Miege, A.; Krause, S.; Sponholz, C.; Donadei, S.; et al. *Integration von Wind-Wasserstoff-Systemen in Das Energiesystem—Abschlussbericht (Final Report)*; PLANET: Essen, Germany; fh lübeck Projekt-GmbH: Lübeck, Germany; Fraunhofer ISI: Karlsruhe, Germany; IFEU: Heidelberg, Germany; KBB: Bannewitz, Germany, 2014.
31. Zivar, D.; Kumar, S.; Foroozesh, J. Underground Hydrogen Storage: A Comprehensive Review. *Int. J. Hydrog. Energy* **2021**, *46*, 23436–23462. [CrossRef]
32. Thema, M.; Bauer, F.; Sterner, M. Power-to-Gas: Electrolysis and Methanation Status Review. *Renew. Sustain. Energy Rev.* **2019**, *112*, 775–787. [CrossRef]
33. Gorre, J.; Ruoss, F.; Karjunen, H.; Schaffert, J.; Tynjälä, T. Cost Benefits of Optimizing Hydrogen Storage and Methanation Capacities for Power-to-Gas Plants in Dynamic Operation. *Appl. Energy* **2020**, *257*, 113967. [CrossRef]
34. Commission of the European Union, Joint Research Centre, Institute for Energy and Transport. *The JRC-EU-TIMES Model: Bioenergy Potentials for EU and Neighbouring Countries*; European Union: Luxembourg, 2015. [CrossRef]
35. Scholz, Y. Renewable Energy Based Electricity Supply at Low Costs: Development of the REMix Model and Application for Europe. Ph.D. Thesis, University of Stuttgart, Stuttgart, Germany, 2012. [CrossRef]
36. *Monitoring Report 2020*; Bundesnetzagentur für Elektrizität, Gas, Telekommunikation, Post und Eisenbahnen: Bonn, Germany; Bundeskartellamt: Bonn, Germany, 2021; p. 503.
37. StromStG 1999: Stromsteuergesetz Vom 24. März 1999 (BGBl. I S. 378; 2000 I S. 147), Das Zuletzt Durch Artikel 207 Der Verordnung Vom 19. Juni 2020 (BGBl. I S. 1328) Geändert Worden Ist. 1999. Available online: http://www.gesetze-im-internet.de/stromstg/ (accessed on 1 February 2022).
38. KWKG 2005: Kraft-Wärme-Kopplungsgesetz Vom 21. Dezember 2015 (BGBl. I S. 2498), Das Zuletzt Durch Artikel 7 Des Gesetzes Vom 8. August 2020 (BGBl. I S. 1818) Geändert Worden Ist. 2015. Available online: http://www.gesetze-im-internet.de/kwkg_2016/ (accessed on 1 February 2022).
39. Husarek, D.; Schmugge, J.; Niessen, S. Hydrogen Supply Chain Scenarios for the Decarbonisation of a German Multi-Modal Energy System. *Int. J. Hydrog. Energy* **2021**, *46*, 38008–38025. [CrossRef]
40. Cerniauskas, S.; Jose Chavez Junco, A.; Grube, T.; Robinius, M.; Stolten, D. Options of Natural Gas Pipeline Reassignment for Hydrogen: Cost Assessment for a Germany Case Study. *Int. J. Hydrog. Energy* **2020**, *45*, 12095–12107. [CrossRef]
41. Van Leeuwen, R.P. Towards 100% Renewable Energy Supply for Urban Areas and the Role of Smart Control. Ph.D. Thesis, University of Twente, Enschede, The Netherlands, 2017.

energies

MDPI

Article

Simulation of Coupled Power and Gas Systems with Hydrogen-Enriched Natural Gas

Yifei Lu [1,2,*], Thiemo Pesch [1] and Andrea Benigni [1,2]

1 IEK-10: Energy Systems Engineering, Institute of Energy and Climate Research, Forschungszentrum Jülich, 52428 Jülich, Germany; t.pesch@fz-juelich.de (T.P.); a.benigni@fz-juelich.de (A.B.)
2 Faculty of Mechanical Engineering, RWTH Aachen University, 52062 Aachen, Germany
* Correspondence: yi.lu@fz-juelich.de

Abstract: Due to the increasing share of renewable energy sources in the electrical network, the focus on decarbonization has extended into other energy sectors. The gas sector is of special interest because it can offer seasonal storage capacity and additional flexibility to the electricity sector. In this paper, we present a new simulation method designed for hydrogen-enriched natural gas network simulation. It can handle different gas compositions and is thus able to accurately analyze the impact of hydrogen injections into natural gas pipelines. After describing the newly defined simulation method, we demonstrate how the simulation tool can be used to analyze a hydrogen-enriched gas pipeline network. An exemplary co-simulation of coupled power and gas networks shows that hydrogen injections are severely constrained by the gas pipeline network, highlighting the importance and necessity of considering different gas compositions in the simulation.

Keywords: gas network simulation; energy system simulation; hydrogen; power-to-gas; multi-energy system

Citation: Lu, Y.; Pesch, T.; Benigni, A. Simulation of Coupled Power and Gas Systems with Hydrogen-Enriched Natural Gas. *Energies* **2021**, *14*, 7680. https://doi.org/10.3390/en14227680

Academic Editor: Bahman Shabani

Received: 10 October 2021
Accepted: 13 November 2021
Published: 16 November 2021

Publisher's Note: MDPI stays neutral with regard to jurisdictional claims in published maps and institutional affiliations.

1. Introduction

The mitigation of climate change requires the rapid decarbonization of all sectors [1]. In recent decades, great progress has been made in the electricity sector by increasing the share of renewable energies [2]. The decarbonization process, however, is not exclusively limited to the electrical energy sector. The focus is therefore now shifting to other sectors, for example, industry and transport, in order to reach the Paris goals on climate change. In the past, different energy sectors mainly operated separately or independently. However, with the increasing use of inter-sectoral flexibility options, different energy sectors are set to be coupled more closely and should therefore be analyzed as a whole [3,4]. An integrated assessment of the electricity and gas sectors is therefore particularly important. In this paper, we establish a simulation tool framework that is capable of solving hydrogen-enriched natural gas flow simulation problems.

1.1. *The Role of Natural Gas (NG) in Future Energy Systems*

Natural gas is an important energy carrier used to generate heat or electricity. In 2019, 10.2% of the total electricity consumption in Germany was covered by natural gas-based generation [5]. In 2020, natural gas storage in Germany amounted to around 23.9 billion cubic meters, which corresponds to roughly 250 TWh of energy [6]. Due to the increasing share of renewable power generation, long-term storage with power-to-gas (PtG) will most likely become necessary and cost-efficient [7]. The gas sector therefore has great potential to provide enormous storage capacity for the power network. Moreover, PtG technologies can provide flexibility to the electrical network [8]. The conversion of electrical energy into chemical energy is an essential part of every PtG technology and is achieved using water electrolysis [9]. In addition to the impacts on the power sector, hydrogen generated from electrolysis can be a key factor in the decarbonization of heavy industries such as steel or chemistry [10].

1.2. Blending Hydrogen into Existing NG Infrastructure

Hydrogen generated using electrolysis can either be directly stored or be transported using pipeline network systems. It can also be further converted into methane using methanation reactions. In order to predominantly utilize the existing gas transmission infrastructure, the feasibility of blending hydrogen into the natural gas network is being intensively investigated [11]. Due to different infrastructure states and regulations, the maximum permitted levels of hydrogen concentration in the natural gas transmission system vary from country to country [12]. German regulations currently allow for up to 10% hydrogen concentration in the natural gas network. In the future, up to 20% is planned, and the German Technical and Scientific Association for Gas and Water (DVGW) estimates that up to 50% hydrogen concentration is feasible [13]. In a recent study, the impact of higher and fluctuating hydrogen concentrations of up to 50% on a variety of industrial combustion systems was investigated [14]. Therefore, in order to simulate future gas networks with uncertain hydrogen concentrations, it is important to take gas mixture properties into consideration.

1.3. Simulation of Hydrogen-Enriched NG Network

Natural gas is typically a mixture of gases, in which the major component is methane. Depending on sources and locations, the compositions of natural gas can vary significantly. In the study [15], it is stated that only around 23% of industrial plants have a real-time natural gas mixture quality measurement, while more than 50% of them only have weekly or monthly data available. In addition, more than 70% of the customers suffer from poor natural gas supply quality about once a month [15]. Natural gas composition must therefore be considered as a factor that has a significant impact on network operation safety. However, in many publications such as [16,17], the properties of natural gas are not listed, which makes it difficult to reproduce their results. When considering the injection of hydrogen into the natural gas network, the impact of gas composition is even greater.

Several research studies were carried out with respect to the steady-state simulation methods for hydrogen-enriched natural gas networks. For example, Abdolahi et al. used several equations of state (EOS) to model the natural gas mixture, whereas the heating values were not considered [18]. In the work by Giulio et al., an analysis was performed to evaluate the gas heating values with respect to the hydrogen injection into the natural gas pipelines, while the impact of the gas mixture properties on the gas flow calculation was not the main focus [19]. Pellegrino et al. established a simulation framework for the hydrogen-enriched natural gas network using the virial EOS (cf. [20]), but the height difference between the pipeline inlet and outlet was not considered [21].

In addition, there are a number of tools available that can be used to analyze the hydrogen-enriched natural gas network. PSS®SINCAL (cf. [22]), SAInt (cf. [23]), and MYNTS (cf. [24]), for example, are well established tools that are capable to run complex simulations for gas pipeline networks. Nevertheless, they are closed-source software and therefore hard to extend with functionalities for the specific requirements of analysis. TransiEnt is a Modelica library that is capable of steady-state and dynamic simulations of coupled-power, gas, and heat networks [25]. However, it is currently only supported in Dymola, which is a commercial software development environment for Modelica. There is a relatively new Python package—so-called pandapipes—that can handle different gas mixture properties for gas network simulation. The nodal difference of gas mixture compositions, however, has not yet been considered [26].

2. Modeling of Gas Pipeline Systems

2.1. Modeling of Gas Pipelines

A gas network typically consists of pipelines, compressor stations, valves, and regulator stations. For simulation purposes, a number of additional virtual components are necessary, for example, short pipes and fictitious resistances. However, the working principle of these components is similar to that of normal pipelines. As the pipeline is the dominant component in a gas network, the modeling of pipelines is explained in detail in this section.

Steady-state gas flows can be calculated using pipeline equations. There are several pipeline equations that can be used to calculate volumetric gas flow rates, although they are mostly derived from the isothermal Euler equation (Equation (1), cf. [27,28]).

$$\frac{\mathrm{d}p}{\mathrm{d}x} - \frac{f}{2D}\rho v|v| - \rho g \sin\theta = 0 \tag{1}$$

After calculating the integral and substituting the constant parameters with real values, the steady-state volumetric flow rate in a pipeline can be calculated as in Equation (2) (cf. [16,29]). Unless stated otherwise, the gas volumetric flow rates will be converted to the ones under standard reference conditions of 15 °C (288.15 K) and 1 bar (101,325 Pa), which correspond to T_{st} and P_{st} in the pipeline equation [30]. The unit of the standardized volumetric flow rate is sm^3/s.

$$Q = C\frac{T_{st}}{P_{st}}D^{2.5}\eta\left(\frac{|P_i^2 - P_j^2 - E|}{LGT_aZf}\right)^{0.5} \tag{2}$$

where $C = \pi\sqrt{\frac{R}{16M_{air}}}$ is a constant, which is around 13.29, D is the pipe diameter, η is the pipe efficiency, f is the friction factor, G is the gas specific gravity, L is the pipe length, P_{st} is the reference pressure, P_i is the inlet pressure, P_j is the outlet pressure, Q is the volumetric flow rate, T_a is the average temperature, T_{st} is the reference temperature, and Z is the compressibility factor. In Equation (2), E represents the potential

$$E = 0.06843G(H_j - H_i)\frac{P_a^2}{T_aZ} \tag{3}$$

where H_i and H_j are the inlet and outlet height, respectively. The average pressure P_a and average temperature T_a can be calculated using the following formulas (cf. [16]):

$$P_a = \frac{2}{3}\left(P_i + P_j - \frac{P_iP_j}{P_i + P_j}\right) \tag{4}$$

$$T_a = T_s + \frac{T_i - T_j}{\ln\left(\frac{T_i - T_s}{T_j - T_s}\right)} \tag{5}$$

The friction factor is a very important variable for calculating gas flow rates. There are various ways to calculate the pipeline friction factor that involve different levels of computational complexity and accuracy. However, the friction factor does not change significantly in the fully turbulent zone [31], which is also the case in this paper. Because the actual friction factor needs to be calculated or calibrated using pipeline efficiency, this paper uses the simplest method for pipeline friction, which is only related to the pipeline diameter D (cf. [16,29]).

$$f = 0.093902D^{-\frac{1}{3}} \tag{6}$$

η is the efficiency of the pipe to convert the theoretical friction factor into an actual one, taking other sources of friction into account, for example, valves and tees. The aging of the pipeline and corrosion or rust inside also contribute additional friction to the gas transmission system. Therefore, the real gas flow rate in a pipe is generally lower than the one calculated by flow equations where $\eta = 1$. To account for such extra flow reductions, the efficiency factor η is usually chosen between 0.75 and 0.95, while experience shows that for an old pipeline it can be reduced to lower than 0.7 [32]. However, Mohitpour et al. suggested η values between 0.92 and 0.97, which are much higher than the ones mentioned above (cf. [33]). In principle, η should be chosen according to actual gas network pipelines. In the remaining part of this paper, all pipeline friction factors are set to 0.85, which is a good assumption for a common pipeline operating status.

By reviewing Equation (2), it can be seen that most constants are preset pipeline parameters. It therefore can be simplified further using algebraic transformations, as shown below in Equation (7):

$$Q = C_{pipe}\left(\frac{|P_i^2 - P_j^2 - E|}{GZ}\right)^{0.5} \tag{7}$$

where C_{pipe} reflects constants and pipeline parameters

$$C_{pipe} = C\frac{T_b}{P_b}D^{2.5}\eta\left(\frac{1}{LT_a f}\right)^{0.5} \tag{8}$$

Typically, as shown in Equation (2), pipeline network simulations are based on volumetric flow rate conservation, which means the sum of all gas volumetric flows injected into or flowing out of one node equals zero. However, considering variant gas compositions in pipelines makes this assumption invalid. To deal with this issue, mass flow conservation is adopted in this work.

2.2. Thermal Formulation

Temperature is another important state variable in a gas network simulation. To calculate the temperature profile alongside a gas pipeline, an extra equation based on the first law of thermodynamics is needed (Equation (9), cf. [27,28]):

$$Q_m\frac{\mathrm{d}\left(h + \frac{v^2}{2}\right)}{\mathrm{d}x} + U_l(T - T_s)\pi D + Q_m g\sin\theta = 0 \tag{9}$$

where Q_m is the gas mass flow rate, h is its specific enthalpy, v is the velocity of gas flow, U_l is the heat transfer coefficient of the pipeline, g is the gravitational acceleration, and $\sin\theta = \frac{H_j - H_i}{L}$.

When considering a steady-state simulation, the change of velocity can be ignored ($\frac{\mathrm{d}v}{\mathrm{d}x} = 0$). By further assuming the gas enthalpy as a function of pressure and temperature, the change of enthalpy $\mathrm{d}h$ can be rewritten as Equation (10) [27]:

$$\mathrm{d}h = \left(\frac{\partial h}{\partial T}\right)_p \mathrm{d}T + \left(\frac{\partial h}{\partial p}\right)_T \mathrm{d}p \tag{10}$$

with $\left(\frac{\partial h}{\partial p}\right)_T = \left(\frac{\partial T}{\partial p}\right)_h\left(\frac{\partial h}{\partial T}\right)_p$, Equation (9) can be rewritten as Equation (11).

$$\frac{\mathrm{d}T}{\mathrm{d}x} + \mu_{JT}\frac{\mathrm{d}p}{\mathrm{d}x} + \frac{U_l}{Q_m c_p}(T - T_s)\pi D + \frac{g\sin\theta}{c_p} = 0 \tag{11}$$

To simplify the calculation, here the potential energy term is ignored. By substituting $\frac{dp}{dx}$ with the help of Equations (1) and (2), Equation (11) can be written in the form of Equation (12) (cf. [16,21]):

$$\frac{\mathrm{d}T}{\mathrm{d}x} = \mu_{JT}\left(\frac{fZ_sRQ_m|Q_m|}{2DPA^2}T + \frac{g\sin\alpha}{\rho}\right) - \frac{U_l\pi D}{Q_m c_p}(T - T_s) \tag{12}$$

After calculating the integral and simplification, the gas pipeline outlet temperature can then be calculated using the following equation:

$$T_j = \frac{\alpha}{\alpha+\beta}\left[T_s - T_s e^{-(\alpha+\beta)L}\right] + T_i e^{-(\alpha+\beta)L} \tag{13}$$

where $\alpha = \frac{U_L}{Q_m c_p}$ and $\beta = \mu_{JT}\frac{ZRfQ_m|Q_m|}{2PaDA^2}$.

2.3. Calculation of Gas Mixture Properties

As we can see in Equation (7), there are two properties that are directly related to the gas composition: the gas specific gravity G and the compressibility factor Z. For the purpose of simulating the gas network, these values are typically obtained using empirical models. In the case of the compressibility factor, for example, Papay's equation [17,21] (Equation (14)) or AGA [34] (Equation (15)) are used, where $p_r = p/p_c$ and $T_r = T/T_c$. These approaches are suitable when calculating the compressibility factor of typical natural gas. However, the critical point conditions (p_c and T_c) used in the formula are constants, assuming that the gas mixture properties do not change during network operation. However, this assumption is no longer valid when considering hydrogen blended into the natural gas network. Therefore, an automated calculation with variant gas composition is not possible using these empirical models.

$$z(p,T) = 1 - 3.52p_r e^{-2.26T_r} + 0.247p_r^2 e^{-1.878T_r} \tag{14}$$

$$z(p,T) = 1 + 0.257p_r - 0.533\frac{p_r}{T_r} \tag{15}$$

Now consider the gas specific gravity and heating value of the gas mixture. According to the DVGW technical regulation [35], natural gas is defined as H-gas and L-gas with respect to its composition. The properties of these two types of gases are shown in Table 1. It can be seen that heating values and specific gravity vary considerably, which makes an accurate calculation more difficult. Therefore, a comprehensive gas mixture property calculation is used in this paper.

Table 1. Properties of natural gas in Germany [30].

Gas Properties	Units	L-Gas	H-Gas
Heating value	MJ/sm^3	30.24–40.5	36–47.16
Specific gravity	-	0.55–0.75	

To calculate the gas mixture properties, the "thermo" python package is used [36]. The gas mixture properties are calculated using the Peng–Robinson EOS (PREOS) for a mixture of any number of compounds. The mathematical formulations can be found in the package repository or in the referenced literature [37,38].

The gas mixture heating values are calculated using another package named "Cantera" [39], based on the combustion data GRI-Mech 3.0 [40]. Limited to the available gas species data in this source, hydrocarbon species with over 3 carbon atoms (e.g., butane, pentane) are considered as methane in the heating value calculation.

$$HV = \sum_{i \in \mathcal{R}} H_i x_i - \sum_{j \in \mathcal{P}} H_j x_j \tag{16}$$

It is first used to balance all the chemical reaction equations of complete combustion based on the composition of the gas mixture. The corresponding heating value (HV) can be subsequently calculated using Equation (16) (cf. [41]), where $\mathcal{R}$ is the set of all reactants (all gas species in the gas mixture and O_2) and $\mathcal{P}$ represents all products. H stands for the enthalpy, and x is the mole fraction of a single gas species in $\mathcal{R}$ or $\mathcal{P}$. In this paper, the higher heating value (HHV) of the gas mixture is used, which assumes all water content in the end product is in a liquid state.

The natural gas composition used in this work is listed in Table A1 [42].

2.4. Solution Flow

In this tool, a gas network is modeled as a set of nodes and pipelines. The network nodes are classified into three types: reference nodes, supply nodes, and demand nodes. For the reference nodes, pressures are known and gas flows need to be calculated to balance network demand. For supply nodes and demand nodes, the gas flows are known and pressures need to be calculated. The passive sign convention system is used in this tool, with negative flows representing supply and positive ones representing consumption.

The simulation method presented in this paper combines conventional pipeline equations with comprehensive modeling of gas mixture and utilizes the Newton–Raphson (NR) method to solve steady-state gas flow calculation problems iteratively. As can be seen in Figure 1, the first step of the NR method is the initialization of network variables, which in this case are the initial estimates of unknown nodal pressures. In an electrical power flow simulation, a "flat-start" initialization is typically applied. In contrast, the pressures at both ends of a pipeline must not be the same; otherwise, this will result in zero flow in the pipeline and thus a poorly conditioned Jacobian matrix. From experience, pressures at pipeline outlets are initialized as 0.98 of those at inlets. To avoid the duplicated assignment of initial estimates, the initialization algorithm assigns pipeline outlet pressures starting from reference nodes until each node has a pressure value. Based on the initial estimates, the physical properties of the gas mixture inside pipelines can be calculated. For each iteration step, the Jacobian matrix is updated, which is later used to calculate the state variables. Afterward, the new gas mixture composition and its properties are calculated. Since the approach presented here is aimed at solving the static flow problem, flows entering a node with different compositions are assumed to be completely mixed at the node. Therefore, the gas mixture composition in a pipeline is always considered to be the same as the one at its inlet. As long as the error after one iteration step is bigger than the set tolerance, the program iterates over to set pressures to the nodes and simultaneously updates the gas mixture properties to be used for the pipelines. If the error remains within the tolerance, the simulation has converged and results are saved.

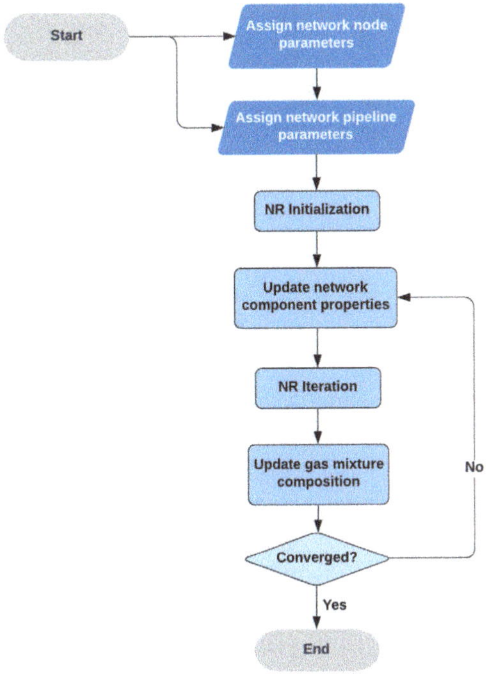

Figure 1. Flow chart of the solving process.

3. Simulation and Results

3.1. Study Case 1: Impacts of Different Calculation Methods on a Single Pipeline with Variable Gas Mixture Composition

Since the pipeline is the dominant component in a gas network, in the first study case the simplest case is considered, which is the analysis of different gas compositions in one single pipeline. The main purpose of this study case is to stress the importance of taking the gas composition into account when performing a pipeline flow calculation with respect to hydrogen blended into natural gas. To achieve this, PREOS is used in this study to calculate the gas mixture properties, which is proved to be one of the most accurate methods to model the natural gas mixture [43]. Since the assumption made for Equation (6) is only valid for systems with higher pressures, only high-pressure gas pipelines are considered in this section. In Germany, the pressure level of high-pressure gas networks is defined as an operational range between 1 and 100 bar [44]. Therefore, in this paper, the analyzed pressure range is also set between 1 and 100 bar. As described in previous sections, multiple pipeline flow equations are available. A comparison of the most popular methods is shown in Figure 2a to give an overview of these different methods. It can be noted that at lower pressures (from 1 to 10 bar), the calculation results of different methods do not differ significantly. Within a higher pressure range (from 10 to 100 bar), the Panhandle B method tends to overestimate the flow rates, while the Weymouth method underestimates them. The flow rates calculated with the method used in this tool lie in the middle range of all analyzed methods, both in the lower pressure range and in the higher pressure range.

Typically, simulations of gas pipeline networks are based on constant gas properties, which is no longer accurate when the properties change significantly, for example, when the operation conditions deviate strongly from the nominal value or when complex systems with different gas mixtures are simulated. As shown in Figure 2b, a comparison between the Papay's equation and the PREOS is made. The calculation results from these two alternatives appear to be quite similar. However, at a higher pressure, the calculated flow rates with Papay's equation, which is based on constant critical conditions, are around 2% lower than those using PREOS which is a more accurate method for calculating gas properties.

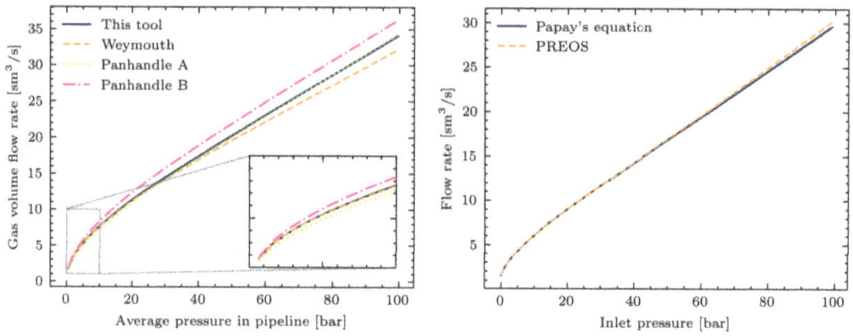

(a) (b)

Figure 2. (**a**) Comparison between different pipeline flow rate calculation methods. (**b**) Comparison of flow rate calculation with constant gas properties and those calculated using PREOS.

The pipeline transmission is generally restricted by the minimum and maximum permitted volumetric flow rates and the maximum permitted pressure. Therefore, it is important to investigate how the flow rate and pressure change when different gas compositions are applied to the pipeline. Now consider a binary gas mixture of CH_4 and H_2. The simplified pipeline equation (Equation (7)) shows that the gas flow rate is dependent on the compressibility factor Z and the specific gravity G of the gas mixture. As shown in Figure 3a, the gas compressibility factor increases in a non-linear fashion with increasing hydrogen concentration. If we consider a 20% hydrogen concentration in the gas mixture, the corresponding compressibility factor is around 2% greater than the one calculated with the linear mole fraction model. When using Papay's equation, it is difficult to calculate the T_c and P_c of the gas mixture; thus, a simple linear mole fraction method is used. As shown in Figure 3a, at low hydrogen concentrations, this method works well, but when the hydrogen concentration is increased further, the error also becomes bigger. Obuba et al. showed that Papay's method tends to underestimate the Z-factor [45], an observation that also corresponds to the results shown in this figure.

Turning now to the gas specific gravity G. As shown in Figure 3b, the gas-specific gravity is linearly related to the hydrogen concentration, and the calculation results obtained from using PREOS and the linear mole fraction model are nearly identical.

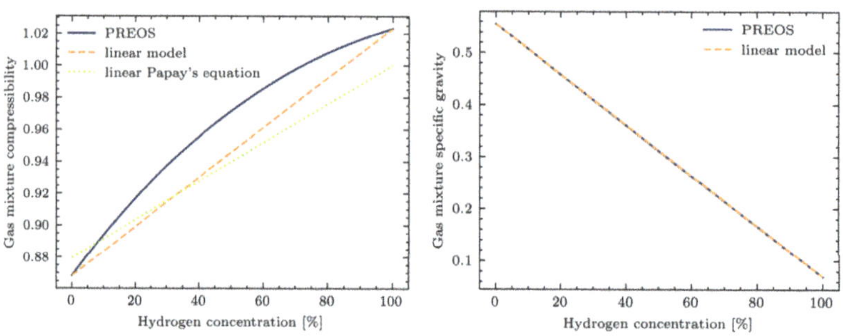

(a) (b)

Figure 3. (**a**) Gas mixture compressibility in the relation to the injection of hydrogen. (**b**) Gas mixture specific gravity in relation to the injection of hydrogen.

After analyzing the gas properties of gas mixtures with different hydrogen concentrations, the pipeline flow calculation results are shown here. In the practical gas network operation, the energy requirement of the demand side has to be met. Therefore, the actual gas flow rate may vary due to variations in the heating values of the gas mixture. In this study case, a constant energy demand at the pipeline outlet is assumed in order to analyze the pressure drops throughout the pipeline with respect to different hydrogen concentrations.

As shown in Figure 4, when considering a certain amount of energy demand, the pressure drop in the pipeline increases with increasing hydrogen concentration. At around 90% hydrogen concentration, however, there is a turning point in the pressure drop curve. Looking back to Equation (7) and Figure 3a, the reason for this turning point is that the increase of the compressibility factor Z slows down when the hydrogen concentration increases.

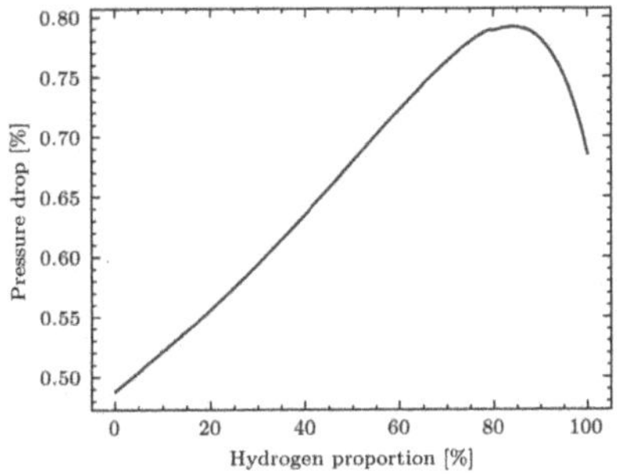

Figure 4. Pressure drop throughout a pipeline in relation to different hydrogen concentrations.

Besides the nodal pressure and the flow rate inside pipelines, the temperature of the transmitted gas mixture is also an important variable for gas network simulation. In Figure 5a, simulations are run to illustrate the temperature profile inside a pipeline with respect to the different inlet temperatures. It can be noted that the gas mixture temperature tends to be the same as its ambient temperature after a long-distance transmission. In Figure 5b, several simulations are run assuming different hydrogen concentration rates inside the pipeline. In the case of natural gas, the gas mixture temperature is assumed to be the same as the ambient temperature if the pipeline is longer than 25 km. However, with the increase of hydrogen concentration inside the pipe, the descent rate of the temperature becomes lower and the temperature difference between the pipe inlet and outlet has to be taken into consideration in certain cases.

Gas transmission in the pipeline network is based on pressure control. The results presented in this section show that in order to keep the same energy flow in one pipeline, gas flow rates and pressure drops also change when the hydrogen concentration changes in the pipeline.

Since the volumetric flow rate is the most important variable to be analyzed, only the volumetric flow will be analyzed and shown in the rest of the paper.

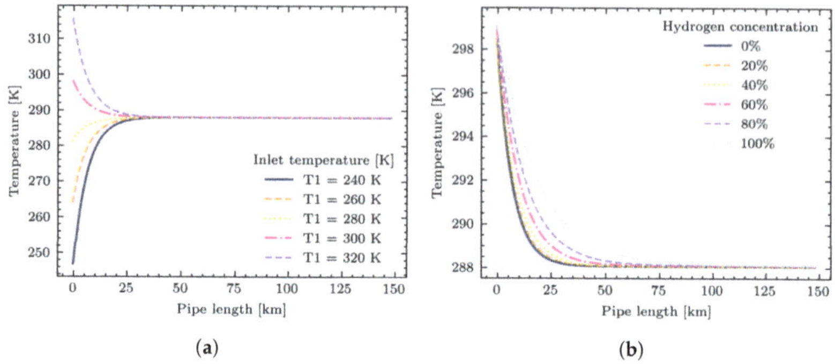

Figure 5. (**a**) Temperature profile throughout a gas pipe with different pipeline inlet temperatures. (**b**) Temperature profile throughout a gas pipe with different hydrogen concentrations.

3.2. Study Case 2: A Simple Gas Network with Hydrogen Injection

In the previous study case, simulations are performed on a simple pipeline model to illustrate the importance and necessity of taking gas mixture properties into account as part of a hydrogen-enriched natural gas network simulation. Considering the possibility of blending hydrogen into the natural gas network, if the hydrogen injection occurs at the demand side, the hydrogen concentration could be gradually increased and therefore cause risky conditions on a local level. In this section, a number of simulations are performed on a hydrogen-enriched natural gas network, to analyze the impacts caused by different hydrogen concentrations in the network or nodal hydrogen injections into the network.

To reproduce the network behavior with the utmost accuracy, a compatible network size for both gas and power grids should be considered. In this work, the gas grid is synthetically generated based on the CIGRE high-voltage transmission benchmark grid [46], ignoring some buses and connections.

The synthetically generated simple gas network model consists of 8 nodes and 8 branches as shown in Figure 6. Node 1 and node 2 are set as reference nodes, where the nodal pressure values are known. At these nodes, the pressures remain constant, while the flow rates are variable to be able to balance the network demands. The other nodes are defined as demand nodes, where the flow rates are predefined. The demand can be also converted into energy demand using the gas mixture heating values.

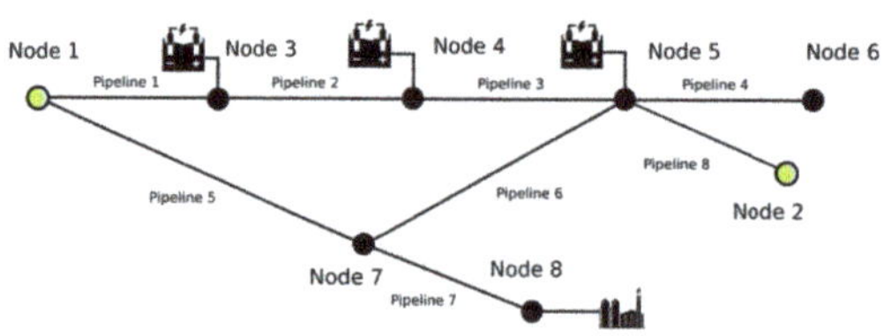

Figure 6. Artificial gas grid.

As shown in Figure 7, when nodal energy demands stay the same, the overall pressure at all demand nodes will be lower when the hydrogen concentration is higher. However, from 80% to 100% hydrogen concentration, the nodal pressures increase slightly, which corresponds to the results shown in Figure 4.

Water electrolysis is one of the most promising electrical flexibility options so far as it can generate hydrogen by electrolyzing water. To analyze the impacts of electrolyzer operation on the natural gas network, it is assumed that a known amount of hydrogen is fed into the natural gas network at the network nodes. In this study case, the distribution of injected hydrogen in the network is analyzed to see if the maximum permitted hydrogen concentration in the pipeline is exceeded.

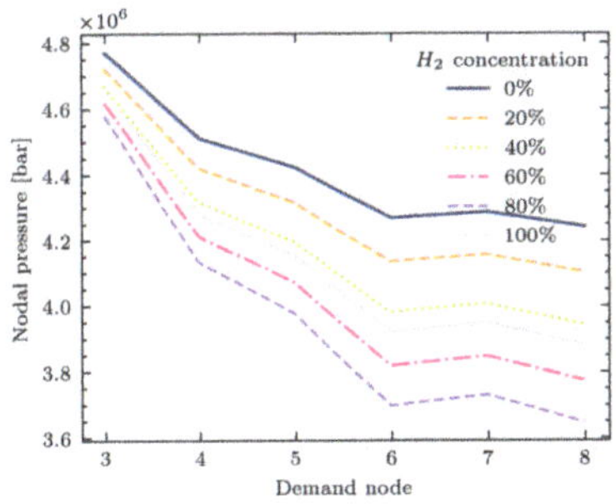

Figure 7. Pressure at demand nodes with respect to different hydrogen concentrations in the network.

Now assume there are three electrolyzers, which are accordingly located at nodes 3, 4, and 5. The energy flow rates of hydrogen blending into the natural gas grid are assumed as 35 MW, which corresponds to around 2.9 sm^3/s. In this case, different gas injections are considered to be fully mixed at conjunction nodes. The corresponding simulation results are shown in Figure 8. As shown in the figure, a relatively high hydrogen concentration of around 29.6% occurs in pipeline 3. In pipelines 2, 4, and 6, the hydrogen concentration also reaches around 13%, which is already critical with respect to current technical regulations. It should be noted, however, that although the assumed hydrogen injection rate is relatively high given the current status of real-world implementation, the total energy delivered with hydrogen covers only around 1.88% of the total energy demand in this system. Therefore, when the total hydrogen share in the gas transmission system is increased, it is important to monitor the hydrogen concentration in pipelines in order to ensure a secure operation.

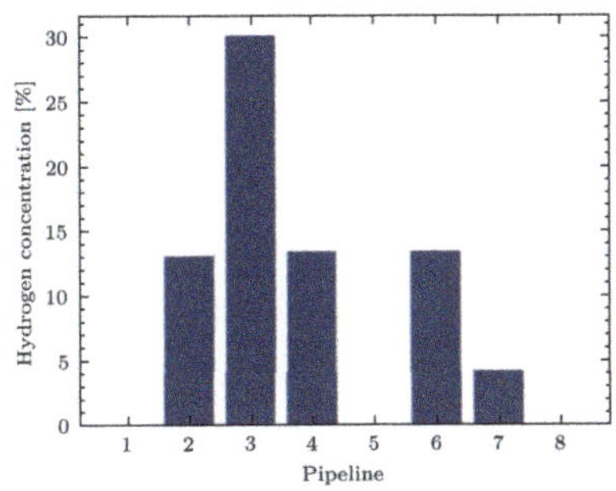

Figure 8. Hydrogen concentration in pipelines with respect to the injection of hydrogen at three nodes.

3.3. Study Case 3: A Simple Example of a Simulation of Coupled Power and Gas Networks

A number of studies analyzed the technical and economical performances of water electrolyzers considering different PV and wind penetration levels [47,48]. However, these studies focused mainly on the electrolyzer and its power consumption rather than on the grid perspective. In this part, a simple analysis is performed by coupling the power grid (Figure 9) and gas grid (Figure 6). To this end, a quasi-dynamic simulation is performed based on a time-series simulation. The energy demands in both networks are calculated using standard demand profiles, which are taken from [49,50] and assume an annual demand of 400 MWh and 600 MWh, based on the scenario defined in the original power system model [46]. To include the generation of renewable energy, weather data from 2015 are used [51]. The wind and PV generation profiles are calculated using the above-mentioned weather data and formulas taken from [52,53]. The total installed generation capacity of wind and PV generation units is set at 500 MW with a wind and PV ratio of 50–50%. Furthermore, renewable energy generation is considered to be equally distributed at each node. In Figure 10, the PV and wind generation are plotted as box plots [54] to show their seasonal behavior. The orange lines in the middle represent the median value of the power generation of each month. The lower and upper bounds of the boxes represent the lower and upper quartiles of monthly generation. The whiskers extend from the box and cover the major range of the data. The outliers refer to the data past the end of the whiskers, which are the peaks and valleys of power generation. It can be seen that although PV has a peak generation in the summer and the maximum total generation of renewable energy usually occurs in warmer seasons, the relatively high continuous generation of renewable energy occurs during winter. Therefore, two scenarios are assumed based on the above-mentioned renewable generation, which correspond to June and December, meaning that the PV-rich and wind-rich seasons can be considered.

To investigate the impact of coupling both networks, the combination of three water electrolyzers and one gas-fired power plant is analyzed (see Figures 6 and 9). The water electrolyzers are considered to be installed at nodes 3, 4, and 5, each with a maximum power consumption rate of 20 MW and an efficiency of 80%. A simple scheduling method is applied to ensure that the electrolyzers are only operated at times when the generation of renewable energies exceeds its nodal power demand. The gas-fired power plant is assumed at node 8, which consumes gas and produces electrical power with an efficiency of 60%. Here, instead of changing settings on the power grid side, a constant volumetric flow rate of gas consumption is assumed so that the impact of the time-variant hydrogen concentration in the natural gas grid can be illustrated.

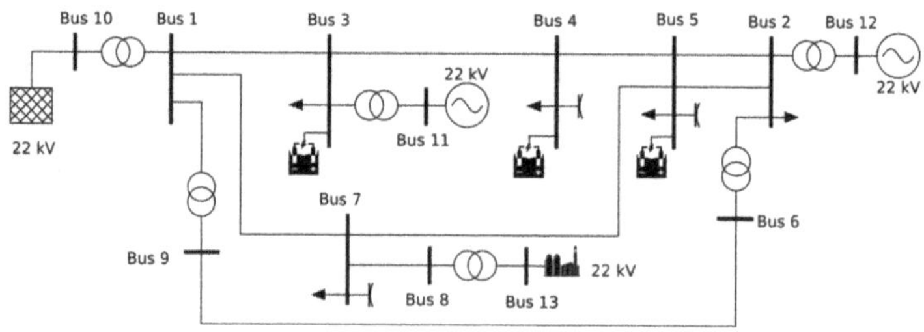

Figure 9. CIGRE high-voltage transmission grid (cf. [46]).

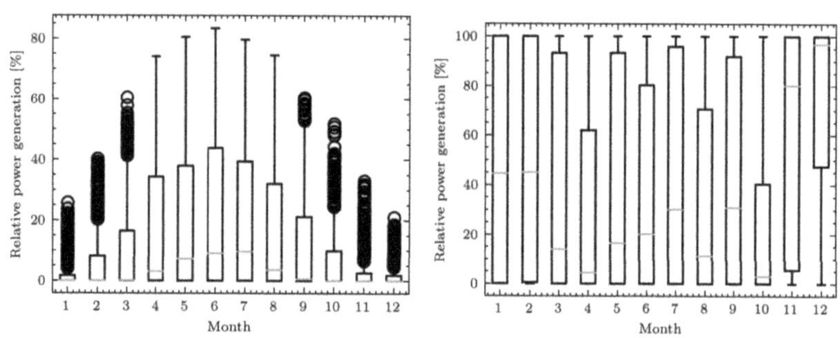

(a) (b)

Figure 10. (**a**) Relative PV power generation in relation to installed generation capacity. (**b**) Relative wind power generation in relation to installed generation capacity.

Two simulations are run separately for June and December 2015 with a sampling time of one hour. The results are shown in Figures 11 and 12. By comparing the results from Figure 11a,b, it is clear that the hydrogen concentrations in June are much higher than those in December. This result can be explained by the fact that the gas demand is usually rather low in summer while generation spikes of renewable energies are rather common. Because natural gas is mostly used for space heating purposes, the gas demand increases during colder days. Although more hydrogen is produced in the winter, the energy supplied with hydrogen does not play a major role.

In Figure 12, the power generation of the gas-fired power plant at node 8 is analyzed. It should be noted that the power generation in June is not stable, which can vary up to 20%. This is caused by the high concentration of hydrogen that occurs in pipeline 7, which directly supplies node 8. This further indicates that directly blending hydrogen into the natural gas network is theoretically impractical during summer. Although the power generation in December is relatively more stable, a difference of over 5% is still possible. Therefore, to ensure a secure operation of the power grid coupled with the hydrogen-enriched natural gas network, the heating values of the gas mixture should be taken into consideration.

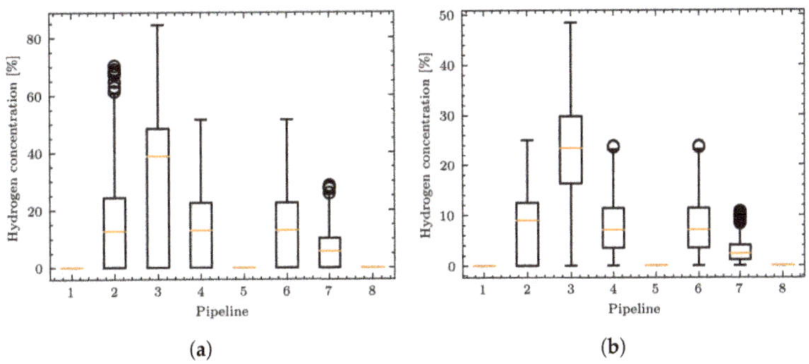

Figure 11. (**a**) Hydrogen concentration in pipelines in June. (**b**) Hydrogen concentration in pipelines in December.

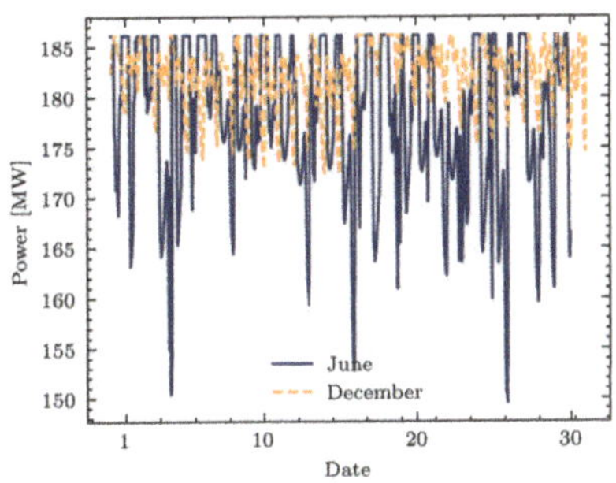

Figure 12. Gas-fired power plant generation with constant volumetric flow consumption.

4. Conclusions

The gas sector can provide seasonal storage capacity and additional flexibility to the power sector in an integrated power and gas system. To accurately analyze the impacts caused by coupling power and gas networks, appropriate modeling and simulation approaches are needed. In this paper, a simulation method and a corresponding tool are presented that are capable to simulate hydrogen-enriched natural gas networks. To achieve this, the Peng–Robinson equation of states is applied to calculate the gas mixture properties.

In the previous section, several study cases were developed and simulated to demonstrate the importance and necessity of modeling variant gas compositions. First, different modeling approaches were analyzed by applying them to a single pipeline. By analyzing the results, it can be concluded that hydrogen injection can pose challenges to the gas pipeline network. The gas mixture composition has a significant impact on the static gas flow calculation, which can represent the long-term gas network operation. With a potentially high penetration of hydrogen in the natural gas network, the gas volumetric flow rate and nodal pressure might vary due to the different hydrogen concentrations in gas pipelines. In addition, when the hydrogen concentration in the natural gas network system increases, the heating value of the transported gas mixture decreases. Therefore, in order to maintain a sufficient energy supply, the volumetric flow rates have to be increased, which in turn leads to increasing pressure drops throughout the pipelines Therefore, local pipeline congestion may occur in the future, which should be analyzed in advance during network planning.

The method is then applied to a simple network, where nodal hydrogen injections take place. The simulation results further confirm the findings from the previous study case. Moreover, the presented method and tool can also be used to calculate hydrogen concentration in pipelines, which is an important state variable in the analysis of hydrogen-enriched natural gas networks. Furthermore, the simulation results also provide information about the heating value of the gas mixture, which is very helpful for real-world applications, where very little real-time measurement data are currently available.

Finally, a coupled power and gas network is simulated and analyzed. This study case shows that the power and gas demand varies greatly on a seasonal level. The feasibility of blending hydrogen directly into the natural gas grid is therefore investigated using the developed simulation tool. Due to the relatively lower gas demand in summer, the permitted amount of synthetic hydrogen in the gas grid is rather limited. Therefore, a more reasonable option is to store the synthetically generated hydrogen separately or together with natural gas in warmer seasons. In winter, the gas demand is higher due to the greater heating demand. Blending hydrogen into the gas grid is therefore more feasible.

Author Contributions: Conceptualization, Y.L., T.P. and A.B.; methodology, Y.L., T.P. and A.B.; software, Y.L.; validation, Y.L.; formal analysis, Y.L.; investigation, Y.L.; resources, Y.L.; data curation, Y.L.; writing—original draft preparation, Y.L.; writing—review and editing, Y.L., T.P. and A.B.; visualization, Y.L.; supervision, T.P. and A.B.; project administration, T.P. and A.B.; funding acquisition, T.P. and A.B. All authors have read and agreed to the published version of the manuscript.

Funding: This research was funded by Operational Program for the promotion of investments in growth and employment for North Rhine-Westphalia from the European fund for regional development (OP EFRE NRW) grant number EFRE-0400111.

Acknowledgments: T.P. gratefully acknowledges funding by the center of excellence "Virtual Institute-Power to Gas and Heat" (EFRE-0400111) by the "Operational Program for the promotion of investments in growth and employment for North Rhine-Westphalia from the European fund for regional development" (OP EFRE NRW) through the Ministry of Economic Affairs, Innovation, Digitalization and Energy of the State of North Rhine-Westphalia.

Conflicts of Interest: The authors declare no conflict of interest.

Abbreviations

The following abbreviations are used in this manuscript:

DVGW	German Technical and Scientific Association for Gas and Water
EOS	Equation of state
HV	Heating value
HHV	High heating value
NG	Natural gas
NR	Newton–Raphson
PREOS	Peng–Robinson equation of state
PtG	Power-to-gas
PV	Photovoltaics

Nomenclature

C	constant, $\pi\sqrt{\frac{R}{16M_{air}}}$, which is around 13.29,
D	pipe diameter (m),
η	pipe efficiency (dimensionless),

f	friction factor (dimensionless),
G	gas specific gravity (dimensionless),
g	gravitational acceleration (m/s^2),
H_i	inlet height (m),
H_j	outlet height (m),
HV	heating value (J/sm^3),
HHV	higher heating value (J/sm^3),
h	specific enthalpy (J/kg),
L	pipe length (m),
P_{st}	reference pressure (Pa),
P_i	inlet pressure (Pa),
P_j	outlet pressure (Pa),
p_c	critical pressure (K),
p_r	pseudo-reduced pressure (dimensionless),
Q	volumetric flow rate (sm^3/s),
Q_m	mass flow rate (kg/s)
T_a	average temperature (K),
T_c	critical pressure (K),
T_r	pseudo-reduced pressure (dimensionless),
T_s	ambient temperature (K),
T_{st}	reference temperature (K),
U_l	heat transfer coefficient ($W/(m^2 \cdot K)$),
v	gas flow velocity (m/s),
Z	compressibility factor (dimensionless),
θ	slope angle (dimensionless)

Appendix A

Table A1. Typical natural gas composition [42].

Gas Species	Typical Analysis (Mole %)	Range (Mole %)
Methane	94.7	87.0–98.0
Ethane	4.2	1.5–9.0
Propane	0.2	0.1–1.5
Iso-butane	0.02	trace–0.3
Butane	0.02	trace–0.3
Iso-pentane	0.01	trace–0.04
Pentane	0.01	trace–0.04
Hexanes plus	0.01	trace–0.06
Nitrogen	0.5	0.2–5.5
Carbon dioxide	0.3	0.05–1.0
Oxygen	0.01	trace–0.1
Hydrogen	0.02	trace–0.05

References

1. Rockström, J.; Gaffney, O.; Rogelj, J.; Meinshausen, M.; Nakicenovic, N.; Schellnhuber, H.J. A roadmap for rapid decarbonization. *Science* **2017**, *355*, 1269–1271. [CrossRef]
2. Wiser, R.; Jenni, K.; Seel, J.; Baker, E.; Hand, M.; Lantz, E.; Smith, A. Expert elicitation survey on future wind energy costs. *Nat. Energy* **2016**, *1*, 16135. [CrossRef]
3. Mancarella, P. MES (multi-energy systems): An overview of concepts and evaluation models. *Energy* **2014**, *65*, 1–17. [CrossRef]
4. Pesch, T.C. *Multiskalare Modellierung Integrierter Energie- und Elektrizitätssysteme*; Schriften des Forschungszentrums Jülich. Energie & Umwelt/Energy & Environment; Verlag des Forschungszentrums Jülich: Jülich, Germany, 2019; Volume 485.
5. Fraunhofer-Gesellschaft. Energy-Charts. Available online: https://energy-charts.info/index.html?l=en&c=DE (accessed on 9 November 2020).
6. Energie Informationsdienst. *Underground Gas Storage in Germany*; DVV Media Group GmbH: Hamburg, Germany, 2020.

7. Thema, M.; Sterner, M.; Lenck, T.; Götz, P. Necessity and Impact of Power-to-gas on Energy Transition in Germany. *Energy Procedia* **2016**, *99*, 392–400. [CrossRef]
8. Clegg, S.; Mancarella, P. Storing renewables in the gas network: Modelling of power–to–gas seasonal storage flexibility in low–carbon power systems. *IET Gener. Transm. Distrib.* **2016**, *10*, 566–575. [CrossRef]
9. Schiebahn, S.; Grube, T.; Robinius, M.; Tietze, V.; Kumar, B.; Stolten, D. Power to gas: Technological overview, systems analysis and economic assessment for a case study in Germany. *Int. J. Hydrog. Energy* **2015**, *40*, 4285–4294. [CrossRef]
10. Dolci, F. Green Hydrogen Opportunities in Selected Industrial Processes. *JRC Tech. Rep.* **2018**. Available online: https://publications.jrc.ec.europa.eu/repository/handle/JRC114766 (accessed on 20 September 2021)
11. Marc W. Melaina.; Olga Antonia.; Michael Penev. *Blending Hydrogen into Natural Gas Pipeline Networks: A Review of Key Issues*; National Renewable Energy Laboratory: Golden, CO, USA, 2013.
12. Fuel Cells and Hydrogen Joint Undertaking. Development of Business Cases for Fuel Cells and Hydrogen Applications for Regions and Cities: Hydrogen Injection into the Natural Gas Grid. Available online: https://www.fch.europa.eu/sites/default/files/FCH%20Docs/171121_FCH2JU_Application-Package_WG5_P2H_Hydrogen%20into%20gas%20grid%20%28ID%202910558%29%20%28ID%202911642%29.pdf (accessed on 20 September 2021)
13. German Bundestag. Critical Values for Hydrogen (H_2) in the Natural Gas Infrastructure. Available online: https://www.bundestag.de/resource/blob/646488/a89bbd41acf3b90f8a5fbfbcb8616df4/WD-8-066-19-pdf-data.pdf (accessed on 8 November 2020).
14. Leicher, J.; Nowakowski, T.; Giese, A.; Görner, K. Power-to-gas and the consequences: Impact of higher hydrogen concentrations in natural gas on industrial combustion processes. *Energy Procedia* **2017**, *120*, 96–103. [CrossRef]
15. Leicher, J.; Giese, A.; Görner, K.; Werschy, M.; Krause, H.; Dörr, H. Natural gas quality fluctuations—surveys and statistics on the situation in Germany. *Energy Procedia* **2017**, *120*, 165–172. [CrossRef]
16. Martinez-Mares, A.; Fuerte-Esquivel, C.R. A Unified Gas and Power Flow Analysis in Natural Gas and Electricity Coupled Networks. *IEEE Trans. Power Syst.* **2012**, *27*, 2156–2166. [CrossRef]
17. Ekhtiari, A.; Dassios, I.; Liu, M.; Syron, E. A Novel Approach to Model a Gas Network. *Appl. Sci.* **2019**, *9*, 1047. [CrossRef]
18. Abdolahi, F.; Mesbah, A.; Boozarjomehry, R.B.; Svrcek, W.Y. The effect of major parameters on simulation results of gas pipelines. *Int. J. Mech. Sci.* **2007**, *49*, 989–1000. [CrossRef]
19. Guandalini, G.; Colbertaldo, P.; Campanari, S. Dynamic modeling of natural gas quality within transport pipelines in presence of hydrogen injections. *Appl. Energy* **2017**, *185*, 1712–1723. [CrossRef]
20. Poling, B.E.; Prausnitz, J.M.; O'Connell, J.P. *The Properties of Gases and Liquids*, 5th ed.; McGraw-Hill: New York, NY, USA; London, UK, 2000.
21. Pellegrino, S.; Lanzini, A.; Leone, P. Greening the gas network – The need for modelling the distributed injection of alternative fuels. *Renew. Sustain. Energy Rev.* **2017**, *70*, 266–286. [CrossRef]
22. Siemens. PSS®SINCAL—Simulation Software for Analysis and Planning of Electric and Pipe Networks. Available online: https://assets.new.siemens.com/siemens/assets/public.1537970929.31ece3a2-e9cc-4528-b9f9-6bf61b613de2.ref-no-69-ps-c-pss-sincal-brochure-hires-intl-sept2018.pdf (accessed on 9 November 2020).
23. Pambour, K.A.; Cakir Erdener, B.; Bolado-Lavin, R.; Dijkema, G.P. SAInt–A novel quasi-dynamic model for assessing security of supply in coupled gas and electricity transmission networks. *Appl. Energy* **2017**, *203*, 829–857. [CrossRef]
24. Merkuryev, Y. (Ed.) *SIMULTECH 2016: Proceedings of the 6th International Conference on Simulation and Modeling Methodologies, Technologies and Applications*; IEEE: Piscataway, NJ, USA, 2016.
25. Andresen, L.; Dubucq, P.; Peniche Garcia, R.; Ackermann, G.; Kather, A.; Schmitz, G. Status of the TransiEnt Library: Transient Simulation of Coupled Energy Networks with High Share of Renewable Energy. In Proceedings of the 11th International Modelica Conference, Linköping Electronic Conference Proceedings, Versailles, France, 21–23 September 2015; Linköping University Electronic Press: Linköping, Sweden, 2015; pp. 695–705.
26. Lohmeier, D.; Cronbach, D.; Drauz, S.R.; Braun, M.; Kneiske, T.M. Pandapipes: An Open Source Piping Grid Calculation Package for the Application in Coupled Multi-Energy Grid Simulations. *Sustainability* **2020**, *12*, 9899. [CrossRef]
27. Chaczykowski, M.; Osiadacz, A.J. Comparative assesment of steady-state pipeline gas flow models / Analiza porównawcza modeli przepływu gazu w rurociągu w stanach ustalonych. *Arch. Min. Sci.* **2012**, *57*, 23–38. [CrossRef]
28. Bales, P.; Kolb, O.; Lang, J. Hierarchical Modelling and Model Adaptivity for Gas Flow on Networks. In *Computational science-ICCS 2009*; Allen, G., Ed.; Lecture Notes in Computer Science; Springer: Berlin, Germany, 2009; Volume 5544, pp. 337–346.
29. Coelho, P.M.; Pinho, C. Considerations about equations for steady state flow in natural gas pipelines. *J. Braz. Soc. Mech. Sci. Eng.* **2007**, *29*, 262–273. [CrossRef]
30. German Technical and Scientific Association for Gas and Water. *Gas Infrastructure-Quality of Gas-Group H*; German version EN 16726:2015+A1:2018; Beuth Verlag GmbH: Berlin, Germany, 2019.
31. Schroeder, D.W., Jr. A tutorial on pipe flow equations. In Proceedings of the PSIG Annual Meeting, Bonita Springs, FL, USA, 11–14 May 2010.
32. Osiadacz, A.J. *Simulation and Analysis of Gas Networks*; Gulf Publishing Company: Houston, TX, USA, 1987.
33. Mohitpour, M.; Thompson, W.; Asante, B. The Importance of Dynamic Simulation on the Design and Optimization of Pipeline Transmission Systems. In Proceedings of the First International Pipeline Conference, Calgary, AB, Canada, 9–13 June 1996; Yoon, M., Mensik, M., Mohitpour, M., Eds.; American Society of Mechanical Engineers: New York, NY, USA, 1996; pp. 1183–1188.

34. Farzaneh-Gord, M.; Khamforoush, A.; Hashemi, S.; Namin, H. Computing Thermal Properties of Natural Gas by Utilizing AGA8 Equation of State. *Int. J. Chem. Eng. Appl.* **2010**, *1*, 20–24. [CrossRef]
35. German Technical and Scientific Association for Gas and Water. *DVGW G 260-Gas Quality*; Wirtschafts- und Verlagsgesellschaft Gas und Wasser mbH: Bonn, Germany, 2013.
36. Caleb Bell and Contributors (2016–2021) Thermo: Chemical Properties Component of Chemical Engineering Design Library (ChEDL). GitHub Repository. Available online: https://github.com/CalebBell/thermo (accessed on 12 September 2021).
37. Peng, D.Y.; Robinson, D.B. A New Two-Constant Equation of State. *Ind. Eng. Chem. Fundam.* **1976**, *15*, 59–64. [CrossRef]
38. Robinson, D.B.; Peng, D.Y.; Chung, S.Y.K. The development of the Peng—Robinson equation and its application to phase equilibrium in a system containing methanol. *Fluid Phase Equilibria* **1985**, *24*, 25–41. [CrossRef]
39. Goodwin, D.G.; Speth, R.L.; Moffat, H.K.; Weber, B.W. Cantera: An Object-Oriented Software Toolkit for Chemical Kinetics, Thermodynamics, and Transport Processes, 2018. GitHub Repository. Available online: https://github.com/Cantera/cantera (accessed on 12 September 2021).
40. Smith, G.P.; Golden, D.M.; Frenklach, M.; Moriarty, N.W.; Eiteneer, B.; Goldenberg, M.; Bowman, C.T.; Hanson, R.K.; Song, S.; Gardiner, W.C., Jr.; et al. GRI-MECH 3.0. Available online: http://combustion.berkeley.edu/gri-mech/version30/text30.html (accessed on 12 September 2021).
41. Klein, S.A.; Nellis, G. *Thermodynamics*; Cambridge University Press: New York, NY, USA, 2012.
42. Enbridge Gas. Chemical Composition of Natural Gas. Available online: https://www.enbridgegas.com/about-enbridge-gas/learn-about-natural-gas (accessed on 2 October 2021)
43. Moiseeva, E.F.; Malyshev, V.L. Compressibility factor of natural gas determination by means of molecular dynamics simulations. *AIP Adv.* **2019**, *9*, 055108. [CrossRef]
44. Bundesnetzagentur. Definitionen der Daten für den Gasbereich. Available online: https://www.bundesnetzagentur.de/SharedDocs/Downloads/DE/Sachgebiete/Energie/Unternehmen_Institutionen/DatenaustauschUndMonitoring/MaStR/DefinitionenDatenGasbereich.pdf?__blob=publicationFile&v=3 (accessed on 15 October 2020).
45. Obuba, J.; Ikiesnkimama, S.S.; Ubani, C. Natural Gas Compressibility Factor Correlation Evaluation for Niger Delta Gas Fields. *IOSR J. Electr. Electron. Eng.* **2013**, *6*, 01–10. [CrossRef]
46. Strunz, K.; Abbasi, E.; Fletcher, R.; Hatziargyriou, N.; Iravani, R.; Joos, G. *TF C6.04.02: TB 575–Benchmark Systems for Network Integration of Renewable and Distributed Energy Resources*; CIGRE: Paris, France, 2014.
47. Davis, W.; Martín, M. Optimal year-round operation for methane production from CO_2 and water using wind and/or solar energy. *J. Clean. Prod.* **2014**, *80*, 252–261. [CrossRef]
48. de Boer, H.S.; Grond, L.; Moll, H.; Benders, R. The application of power-to-gas, pumped hydro storage and compressed air energy storage in an electricity system at different wind power penetration levels. *Energy* **2014**, *72*, 360–370. [CrossRef]
49. Bundesverband der Energie- und Wasserwirtschaft. Standardlastprofile Strom. Available online: https://www.bdew.de/energie/standardlastprofile-strom/ (accessed on 16 September 2021).
50. Bundesverband der Energie- und Wasserwirtschaft. Standardlastprofile Gas. Available online: https://www.bdew.de/energie/standardlastprofile-gas/ (accessed on 16 September 2021).
51. Pfenninger, S.; Staffell, I. Long-term patterns of European PV output using 30 years of validated hourly reanalysis and satellite data. *Energy* **2016**, *114*, 1251–1265. [CrossRef]
52. Quaschning, V. *Understanding Renewable Energy Systems*; revised edition ed.; Routledge Taylor & Francis Group Earthscan from Routledge: London, UK; New York, NY, USA, 2016.
53. Zhou, W.; Yang, H.; Fang, Z. A novel model for photovoltaic array performance prediction. *Appl. Energy* **2007**, *84*, 1187–1198. [CrossRef]
54. McGill, R.; Tukey, J.W.; Larsen, W.A. Variations of box plots. *Am. Stat.* **1978**, *32*, 12–16.

energies

MDPI

Article

Current Legislative Framework for Green Hydrogen Production by Electrolysis Plants in Germany

Lena Maria Ringsgwandl [1,*], Johannes Schaffert [2,*], Nils Brücken [2,*], Rolf Albus [2] and Klaus Görner [2]

[1] ALBA Group, 10719 Berlin, Germany
[2] Gas-und Waerme-Institut Essen e.V. (GWI), Hafenstrasse 101, 45356 Essen, Germany; rolf.albus@gwi-essen.de (R.A.); klaus.goerner@gwi-essen.de (K.G.)
* Correspondence: lena.ringsgwandl@albagroup.de (L.M.R.); johannes.schaffert@gwi-essen.de (J.S.); nils.bruecken@gwi-essen.de (N.B.)

Abstract: (1) The German energy system transformation towards an entirely renewable supply is expected to incorporate the extensive use of green hydrogen. This carbon-free fuel allows the decarbonization of end-use sectors such as industrial high-temperature processes or heavy-duty transport that remain challenging to be covered by green electricity only. However, it remains unclear whether the current legislative framework supports green hydrogen production or is an obstacle to its rollout. (2) This work analyzes the relevant laws and ordinances regarding their implications on potential hydrogen production plant operators. (3) Due to unbundling-related constraints, potential operators from the group of electricity transport system and distribution system operators face lacking permission to operate production plants. Moreover, ownership remains forbidden for them. The same applies to natural gas transport system operators. The case is less clear for natural gas distribution system operators, where explicit regulation is missing. (4) It is finally analyzed if the production of green hydrogen is currently supported in competition with fossil hydrogen production, not only by the legal framework but also by the National Hydrogen Strategy and the Amendment of the Renewable Energies Act. It can be concluded that in recent amendments of German energy legislation, regulatory support for green hydrogen in Germany was found. The latest legislation has clarified crucial points concerning the ownership and operation of electrolyzers and the treatment of green hydrogen as a renewable energy carrier.

Keywords: hydrogen; power-to-hydrogen; power-to-gas; energy law; energy regulation; renewable energy; legal framework; energy; energy transition; electrolysis

Citation: Ringsgwandl, L.M.; Schaffert, J.; Brücken, N.; Albus, R.; Görner, K. Current Legislative Framework for Green Hydrogen Production by Electrolysis Plants in Germany. *Energies* **2022**, *15*, 1786. https://doi.org/10.3390/en15051786

Academic Editor: Attilio Converti

Received: 31 January 2022
Accepted: 20 February 2022
Published: 28 February 2022

Publisher's Note: MDPI stays neutral with regard to jurisdictional claims in published maps and institutional affiliations.

1. Introduction

Hydrogen as an energy carrier can be used as a fuel that reacts without causing harmful emissions. When hydrogen is reacted with pure oxygen, the reaction product is water only, making hydrogen a unique fuel. The broad use of pure hydrogen is a promising future scenario for various end-use sectors, especially those that cannot easily cover their energy demands by renewable electricity. Prominent examples are in heavy-duty transportation, such as trucks, ships, or trains, and industrial high-temperature processes. Finally, the admixture of hydrogen into natural gas is also discussed as an option [1] to partly decarbonize all end-use sectors, including all gas-fired domestic and commercial technologies, before a mass rollout of dedicated pure hydrogen technologies can be realized.

Besides its unique characteristics as a fuel, hydrogen's second essential advantage is that it can be produced using only renewable electricity and water. The process is often referred to as "power-to-hydrogen", or more generally "power-to-gas" (PtG). The resulting renewable hydrogen—in the following referred to as green hydrogen—can be transported and stored in analogy to the proven natural gas technologies [2] and, thus, serve as a large-scale option to convert and store renewable electricity to cover renewable energy demands all around the year.

Academia, industry, and governments are constantly developing strategies to roll out hydrogen technologies, some of which have reached high technology readiness and are all set for mass production. However, in the existing energy systems, with all the interdependent entanglements of existing laws, regulations, infrastructures, and end-use applications, the market for hydrogen technologies is not developing swiftly. Until recently, decision-makers in Germany faced a lack of planning reliability due to missing legal and regulatory frameworks in combination with high technology cost. In addition, unclear future prospects, e.g., concerning prices, fees, and tariffs levied for the electricity needed for green hydrogen production, acted as barriers to investment decisions. The regulatory framework in Germany has been improved in 2021, providing more reliability. However, business models are still missing, and some aspects that appear crucial from the societal perspective, such as the long-term storage option of renewable energy in gaseous form, are not met by today's regulatory framework in Germany.

The situation is even more complex since different qualities of hydrogen could potentially be available from domestic production or imports, which could diverge dramatically in terms of environmental footprints. Depending on the environmental footprints and prices of available hydrogen on the market, different consumer groups are expected to show varying levels of acceptance.

This work reviews, analyses, and discusses the current situation for the case of Germany in 2021 with a focus on the two following research questions: Do German laws and regulations support hydrogen produced from renewable electricity (green hydrogen) in its competition with grey, blue, turquoise, or pink hydrogen? Do German laws and regulations support green hydrogen in its potential role as a long-term renewable energy storage option?

2. Materials and Methods

This work reviews and assesses the current legal framework for hydrogen in Germany. The corresponding central laws and acts as well as the national hydrogen strategy are discussed:

- Energy Industry Act (EnWG, 2021) [3];
- Renewable Energies Act 2021 (EEG, 2021) [4];
- Renewable Energies Ordinance (EEV, 2021) [5];
- National Hydrogen Strategy (NHS, 2020) [6].

3. Results

3.1. Hydrogen—Technology Options for Its Production and the Resulting "Color Scale"

In this section, the different possibilities to produce hydrogen are presented. Figure 1 shows the processes, educts, by-products, and "color" of the product hydrogen. The color theory is intended to reflect the degree of sustainability of the hydrogen produced. While hydrogen from fossil natural gas is turquoise, blue, or grey, hydrogen produced from biomass and biogas is green. Hydrogen from the electrolysis process, which is the focus of this paper, is also considered green hydrogen when renewable electricity sources are used, yellow when the electricity mix is used, and pink when nuclear electricity is used. The different production processes and the respective technology readiness levels (TRLs) are described in more detail below.

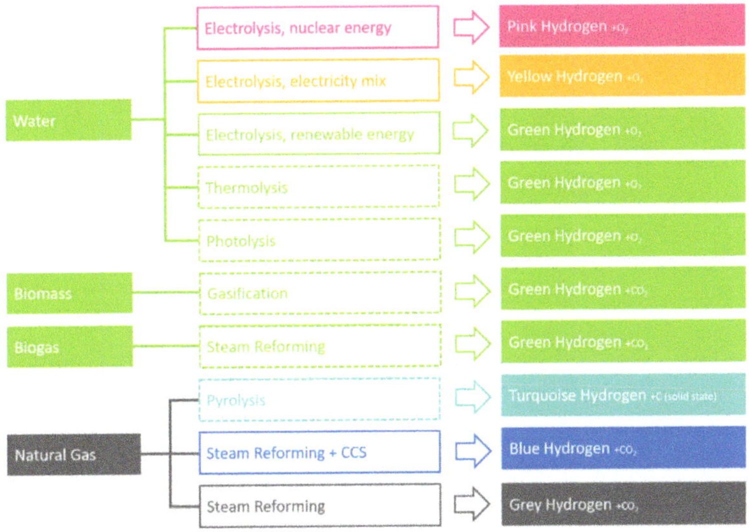

Figure 1. Overview of possible hydrogen production pathways using various technology options and electricity sources. Dashed boxes indicate technologies that are not yet available on an industrial scale.

In the electrolysis process, water is split into its components hydrogen and oxygen by the addition of electrical energy. Alkaline electrolysis (AEL) (TRL 8-9) [7], proton exchange membrane electrolysis (PEMEL) (TRL 8-9) [7], and high-temperature electrolysis (HTE) (TRL 5) [8] are known as common electrolysis technologies. While AEL and PEMEL have already been in operation for many years, HTE plants are still in the development stage, and HTE is characterized in particular by a high operating temperature of 700–1000 °C [8]. The main difference between AEL and HTE is the high temperature of the electrolyte. Essentially, AEL differs from the solid electrolyte of PEMEL by its additional caustic circuit, which transports the liquid electrolyte. While AEL has been used on a large scale for many decades, PEMEL has only been installed in the MW range for a few years. One advantage of AEL is the comparatively low specific plant costs. PEMEL is characterized by a more compact design and better dynamics, which means more flexible operating point changes, especially when renewable electricity is purchased. In addition to hydrogen, there is the possibility of removing the process heat and oxygen as by-products. As can be seen in Figure 1, the color of the hydrogen depends on the electricity used. Thus, pink hydrogen can be produced by using nuclear energy, yellow hydrogen by using the public electricity mix, and green hydrogen by using renewable electricity.

Thermolysis describes the process at temperatures above 1500 °C [9] with which the thermal decomposition of water into its components hydrogen and oxygen begins. Thus, it is theoretically possible to obtain hydrogen directly from water vapor at a very high temperature level. Challenges include technically controlling the working temperature, separating hydrogen, and avoiding direct recombination with oxygen back to water. Lowering the temperature of thermal water splitting can be achieved through coupled chemical reactions based on so-called metal oxide redox systems. The TRL for thermolysis is currently about 2 to 5 [8–10].

In photolysis, sunlight is used with the aid of a catalyst to effect direct water splitting. In 1972, Japanese scientists Fujishima and Honda discovered that titanium dioxide was suitable as such a catalyst [11]. Although worldwide efforts have been made since then to better understand the process flow as well as how the catalyst works, there is still a lack of understanding to achieve major efficiency improvements. Photolysis, like thermolysis, is at a low TRL (about 3) [10].

Biomass gasification describes a process in which biomass is supplied with heat, steam, and oxygen in a controlled manner and converted into hydrogen as well as other products. The biomass gasification process runs at temperatures of >700 °C without combustion. Organic or fossil carbonaceous materials are converted into carbon monoxide, hydrogen, and carbon dioxide with controlled amounts of oxygen and steam. One of the major challenges in hydrogen production using biomass gasification is the reduction in investment costs for the plant as well as for the biomass feedstock. The dual fluidized bed gasification (DFB) concept from the Vienna University of Technology is a promising approach to biomass gasification. The TRL for the DFB is currently 5 [12].

Steam methane reforming (TRL 9) [13] is today's standard in large-scale hydrogen production, usually found in industrial contexts, such as the chemical industry. Steam reforming uses methane (natural gas) which is thermally decomposed into CO and H_2 by hot, pressurized water steam in presence of a suitable catalyst. The water–gas shift reaction then leads to more H_2 output and CO_2 formation before, in a final step, the two are separated, e.g., by a pressure swing adsorption. The resulting hydrogen from today's large-scale plants is "grey hydrogen". However, renewable methane from biogenic sources that are connected to a large-scale gas processing plant by collection lines could be a technology option to produce "green hydrogen" via steam reforming as well. Steam reforming is therefore represented twice in Figure 1. The carbon dioxide emissions resulting from steam reforming plants could be reduced by integrating carbon capture technologies, which are not yet state-of-the-art (TRLs 7-8) [13,14].

Finally, one more alternative approach for hydrogen production is the technology class of pyrolysis. Here, hydrogen is provided by separating the carbon components from the methane molecules. Since the process is endothermic, heat must be added, for example in processes that use hot liquid metal baths to crack the molecules [15,16] at technology readiness levels in the range of 3-5 [13]. Other processes work, for example, with electron beam plasmas to split the methane molecules but are at a very early stage of development (TRLs 2-3) [14]. Pyrolysis technologies could be options for very large scale hydrogen production plants in the hundreds of megawatts order of magnitude. Due to their suitability for large-scale application, pyrolysis is combined with natural gas conversion in Figure 1, producing the so-called "turquoise hydrogen". This fossil-based hydrogen variant is to be distinguished from the "grey hydrogen" since its by-product carbon accumulates in solid form. Unlike gaseous CO_2, for example, the chemically inert solid carbon is prevented from reaching the atmosphere and acting as a greenhouse gas. In contrast, solid carbon products such as carbon black can be sold as dyes for specific industries.

3.2. *Current Techno-Economic Situation of Green Hydrogen Production*

The production of green hydrogen—for energy storage or other uses—is not yet economically viable or competitive in the market. The economic competitiveness of green hydrogen production mainly depends on the cost of electrolyzers, the cost of renewable electricity used in the process, the load factor (operating hours per year), and the plant scale [17–20]. With regard to the German legal situation, the price of electricity from renewable energy sources (RESs) is most interesting. The cost of RES power has generally decreased in recent years [21], meaning the production costs of green hydrogen could also drop. Currently, however, Germany has one of the highest electricity prices worldwide [22], which is a hindrance to the economic feasibility of green hydrogen production. Taxes and levies are a large part of the electricity price in Germany (accounting for approximately 50% of the electricity price for household consumers) [23]. Thus, any exceptions to any of the imposed levies are a major concern for PtG plant operators.

3.3. *Introduction: German Energy Industry Act (EnWG) and Renewable Energies Act (EEG)*

To a large extent, German energy law is formed by European (i.e., European Union) law. While we will not examine European energy law in itself, it plays an important role in German legislation and the interpretation of German energy law. While the European treaties were important in setting ground rules for a common European energy market, the EU's directives and regulations are most important in providing common regulation for the EU's member states. Both kinds of regulation are binding for the member states; directives, however, are not binding for European citizens until the member state has transposed them into national law (which they are obligated to do). Regulations, on the other hand, are directly binding for member states and citizens, much like national laws.

The two central laws for the development of green hydrogen in German energy legislation are the Renewable Energy Sources Act (EEG 2021) and the Energy Industry Act (EnWG).

The EEG 2021 (with predecessing laws going back to the 1990s [24]) governs the promotion of renewable energy sources for electricity production. It is meant to extend electricity production from renewable sources and to keep the overall cost of energy at a low level (§ 1 (1) EEG 2021). To this end, it provides two main instruments:

- An obligation of the transmission system operators (TSOs) to provide network access;
- An obligation of TSOs to pay either market premiums or feed-in tariffs to the producers for renewable electricity fed into the system.

Both instruments are designed to offer renewable energy producers a reliable legal and financial basis for their investments.

The market premiums or feed-in tariffs are financed via the "EEG surcharge". It is charged as part of the electricity price and is thus paid for by the electricity consumers.

The EnWG, on the other hand, is not limited to energy from renewable sources. It provides an overall framework for a functioning, competitive energy market (§ 1 EnWG). Among other things, it regulates the operation of energy networks (electricity and gas) and the tasks of TSOs and distribution system operators (DSOs). It further includes rules on the unbundling of network operators, storage facility operators, and suppliers, as well as rules for grid connection and the powers of the regulatory authority.

The EnWG also is the basis for a number of ordinances regarding grid access charges. We will also discuss the unbundling regulations as they pertain to the question of which players in the market are allowed to run an electrolyzer/PtG plant under German law.

3.4. *Applicability of the EnWG to Hydrogen*

German legislation has recently moved forward with regard to hydrogen production. Its objective in § 1 EnWG now mentions the supply of hydrogen. Hydrogen is also mentioned as one type of energy (see definition in § 3 No. 14), albeit only as far as it is used in "grid-bound" energy supply.

Hydrogen falls under the EnWG's definition of gas according to § 3 No. 19a EnWG; however, the definition only applies to hydrogen if it is made from water electrolysis and injected into the gas grid. § 3 No. 19a makes no mention of the electricity source used for the electrolysis process, meaning that not only green hydrogen is encompassed. However, neither grey nor blue hydrogen is included.

Hydrogen also falls under the EnWG's definition of biogas (§ 3 No. 10c EnWG), provided it is produced via water electrolysis using electricity predominantly from renewable energy sources [25–27].

As for energy storage via green hydrogen, the amended EnWG now defines "hydrogen storage plants" in § 3 No. 39b. They are defined as plants for the storage of hydrogen in ownership of or in operation by an energy supply undertaking. Hydrogen storage plants of hydrogen grid operators for their tasks are not encompassed by this definition and are not hydrogen storage plants according to § 3 No. 39b EnWG.

3.4.1. Ownership and Operation of Power-to-Hydrogen Plants: Unbundling Regulations

The energy market is highly regulated in Germany. There are strict rules on which players can or cannot—for example—operate electricity generation facilities, operate energy networks, or sell/supply energy. How these regulations apply to hydrogen production/PtG facilities is important to potential operators, investors, and network operators alike.

Unbundling regulations are to be found in §§ 6–10 EnWG. They transpose binding European unbundling legislation into German national law [28]. Unbundling generally is the idea that the operation of gas/electricity networks (distribution and transmission networks) has to be separated from electricity/natural gas generation and supply. The reason behind this is that network operation is a natural monopoly while generation and supply are competitive activities. Therefore, DSOs and TSOs must not be in a position to exploit their position as monopolists and unfairly influence the generation/supply market [29].

According to § 6 (1) EnWG, any network operator (NO), i.e., TSO or DSO, has to be independent from any kind of electricity/gas generation or supply. The EnWG does not allow the operation for NOs. Instead, it prescribes ownership unbundling for these activities [30].

The latest amendment to the EnWG now includes a definition of energy storage facilities (§ 3 No. 15d), which was missing from the EnWG before. Are PtG plants energy storage facilities and, if so, is this of any consequence for hydrogen energy storage? § 3 No. 15d EnWG now defines energy storage facilities. They are "facilities consuming electrical energy for electrical, chemical, mechanical or physical intermittent storage and reproduce it as electrical energy or in another form of energy". This wording includes PtG plants producing green hydrogen for energy storage. The official justification for the new amendment (BR-Drs. 165/21 v. 12.02.2021) [31] explicitly mentions PtX (power-to- … technologies, including PtG).

There has been some discussion about whether or not NOs should be allowed to own or operate electrolyzers [16,25,27,29]. The latest amendment of the EnWG, however, has largely rendered this question moot as far as TSOs and electricity DSOs are generally concerned.

According to § 7 (1) EnWG, electricity DSOs are not allowed to own an energy storage facility or to build, manage, or operate one. According to § 8 (2) EnWG, electricity TSOs are not allowed to own an energy storage facility or to build, manage, or operate one. § 10b (3) EnWG holds a similar regulation for TSOs within a VIU. Notice it is not gas NOs that are forbidden from engaging in energy storage facilities. The official justification merely states that the rules for gas NOs in this regard are "the general regulations" [31]. Natural gas network operators also have to abide by the unbundling regulations according to §§ 6 ff. EnWG; hence, the reasonable legal assessment is that they are also not allowed to engage in the ownership or operation of an electrolyzer [19,30].

There is an exception to the prohibition for electricity network operators: § 11a (1) EnWG says that electricity NOs are allowed to build, manage and operate energy storage facilities under certain circumstances. The energy storage facility has to be owned by a third party. The construction, management, and operation have to be tendered in an open and fair procedure. All of this only applies if the energy storage facilities are necessary for the NO to fulfil its duties according to § 11 (1) EnWG. These are to run a secure, reliable, and effective network in a non-discriminatory way and to maintain it, optimize it according to demand, and augment and expand it, as far as is economically reasonable.

§ 11b EnWG states that in certain cases, an electricity NO may own energy storage facilities or operate, build or manage them, if the regulatory authority gave permission to do so or if the regulatory authority allowed it for several or all NOs according to § 29 EnWG. This applies only to energy storage facilities that are fully integrated network components. Both § 11a and § 11b EnWG only apply to energy storage facilities that can produce electric energy. NOs could only make use of electrolyzer exceptions if they also re-electrify the hydrogen.

An example that the (green hydrogen) market has already tried these regulations is the decision made by the Federal Network Agency regarding the project "ELEMENT EINS" [32]. The planned project was to build a large-scale electrolyzer for the production of green hydrogen. The project partners were three network operators (one electricity TSO, two natural gas TSOs). The two natural gas TSOs were to build and operate the electrolyzer, and all three applicants would offer network access to their respective networks. The application pertained to the investments necessary for the project and their financing via network fees. After long consideration, the Federal Network Agency denied the application for the project because the operation of such a plant was not within the original tasks of network operators. The agency found among other considerations that the construction of the electrolyzer was not necessary for the expansion of the supply network according to § 11 EnWG. § 11 (1) EnWG obliges NOs to run and also optimize or expand their networks as needed. The Federal Network Agency did not see the construction of the electrolyzer as part of the applicants' duties. Furthermore, the agency saw the possibility that the operation of the electrolyzer by an NO might be a risk to the network operation itself. Since the successful operation of the electrolyzer depends on congestions in the network, the NO might therefore have less of an incentive to secure an efficient and safe network in the first place. It should be kept in mind that the application was aimed at the financing of the electrolyzer investment costs via network fees. According to the Agency, this could discriminate against other potential electrolyzer operators, which do not have the possibility of offsetting their costs via network charges.

Since the time the Federal Network Agency issued the decision, the unbundling legislation in the EnWG was amended, as seen above for §§ 11 a, b EnWG. This begs the following question: would the Federal Network Agency issue a different decision today? In its decision, the Federal Network Agency stressed specifically that the legislator had not assigned the operation of an electrolyzer to an NO. This has changed to some degree as the legislator has now specifically prohibited the ownership and operation of energy storage facilities for electricity NOs (§ 7 EnWG). If the exemptions of §§ 11 a and b EnWG do not apply, there is still no way for an electricity NO to operate an electrolyzer. For natural gas NOs, there has been no change in legislation, and the decision would probably be issued in the same way now.

An overview of the current regulatory status of potential electrolyzer owners and operators in Germany is given in Table 1.

Table 1. Regulation of ownership and operation of electrolyzers in Germany for potential stakeholders.

Stakeholders	Ownership or Operation of Electrolyzers	Further Explanations
TSOs (natural gas)	No explicit regulation in German law; most probably not admissible.	See Section 3.4.1
DSOs (natural gas)	No explicit regulation in German law; most probably not admissible.	See Section 3.4.1
TSOs (electricity)	No ownership or operation of electrolyzers permitted: §§ 6, 8 EnWG. Exceptions possible: § 11 a and b EnWG.	See Section 3.4.1
DSOs (electricity)	No ownership or operation of electrolyzers permitted: §§ 6, 7 EnWG. Exceptions possible: § 11 a and b EnWG.	See Section 3.4.1
Third parties, e.g., plant manufacturers	Ownership/operation possible.	For exceptions from charges see Section 3.5.3

3.4.2. Grid Tariffs

Energy storage facilities are generally considered "final customers" in terms of electricity consumption [28]. The EnWG defines final consumers as "natural or legal persons purchasing energy for their own use" [28]. In this sense, PtG plant operators are final consumers since they purchase energy (electricity) for their own use. However, subjecting them to all the regular levies also means that the costs for running a PtG plant for storage might be high. There is also a definition of wholesale customers (§ 3 No. 21 EnWG) defining wholesale customers as "a natural or legal person purchasing electricity for the purpose of resale inside or outside the system where it is established" [28]. This definition seems to be a little closer to the business model of an energy storage plant than the definition of a final consumer [28]; however, the wording does not fit—electrolyzer operators might not buy electricity primarily to resell it but actually to store it. The prevailing legal opinion is that energy storage plant operators are to be classified as final consumers [28].

There is a general obligation to pay network fees according to § 21 (1) EnWG. Usage of the grid occurs when taking power from the grid as well as feeding it into the grid. There is, however, an exception to this rule which applies to electrolyzer plants. § 118 (6) EnWG provides that plants constructed after 2008 and beginning operation on or after 4 August 2011 are exempted from network access fees regarding the electricity taken for storage. This expressly includes PtG plants [33]. § 118 (6) EnWG does not include an exception from feed-in fees for feeding electricity into the grid (from a PtG plant); this, however, is covered by § 15 (1) StromNEV (electricity network fee regulation ordinance): there are no charges for the injection of electricity. It should be noted that § 15 StromNEV applies to electricity of any source, not just renewable sources. Therefore, § 15 StromNEV does not constitute a privilege.

3.5. Green Hydrogen Production under the EEG 2021

In the 2021 amendment of the Renewable Energies Act, hydrogen is taken into account to a larger degree. Before, the electricity supply for green hydrogen production was not exempted from surcharges. The amendments will be described in detail in the following subsections.

3.5.1. Applicability of the EEG 2021 to Hydrogen

The EEG 2021 governs the promotion of energy from renewable sources. One of its objectives is the system integration of electricity from renewable sources (RES) (§§ 1, 2 EEG 2021). The EEG contains support schemes for RES, which directly affect the cost of electricity consumed for green hydrogen production and will be assessed in the following.

Does green hydrogen qualify as renewable energy according to § 3 No. 21 EEG 2017? "Renewable energy" according to this provision is, among others, "energy from biomass including biogas, bio methane (...)" (see § 3 No. 21 in [34]). Hydrogen of any kind is not named in the provision. It does not fit under this regulation because it is not derived from biomass, but also not biogas. § 3 No. 11 EEG 2021 defines biogas as "any gas obtained from anaerobic fermentation of biomass". Biomass is not actually defined in the EEG 2021 [35]. However, the official justification of the EEG 2009 gives an indication: Biomass encompasses "biogenic energy carriers in solid, liquid or gaseous form" [36], which have to be biodegradable and generally derived from plant or animal origin [36]. Hydrogen is none of these things and therefore does not fall under this definition of "renewable energy". It has been argued that one could draw an analogy and thus subsume hydrogen under the term "biogas", with the reasoning being as follows: the EEG's objective is to promote climate and environmental protection, and, to that end, to promote the use of renewable electricity and the development of suitable technologies (see § 1 (1) in [33,34]); because of this, green hydrogen should also be subsumed under biogas since it has all the attributes and functions of biogas [33]. While this argumentation may appear preferable for the development of green hydrogen production in Germany, the reasoning lacks justification. In German law, an analogy requires an accidental gap in regulation, in other words, that the legislator would have included hydrogen if he had only remembered to do so. This might have been the case for the EEG 2017; however, the legislator for the EEG 2021 clearly made an amendment to include green hydrogen in other regulations (see § 69b EEG 2021). Therefore, hydrogen does not qualify as biogas in the sense of the EEG 2021.

It is important to underline the difference between the two relevant laws since in contrast to the EEG 2021, the EnWG does qualify hydrogen as biogas. This is due to different definitions of the term "biogas". Hence, it is important to apply this term only with clear reference to one law and not to assume that the EnWG and the EEG 2021 employ the same definitions.

3.5.2. The EEG Surcharge

Since the operating expenditures are a crucial and potentially hindering factor in the economic viability of hydrogen production, we should take a closer look at the charges and levies that apply for green electricity supply.

The EEG surcharge serves the purpose of promoting renewable energies in Germany. It is part of the electricity price and as such has to be paid by the electricity consumers. The revenues created from the EEG surcharge are used to pay certain remunerations for operators of renewable energy, i.e., to electricity generators. The plant operators feed electricity into the power grid and receive fixed remunerations from the TSO. The TSO in turn is reimbursed via the revenues from the EEG surcharge [37].

Since electrolyzers draw power from the grid, they are considered "final consumers" (see § 3 No. 33 EEG 2021). Consequently, they fall under the scope of § 61 (I) EEG 2021 and thus have to pay the EEG surcharge as part of the electricity price. This does not seem reasonable for plant operators, since they argue their purpose is the storage of electricity, not consumption. The last amendment of the EEG has improved this situation, however. § 61 (II) EEG 2021 stipulates that under certain circumstances, the EEG surcharge can be reduced or be omitted completely. For electrolyzers, it refers to § 69b EEG 2021 (titled "production of green hydrogen"). This exemption will be explored in the following section.

3.5.3. Exemption from the EEG Surcharge

§ 69b EEG 2021: Green hydrogen production

§ 69b EEG 2021 exempts PtG plants from the duty of paying the EEG surcharge. There are several requirements to be met in order to qualify for the exemption. First, it only applies to the electricity used for the production of green hydrogen. This begs the following question: what exactly qualifies as green hydrogen? The EEG 2021 itself does not define green hydrogen. § 69b stipulates that the exemption is only applicable when an ordinance has given detailed requirements for the qualification of green hydrogen. As of July 2021, an (amended) ordinance is in force: the "Ordinance for the implementation of the renewable energies act and the wind power at sea act" (also renewable energies ordinance, EEV, [5]). §§ 12h-12l EEV apply to green hydrogen. The objective of the new regulations is to exempt certain hydrogen production plants from the EEG surcharge, thus making electricity cheaper to them and eventually rendering the electrochemical production of green hydrogen more economically feasible and competitive [5].

Green hydrogen according to § 69b EEG 2021 is hydrogen made within the first 5000 full-load hours of the year (calendar year) in a green hydrogen production facility. According to § 12i EEV, it is required that the production uses exclusively electricity that:

- Verifiably stems from renewable energy production plants (§ 3 No. 21 EEG 2021);
- Stems verifiably at least 80% from plants within the German bidding zone; 20% at the most may come from plants within a bidding zone electrically connected to the German bidding zone;
- Was not produced receiving any payments or support according to the EEG, the EEV, or the Combined Heat and Power Act (CHP Act) [38].

§ 12i (2) EEV gives further definition regarding electricity stemming from renewable energy plants. Renewable energies according to § 3 No. 21 EEG 2021 are, among others, hydropower, solar power, energy from biomass, and geothermal energy. § 12i (3) EEV gives instructions for the calculation of full-load hours. It is irrelevant for which purpose the hydrogen is produced. Furthermore, § 69b EEG 2021 only applies to facilities taken into operation before 1 January 2030. According to § 12 h EEV, exemptions apply to electricity used from 1 January 2022 on [5].

§ 64a EEG 2021: Limitation for undertakings with intensive electricity costs

There is another option to reduce the EEG surcharge for green hydrogen production. §§ 63 No. 1 a, 64a EEG 2021 offers a limitation of the EEG surcharge to be paid by PtG plant operators. It applies to hydrogen production plants qualifying as undertakings with intensive electricity costs.

A short digression in order to explain the notion of "undertakings with intensive electricity costs" is as follows: The limitation of the EEG surcharge is an established instrument within the EEG's regime. Companies in certain industries (e.g., railroad, production of goods, mining, see annex 4 to EEG 2021) can apply for the limitation, provided they meet certain requirements. The objective is to keep these companies competitive in the market and to prevent them from moving their operation abroad.

According to §§ 63 No. 1a, 64a EEG 2021, companies producing hydrogen electrochemically are also eligible for this kind of EEG surcharge limitation. The limitation applies to hydrogen production for energy storage since § 64a (1) EEG 2021 states that the regulation applies regardless of the intended usage of the hydrogen. It is interesting that § 64a EEG 2021 does not make any demands regarding the source of the electricity used. As long as the hydrogen is produced electrochemically, the electricity could also stem from nuclear, coal, or other nonrenewable power sources.

Not any company involved in hydrogen production is eligible for the limitation. § 64a EEG 2021 only applies to companies active in the sector "production of industry gases" (§ 64a EEG 2021, annex 4, No. 78). Additionally, the production has to constitute the largest portion of the total added value of the company.

In case an undertaking is eligible for limitation or exemption according to both § 64a and § 69b EEG 2021, they have to decide for which one to apply. Both regulations are mutually exclusive (§ 12i (1) EEV). Only one of the regulations can be applied throughout the year. It is albeit possible to switch to the other one in the following year, and back the year after that.

§§ 61a ff. EEG 2021: Exemptions for self-provision

Another Possibility for an Exemption from the EEG Surcharge Is Offered by §§ 61a ff. EEG 2021.The exemptions apply to entities operating their own power generating facilities, without the electricity passing through the power grid (§ 3 No. 19 EEG 2021). In terms of green hydrogen production, these exemptions would benefit hydrogen projects if they produce their own electricity with renewable sources. The requirement that the electricity has to be self-produced and may not have been passed through the power grid means that the exemptions of §§ 61a ff. EEG 2021 cannot be met by many companies [39]. In case an electrolyzer is operated by an industrial hydrogen consumer, partial savings on electricity cost may be achieved by using renewable energy from their own renewable energy source production plants.

3.6. Political Developments in Germany: The National Hydrogen Strategy

The growing attention for hydrogen as an energy carrier is reflected by the fact that the German Federal Government has released its own hydrogen strategy, the National Hydrogen Strategy [6]. While it is a communication that is not legally binding, it does reflect the governments' goals which can be indicative of coming legislation. The National Hydrogen Strategy was issued in June 2020. It outlines its goals within the context of the energy transition and the German 2030 Climate Action Plan as well as the fact that Germany is committed to achieving greenhouse gas (GHG) neutrality [6]. The goal for GHG neutrality has recently been set higher (from 2050 to 2045) through § 3 (2) of the Federal Climate Protection Act [40] and is not reflected in the current version of the NHS, which as of now cites the 2030 Climate Action Plan with GHG neutrality for 2050 [41]. The Strategy further states that fossil fuels have to be replaced by alternative energy sources and that hydrogen will play a key role in this endeavor [6]. The Strategy in this respect names uses of hydrogen as an energy carrier (e.g., for fuel-cell-powered vehicles); as an instrument in sector coupling, especially in energy applications that cannot be electrified; and as an energy storage medium. The strategy also states the goal of replacing grey hydrogen with green hydrogen in industrial processes that use grey hydrogen today [6]. The last part relates to an important factor of the German hydrogen strategy: It focuses exclusively on green hydrogen [6]. It can be said that it regards only green hydrogen to be *"sustainable in the long term"* [6], although turquoise hydrogen is also mentioned [6]. Within these conditions, the strategy aims to make hydrogen a "competitive option", to develop a "domestic market for hydrogen technology" in Germany, "establishing hydrogen as an alternative for other energy sources" and making it "a sustainable base for the industrial sector" [6]. Via the strategy, the federal government also announces wide-ranging financial support for the promotion and deployment of green hydrogen technology [6].

4. Discussion

In the following, we discuss the role of green hydrogen in the German energy system. Is green hydrogen supported by the current legal framework (Section 4.1), i.e., the Energy Industry Act and the Renewable Energies Act including its amendments? Which future developments can be drawn from the political National Hydrogen Strategy (Section 4.2)?

4.1. Hydrogen in the Current Legal Framework

The current legal framework for (green) hydrogen is the result of some recently adopted changes in legislation: The EEG 2021 was adopted (and in its wake, the EEV was amended) and the EnWG was amended as well in 2021. These laws now explicitly mention hydrogen. This is remarkable as, until 2021, there were no regulations directly addressing hydrogen or electrolyzers. Lawyers mostly had to interpret existing regulations, for the qualification of hydrogen; e.g., "Does hydrogen qualify as natural gas/biogas?", "Under which conditions is hydrogen a renewable energy carrier?". The characteristics of hydrogen as an energy carrier and a means of storing and converting electricity did not fit into some regulations. This was especially discussed when it came to questions of unbundling: Since green hydrogen is a gas that is produced from electricity, it is technically associated somewhere between the natural gas sector and the electricity sector, and it was for a long time ambiguous which sector it "belonged" to. The recent legislation is a step forward as it addresses hydrogen and hydrogen production directly. This provides legal certainty for network operators, investors, and regulatory authorities alike. This certainty is especially important for investors and potential operators or electrolyzers.

The legal framework does not, however, provide a coherent, overall regulation for hydrogen production [42]. As shown above, any definition of "green hydrogen", or hydrogen as biogas, is only relevant within the scope of the respective act. For example, the definition of green hydrogen as given in the EEV is only relevant for the exemption from the EEG surcharge according to § 69b EEG 2021. So far, there is no single and overall valid legal definition of what green hydrogen exactly is and what it qualifies for.

The aforementioned exemptions from grid access charges (§ 118 EnWG) and from the EEG surcharge (§ 64a and § 69b EEG 2021) provide support for the production of green hydrogen in Germany. Another factor for the competitiveness of green hydrogen production is carbon pricing [42]. While the price of carbon does not directly affect the hydrogen production process, an increase in the carbon price could level the playing field between grey and green hydrogen and would make green hydrogen financially more interesting for industrial companies. The German legislator will have to examine whether the current way of carbon pricing is efficient or whether it has to be adjusted in scope and price, as some have suggested that the carbon price in Germany is not high enough [42,43]. This can also be addressed in the European Union's emissions trading system at the same time.

The legislation passed in 2020 and 2021 shows that the German legislator regards hydrogen, especially green hydrogen, as a relevant factor for the German energy industry. As a result, the current legislation is overall more supportive of green hydrogen than in recent years.

4.1.1. Hydrogen in the Energy Industry Act

The EnWG has had several amendments adopted in 2021. It now addresses hydrogen and regulates its use in the German energy system. This in itself is an important step forward. The new definition of energy storage facilities (§ 3 No. 15d) encompasses electrolyzers. This definition as well as the new definition of (natural) gas, energy, and biogas (§ 3 No. 10f, 14 and 19a EnWG) now addressing hydrogen open the door for regulation of (green) hydrogen and its production. The EnWG thus regulates not only green hydrogen, but also other kinds of hydrogen. The regulations differ, however, since the general regulations regarding hydrogen apply to any "color" of hydrogen. Only hydrogen produced through electrolysis using electricity from renewable sources qualifies as biogas. Consequently, any regulation within the EnWG regarding biogas applies to green hydrogen. Whether or not the regulations are in themselves beneficial, the EnWG now provides a clearer legal framework for any entity planning to produce hydrogen. With the technique of encompassing hydrogen into already existing terms (natural gas, biogas), the legislator also was able to employ the already existing regulation as opposed to creating an overall new framework.

One criticism pertaining to the new definitions of natural gas and energy could be that they only apply to hydrogen as far as it is fed into the grid or used for grid-bound energy supply (§ 3 No. 14, 19a EnWG). This is consistent within the framework of the EnWG because this prerequisite also applies to natural gas and other energy carriers [15]; however, one interesting feature of hydrogen is that it is not dependent on the gas grid, but can be transported in other ways.

The EnWG now holds several important regulations with regard to unbundling (see above), giving potential operators of electrolyzers the much-needed legal certainty for their business. Generally, the fact that network operators are not allowed to own or operate electrolyzers is probably going to be supportive of a market launch for hydrogen production. There has been some criticism of the exceptions to those rules. Any employment of electrolyzers can potentially interfere with the energy market.

Generally, while there are good arguments for allowing NOs to operate PtG plants regarding grid safety, it would be inconsistent with the current legal framework. Since the delineation between competitive and system-based measures would be blurred, there would be a certain risk of market distortion. The benefits of PtG for grid operation could probably also be reaped if the plants were operated not by NOs, but by competitive undertakings.

4.1.2. Hydrogen in the Renewable Energies Act and Its Amendment

The amendments of the EEG 2021/Renewable Energies Act could well make the production of green hydrogen economically more viable. As stated above, a major concern was (or is) the high cost of electricity in Germany, mainly due to a number of publicly imposed charges. This problem is partially solved by the regulations in § 69b and also § 64a EEG 2021 as they give a possibility to limit or be exempted from the EEG surcharge. Both state no conditions for the usage of the hydrogen produced and thus also apply to hydrogen produced for electricity/energy storage. Both regulations are relatively new and a necessary step towards promoting hydrogen as a means for energy storage.

It is worth analyzing the regulation in the EEV. Many of the detailed requirements regarding § 69b EEG 2021 are left to the EEV. The overall possibility of an exemption from the EEG surcharge is beneficial. The detailed instructions and the definition as to what green hydrogen is within the scope of the EEG 2021 are also positive: they provide legal certainty for PtG plant operators, investors, and government agencies. This being said, the requirements are quite restrictive since only the first 5000 full-load hours per year are exempted. Any electricity used after/beyond the first 5000 full-load hours is charged with the full EEG surcharge.

4.2. The National Hydrogen Strategy

Since the National Hydrogen Strategy (NHS) is a political strategy, it is not legally binding and does in itself not yield any consequences for entities dealing with hydrogen. As mentioned above, the strategy focuses on green hydrogen, as the German Federal Government only sees green hydrogen as sustainable [29]. It is coherent that the strategy then focuses on green hydrogen since it sees hydrogen as one possibility of decarbonization and of reducing Germany's greenhouse gas emissions [29]. The NHS addresses some regulatory barriers in the promotion of green hydrogen production. By the time this paper was written, some of the regulation which was in place when the NHS was issued had already been amended.

The NHS acknowledges the problem of the current high cost of hydrogen production. The high cost of electricity—especially the price components induced by regulation—is addressed. The NHS names as one of several measures "a fair design of price components induced by the state" [29]. The NHS promises an analysis as to how the production of green hydrogen can be exempted from taxes, levies, or charges. The exemption from the EEG surcharge is specifically mentioned as one goal. Here, the strategy addresses one central regulatory problem regarding hydrogen production and has specific measures to solve it.

Another measure is the exploration of model projects of cooperations between operators of electrolyzers and network operators [29]. Since these have to be in line with the unbundling regulations, the NHS states that the government will explore the need for possibly amending the regulatory framework. The aim here is to *"ease the burden"* on the grid. A general clarification of the status of electrolyzers in the unbundling regulations is not the goal here; this could also be a sign that the government sees no need for clarification. This proposal is on the one side supportive of green hydrogen since it aims to enable the production of hydrogen. On the other hand, the operation of electrolyzers by network operators always involves the danger of market distortion. Therefore, this measure could not be supportive for the market launch of green hydrogen in the long run. This measure is also one of the few instances where the NHS indirectly mentions green hydrogen as an energy storage medium. Remarkably, the strategy almost does not mention the possibility of green hydrogen as an energy storage medium. This is on the one hand surprising because energy storage is a crucial challenge in the promotion of renewable energy sources. Green hydrogen could be an important storage medium here. On the other hand, the technology and the financial side of hydrogen as an energy storage medium are not yet on a level where it could be widely employed for storage.

The Strategy promises several measures to foster the development of a hydrogen market, some of which could be beneficial for green hydrogen production (e.g., introducing carbon dioxide pricing for fossil fuels and possible exemption from taxes, levies, and surcharges for green hydrogen production) [6]. Measures such as the support of electrolyzers for industrial procedures could indirectly also be beneficial for the economic viability of hydrogen energy storage since they might bring down prices for electrolyzers by promoting their production on a mass scale. Overall, the NHS addresses regulatory barriers for green hydrogen and suggests concrete and probably promising measures.

5. Conclusions and Outlook

German legislation has made significant progress in the last two years in terms of the regulation of green hydrogen as an energy carrier. Two central acts in German energy law —the EnWG and the EEG 2021—now explicitly address and categorize hydrogen, providing legal certainty for plant operators and network operators. One hindrance in the production of green hydrogen is the high electricity price. This has been addressed through exemptions from grid tariffs and the EEG. The regulatory framework is confined to the scope of the respective acts, meaning there is no comprehensive regulation of green hydrogen production in German law yet.

The newly formed German federal government includes green hydrogen in its coalition agreement [44]. In line with the current NHS, the focus remains on green hydrogen. Among other goals, the new government aims to make Germany a lead market for hydrogen technologies and promote the production of green hydrogen. The coalition agreement does not make any direct announcements regarding new regulation, apart from promising purchasing quotas for public agencies [44]. The coalition agreement also promises to achieve competitive electricity prices for German companies; this would have to involve some legislation as the electricity prices in Germany are in a large part regulation-driven by charges, levies, and taxes [17].

The European Hydrogen Strategy (EHS) [45] will also further play a role as it shapes European legislation which, in turn, informs German national legislation. It is worth pointing out one difference in their terminology: The NHS focuses on green hydrogen and uses the terminology described in Section 3.1. The EHS however uses the following two terms: (a) "renewable" or "clean" hydrogen that is produced via water electrolysis using electricity from renewable sources and (b) "low-carbon" hydrogen which is fossil-based hydrogen with carbon capture or electricity-based hydrogen regardless of the electricity source [45]. Why the Commission chose this terminology is not entirely clear. It could be suggested that the aim is to shift policies in a new direction—possibly to "blur the line between 'green' and 'blue' hydrogen, essentially defining both as 'clean'" [46]. In this case, the German and the European strategies differ in a key point, and it remains open how this will shape the future German legislation regarding hydrogen.

New legislation is also to be expected from the European Union—and subsequently on the German level where it will have to be transposed. The European Commission has issued a draft to amend the current Renewable Energy Directive (RED II) [47]. The proposed amendment for the so-called RED III [48] was issued by the Commission in July 2021; it includes a quota for the member states for hydrogen used in industry: by 2030, 50% of the hydrogen used for industry purposes has to be contributed by renewable fuels of nonbiological origin [48]. If and when this proposal is passed, this quota will inform the German legislation on hydrogen and possibly be a promotor of green hydrogen in the future.

Author Contributions: Conceptualization and methodology, L.M.R., J.S., N.B.; formal analysis, L.M.R.; investigation, resources, visualization, L.M.R., J.S., N.B.; writing—original draft preparation, L.M.R.; writing—review and editing, J.S., N.B.; supervision, R.A., K.G. All authors have read and agreed to the published version of the manuscript.

Funding: This research received no external funding.

Institutional Review Board Statement: Not applicable.

Acknowledgments: The authors would like to thank Eadbhard Pernot, Christian Hampel, Elina Martens, Manfred Lange, and Frank Burmeister.

Conflicts of Interest: The authors declare no conflict of interest.

References

1. Leicher, J.; Schaffert, J.; Cigarida, H.; Tali, E.; Burmeister, F.; Giese, A.; Albus, R.; Görner, K.; Carpentier, S.; Milin, P.; et al. The Impact of Hydrogen Admixture into Natural Gas on Residential and Commercial Gas Appliances. *Energies* **2022**, *15*, 777. [CrossRef]
2. Creos; DESFA; Elering; Enagás; Energinet; Eustream; FGSZ; Fluxys Belgium; Gasgrid Finland; Gasunie; et al. Extending the European Hydrogen Backbone-A European Hydrogen Infrastructure Vision Covering 21 Countries 2021. Available online: https://gasforclimate2050.eu/wp-content/uploads/2021/06/European-Hydrogen-Backbone_April-2021_V3.pdf (accessed on 20 November 2021).
3. *Energiewirtschaftsgesetz Vom 7. Juli 2005 (BGBl. I S. 1970, 3621), Das Zuletzt Durch Artikel 84 Des Gesetzes Vom 10. August 2021 (BGBl. I S. 3436) Geändert Worden Ist; [Energy Industry Act of 7 July 2005 (Federal Law Gazette I, p. 1970, 3621), as Last Amended by Article 84 of the Act of 10 August 2021 (Federal Law Gazette I, p. 3021)]*. Available online: https://www.gesetze-im-internet.de/enwg_2005/BJNR197010005.html (accessed on 24 January 2022).
4. *Erneuerbare-Energien-Gesetz Vom 21. Juli 2014 (BGBl. I S. 1066), Das Zuletzt Durch Artikel 11 Des Gesetzes Vom 16. Juli 2021 (BGBl. I S. 3026) Geändert Worden Ist; [Renewable Energies Act of 21 July 2014 (Federal Law Gazette I, p. 1066), as Last Amended by Article 11 of the Act of 16 July 2021 (Federal Law Gazette I, p. 3026)]*. 2021. Available online: https://www.gesetze-im-internet.de/eeg_2014/BJNR106610014.html (accessed on 20 November 2021).
5. *Erneuerbare-Energien-Verordnung Vom 17. Februar 2015 (BGBl. I S. 146), Die Zuletzt Durch Artikel 87 Des Gesetzes Vom 10. August 2021 (BGBl. I S. 3436) Geändert Worden Ist); [Renewable Energies Ordinance of 17 February 2015 (Federal Law Gazette I, p. 146), as Last Amended by Article 87 of the Act of 10 August 2021 (Federal Law Gazette I, p. 3436)]*. Available online: https://www.gesetze-im-internet.de/ausglmechv_2015/BJNR014610015.html (accessed on 24 January 2022).

6. Federal Ministry for Economic Affairs and Energy. *The National Hydrogen Strategy*; Federal Ministry for Economic Affairs and Energy: Berlin, Germany, 2020.
7. Taibi, E.; Blanco, H. *Hydrogen Series-Part 2: Green Hydrogen Cost Reduction: Scaling up Electrolysers to Meet the 1.5 °C Climate Goal*; IRENA: Abu Dhabi, United Arab Emirates, 2021.
8. Pinsky, R.; Sabharwall, P.; Hartvigsen, J.; O'Brien, J. Comparative Review of Hydrogen Production Technologies for Nuclear Hybrid Energy Systems. *Prog. Nucl. Energy* **2020**, *123*, 103317. [CrossRef]
9. Boretti, A. Technology Readiness Level of Solar Thermochemical Splitting Cycles. *ACS Energy Lett.* **2021**, *6*, 1170–1174. [CrossRef]
10. Albrecht, U.; Altmann, M.; Barth, F.; Bünger, U.; Fraile, D.; Lanoix, J.; Pschorr-Schoberer, E.; Vanhoudt, W.; Weindorf, W.; Zetra, M.; et al. *Study on Hydrogen from Renewable Resources in the EU-Final Report*; Ludwig-Bölkow-Systemtechnik (LBST): Ottobrunn, Germany; Brussels, Belgium, 2015.
11. Fujishima, A.; Honda, K. Electrochemical Photolysis of Water at a Semiconductor Electrode. *Nature* **1972**, *238*, 37–38. [CrossRef] [PubMed]
12. Binder, M.; Kraussler, M.; Kuba, M.; Luisser, M. Hydrogen from Biomass Gasification. Available online: https://www.ieabioenergy.com/wp-content/uploads/2019/01/Wasserstoffstudie_IEA-final.pdf (accessed on 20 November 2021).
13. Parkinson, B.; Balcombe, P.; Speirs, J.F.; Hawkes, A.D.; Hellgardt, K. Levelized Cost of CO_2 Mitigation from Hydrogen Production Routes. *Energy Environ. Sci.* **2019**, *12*, 19–40. [CrossRef]
14. Kerscher, F.; Stary, A.; Gleis, S.; Ulrich, A.; Klein, H.; Spliethoff, H. Low-Carbon Hydrogen Production via Electron Beam Plasma Methane Pyrolysis: Techno-Economic Analysis and Carbon Footprint Assessment. *Int. J. Hydrog. Energy* **2021**, *46*, 19897–19912. [CrossRef]
15. Abánades, A.; Rubbia, C.; Salmieri, D. Thermal Cracking of Methane into Hydrogen for a CO_2-Free Utilization of Natural Gas. *Int. J. Hydrog. Energy* **2013**, *38*, 8491–8496. [CrossRef]
16. Dagle, R.A.; Dagle, V.; Bearden, M.D.; Holladay, J.D.; Krause, T.R.; Ahmed, S. *An Overview of Natural Gas Conversion Technologies for Co-Production of Hydrogen and Value-Added Solid Carbon Products*; Pacific Northwest Natl Lab(PNNL): Richland, WA, USA, 2017.
17. International Renewable Energy Agency (IRENA). *A Renewable Energy Perspective*; IRENA: Abu Dhabi, United Arab Emirates, 2019.
18. Böhm, H.; Zauner, A.; Rosenfeld, D.C.; Tichler, R. Projecting Cost Development for Future Large-Scale Power-to-Gas Implementations by Scaling Effects. *Appl. Energy* **2020**, *264*, 114780. [CrossRef]
19. Schmidt, O.; Gambhir, A.; Staffell, I.; Hawkes, A.; Nelson, J.; Few, S. Future Cost and Performance of Water Electrolysis: An Expert Elicitation Study. *Int. J. Hydrog. Energy* **2017**, *42*, 30470–30492. [CrossRef]
20. Thema, M.; Bauer, F.; Sterner, M. Power-to-Gas: Electrolysis and Methanation Status Review. *Renew. Sustain. Energy Rev.* **2019**, *112*, 775–787. [CrossRef]
21. Van Hulst, N. *The Clean Hydrogen Future Has Already Begun*; International Energy Agency: Paris, France, 2019; Available online: https://www.iea.org/commentaries/the-clean-hydrogen-future-has-already-begun (accessed on 12 October 2021).
22. Borning, M. Bringt die Nationale Wasserstoffstrategie endlich die passende Regulierung? *EnergieRecht* **2020**, *04/20*, 135–143.
23. Federal Ministry for Economic Affairs and Energy (BMWi). Energy Prices and Transparency for Consumers. 2021. Available online: https://www.bmwi.de/Redaktion/EN/Textsammlungen/Energy/electricity-prices.html (accessed on 22 October 2021).
24. Federal Ministry for Economic Affairs and Energy (BMWi). Das Erneuerbare-Energien-Gesetz. Available online: https://www.erneuerbare-energien.de/EE/Redaktion/DE/Dossier/eeg.html (accessed on 4 November 2021).
25. *Directive2009/28/EC of the European Parliament and of the Council of 23 April 2009 on the Promotion of the Use of Energy from Renewable Sources and Amending and Subsequently Repealing Directives 2001/77/EC and 2003/30/EC (Text with EEA relevance), OJ L 140, 5.6.2009, p. 16–62*. Available online: https://eur-lex.europa.eu/LexUriServ/LexUriServ.do?uri=OJ:L:2009:140:0016:0062:en:PDF (accessed on 24 January 2022).
26. Elspas, M.; Graßmann, N.; Rasbach, W. *Energiewirtschaftsgesetz (EnWG)*; Erich Schmidt Verlag (ESV): Berlin, Germany, 2018.
27. *Gesetzentwurf der Fraktionen der CDU/CSU und FDP-Entwurf eines Gesetzes zur Neuregelung energiewirtschaftsrechtlicher Vorschriften*; BT-DRs. 17/6072; Deutscher Bundestag: Berlin, Germany, 2011.
28. Kreeft, G. *Legislative and Regulatory Framework for Power-to-Gas in Germany, Italy and Switzerland*; STORE&GO Project: Groningen, The Netherlands, 2018; pp. 22–28.
29. European Comission Third Energy Package. Available online: https://ec.europa.eu/energy/topics/markets-and-consumers/market-legislation/third-energy-package_en#unbundling (accessed on 15 June 2021).
30. Borning, M. Der Schlüssel für die Energiewende Wartet auf die Passende Regulierung. *EnergieRecht* **2020**, *03/20*. Available online: https://erdigital.de/field/Autoren/q/%22Martin+Borning%22/truncation/0/suche.html (accessed on 24 January 2022).
31. *Entwurf Eines Gesetzes zur Umsetzung Unionsrechtlicher Vorgaben und zur Regelung Reiner Wasserstoffnetze im Energiewirtschaftsrecht*; Gesetzesentwurf der Bundesregierung, BT-Drs. 19/27453; Bundesregierung: Berlin, Germany, 2021.
32. Bundesnetzagentur (National Grid Agency). *Beschlusskammer 4, Beschluss Vom 27.01.2021, Az. BK4-19-049*; Bundesnetzagentur: Cologne, Germany, 2021. Available online: https://www.bundesnetzagentur.de/DE/Beschlusskammern/1_GZ/BK4-GZ/2019/BK4-19-0049/BK4-19-0049_Beschluss_download_bf.pdf?__blob=publicationFile&v=2 (accessed on 9 November 2021).
33. Sieberg, C.; Cesarano, C. Rechtsrahmen für eine Wasserstoffwirtschaft. *Recht Energ.* **2020**, *6*, 238.

34. *Erneuerbare-Energien-Gesetz vom 21. Juli 2014 (BGBl. I S. 1066), das zuletzt durch Artikel 11 des Gesetzes vom 16. Juli 2021 (BGBl. I S. 3026) geändert worden ist" / Englisch: Renewable Energies Act of 21 July 2014 (Federal Law Gazette I, p. 1066), as last amended by Article 11 of the Act of 16 July 2021 (Federal Law Gazette I, p. 3026).* Available online: https://www.bgbl.de/xaver/bgbl/start.xav?start=%2F%2F*[%40attr_id%3D%27bgbl114s1066.pdf%27]#__bgbl__%2F%2F*%5B%40attr_id%3D%27bgbl114s1066.pdf%27%5D__1646034210817 (accessed on 24 January 2022).
35. Bundesministerium für Justiz. Biomasseverordnung vom 21. Juni 2001 (BGBl. I S. 1234), die uletzt durch Artikel 8 des Gesetzes vom 13. Oktober 2016 (BGBl. I S. 2258) geändert worden ist. 2016. Available online: https://Www.Gesetze-Im-Internet.de/Biomassev/ (accessed on 24 January 2022).
36. *Gesetz zur Neuregelung des Rechts der Erneuerbaren Energien im Strombereich und zur Änderung damit Zusammenhängender Vorschriften vom 25. Oktober 2008.* BGBl. I S. 2074. Available online: https://www.bgbl.de/xaver/bgbl/start.xav?startbk=Bundesanzeiger_BGBl&start=//*%255B@attr_id=%27bgbl108s2074.pdf%27%255D#__bgbl__%2F%2F*%5B%40attr_id%3D%27bgbl108s2074.pdf%27%5D__1646034140825 (accessed on 24 January 2022).
37. Federal Ministry for Economic Affairs and Energy (BMWi). State-Imposed Components of the Electricity Price. Available online: https://www.bmwi.de/Redaktion/EN/Artikel/Energy/electircity-price-components-state-imposed.html (accessed on 9 November 2021).
38. *Kraft-Wärme-Kopplungsgesetz vom 21. Dezember 2015 (BGBl. I S. 2498), das Zuletzt Durch Artikel 88 des Gesetzes vom 10. August 2021 (BGBl. I S. 3436) geändert worden ist; [Combined Heat and Power Act of 21 December 2015 (Federal Law Gazette I p. 2498), as Last Amended by Article 88 of the Act of 10 August 2021 (Federal Law Gazette I p. 3436)].* Available online: https://www.bmwi.de/Redaktion/DE/Downloads/G/gesetzesentwurf-zur-aenderung-der-bestimmungen-zur-stromerzeugung-aus-kwk-und-eigenversorgung.pdf?__blob=publicationFile&v=4 (accessed on 28 January 2022).
39. Buchmüller, C. Die energie- und regulierungsrechtlichen Baustellen auf dem Weg zur Wasserstoffwirtschaft. *Zeitschrift für Umweltrecht.* **2021**, 4, 195. Available online: https://www.zur.nomos.de/archiv/2021/heft-4/ (accessed on 28 January 2022).
40. *Bundes-Klimaschutzgesetz vom 12. Dezember 2019 (BGBl. I S. 2013), das durch Artikel 1 des Gesetzes vom 18. August 2021 (BGBl. I S. 3905) geändert worden ist 2019.* Available online: https://Www.Gesetze-Im-Internet.de/Ksg/BJNR251310019.Html (accessed on 28 January 2022).
41. German Federal Government. *Klimaschutzprogramm 2030 Der Bundesregierung Zur Umsetzung Des Klimaschutzplans 2050*; German Federal Government: Berlin, Germany, 2019. Available online: Https://Www.Bundesfinanzministerium.de/Content/DE/Downloads/Klimaschutzprogramm-2030-Der-Bundesregierung-Zur-Umsetzung-Des-Klimaschutzplans-2050.Pdf?__blob=publicationFile&v=2 (accessed on 24 January 2022).
42. Hoffmann, B.; Halbig, A.; Senders, J.; Nysten, J.V.; Antoni, O.; Müller, T. Auf dem Weg zum Wasserstoffwirtschaftsrecht? In *Stift. Umweltenergierecht, WürzburgerStudien zum Umweltenergierecht, No. 21, 2021*; 2021; p. 54. Available online: https://stiftung-umweltenergierecht.de/wp-content/uploads/2021/07/Stiftung_Umweltenergierecht_Wuerzburger-Studie-zum-Umweltenergierecht_21_Wasserstoffwirtschaftsrecht_2021-07-12-2.pdf (accessed on 24 January 2022).
43. McKinsey & Company, Hydrogen Council: Hydrogen Insights a Perspective on Hydrogen Investment, Market Development and Cost Competitiveness: February 2021. Available online: https://hydrogencouncil.com/wp-content/uploads/2021/02/Hydrogen-Insights-2021-Report.pdf (accessed on 16 January 2022).
44. Mehr Fortschritt Wagen-Bündnis für Freiheit, Gerechtigkeit und Nachhaltigkeit. *Koalitionsvertrag 2021–2025*; Zwischen der Sozialdemokratischen Partei Deutschlands (SPD); Bündnis 90/Die Grünen und den Freien Demokraten (FDP): Berlin, Germany, 2021.
45. European Commission. *Communication from the Commission to the European Parliament, the Council, the European Economic and Social Committee and the Committee of the Regions: A Hydrogen Strategy for a Climate-Neutral Europe*; Brussels, Belgium, 2020. Available online: https://ec.europa.eu/energy/sites/ener/files/hydrogen_strategy.pdf (accessed on 16 February 2022).
46. Fleming, R. Clean or Renewable–Hydrogen and Power-to-Gas in EU Energy Law. *J. Energy Nat. Resour. Law* **2020**, *39*, 43–63. [CrossRef]
47. *Directive (EU) 2018/2001 of the European Parliament and of the Council of 11 December 2018 on the Promotion of the Use of Energy from Renewable Sources (Text with EEA Relevance).* 2018, Volume OJ L 328, pp. 82–209. Available online: https://Eur-Lex.Europa.Eu/Legal-Content/En/TXT/PDF/?Uri=CELEX:32018L2001 (accessed on 23 January 2022).
48. Proposal for a Directive of the European Parliament and of the Council, Amending Directive (EU) 2018/2001 of the European Parliament and of the Council, Regulation (EU) 2018/1999 of the European Parliament and of the Council and Directive 98/70/EC of the European Parliament and of the Council as Regards the Promotion of Energy from Renewable Sources, and Repealing Council Directive (EU) 2015/652, Brussels, 14 July 2021, Volume COM(2021) 557 Final. Available online: https://eur-lex.europa.eu/legal-content/EN/TXT/?uri=CELEX%3A52021PC0557 (accessed on 20 January 2022).

MDPI
St. Alban-Anlage 66
4052 Basel
Switzerland
Tel. +41 61 683 77 34
Fax +41 61 302 89 18
www.mdpi.com

Energies Editorial Office
E-mail: energies@mdpi.com
www.mdpi.com/journal/energies

www.ingramcontent.com/pod-product-compliance
Lightning Source LLC
Chambersburg PA
CBHW061815210926
51850CB00006B/712
* 9 7 8 3 0 3 6 5 7 5 5 7 5 *